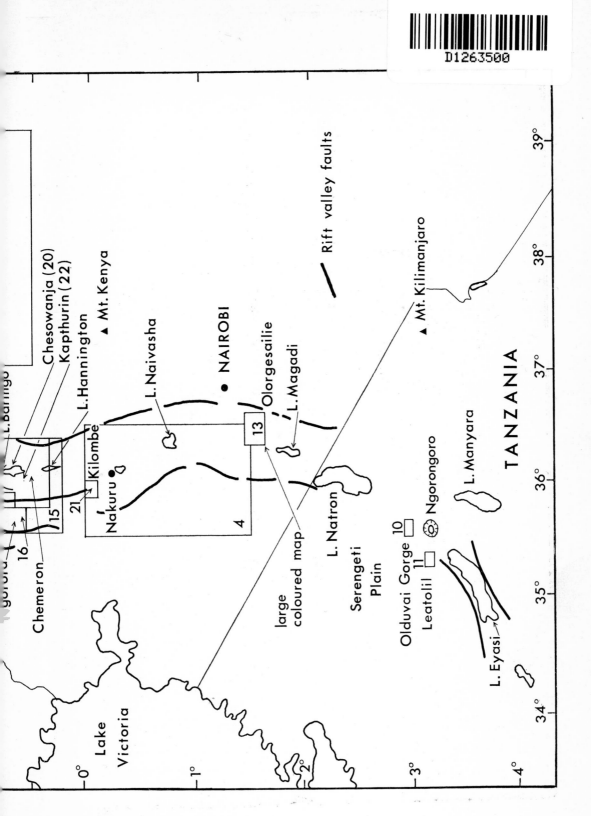

Chesowanja (20)
Kapthurin (22)
L.Hannington
▲ Mt. Kenya
L. Baringo
L. Naivasha
Chemeron
16
15
21
Kilombe
Nakuru ●
NAIROBI
Olorgesailie
L. Magadi
13
4
large
coloured map
L. Natron
Serengeti
Plain
Olduvai Gorge 10
Leatolil 11
Ngorongoro
L. Manyara
TANZANIA
L. Eyasi
Lake
Victoria
Rift valley faults
▲ Mt. Kilimanjaro
0°
1°
2°
3°
4°
34°
35°
36°
37°
38°
39°

GEOLOGICAL BACKGROUND TO FOSSIL MAN

Recent research in the Gregory Rift Valley, East Africa

W. W. Bishop.

GEOLOGICAL BACKGROUND TO FOSSIL MAN

Recent research in the Gregory
Rift Valley, East Africa

Edited by
WALTER W. BISHOP

1978

Published for
The Geological Society of London

by

SCOTTISH ACADEMIC PRESS
UNIVERSITY OF TORONTO PRESS

Published by

SCOTTISH ACADEMIC PRESS LTD
33 Montgomery Street
Edinburgh, EH7 5JX

First published 1978

ISBN 7073 0143 2

First published 1978 in Canada and the United States
by University of Toronto Press
Toronto   Buffalo   London

ISBN 0-8020-2302-9

Printed in Great Britain by
Western Printing Services Ltd, Bristol

In Memoriam

WALTER WILLIAM (BILL) BISHOP was born 4 May 1931 and died suddenly on 20 February 1977 in his forty-sixth year. His activities and achievements during this relatively short period of time were far greater than those of many who have survived a full life span. His potentialities and promise for the future were prodigious; it is a tragedy to geology that they were not to be fulfilled.

The Geological Society remembers him with gratitude, for he stepped in as Scientific Editor at a time (1969) when the situation regarding its publications was at a crisis; he not only cleared up an intolerable backlog, but by insight and imagination, widened the scope, improved the style of presentation and increased the number of issues of the Journal, thereby greatly enhancing its reputation. His advice and understanding were so extensive and valued that he continued as Secretary until 1975.

His career was varied, his experience wide, and for his age he had an almost unparalleled list of distinctions, credits and publications. A citation of his curriculum vitae would testify to the facts, but I shall confine attention to Bill as a scientist and as a person.

He obtained a B.Sc. Honours degree in Geography at Birmingham University in 1952, continuing to his Ph.D. in 1956. He served as a geologist on the Uganda Geological Survey from 1956 to 1959; it was here that I first met him, appropriately, at his camp near Napak, where he was making wonderful discoveries of fossil vertebrates in tuffs of the spectacular volcano that I had mapped in 1939. Thereafter, from 1959 to 1965 much of his work was concerned with museum curating, first in the Hunterian Museum, Glasgow University, and then as director of the Uganda National Museum, Kampala. Here he showed his supreme ability in organisation and acquiring enthusiastic support for his ventures and ideas. Bill had already, while in Uganda, gained a considerable reputation for his work in the field and on expeditions, concerned with Ugandan geomorphology, with Cenozoic geology in general and in vertebrate palaeontology. Teaching at Makerere University was associated with his museum activities; and he then, as subsequently, proved himself an inspiring lecturer and a person greatly concerned with the welfare of students.

Although his main interest was with Cenozoic geology, he pursued this in all its aspects, stratigraphical, sedimentational and palaeontological, using radiometric and palaeomagnetic methods of age determination and, as a keen observer of the behaviour and fate of existing vertebrates, he instituted comparative studies by his research workers of the patterns of dispersal of present day mammalian remains.

My concern to get him appointed to the staff at Bedford College was motivated not only because of his intrinsic merit, but also the great contribution that he could make to systematic and detailed sedimentological and palaeontological studies in the Kenya Rift Valley project. This, indeed, worked exceedingly well, since, for almost the first time, such studies were conducted within an already well-established stratigraphical and radiometrically dated framework. These common interests continued with his appointment to the Chair and Headship of the Department of Geology at Queen Mary College, from where he initiated comparable studies in Pakistan.

The numerous symposia that he attended as an invited speaker, the many special lectures that he was invited to give and the colloquia of various kinds that he organised testify to his world-wide recognition in his field. All of these numerous commitments he took in his stride, giving the appearance of completely unruffled competence, enthusiasm and organisational ability. It is a clear reflection of the esteem with which he and his work were regarded internationally that he was invited to the Directorship of the Peabody Museum in the University of Yale, with a Chair in the Department of Geology, an appointment that he was due to take up in July 1977.

Perhaps, above all else he had a gift for acquiring the respect and affection from all of those with whom he came in contact. He was sympathetically approachable by everyone; indeed, his concern with student affairs stemmed from his Birmingham days where he was President of the Guild of Undergraduates. Despite his self-imposed, exacting schedule he had time for everyone, received and wrote individual letters to persons all over the world. He founded in Bedford College a Light Opera Group (BLOG), in which he initially participated, for he was an actor and singer of some merit; this has now become an institution in the College and he continued his operatic interests at Queen Mary College and indeed a week before his death, he sang an end to a lecture at Glasgow University in his usual effortless style.

Bill had a great concern for adult education in Glasgow and particularly for the training of young persons from all walks of life. After several years of serving in a voluntary capacity on the Advisory Committee of the Brathay Exploration Group, based on Ambleside, Cumbria, he was recently elected Chairman of the Committee. Indeed, he participated in organising expeditions in East Africa in which Brathay students actively took part. Above all else, he will be remembered for his wide sense of humanity and true social values.

This book is a most fitting and lasting tribute to the memory of Bill Bishop. Based on a major symposium which he organised and, from the breadth of his knowledge and acquaintance, selected the contributors, he has arranged, advised on, and edited all of the many articles contained in this volume. It is a monumental work and one of true devotion both on his part and that which he excited among others. It may appear invidious to cite particular persons, but I should like to record some of those who both inspired and were inspired by Bill: Louise, Mary and Richard Leakey, Bob and Shirley (Coryndon) Savage and Glynn Isaac.

And finally, let me say, that despite all his preoccupations he was a man with a family of which he was intensely proud and concerned; he bestowed on them his first consideration, but it is only fair to add that in his wife, Sheila, he received always the staunchest and most unstinting support that any man could ever have.

B. C. KING

Contents

In Memoriam   v

Acknowledgements   xi

WALTER W. BISHOP. Introduction   xii

SIR PETER KENT, F.R.S. Historical background: Early exploration in the East African Rift—The Gregory Rift Valley   1

PART I. FRAMEWORKS: STRUCTURAL—VOLCANIC—GEOPHYSICAL

1. E. RONALD OXBURGH. Rifting in East Africa and large-scale tectonic processes   7

2. ROBERT M. SHACKLETON. Structural development of the East African Rift System   19

3. BASIL C. KING. Structural and volcanic evolution of the Gregory Rift Valley   29

4. LAURENCE A. J. WILLIAMS. Character of Quaternary volcanism in the Gregory Rift Valley   55

5. M. AFTAB KHAN & CHRISTOPHER J. SWAIN. Geophysical investigations and the Rift Valley geology of Keyna   71

PART II. BACKGROUND: PALAEONTOLOGICAL AND ARCHAEOLOGICAL PROBLEMS

6. ANDREW HILL. Taphonomical background to fossil man—problems in palaeo-ecology   87

7. R. T. SHUEY, FRANK H. BROWN, G. G. ECK & F. CLARKE HOWELL. A statistical approach to temporal biostratigraphy   103

8. BERNARD A. WOOD. Allometry and hominid studies   125

9. GLYNN LL. ISAAC. The first geologists—the archaeology of the original rock breakers   139

PART III. REGIONAL STUDIES IN THE GREGORY RIFT VALLEY

A. OLDUVAI GORGE AND LAETOLIL, TANZANIA; OLORGESAILIE, KENYA

10. MARY D. LEAKEY. Olduvai Gorge 1911–1975: a history of the investigations   151

11. MARY D. LEAKEY, R. L. HAY, C. H. CURTIS, R. E. DRAKE, M. K. JACKES, & T. D. WHITE. Fossil hominids from the Laetolil Beds, Tanzania   157

12. ROBERT M. SHACKLETON. Geological Map of the Olorgesailie Area, Kenya   171

13. GLYNN LL. ISAAC. The Olorgesailie Formation: Stratigraphy, tectonics and the palaeogeographic context of the Middle Pleistocene archaeological sites   173

B. THE LAKE BARINGO BASIN, KENYA

14. GREGORY R. CHAPMAN & MAUREEN BROOK. Chronostratigraphy of the Baringo Basin, Kenya   207

15. PETER DAGLEY, ALAN E. MUSSETT & H. C. PALMER. Preliminary observations on the palaeomagnetic stratigraphy of the area west of Lake Baringo, Kenya   225

16. MARTIN H. L. PICKFORD. Geology, palaeoenvironments and vertebrate faunas of the mid-Miocene Ngorora Formation, Kenya   237

17. MARTIN H. L. PICKFORD. Stratigraphy and mammalian palaeontology of the late-Miocene Lukeino Formation, Kenya   263

18. SHIRLEY CAMERON CORYNDON. Fossil Hippopotamidae from the Baringo Basin and relationships within the Gregory Rift, Kenya   279

19. ALAN W. GENTRY. Fossil Bovidae of the Baringo Area, Kenya   293

20. WILLIAM BISHOP, ANDREW HILL & MARTIN PICKFORD. Chesowanja: A revised geological interpretation   309

21 (a) WALTER W. BISHOP. Geological framework of the Kilombe Acheulian archaeological site, Kenya   329

    (b) JOHN A. J. GOWLETT. Kilombe—an Acheulian site complex in Kenya   337

22. PETER W. J. TALLON. Geological setting of the hominid fossils and Acheulian artifacts from the Kapthurin Formation, Baringo District, Kenya   361

C. THE LAKE TURKANA (RUDOLPH) BASIN, KENYA AND ETHIOPIA

23. ROBERT J. G. SAVAGE & PETER G. WILLIAMSON. The early history of the Turkana depression   375

24. CARL F. VONDRA & BRUCE E. BOWEN. Stratigraphy, sedimentray facies and palaeoenvironments, East Lake Turkana, Kenya   395

25. IAN C. FINDLATER. Isochronous surfaces within the Plio-Pleistocene sediments east of Lake Turkana   415

26. ANNA K. BEHRENSMEYER. Correlation of Plio-Pleistocene sequences in the northern Lake Turkana Basin: a summary of evidence and issues   421

27. FRANK J. FITCH, PAUL J. HOOKER & JOHN A. MILLER. Geochronological problems and radioisotopic dating in the Gregory Rift Valley   441

28. G. H. CURTIS, R. E. DRAKE, T. E. CERLING, B. W. CERLING & J. H. HAMPEL. Age of KBS Tuff in Koobi Fora Formation, East Lake Turkana, Kenya   463

29. ANDREW BROCK. Magneto-stratigraphy east of Lake Turkana and at Olduvai Gorge: a brief summary   471

30. F. H. BROWN, F. CLARK HOWELL & G. G. ECK. Observations on problems of correlation of late Cenozoic hominid-bearing formations in the North Lake Turkana Basin   473

31. YVES COPPENS. Evolution of the hominids and of their environment during the Plio-Pleistocene in the lower Omo Valley, Ethiopia   499

32. PETER G. WILLIAMSON. Evidence for the major features and development of Rift Palaeolakes in the Neogene of East Africa from certain aspects of lacustrine mollusc assemblages   507

33. JOHN W. K. HARRIS & INGRID HERBICH. Aspects of early Pleistocene hominid behaviour east of Lake Turkana, Kenya   529

D. THE AFAR AREA, ETHIOPIA

34. DON C. JOHANSON, MAURICE TAIEB, B. T. GRAY & YVES COPPENS. Geological framework of the Pliocene Hadar Formation (Afar, Ethiopia) with notes on palaeontology including hominids   549

List of authors and addresses   565

Index of Subjects   569

Index of Authors   579

ENDPAPERS

The locations of the areas discussed and the general geological framework in which they lie are indicated in the endpapers; these areas are numbered the same as the relevant chapters in the book.

Acknowledgements

It was with tragic suddenness that Bill Bishop died before his final revision of the proofs of this book was finished. The brief introductions he had planned to write for each of the parts of the book and the final summary were not completed before his death and have had to be omitted.

The publishers extend grateful thanks to his contributors, particularly to those who have assisted them—in one way or another—to complete the press revision, and to answer queries which appeared to be still unresolved.

But above all gratitude must be expressed to Mrs. G. W. Flinn of Liverpool, who has prepared the Index and has noted a number of inconsistencies of the spellings of proper names and geological terms. These have now been standardised as far as may be possible within individual articles, but without going to the further expense of redrawing several of the line illustrations. All concerned, both authors and readers, have cause to be grateful to Mrs. Flinn for her most careful and painstaking contribution to this work.

Finally, the Geological Society thanks and acknowledges the extraordinary time and effort expended by Mr. Douglas Grant and Mr. T. L. Jenkins of the Scottish Academic Press, without which the book could not have appeared.

Introduction

Since the finding of the skull of *Australopithecus boisei*, or 'Zinj', at Olduvai Gorge in 1959 by Dr. Mary Leakey there has been increasing involvement by geologists in the elucidation of the time-framework and environmental setting of man's fossil ancestors. The pioneering enthusiasm of Louis and Mary Leakey encouraged many anthropologists, and scientists in associated disciplines, to continue the search for better preserved fossils of early man and the animals who were his contemporaries.

Over 550 hominid fossils have been discovered during the last 18 years from the Gregory Rift Valley. The Rift may be considered as a field laboratory, varying from 40 to 80 kilometres in width but over 2000 kilometres in length. It contains unique fossil localities in the Afar region of northern Ethiopia and the Omo valley in the south, round Lake Turkana (formerly called Lake Rudolf) and several other fossiliferous sedimentary basins in Kenya, and in the Laetolil/Olduvai Gorge region of northern Tanzania.

This volume stems from a 3 day symposium held in February 1975 at the Geological Society, London. It contains 35 papers by 50 authors and presents the detailed geological context for the major hominid finds established in the Gregory Rift Valley during the last eighteen years.

Part I describes the broad structural, volcanic and geophysical setting and the history of development of the Gregory Rift.

Part II discusses palaeontological problems, including those involved in reconstructing former living communities from remnant fossil assemblages. A new statistical approach to temporal biostratigraphy is outlined together with the importance of allometry in hominid studies. A guide is provided to the problems of interpreting the debris of broken stone left by those 'earliest geologists', our palaeolithic ancestors.

Part III contains detailed regional studies for four sections of the Gregory Rift: **A.** The history of research at Olduvai Gorge is described together with the context of exciting new finds from Laetolil, Tanzania. The Olorgesailie palaeolithic sites in Kenya are placed into their geological setting. **B.** Numerous authors contribute papers on the lithological succession, chronology, magnetostratigraphy, palaeontology and archaeology of Baringo District, Keyna investigated since 1965. **C.** Remarkable hominid discoveries from the Lake Turkana Basin (Keyna) and the Omo Valley (Ethiopia) made by international research teams are described with reference to their geological, palaeontological and archaeological context. Chronostratigraphic and palaeomagnetic sequences are outlined and some of the problems relating to isotopic dating and biostratigraphy are discussed. **D.** The geological setting is outlined for the latest hominid discoveries from the Afar region of the Ethiopian sector of the Rift Valley.

W. W. Bishop

SIR PETER KENT

Historical background: Early exploration in the East African Rift — The Gregory Rift Valley

In relation to modern lines of communication it seems surprising that the Gregory Rift Valley was the last part of the system to become known. Much of the earlier exploration had however been centred on the problem of the sources of the Nile, and in consequence the Western or Albertine Rift was explored by Samuel Baker as early as 1862/63 (Baker 1866). Additionally there was a strong tendency to use the convenient base at Zanzibar Island for journeys inland by the Arab slave trading routes from Pangani and Bagamoyo; these led to the Tanganyika Rift and Nyasaland rather than to the area of modern Kenya. The first penetrations into the Gregory Rift area were in 1883; Joseph Thomson made an extensive journey into Central Kenya which he described in his book of 1887, 'Through Masai Land' which had as a subtitle, 'a journey of exploration among the snowclad volcanic mountains and strange tribes of Eastern Equatorial Africa—being the narrative of the Royal Geographical Society's Expedition to Mount Kenya and Lake Victoria Nyanza 1883–84'.

In his classic journey Thomson practically encircled the lower slopes of Mount Kilimanjaro and reached the Gregory Rift wall near the Ngong Hills. He then went north to Lake Baringo and westwards to Lake Victoria, before returning to his starting point at Mombasa. His observations on the geology were of good standard for the time. For example he referred to 'enormous masses of porphyritic sanidine rock forming a lava cap to the underlying metamorphic rocks' on the western side of the Rift.

About the same time a German naturalist Dr Gustav Fischer visited the southern part of the Rift Valley in what is now Tanzania, reaching Naivasha and overlapping with the southern part of Thomson's traverse and completing the traverse of the southerly Gregory rift. His mapping was however of modest standard, and some of his localities were only rediscovered decades later.

The next major contribution was from Ludwig von Hohnel, who in 1894 published in English an account of 'Discovery of Lakes Rudolf and Stefanie—a narrative of Count Samuel Teleki's exploring and hunting expedition in eastern Equatorial Africa in 1887 and 1888.' This account included an excellent scientific record and maps of the Rift from Baringo northwards to Lake Rudolf,[1] and discovery of an active volcano, named after Count Teleki. The geology was recorded along the route of the expedition and former high levels of the various lakes were observed and recorded.

[1] Now renamed Lake Turkana.

A major advance came in 1896 when J. W. Gregory published 'The Great Rift Valley—being the narrative of a journey to Mount Kenya and Lake Baringo—with some account of the Geology, Natural History, Anthropology and future prospects of British East Africa.' This described a journey in 1892–3 from the Kenya coast at Mombasa to the Rift Valley, climbing Mount Kenya en route and returning to the Coast. Gregory's work marked the beginning of scientific understanding of the Rift Valley. He had already travelled extensively in other parts of the world and he recognised the Rift was a true graben (fault trough), and discussed its origin. He observed that much of the faulting was very recent, that the closely spaced block-and-trough strips were relatively shallow features, due to crustal extension. He worked out (correctly) the order of superposition of the members of the volcanic sequence, although modern ideas of chronology are somewhat different from his. Some of his observations still remain to be followed up. For example, he drew attention to 'bastions' where the Rift Valley wall was particularly abrupt and steep, separating stretches with more highly developed step faulting. (It might be suggested that the step fault sequences represent a degree of gravity collapse of the Rift wall into the relatively incompetent valley filling: in fact that this particular type of faulting is to a large degree a superficial gravity effect. This is a matter which might be further investigated).

J. W. Gregory returned to East Africa much later, in 1919, at the invitation of the Government. By then the country had been largely opened up and mapped, although there was still no road for wheeled traffic from Nairobi into the Rift Valley. He was able to draw on the work of a range of government administrators who were well aware of the importance of scientific discoveries. Notable among them was C. W. Hobley (commemorated in the name of the Miocene *Deinotherium hobleyi* Andrews), who was responsible for the first recognition of vertebrate bearing Miocene, described by F. Oswald (1913, 1914) at Karungu on the shores of Lake Victoria, and who himself made extensive contributions on the geology and archaeology of the Gregory Rift Valley from 1894 onwards. The Geological Society has a direct link with Gregory, for he was awarded the Bigsby Medal in 1905 and was President from 1928–30.

In 1919 E. J. Wayland set up the first Geological Survey in East Africa, at Entebbe in Uganda. In the subsequent fifteen years Wayland made an outstanding contribution to knowledge of rift and basin development in the region, recognising for example the late river reversals in the Lake Victoria basin, and he contributed also to the geology of areas further east including Olduvai and Kavirondo. He was deeply impressed by the straightness and structural transgressiveness of the fractures in the Western Rift, with the uplift of the Ruwenzori massif, and put forward a general theory of origin of the rift valley system by lateral compression, a theory which was strongly supported by other authors. In 1933 he was awarded the Bigsby Medal of the Geological Society for his East African work. Wayland was followed by a notable sequence of geologists of international reputation who between them made Uganda one of the geologically better known parts of Africa, and the experience of this led in due course to the setting up of corresponding bodies in Tanganyika and eventually in Kenya. Except in Uganda however geological work was concentrated on the rocks with direct economic mineral potential; the rifts and the volcanic spreads were largely neglected. It has remained for the present generation of geologists to fill this very large gap.

Formal geophysical observations began early in Tanzania with the setting up of a series of gravity stations by Kohlshutter in 1899, but it was not until 1933 that the Gregory and Western Rift Valleys were investigated, by E. C. Bullard, who found a marked gravity deficiency in the rift and deduced that the rift floors must be held down by overthrusting of the margins (1936). In this he was following the views of Wayland, and the compressional hypothesis was endorsed in part by Bailey Willis (1936), but other explanations are now preferred.

Serious work on the archaeology and later geology of the Gregory Rift valley began in the later 1920s when L. S. B. Leakey returned to his birthplace from Cambridge, and began the long series of investigations which have made the area a classic one in relation to early man and his development. Leakey supported by Solomon and others worked on the high level 'Gamblian' lakes in the Rift valley basins, and this work was extended northwards by Fuchs around Lake Rudolf and by Nilsson in Ethiopia. The early interpretation of their 'pluvials' as the tropical equivalent of the northern ice ages is no longer accepted, but the evidence of high level lake waters and their relation to the activity of early man still remains. Leakey with his collaborators made Olduvai one of their main areas of activity, and it is now one of the world's classic areas for Pleistocene geology and the history of early man. Leakey extended his activities into the Kavirondo Rift, to the Pleistocene of Kanam and Kanjera, and to the Miocene rocks of the Legetet and Rusinga areas, with their important pre-hominid anthropoids.

At the present day the Rift system is widely regarded as a landward continuation of the mid-ocean rises, and by some theorists as belts where the crust has opened by miles or tens of miles. These concepts need careful analysis, for the link with the oceanic system looks less good on a large-scale map than it does on the map of the world; the chronology of rift development is quite complex regionally and fails to agree in detail with that of the ocean; finally the whole width of the rift is, over much of its length, floored with continental sialic basement rocks. We need more information on the dating of the different parts of the rift; more information on the dating of the broad swell of which the Gregory Rift Valley forms the crestline; more information about the floor of the rift beneath the volcanics. Fortunately there are now many competent people producing the critical geological and geophysical data as a basis for informed syntheses.

This Volume records progress in dealing with these problems, important not only in the content of the continental development but also in relation to the history, evolution and ecology of what Louis Leakey called 'Adams Ancestors'.

References

BAKER, S. W. 1866. *The Albert Nyanza, Great Basin of the Nile*, Vols. 1 and 2, Macmillan London, 395 pp. and 384 pp.

BULLARD, E. C. 1936. Gravity Measurements in East Africa. *Phil. Trans. Roy. Soc. A.* **235**, 445–531.

GREGORY, J. W. 1896. *The Great Rift Valley*, Seeley Service, London.

HOHNEL, L. VON, 1894. *Discovery of Lakes Rudolf and Stefanie*, 2 vols, (English translation of 1892 German Edition). 435 pp., 397 pp.

OSWALD, F. 1913. The Miocene Beds of the Victoria Nyanza, *Journ. E. African Nat. Hist. Soc.* **3**, 2–8.

—— 1914. The Miocene Beds of the Victoria Nyanza. *Q. Jl geol. Soc. Lond.* **70**, 128–62.

THOMSON, J. 1887. *Through Masai Land*, 364 pp.

WILLIS, B. 1936. *East African Plateaus and Rift Valleys*, Carnegie Institute, Washington.

Part I

FRAMEWORKS:
Structural – Volcanic – Geophysical

1

E. RONALD OXBURGH

Rifting in East Africa and large-scale tectonic processes

The East African Rift System has many of the geophysical characteristics of a mid-ocean ridge and is laterally continuous with the Red Sea and Gulf of Aden spreading centres. Although all three were initiated simultaneously on continental crust in pre-Miocene times, that in East Africa did not evolve oceanic crust.

Seismic, thermal and geochemical evidence suggest that the lithosphere under much of Africa is about 300 km thick, whereas under the rift it is today 40–50 km. Thin lithosphere can result either from mechanical necking or from the upward migration of the isotherm defining the lithosphere/asthenosphere boundary. Both processes are important in the evolution of a continental rift into a spreading ocean, but the limited extension in East Africa indicates that here the latter must dominate. It is, however, difficult to transfer sufficient heat into the lithosphere from below in the time available; conductive processes are too slow, and transfer by magma, although fast enough, requires more mass addition to the lithosphere than is compatible with the limited surface extension. Hot volatiles might transfer heat to the lithosphere from below and depress melting temperatures without adding significantly to its mass.

Rifts have been attributed to convective mantle plumes or to membrane stresses in the lithosphere. If the lithosphere is thick, mantle plumes are unlikely to generate rifts in the time available.

Introduction

The East African Rift System of which the Gregory Rift is a part belongs to a class of tectonic phenomena which have been loosely grouped together under the name of 'mid-plate tectonics'. The members of this group have one main feature in common, that they seem to involve magmatic, seismic or deformational activity which is not directly attributable to the interactions between plate margins which have been used so successfully to interpret a wide range of other tectonic phenomena.

There have been two broadly contrasting approaches to mid-plate phenomena. The first assigns an essentially passive role to the plate, and the various mid-plate phenomena are seen as the expression of some local special condition in the underlying asthenosphere. Following this approach Morgan (1971, 1972a, b) has proposed a system of mantle convective plumes. Alternatively, the mantle may be regarded as being entirely passive and simply responding to conditions developed in the overlying plates. This approach depends upon the behaviour of plates as stress guides and their response to stresses resulting from thermal contraction, their motion across the surface of a non-spherical earth, and marginal interactions with neighbouring plates (e.g. Turcotte & Oxburgh 1973).

In the discussion which follows we shall attempt to establish whether either of these tectonic hypotheses provides a satisfactory explanation for the phenomena observed.

Aspects of rift geology

In so far as there have been a number of extensive recent reviews of the crustal characteristics of the East African Rift System (e.g. McConnell 1972; Baker *et al.* 1972; Clifford & Gass 1970) we shall touch on these only to a small extent and concentrate on their role in the larger scale plate tectonics processes.

Fig. 1:1 shows the position of the Gregory Rift in relation to the rest of the rift

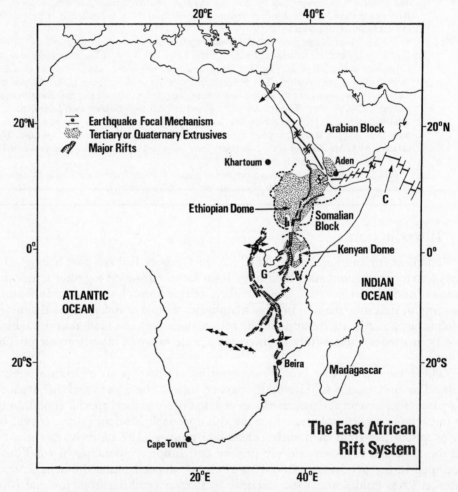

FIG. 1:1. A sketch map of the East African Rift System; dashed lines indicate the shape of the Kenyan and Ethiopian Domes, and arrows the nature of the fault plane solutions. C—Carlsberg Ridge; G—Gregory Rift.

system and the Red Sea and Gulf of Aden. The central parts of both the Red Sea and the Gulf of Aden are today occupied by oceanic crust which has evidently formed by spreading from an axial ridge system. In contrast, the crust within which the East African Rift System is developed, is of continental thickness and has an uncertain amount of separation which is nowhere more than 30 km, in contrast to the hundreds of kilometres separation on the other two arms.

At the largest scale, therefore, we may ask whether in the rift system we see the first stages of the generation of a new ocean or whether it represents a 'failed arm'—a line of rupture formed as part of the same fracture system which gave rise to the Red Sea and Gulf of Aden, but for some reason never developing into an ocean.

Movement along all three arms began in the Miocene and in the Red Sea and Gulf of Aden arms there was crustal subsidence with the formation of marine basins of restricted circulation within which thick evaporites were deposited. Such a phase of pre-separation crustal subsidence (presumably isostatic, resulting from lithospheric necking) is now recognised in a number of oceans (e.g. Kinsman 1975). In eastern Africa it seems that there was continuous localised differential vertical movement without the development of one clearly defined zone of crustal separation. Instead there developed a broad north-south trending zone within which there was a polygonal pattern of normal faulting; locally pairs of faults, or fault groups, with opposed throw gave rise to well-defined graben structures of which the Gregory rift is probably the best example.

Basaltic volcanism began in the early Tertiary along many parts of the rift system. Baker *et al.* (1972) propose an Eocene–Oligocene age for the onset of extrusion in Ethiopia but reliable age data are rather sparse; there is some suggestion that volcanicity begins progressively later in a southerly direction (Oxburgh & Turcotte 1974).

The greatest volumes of extrusives were spatially associated with the Kenyan and Ethiopian domes (Fig. 1:1). Both are regional upwarpings of the basement and are, in detail, somewhat complicated; they were produced by several superimposed and separate phases of vertical movement which differed in amount and did not exactly correspond geographically; post-Cretaceous, Miocene and late Tertiary phases are recognised giving a net surface flexuring with a length of the order of 1000 km and an amplitude of around 10 km in Ethiopia and perhaps half that in Kenya (Baker *et al.* 1972).

The continental lithosphere

We now consider the nature of the upper mantle under the Afro-Arabian block prior to rifting. The concept of plates which are internally rigid and comprise a relatively thin crustal layer which overlies a thicker and mechanically coupled layer of upper mantle material, is well established for the oceans and fairly well understood; seismic, thermal and petrological observations provide an internally consistent model. Continental crust forms by a different process and there is no reason to suppose that continental lithosphere should have the same thickness and properties as oceanic plates; indeed, the fact that continental crust is significantly thicker and less dense than oceanic crust, and contains a sufficient concentration of heat-producing elements for their distribution to be a major influence in determining the conductive surface heat-flux, all suggest that different considerations from those applicable in oceanic areas must determine the thermal structure and thus the thickness of the continental lithosphere.

We use the term lithosphere to signify the cool and therefore mechanically coherent crust and upper mantle system which moves as a unit in the directions indicated by movements between plates at the surface. We shall not deal with the detailed arguments here, but there are now three independent lines of evidence which suggest that the continental lithosphere is significantly thicker than that in the oceans. Seismological observations (Jordan & Frazer 1975) indicate that the velocity structure under continents is different from that in oceanic regions down to about 400 km; whether this difference is compositional or thermal in origin, it implies that 400 km or so of upper mantle are coupled to the overlying continent and move with it. The values derived for the mantle contribution to the conductive continental heat flow in old stable areas, are so low that a thick, (200–400 km) relatively cool region with a very low thermal gradient, seems to have underlain shield areas for times $> 10^9$ m.y. Isotopic and petrological considerations (e.g. Armstrong & Hein 1973; Boyd 1974) seem to require a coupled thickness of sub-continental upper mantle which is at least 200 km.

This discussion has been presented in order to show that the coherent unit which must undergo extensional strain during continental rift development must be at least several hundred kilometres thick.

Geophysical aspects of the Rift System

Direct geophysical studies in different parts of the Rift System have provided considerable insight into the extent of the modification of the lithosphere beneath the rift. These have been reviewed by Baker *et al.* (1972) and Darracott *et al.* (1973). The seismicity of the Red Sea and Gulf of Aden continues into Ethiopia and thence southward, with a somewhat diffuse distribution, along the general trend of the Rift System. A number of fault plane solutions have been published for earthquakes within this zone and show either extensional or strike slip movements on the active faults from which they are recorded (Fig. 1:1). This zone of present-day shallow-depth seismicity is about 40 km thick (Maasha 1975) and approximately overlies a region in the upper mantle with a number of features similar to those observed beneath mid-ocean ridges. There appear to be delays in the propagation of both P waves and S_n beneath the rift (Fairhead & Girdler 1971; Gumper & Pomeroy 1970). Together these observations suggest a relatively thin lithosphere in the vicinity of the rift, with a transition to relatively more ductile upper mantle ('asthenosphere') at a much shallower depth than to either side. Such a local increase in ductility is attributable to a high value for θ,[1] the

---

[1] $\theta = T/Tm$, where T is the mantle temperature and Tm is the temperature of onset of melting, both in °K; for values of $\theta > 0.75$ silicate materials are expected to behave in a ductile fashion on geological time scales and Oxburgh & Turcotte (1976) have used this value of θ to define a base to the mechanical lithosphere.

normalised temperature of the upper mantle. High values of θ may occur simply because the temperature at any particular depth is anomalously high, but equally could result from temperatures of beginning of melting which had been locally depressed by anomalous pressures of volatiles.

We now consider the results of gravity surveys made over the rifts and adjacent areas. As pointed out by Girdler & Searle (1975) there is typically a broad ($\sim$ 1000 km) negative Bouguer anomaly roughly centred on the rift system as a whole, with a superimposed smaller scale ($\sim$ 100 km) positive anomaly centred on the Eastern Rift. The negative anomaly has been interpreted in terms of a thin lithosphere. The positive anomaly is thought to indicate limited intrusion of asthenospheric ultramafic material into the continental crust.

Darracott *et al.* (1972) propose a gravity profile (Fig. 1:2) in which a 90 km lithosphere (density 3.34) thins to about 55 km across the rift zone within which asthenospheric material (density 3.24) occurs at a shallower depth than on either side. Clearly this is a highly generalised regional model as the authors would, no doubt, be the first to emphasise; it is, however, the case that for the proposed lateral density contrast ($\sim$ 3 per cent) to be the result of thermal expansion, that temperature contrast producing it would have to be considerably in excess of 1000°C. Melting would occur before such temperature differences were reached and would reduce the density of the lithosphere. However, roughly 30 per cent melting would be required to explain the proposed density difference. These difficulties have also been discussed by Bailey (1972, 1974b).

We conclude, therefore that either (1) the proposed deep structure with its 3 per cent density contrast results not solely from a temperature difference, but a difference in chemical composition, i.e. low density material of different chemical composition has been physically introduced from elsewhere into a lithosphere which has somehow been thinned; or that (2) at least part of the density contrast results from solid-solid, or solid-liquid, phase changes and that the asthenosphere may retain a high degree partial melt; or (3), that the proposed structure is over-

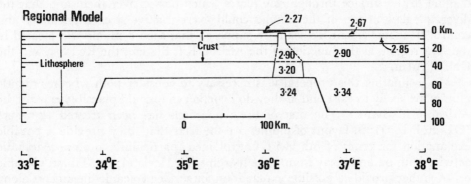

FIG. 1:2. The lithosphere/asthenosphere model proposed for the East African Rift System just south of the equator by Darracott *et al.* (1972) satisfying gravity observations. Figures on the diagram indicate densities.

generalised, and that in particular a much thicker lithosphere with a deeper and smaller density contrast and of different geometry is required.

The first possibility is not considered probable because the amount of 'necking' of the lithosphere required to allow an upwelling of material from below on the scale suggested by the model gravity profiles would correspond to about 100 km of extension at the surface and this is precluded by geological observations. The writer favours a combination of the second and third possibilities; the density contrast under the rift system probably is somewhat less than shown in Fig. 1:2 and extends to a depth of more than several hundred kilometres but in order to satisfy the observed width of the negative anomaly, must be much narrower, particularly at depth.

We therefore require a large-scale tectonic process which can produce limited separation of continental lithosphere more than several hundred kilometres thick, giving rise to the rift system at the surface and an underlying zone of high values of θ and some partial melting. We may also consider the evolution of such a structure into a spreading ocean.

Tectonic hypotheses

No attempt will be made to review all the hypotheses which have been advanced for the origin of the rift system (for reviews and other models see Bailey 1964, 1974, 1975; Baker *et al.* 1972; Gass 1970, 1973; King 1970; Harris 1970). We here restrict discussion to 'mantle plume tectonics' and 'membrane tectonics'.

A. *Mantle plumes*

A world-wide system of mantle plumes was proposed by Morgan (1971; 1972a, b). He suggested that a limited number of localities characterised by particularly profuse magmatic activity were underlain by mantle plumes— localised, pipe-like zones of rapidly ascending mantle flow. These plumes originated at or close to the core-mantle boundary and remained relatively fixed with respect to each other in space and time. It was thought that such plumes could deliver magma to the surface through any plates which passed over them and that the diverging flow at top of the plumes could exert tractive stresses on plates and control their movement. It has been supposed that such a plume was located in the Afar region, at the junction of the African Rift System, the Red Sea and the Gulf of Aden.

In evaluating this proposal it is necessary to consider both whether mantle plumes are likely to exist and, if they do, whether they could plausibly explain the African Rift System. The first of these questions has been treated elsewhere (Oxburgh 1974); in favour of plumes are the facts that, they provide a possible explanation for zones of otherwise unexplained, particularly profuse magmatic activity both on and away from plate margins (e.g. Iceland; the Tibesti volcanic centre); they provide a possible explanation for surface volcanic lineaments along which the age of onset of volcanism seems to have changed regularly with time (e.g. the Hawaiian Chain, Dalrymple *et al.* 1973). The physical basis for the existence of mantle plumes has, however, been seriously constrained (Parmentier *et al.* 1975). Detailed analysis of plate motions and the geometry of possible plume traces (Molnar & Atwater. 1973) shows that if plumes are responsible for the features attributed to them, they must be able to migrate with respect to each other with velocities of a few cm/yr.

On the second question, the results of Parmentier *et al.* (1975) show that plume-induced shear stresses at the base of the lithosphere, of more than ten bars are rather improbable. Such stresses are low by comparison with the strength of the lithosphere. If temperatures at the top of the plume are abnormally high, that part of a plate which passed over a plume would be heated; it might be suggested that such local heating of the base of the plate could induce thermal stresses which might both dome and rupture the brittle upper part of the plate. Taking a mean plate velocity of 4 cm/yr and assuming a characteristic plume width of 400 km, a point on the bottom of the plate, which passed directly over the centre of the plume, would be exposed to the heat source for about 10 m.y. This would allow conductive heating of about the bottom 20 km of the plate. Such heating would, however, only be important, if the top of the plume were significantly hotter than the ambient mantle temperature at that depth. Temperatures in the plume, as outside, are ultimately liable to buffering by partial melting and are unlikely to be more than 300°C higher than the ambient temperature. Thus conductive heating from an underlying plume could be expected to give a rather small temperature increase over a depth interval which was small by comparison with the thickness of the plate.

If, however, the plume was able to discharge magma and/or volatiles into the passing plate the situation might be different. Convective transfer of heat from the plume to the plate would then take place and the only limit on the amount of heating is provided by the limits thought reasonable for the rates and total amounts of fluid transferred from the plume to the lithosphere.

The chemistry of the magmas observed at the surface in rift zones (Bailey 1966, 1974a, b) presents problems if they are derived from plumes; it suggests that they have either been generated or at least undergone a complete re-equilibriation with a mineral assemblage stable at depths of less than 200 km or so; unless plume magmas never reach the surface or invariably pause during their ascent to re-equilibriate in the upper part of the lithosphere, it seems that their role is unimportant in heating the lithosphere. It is not difficult to show that unless the water content of plume material is much greater than 0.1 per cent or the mean upward velocity of material in the plume is very much greater than 20 cm/yr heating by the transfer of hot water to the lithosphere would be of the order of a few tens of degrees C.

In fact, if it were possible for a plume to discharge only a water-rich fluid phase into an overlying plate, the main effect would be to depress the melting temperature of the plate and to produce intra-plate melting without any significant elevation of the rock temperature. Volume increases associated with the melting could perhaps produce surface elevation and arching and the dilational features observed. Bailey (1970, 1974a, b) has emphasised the possible effects of local concentrations of volatiles of this kind.

B. *Membrane tectonics*

The approach of membrane tectonics is entirely different. It is supposed that no special conditions prevail in the mantle to initiate rifting. The break-up of continental plates is attributed to stresses generated within them. The essential idea is illustrated in Fig. 1:3 It is known from a variety of lines of evidence that plates have undergone substantial changes in latitude; as they do so, however, it is necessary for them to accommodate their shape to the change in curvature of the geoid between the equator and the poles. Although for many geological purposes the amount of this curvature change is unimportant, it is very significant for large plates. As an illustration, if the African plate were instantaneously transported from its present position and came to rest near one of the poles without change in curvature, there would be a vertical mismatch (i.e. a gap) of several tens of kilometres between the centre of the plate and the underlying geoid! The figure of the earth is now known so well that it can be regarded as firmly established that all plates do change their curvature continuously as they change latitude. This problem has been analysed theoretically (Turcotte, 1974) and it has been shown that a plate moving towards the equator should have its margin in compression and its central part in tension; for motion away from the equator the tensional and compressional zones are reversed (Fig. 1:3). The magnitudes of the stresses for plates the size of the African plate undergoing substantial latitude changes are in the kilobar range (i.e. of the same order as the strength of the plates) and should be sufficient to rupture the plate. The theory has been applied to the African plate (Oxburgh & Turcotte 1974) as an explanation of the rift system.

The membrane theory depends upon the assumption that the plates, or at

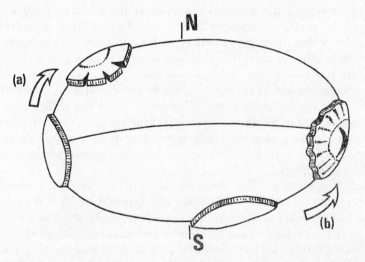

FIG. 1:3. Membrane Tectonics: a plate moving away from the equator (a) undergoes an increase in radius of curvature and is extended at its margin and compressed in its interior. The opposite effects are found in a plate moving towards the equator (b).

any rate, their uppermost parts behave in brittle fashion; if the plates were completely ductile there would be no membrane effect because the plate would continuously yield homogeneously to take up a new curvature. Insofar as the membrane effect and the magnitude of the stresses do not depend on the thickness of brittle material, it is sufficient if only the cold, outer few tens of kilometres behave in brittle fashion. Evidence that the lithosphere does behave in this way on time scales of geological interest is provided by the 'forebulge' observed in oceanic plates as they bend downwards into a subduction zone (Oxburgh & Turcotte 1976).

Membrane theory does not provide any direct explanation for the magmatic activity which is found associated with every rift system. It is necessary to postulate either that the asthenosphere contains small amounts of melt which are permitted to rise to the surface by the development of an extensional stress regime in the overlying lithosphere; or, that melts are developed *within* the lithosphere. In this latter case melting may occur in localised zones of strain working by processes similar to those described by Anderson & Perkins (1974). It should be emphasised that it is only the melts at lower degrees of lithospheric extension which need be derived in this way; as a rift opens into a proto-ocean, presumably the lithosphere necks progressively, initiating and localising a zone of diapiric uprise of the asthenosphere which eventually may develop into a steady flow feeding a spreading ocean ridge. Once such a process has been established melts may presumably be derived by the same process of adiabatic decompression as occurs under oceanic ridges (Oxburgh 1965). This is essentially the sequence of events proposed by Gass (1970; 1973).

The scale and pattern of movements of the Kenyan and Ethiopian domes allow some constraints to be placed upon their mode of formation. The speed and spasmodic character of the uplift indicate that movement of material must have occurred within the lithosphere and in view of the association of both features with volcanic activity it is tempting to associate both with the lateral migration of magma.

Discussion and conclusions

The surface geology suggests that events leading to the formation of the triple rift system which split the Afro-Arabian block began in pre-Miocene times. There was, however, presumably no spreading or major surface rifting until the Miocene when the Carlsberg ridge seems to have propagated northwards from the oceanic crust of the Indian Ocean into the continental Afro-Arabian block. Previously it must have terminated in oceanic crust at a transform fault (Owen Fracture Zone) sub-parallel to the Afro-Arabian coast line (Laughton *et al.* 1972).

Between the Miocene and the present, Africa has undergone considerable movements. Palaeomagnetic evidence (McElhinny 1973) shows that in that time it has undergone a northward latitude change of about 10°; in addition it may have experienced longitude changes, situated as it is between two spreading ridges. It is therefore not easy to see how a fixed mantle plume could both have initiated the Afar triple junction break, and continue to underly it today (for a contrary view see Burke & Wilson 1972).

The problem with the membrane hypothesis is rather different. The theoretical basis for membrane stresses is sound and it is hard to see why they should not both exist in the lithosphere and be of the right magnitude to cause extensional failure of the kind observed. There is, however, no conclusive evidence which links rift formation with such stresses.

The thinning of the lithosphere which is evident under the rift system can come about in one of two ways. Insofar as the lithosphere is the cool, and thus relatively strong and mechanically coherent part of the upper-mantle and the coupled overlying crust; the lithosphere may be thinned simply by local heating and elevation of the geotherms. Alternatively, it may be thinned mechanically by a necking process during extensional deformation. The relatively low upper limits of a few tens of kilometres which may be placed on extension across the rift on the basis of the surface geology, mean that necking can play relatively little part in the thinning of the lithosphere under the rift although it may well have been important for the Red Sea and Gulf of Aden.

From a thermal point of view, it is essential to make a clear distinction between conductive and convective processes in considering the evolution of the rifted lithosphere. They have quite different characteristic rates associated with them. Over a period of 50 m.y. significant heating of rocks by conduction in response to an applied heat source occurs only over a distance of 40 km or so. This is a small fraction of the thickness of the lithosphere. Much more rapid heating can be achieved if heat is transferred by convection, i.e. if magma or hot fluids penetrate the lithosphere. The difficulty here is that the amount of heating which can be accomplished is limited by the volume of new material which may be added to the lithosphere without giving rise to more extension than is observed at the surface.

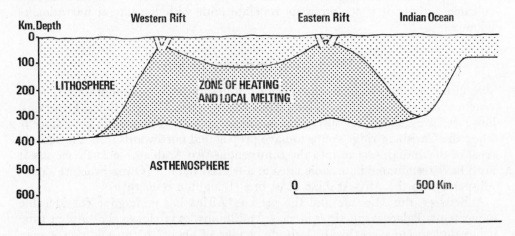

Fig. 1:4. A sketch of the proposed lithospheric configuration under the East African Rift System. The pre-rifting lithosphere has been partly modified in thickness by necking but predominantly by heating and partial melting. The Zone of modified lithosphere (heavy stipple) would presumably be geophysically indistinguishable from asthenopshere although it might differ chemically.

At present it seems most likely that several mechanisms contribute to the high temperatures under the rift system and the consequently thin present day lithosphere; a combination of heating by strain working and the lowering of lithospheric melting temperatures by the rise of hot volatiles from the lower part of the lithosphere to the asthenosphere, could give rise to the widespread zone of partial melting or near partial melting under the rift which seems indicated by the geophysical evidence (Fig. 1:4). Both of these phenomena could arise from membrane stresses. On the question of whether the rift system will evolve into an ocean, the writer is inclined towards the view of Burke & Whiteman (1973) that whether lithospheric breaks evolve into spreading oceans depends on the pattern of worldwide plate interaction at the time the break is formed and that favourably oriented breaks may evolve in this way while others may not. All the breaks would, however, have undergone the same evolutionary sequence of events before spreading. It is hard to see how plumes of the type suggested by Morgan could generate rifts but slowly migrating plume-like mantle flows with different and special characteristics, could conceivably do so.

References

ANDERSON, O. L. & PERKINS, P. C. 1974. Runaway temperatures in the Asthenosphere resulting from viscous heating. *J. Geophys. Res.* **79**, 2136–8.

ARMSTRONG, R. L. & HEIN, S. M. 1973. Computer simulation of Pb and Sr isotope evolution of the earth's crust and Upper Mantle. *Geochim. Cosmochim. Acta.* **37**, 1–18.

BAILEY, D. K. 1964. Crustal warping—a possible tectonic control of Alkaline magmatism. *J. Geophys. Res.* **69**, 1103–11.

—— The system $Na_2O–Al_2O_3–Fe_2O_3–SiO_2$ at 1 Atmosphere and the Petrogenesis of Alkaline Rocks. *J. Pet.*, **7**, 114–70.

—— 1970. Volatile flux, heat focussing and the generation of magma, in *Mechanism of Igneous Intrusion, Geol. J.* Special Issue No 2, 177–86.

—— 1972. Uplift, rifting and magmatism in continental plates. *J. Earth Sci.* (Leeds), **8**, 225–39.

—— 1974a. Melting in the deep crust: 436–42 in *The Alkaline Rocks* (Ed. H. Sørensen), New York (Wiley). pp 622.

—— 1974b. Continental Rifting and Alkaline Magmatism: 148–59 in *The Alkaline Rocks* (Ed. H. Sørensen) New York (Wiley). 622 pp.

BAKER, B. H., MOHR, P. A. & WILLIAMS, L. A. J. 1972. Geology of the Eastern Rift System of Africa. *Geol. Soc. Amer. Special Paper*, **136**, 67 pp.

BOYD, F. R. 1974. Ultramafic nodules from the Frank Smith Kimberlite pipe, South Africa. *Carnegie Inst. Washington Yearbook*, **73**, 285–94.

BURKE, K. & DEWEY, J. F. 1974. Two plates in Africa during the Cretaceous? *Nature, Lond.*, **249**, 313–16.

—— & WHITEMAN, A. J. 1973. Uplift, Rifting and the Break-up of Africa, 735–53, in *Implications of Continental Drift to the Earth Sciences*, **2**, London (Academic Press). 1184 pp.

—— & WILSON, J. T. 1972. Is the African plate Stationary? *Nature, Lond.*, **239**, 387–90.

CLIFFORD, T. N. & GASS, I. G. 1970. Eds. *African Magmatism and Tectonics*. Edinburgh (Oliver & Boyd). 461 pp.

DALRYMPLE, G. B., SILVER, E. A. & JACKSON, E. D. 1973. Origin of the Hawaiian Islands. *American Scientist*, **61**, 294–308.

DARRACOTT, B. W., FAIRHEAD, J. D. & GIRDLER, R. W. 1972. Gravity and magnetic surveys in Northern Tanzania and Southern Kenya. *Tectonophysics*, **15**, 131–41.

FAIRHEAD, J. D., GIRDLER, R. W. & HALL, S. A. 1973. The East African Rift System, 757–66, in *Implications of Continental Drift to the Earth Sciences*, **2**, London (Academic Press). 1184 pp.

FAIRHEAD, J. D. & GIRDLER, R. W. 1971. The Seismicity of Africa. *Geophys. J. Roy. Astr. Soc.*, **24**, 271–301.

GASS, I. G. 1970. Tectonic and magmatic evolution of the Afro-Arabian dome, 285–97, in *African Magmatism and Tectonics*. Edinburgh (Oliver & Boyd). 461 pp.

—— 1973. The Red Sea Depression: Causes and Consequences, 779–85, in *Implications of Continental Drift to the Earth Sciences*, **2**, London (Academic Press). 1154 pp.

GIRDLER, R. W. & SEARLE, R. C. 1975. Continental Rift Systems, 33–9, in *Geodynamics Today*. London (The Royal Society). 197 pp.

GUMPER, F. & POMEROY, P. W. 1970. Seismic waves and earth structure of the African Continent. *Bull. Seism. Soc. Amer.*, **60**, 651–8.

HARRIS, P. G. 1970. Convection and magmatism with reference to the African Continent, 419–38, in *African Magmatism and Tectonics*. Edinburgh (Oliver & Boyd). 461 pp.

JORDAN, T. H. & FRAZER, L. N. 1975. Crustal and Upper Mantle Structure from Sp Phases, *J. Geophys. Res.*, **80**, 1504–18.

KING, B. C. 1970. Vulcanicity and Rift Tectonics in East Africa, 263–85, in *African Magmatism and Tectonics*. Edinburgh (Oliver & Boyd). 461 pp.

KINSMAN, D. J. J. 1975. Salt Floors to Geosynclines. *Nature, Lond.* **255**, 375–78.

LAUGHTON, A. S., SCLATER, J. G. & McKENZIE, D. P. 1972. The Structure and Evolution of the Indian Ocean, 201–12, in *Implications of Continental Drift to the Earth Sciences*. London (Academic Press). 622 pp.

MAASHA, N. 1975. The Seismicity of the Ruwenzori Region in Uganda. *J. Geophys. Res.*, **80**, 1485–96.

McCONNELL, R. B. 1972. Geological Development of the Rift System of Eastern Africa. *Geol. Soc. of Amer. Bull.*, **83**, 2549–72.

McELHINNY, M. W. 1973. *Palaeomagnetism and plate tectonics*. Cambridge (Cambridge University Press). 358 pp.

MOLNAR, P. & ATWATER, T. 1973. On the relative motion of 'Hot-Spots'. *Nature, Lond.* **246**, 288–91.

MORGAN, W. J. 1971. Convection Plumes in the lower mantle. *Nature, Lond.* **230**, 42–3.

—— 1972a. Deep Mantle Convection Plumes and Plate Motions. *Am. Ass. Petrol. Geol. Bull.*, **56**, 203–13.

—— 1972b. Plate Motions and deep Mantle Convection. *Geol. Soc. Amer. Mem.*, **132**, 7–32.

OXBURGH, E. R. 1965. Volcanism and mantle convection. *Phil. Trans. Roy. Soc., Lond.*, A. **258**, 142–4.

—— 1974. The plain man's guide to plate tectonics. *Proc. Geol. Ass.*, **85**, 299–359.

—— & TURCOTTE, D. L. 1974. Membrane Tectonics and the East African Rift. *Earth Plan. Sci. Letters.*

—— & TURCOTTE, D. L. 1976. The Physico-Chemical Behaviour of the Descending Lithosphere. *Tectonophysics*, **32**, 107–28.

PARMENTIER, E. M., TURCOTTE, D. L. & TORRANCE, K. E. 1975. Numerical experiments on the Structure of Mantle Plumes, *J. Geophys. Res.*, **80**, 4417–24.

TURCOTTE, D. L. 1974. Membrane Tectonics. *Geophys. J. Roy. Astr. Soc.*, **36**, 33–42.

—— & OXBURGH, E. R. 1973. Mid-Plate tectonics. *Nature, Lond.*, **244**, 337–9.

2

ROBERT M. SHACKLETON

Structural development of the East African Rift System

As the purpose of this volume is to outline the background to hominid evolution in East Africa, this paper emphasises the influence of structural development on palaeo-environments rather than the mechanism or causes of rifting.

Late Mesozoic Stages

During the late Mesozoic, eastern Africa was a lowland area, flooded from the opening Indian Ocean as far westward as a shoreline which migrated back and forth across Ethiopia, Kenya and Tanzania, with a western limit at about 38°E. Subsidence of the Gulf of Aden Rift and associated uplift of the Horn of Africa began in the Jurassic (Azzaroli & Fois 1964). The future southern Red Sea area was subsiding but thicknesses and facies of Jurassic and Lower Cretaceous successions show that the Afar depression had not begun to subside. The Jurassic and Lower Cretaceous marine sedimentation of limestones, shales and sandstones, in Ethiopia and Kenya, suggests low to moderate relief in the source areas to the west. The regions lay about 20° further south than at present. Small basalt eruptions occurred in Somalia (Jurassic) and in northern Ethiopia.

Late in the Cretaceous the surface of low relief which had developed across the interior of East Africa was unevenly uplifted. The greatest elevation (to about 500 m) was in central Kenya and in Ethiopia (Saggerson & Baker 1965). Meanwhile, the coastal regions of East Africa were subsiding. Because the uplifted areas were just those whose subsequent further uplift was clearly associated with rifting, the late Cretaceous uplifts are interpreted as the first stage in the rifting process. The only evidence of subsidence in the rift zone at this time is in Turkana, where Cretaceous dinosaur bones have been reported from the Turkana Sandstone near Lokitaung (Arambourg & Wolff 1969). The Turkana Sandstone accumulated in a depression centred slightly west of the present Lake Turkana and limited to the west by a basement rise, thought to be the result of monoclinal warping and local faulting along the Kenya–Uganda border (Baker & Wohlenberg 1971). At Losodok, Turkana Sandstones dipping eastwards are overlain unconformably by basalts dipping west and at the base of the basalts is a sedimentary horizon with Lower Miocene (Burdigalian) vertebrates (Arambourg & Wolff 1969). In south-west Ethiopia sandstones, a few metres thick, interpreted as sheet flood deposits on a pedeplained basement surface, underlie Tertiary basalts, sometimes unconformably (Davidson *et al.* 1973). These sandstones might be Cretaceous although a Tertiary age seems more probable in view of their regional near-conformity with the basalts. In the Loperot area of southern Turkana, sandstones also classified as Turkana Grits contain a Miocene fauna (Dixey 1945; Joubert 1966).

Early Tertiary Stages, 65–20 m.y. ago

During the early Tertiary, prolonged stability allowed the reduction of large areas of East Africa to a pedeplain. Inselbergs and hill ranges preserving remnants of the late Cretaceous (and older) surfaces stood above the early Tertiary surface. The watershed between the Indian and Atlantic oceans was situated near the western side of the present Gregory Rift.

At least 30 m.y. ago[1] (Baker *et al.* 1971) and perhaps considerably earlier (Zanettin & Justin Visentin 1974 and 1975; D. Rex pers. comm.), basalts were erupted in southwest Ethiopia and northwest Kenya.

By 25 m.y. ago, much of Ethiopia had been flooded by basalts (Trap Series) to a thickness of over 2500 m in southern Afar and 2000 m over parts of the western plateau. These basalts were apparently erupted from fissures in Afar and the Ethiopian Rift. Afar was subsiding under continuing tension but the basalts flowed out fast enough to flood far beyond the limits of the future rift zones. They flooded a smooth surface cut on Mesozoic sediments. The lava surfaces themselves were mostly smooth and red soils developed on some of them between eruptions.

An angular unconformity within the basalts of the Ethiopian plateau, seen at Maichew (12°45′N, 38°30′E) suggests an early phase of deformation followed by erosion but the regional significance of this unconformity is not clear. Mapping in the Ethiopian highlands northwards from Addis Ababa, together with radiometric dating, is thought to show that an older, pre-Oligocene series of basalts (Ashangi Basalts) suffered large-scale warping, followed by the erosion of a level surface (Ashangi peneplain) before being unconformably overlain, throughout a region extending 500 km north of Addis Ababa, by the Aiba Basalts and Alaji Rhyolites about 32–28 m.y. old. (Zanettin *et al.* 1974; Morbidelli *et al.* 1975). About 25–22 m.y. ago, alkali granites were emplaced, probably as subvolcanic complexes, in the Danakil Alps, central Afar and at the western margin of Afar. Rhyolites and ignimbrites were erupted at about this time. The Afar rifting is thought to have been intitiated between 25 and 23 m.y. ago (Barberi *et al.* 1975) although the Red Sea model of Girdler and Styles (1974) implies that three-quarters to four-fifths of the present width of Afar must have been formed during the early Tertiary spreading episode between 41 and 43 m.y. ago (Mohr 1975). There is no geological evidence known from Afar to prove or disprove this.

The Kisumu (formerly Kavirondo) Rift in western Kenya must have been initiated as early as 22 m.y. ago; this is the age of the earliest dated tuff erupted from the Kisingiri volcano at the southwestern extremity of the Rift (Baker *et al.* 1971). This graben controlled the location of volcanoes and sedimentation; movements on faults within the Rift led to slumping of tuffaceous sediments and to

---

[1] Dates quoted have in most cases been reduced to the nearest whole number and error limits omitted; all are KAr– dates unless otherwise stated.

unconformities within the Lower Miocene sequences on Rusinga Island (Shackleton 1951). Near the eastern end of the Kisumu Rift, the Tinderet volcano began to erupt more than 20 m.y. ago: one of the early lavas is dated at 20.5 m.y. (Baker *et al.* 1971). Early Miocene lake deposits in the Rift reflect tectonic depression although volcanic damming may have been partly responsible for the lakes. The Western (Albertine) Rift was also subsiding during the Miocene (Bishop 1965). But although the Afar, Gregory Rift, Kisumu and Western Rifts were initiated in the Miocene they must then have been very shallow depressions. There is no definite evidence that 20 m.y. ago the Gregory Rift extended south of the Nakuru area just south of the Equator.

Stages from 20–13.5 m.y. ago
(Lower to Mid–Miocene)

In Ethiopia the eruption of basalts of the Trap Series continued into the Middle Miocene. The earliest direct evidence of major structures developing at the margins of the rifts is in southern Afar, where arrays of tilted fault blocks of trap basalts and older rocks, with dominantly NNW trending faults, are overlain unconformably by horizontal rhyolites dated at 15.3 m.y. (Black *et al.* 1975). This phase of predominantly acid volcanism in south Afar appears to have lasted until about 10 m.y. ago (Chessex *et al.* 1975). Along the western margin of Afar there is a wide zone of eastward-tilted blocks with antithetic faults, where the tilted basalts are unconformably overlain by basalts and rhyolites yielding ages of 10–13 m.y. (Zanettin & Visentin 1974). By 15 m.y. ago, the Afar depression was already separated from the surrounding uplifted volcanic highlands to the south, east and west, by zones of tilted blocks. After this time, lavas erupted within Afar were confined to the depression instead of spreading far over the surrounding areas. Big shield volcanoes were being built up on the western Ethiopian plateau while Afar was subsiding.

The main Ethiopian Rift, which extends about 450 km south-southwest from Afar before dying out in a series of splayed tilt blocks, may have begun developing during the Miocene; the fault scarps in the south are deeply eroded and old-looking but precise data are lacking, and also for the more recent-looking Stephanie Rift which is offset en echelon between the Main Ethiopian Rift and the Omo–Lake Turkana extremity of the Gregory Rift.

In the Turkana depresson west of Lake Turkana, eruption of basalts continued, dates ranging from 32–14 m.y. (Baker *et al.* 1971). The Samburu basalts were being erupted farther south in the Gregory Rift during a similar interval, with dates of 23 to 18.5 m.y. (Baker *et al.* 1971). These basalts thicken from Laikipia towards the rift (Shackleton 1946b) but seem to have been restricted within it on the west by the Elgeyo escarpment (Walsh 1969). The Samburu basalts were tilted, faulted and eroded before being overlain unconformably by another series, the Elgeyo basalts, in the western part of the Gregory Rift just north of the equator (Walsh 1969). One of these Elgeyo basalts is dated at 15 m.y. (Baker *et al.* 1971) indicating that the preceding phase of faulting occurred at about the same time as that in southern Afar.

Outside the Kisumu and Gregory Rifts, isolated volcanoes (Elgon, Moroto, Napak, Kadam) were built up, from about 20 m.y. ago (Baker *et al.* 1971) over a complex early Miocene topography (Bishop & Trendall 1967). Under the Elgon volcanics there is a buried slope trending ENE and about 20 to 50 km wide, which leads down from the Kitale surface at 1750 m–1830 m to the Kyoga surface at 1060 m–1220 m. This slope is interpreted (Bishop & Trendall 1967) as an erosion scarp separating two surfaces of different ages, although its lower part is evidently tectonic since it forms the southern limb of the North Elgon depression. A scarp beneath Napak is likewise associated with a subvolcanic depression. The Kitale and Kyoga surfaces were regarded as one and the same, and the slope connecting them as tectonic, by Shackleton (1951), Pulfrey (1960) and Baker and Wohlenberg (1971) whereas Dixey (1948) and Bishop & Trendall, (1967) regarded the Kitale surface as older: the latter authors correlated the Kyoga surface with the Kasubi surface of Central Uganda, where it is clearly at a lower level and younger than the Buganda surface. The Turkana escarpment north of Moroto is the result of a monocline along which lavas from Moroto and fossiliferous sub-volcanic sediments derived from the northeast, were tilted 3° to the ENE (Varne 1965).

Within the Kisumu Rift, in the long interval between 22 and 14 m.y. ago, only about 200 m of mostly subaerial agglomerates, tuffs and sediments accumulated at the western side of the Tinderet volcano. In southwest Kenya the westward tilted south Kavirondo block is bordered to the south by the Siria and Tarime faults and to the north by the Kaniamwia and Sondu faults and flexures. Along the southern side of this block the flood phonolites rest on a smooth surface which slopes from 1850 m in the east to 1500 m near Tarime (in N. Tanzania). If projected westwards, this uniform slope would meet the sub-volcanic Lower Miocene sediments of Karungu. These sediments, about 50 m thick, with laterite at the base, are succeeded without apparent break by the lavas and tuffs erupted from the Kisingiri volcano. The sediments are dated at about 22 m.y. (Baker *et al.* 1971). Thus sediments more than 22 m.y. old rest on virtually the same surface as that which 50 km away is covered by phonolites between 11 and 13.5 m.y. old. In 10 m. years the surface was not appreciably lowered. Similarly the Kitale surface is covered in the east by the Uasin Gishu phonolites about 12 m.y. old and in the west by the eastern part of the Elgon volcanics about 20 m.y. old. One might suppose that such slow lowering of the surface implies prolonged stability near sea level but the evidence of Bishop & Trendall (1967) shows that throughout this time the Kitale surface must have formed a plateau at an elevation of at least 610 m above the Kyoga surface to the west. In Kenya it was recognised (Saggerson & Baker 1965) that 'dissection of the sub-Miocene plain by the end-Tertiary erosion cycle was well advanced before the eruption of the late Miocene phonolites'. The late Miocene phonolites must therefore have flooded out from the rift over plains which stood at an elevation of over 600 m near the rift but sloped eastward to the sea, and were being eroded from both sides to lower levels.

Stage from 13.5–11 m.y. ago (Middle Miocene)

In Kenya an important event was the eruption, between 13.5 and 11 m.y. ago of about 2500 km³ of flood phonolites (Baker *et al.* 1972). These came from sources within the Gregory Rift (suggestions of more remote local sources are unconvincing) but they flooded far beyond the limits of the Rift, reaching over 300 km southeastwards down a shallow valley to the Yatta Plateau and 200 km southwest and west to Tarime (Tanzania) and the Gulf of Kisumu. At this time the Gregory Rift was not deep enough to contain the phonolites which reached a maximum thickness of about 700 m (McCall 1967) just north of the equator, where one group of phonolites can be traced from Laikipia east of the Rift across the floor to the Uasin Gishu plain west of the Rift (McCall 1967; Walsh 1969). East of the Rift, in Laikipia, these phonolites (Rumuruti phonolites) flowed eastwards through a gap between hills, covering lacustrine sediments (dammed by lavas?) in valleys, and onto the flat sub-Miocene surface (Shackleton 1946). West of the Rift the equivalent Uasin Gishu phonolites flowed over the Kitale surface and farther south the phonolites flowed across the sub-Miocene surface above which stood hills surmounted by the end-Cretaceous surface. This phonolite flood allows correlation of surfaces across the Rift. Unless different surfaces had been coincidentally uplifted to the same elevations on opposite sides of the Rift, it must be concluded that the Kitale surface cannot be equivalent to the end-Cretaceous surface as believed by Dixey (1946) and Pulfrey (1960) but must correspond to the sub-Miocene surface east of the Rift (Shackleton 1951; Bishop & Trendall 1967).

Stage from 11–8 m.y. ago (Upper Miocene)

In southern Afar, an important phase of down-faulting was followed by the eruption of fissure basalts (11–10 m.y.), rhyolites, ignimbrites and pumice (10–9 m.y.) in which lacustrine sediments (Chorora Formation) are intercalated (Christiansen *et al.* 1975). These diatomaceous lake beds lie at the foot of the southeast Afar escarpment where a lake was impounded by the Assabot volcano (Sickenberg & Schönfeld 1975). Further downwarping, faulting and slight tilting occurred along the south Afar margin between 9 and 8 m.y. ago (Christiansen *et al.* 1975).

In the Turkana basin, widespread and thick ignimbrites and rhyolites, with mugearites and trachytes, were erupted, probably in late Miocene or early Pliocene times. Farther south along the Gregory Rift, in the Baringo area, faulting contemporaneous with sedimentation is shown by abrupt changes in thickness of members of the Tugen Hills Group, about 10 m.y. old, as they are traced across faults (Bishop & Pickford 1975). The impressive Samburu monocline (Shackleton 1946b) along which the Samburu basalts and overlying Rumuruti phonolites dip west towards the Rift, probably originated late in the Miocene or early in the Pliocene.

Along the northern side of the Kisumu Rift, north of Koru, tuffs and sediments deposited between 20 and 14 m.y. ago were apparently displaced about

500 m vertically by a monoclinal flexure and subsidiary faults, between 14 and 10 m.y. ago, forming an escarpment against which were banked the Upper Tinderet volcanics (Binge 1962; Jennings 1964; cf. Shackleton 1951; Bishop & Trendall 1967). The upper Tinderet volcanics give dates ranging from 9.9 to 5.6. m.y. (Baker *et al.* 1971).

Stage from 8–5 m.y. ago (Late Miocene)

Between 8 and 6.5 m.y. ago the Dalha Series and correlative flood basalts with subordinate ignimbrites were erupted around the margins and probably across the floor of the Afar Depression (Barberi *et al.* 1975), and south of the Tadjurah Gulf. These basalts were in turn faulted, between 5 and 4 m.y. ago and deeply eroded before being covered by further volcanics. Basalts were erupted in South Turkana (Lothagam basalts) at about the same time (8.3 m.y.) as the Dalha Series in Afar (Baker *et al.* 1971). The predominantly basaltic Aberdare range at the eastern edge of the Gregory Rift was mainly built between about 6.5 and 5 m.y. ago (Baker *et al.* 1971).

In the central part of the Gregory Rift, voluminous trachytes, trachytic tuffs and phonolites were erupted between 7 and 5 m.y. ago. They include the Kabarnet trachyte (7 m.y.) in Kamasia, the Thomson's Falls phonolites (6.5 m.y.), the Mbagathi phonolitic trachyte, Nairobi phonolite (5.2 m.y.) and Kandizi phonolite of the Nairobi area. Farther south the volcanoes of Esayeti (6.7–3.6 m.y.) and Olorgesailie (5.8 m.y.) were built up at about this time (Baker *et al.* 1971). Lavas from these volcanoes flowed eastwards beyond the eastern boundary faults of the rifts, over areas which are now at a much higher elevation than the rift floor on which these old volcanoes stand, but at the western side, the Nguruman fault scarp contained the Kirikiti basalts, about 5 m.y. old (Baker *et al.* 1971), within the Rift. Strong faulting in the southern sector of the Gregory Rift occurred between the building of the Olorgesailie and Esayeti volcanoes and the eruption of the succeeding Limuru trachytes.

Stage from 5–2 m.y. (Pliocene)

In southern and central Afar, a stratiform volcanic series, mainly fissure basalts, ranging in age from 5–2 m.y. and rhyolite domes about 3–5 m.y., are associated with lacustrine clays, sands and gravels dated at about 4 to 3 m.y. old; very intensive internal faulting affected the whole of the Afar depression after the eruption of this assemblage (Barberi *et al.* 1975). A series of stratoid trachytes, rhyolites, ignimbrites, tuffs and pumice (Nazareth Series) forming the floor of the northern part of the Main Ethiopian Rift has an age range of 5–2 m.y. (Meyer *et al.* 1975). Faulting occurred contemporaneously with the eruption of this series. In the Gregory Rift between 5 and 2 m.y. ago, sedimentation prevailed in the Omo basin and Turkana with minor eruptions such as the Kanapoi basalt (2.9 m.y.) but farther south, especially in the Nakuru basin, thick ignimbrite sheets and trachytes (Kinangop tuffs) with intercalated diatomaceous lake beds, covered a large area of the rift and the area to the east. A date of 3.3 m.y. was obtained from a feldspar from a trachyte in the Kinangop tuffs (Baker *et al.* 1971). The original extensive smooth surface of the ignimbrite sheets has subsequently been dislocated by an array of faults, as a result of which the surface now ranges from below the floor of the rift near Nakuru to high on the eastern flank (1300 m). The main topographic features of the existing rift are the result of this young, probably end-Pliocene, faulting.

Farther south, the Limuru trachytes lap round the faulted and eroded Esayeti volcano. They flowed eastwards from the rift as far as Nairobi. Dates of 1.5, 1.5, and 1.7 m.y. (Baker *et al.* 1971) are probably too young but in this sector of the rift also the main faulting responsible for the present topography is late Pliocene. The Singiriani basalts, dated at 2.3 m.y. (Baker *et al.* 1971) flooded round eroded fault ridges of Limuru trachyte and were confined to the floor of the rift by pre-existing fault scarps. They were themselves dislocated by reactivation of pre-existing faults soon after being erupted.

The End-Tertiary surface of Kenya, on which Upper Pliocene marine sediments were deposited near the coast, was uplifted during the Upper Pliocene and probably Lower Pleistocene. This uplift raised the surface to elevations of between 1200 and 1500 m along the eastern side of the Gregory Rift (Saggerson & Baker 1965).

Stage from 2–0 m.y. (Pleistocene)

During the last 2 m.y. a drastic change occurred in Afar and the Main Ethiopian Rift. Mainly between 1.8 and 1.6 m.y. ago, rapid and intense faulting (Nazareth Phase) restricted to a narrow central zone, gave rise to the Wonji fault belt (Mohr 1960; 1962). Since about 1.3 m.y. ago a transition to oceanic crust has taken place. Faulting and the formation of open fissures has been restricted mainly to spreading ridges, along which there are basaltic fissure eruptions and complex strato-volcanoes, forming Central Volcanic Ranges, exemplified by Erta–Ale. In the Asal Rift spreading centre at the western end of the Tadjurah Gulf, an axial graben, 10 km wide and topographically comparable to a median valley, has along its axis a narrower zone of fissured recent basalts (Harrison *et al.* 1975).

In the northern part of the main Ethiopian Rift, the Nazareth Phase of intense but localised faulting was followed by eruption of basalts, and after further faulting, the eruption of a string of volcanoes partly basaltic, partly acid. These in turn are cut by fissures (Meyer *et al.* 1975). Open fissures are also seen further south in the Main Ethiopian Rift.

In the Gregory Rift during the last 2 m.y. eruptions mainly of basalts, trachyte and phonolite have been most intense between Lakes Baringo and Magadi. A close network of faults has dislocated the lava sheets. A series of volcanoes and calderas formed along the axis of the rift.

Middle Pleistocene lake beds, as at Olorgesailie (Legemunge) dated at 0.42 m.y. and at Kariandusi permit an estimate of the extent of mid-Pleistocene faulting. Faults with throws of up to 50 km occur both at Olorgesailie and Kariandusi. At both localities the lacustrine beds rest on eroded fault complexes of underlying lavas, the youngest of which at Olorgesailie gives dates of 1.25 to 0.63

m.y. (Baker *et al.* 1971). Upper Pleistocene deposits at these localities are not faulted but north and south of the Menengai caldera, Upper Pleistocene deposits are faulted, resulting in a relative depression of the rift floor in the Nakura basin by 30 m. North and south of Menengai there are lines of open fissures (McCall 1967) but there is no transition to oceanic crust as in Afar.

West of the Gregory Rift, renewed tilt away from the shoulders of the Albertine Rift in mid-Pleistocene times reversed the drainage in western Uganda and formed Lakes Kyoga and Victoria.

Conclusions

1. Episodic uplift of the Kenya dome, Ethiopia and Western Uganda is expressed by uplifted erosion surfaces between which the spacing increased towards the rift. Uplift began in the Mesozoic.

2. Volcanism associated with the rifting probably began in the Oligocene in parts of Ethiopia, in the Miocene in Kenya and still later in Tanzania.

3. Disconnected rifts, with triple junctions at the crests of domal uplifts, developed by monoclinal downwarps and faulting, from Miocene onwards. There is no definite evidence of systematic southward propagation of fractures; there is not even yet a continuously connected series of rifts.

4. A general structural sequence can be recognised. (i) domal uplifts of large amplitude; (ii) monoclinal bends, down towards the future rifts; (iii) major boundary faults; (iv) closely spaced faults across the whole width of the rift floor; (v) faulting in a narrow central belt; (vi) fissuring in a still narrower belt; (vii) transition to oceanic crust. The last is only seen in Ethiopia.

5. Movements on faults have recurred so frequently that it is generally misleading to separate distinct phases but the rate of movement has fluctuated, on the whole accelerating from Miocene to Pleistocene. The largest vertical movements in the Gregory Rift have occurred during the last 2 m. years.

6. Because movement on most faults has been repeated, the throw increases downwards in successive formations and the throw at the surface is no indication of the total throw on the faults or of the extension indicated by them.

7. The principal topographic and environmental effects of the rifting have been to convert subdued plains on basement, with inselbergs, into volcanic highlands transected by deep grabens; to rejuvenate, divert, reverse and pond rivers; to form volcanoes which were probably forested and from which streams radiated; to form lakes at the foot of fault scarps, in plunge depressions within fault strips and as the result of volcanic damming.

References

ARAMBOURG, C. & WOLFF, R. G. 1969. Nouvelles données paléontologique sur l'age des grès de Lubur ('Turkana Grits') a l'ouest du lac Rudolf. *C.R. Soc. Geol. France*, **6**, 190–2.
AZZAROLI, A. & FOIS, V. 1964. Geological outlines of the northern end of the Horn of Africa. *Intern. Geol. Congr. Rep. 23rd Sess.*, New Delhi, Pt. **IV**, 293–314.
BAKER, B. H., MOHR, P. A. & WILLIAMS, L. A. J. 1972. Geology of the Eastern Rift System of Africa. *Geol. Soc. Am. Spec. Paper* **136**, 1–67.

——, WILLIAMS, L. A. J., MILLER, J. A. & FITCH, F. J. 1971. Sequence and geochronology of the Kenya Rift volcanics. *Tectonophysics*, **11**, 191–215.

—— & WOHLENBERG, J. 1971. Structure and evolution of the Kenya Rift valley. *Nature, Lond.* **229**, 538–42.

BARBERI, F., FERRARA, G., SANTACROCE, R. & VARET, J. 1975. Structural evolution of the Afar triple junction. *In:* Afar Depression of Ethiopia, ed. Pilger, A., & Rosler, A., Stuttgart, 38–45.

BINGE, F. W. 1962. Geology of the Kericho Area. *Rept. Geol. Surv. Kenya.* **50.**

BISHOP, W. W. 1965. Quaternary Geology and Geomorphology in the Albertine Rift Valley, Uganda. *Geol. Soc. Am. Spec. Paper*, **84**, 293–321.

—— & PICKFORD, M. 1975. Geology, fauna and palaeo environments of the Ngorora Formation, Kenya Rift Valley. *Nature, Lond.* **254**, 185–92.

—— & TRENDALL, A. F. 1967. Erosion-surfaces, tectonics and volcanic activity in Uganda. *Q. Jl. Geol. Soc. Lond.*, **122**, 385–420.

BLACK, R., MORTON, W. H. & REX, D. C. 1975. Block tilting and volcanism within the Afar in the light of recent K/Ar age data. *In:* Afar Depression of Ethiopia, ed. Pilger, A., & Rosler, A., Stuttgart, 296–300.

CHESSEX, R., DELALOYE, M., MULLER, J. & WEIDMANN, M. 1975. Evolution of the volcanic region of Al Sabieh (T.F.A.I.) in the light of K/Ar determinations. *In:* Afar Depression of Ethiopia, ed. Pilger, A. & Rösler, A., Stuttgart, 221–7

CHRISTIANSEN, T. B., SCHAEFER, H.-U. & SCHÖNFELD, M. 1975. Geology of Southern and Central Afar, Ethiopia. *In:* Afar Depression of Ethiopia, ed. Pilger, A. & Rösler, A., Stuttgart, 259–77.

DAVIDSON, A., MOORE, J. M. & DAVIES, J. C. 1973. Preliminary report on the geology and geo-chemistry of parts of Sidamo, Gemu Gofa and Kefa Provinces, Ethiopia. *Imp. Ethiopian Govt. Ministry of Mines*, Omo River Project Rept, I, 1–21.

DIXEY, F. 1945. Miocene sediments in South Turkana. *Journ. E. A. Nat. Hist. Soc.*, **18**, 13–14.

—— 1948. Geology of Northern Kenya. *Rep. Geol. Surv. Kenya*, **26.**

GIRDLER, R. W. & STYLES, P. 1974. Two stage Red Sea Floor spreading. *Nature, Lond.* **247**, 7–11.

HARRISON, C. G. A., BONATTI, E. & STIELDJES, L. 1975. Tectonism of axial valleys in spreading centers: data from the Afar Rift. *In:* Afar Depression of Ethiopia, ed. Pilger, A. & Rosler, A., Stuttgart, 178–98.

JENNINGS, D. J. 1964. Geology of the Kapsabet–Plateau Area. *Rept. Geol. Surv. Kenya*, 63.

JOUBERT, P. 1966. Geology of the Loperot Area. *Rept. Geol. Surv. Kenya*, **74.**

McCALL, G. J. H. 1967. Geology of Nakuru–Thomson's Falls—Lake Hannington Area. *Rept. Geol. Surv. Kenya*, **78.**

MEYER, W., PILGER, A., ROSLER, A. & STETS, J. 1975. Tectonic evolution of the northern part of the Main Ethiopian Rift in southern Ethiopia. *In:* Afar Depression of Ethiopia, ed. Pilger, A. & Rosler, A., Stuttgart, 352–62.

MOHR, P. A. 1960. Report on a geological excursion through southern Ethiopia. *Bull. Geophys. Obs., Univ. Addis Ababa*, **3**, 9–20.

—— 1962. The Ethiopian Rift System. *Bull. Geophys. Obs., Univ. Addis Ababa*, **5**, 33–62.

—— 1975. Structural setting and evolution of Afar. *In:* Afar Depression of Ethiopia, ed. Pilger, A. & Rösler, A., Stuttgart, 27–37.

MORBIDELLI, L., NICOLETTI, M., PETRUCCIANI, C. & PICCIRILLO, E. M. 1975. Ethiopian South-Eastern Plateau and related escarpments: K/Ar ages of the main volcanic events (Main Ethiopian Rift from 8° 10' to 9.00 lat. North). *In:* Afar Depression of Ethiopia. ed. Pilger, A. & Rösler, A. Stuttgart, 362–9.

PULFREY, W. 1960. Shape of the sub-Miocene erosion bevel in Kenya. *Bull. Geol. Surv. Kenya*, **3.**

SAGGERSON, E. P. & BAKER, B. H. 1965. Post-Jurassic erosion-surfaces in eastern Kenya and their deformation in relation to rift structure. *Q. Jl. Geol. Soc. Lond.*, **121**, 51–72.

SHACKLETON, R. M. 1946a. Geology of the Migori Gold Belt and adjoining areas. *Rept. Geol. Surv. Kenya*, **10.**

—— 1946b. The Geology of the Country between Nanyuki and Maralal. *Rept. Geol. Surv. Kenya*, **11.**

—— 1951. A contribution to the geology of the Kavirondo Rift Valley. *Q. Jl. Geol. Soc. Lond.*, **106**, 345–92.

WALSH, J. 1969. Geology of the Eldama Ravine—Kabarnet area. *Rept. Geol. Surv. Kenya*, **83.**

VARNE, R. 1965. The Geology of Moroto Mountain, Karamoja, north-eastern Uganda. Unpub. Ph.D. thesis, Univ. Leeds.

ZANETTIN, B., GREGNANIN, B., JUSTIN-VISENTIN, E., MEZACASA, G. & PICCIRILLO, E. M. 1974. Petrochemistry of the volcanic series of the Central eastern Ethiopian plateau and relationships between tectonics and magmatology. *Cons. Naz. Della Ric.*, Padova, 1–35.

——— & JUSTIN-VISENTIN, E. 1974. The volcanic succession in Central Eritrea: *In:* The Volcanics of the Western Afar and Ethiopian Rift Margins. *Cons. Naz. della Ric.*, Padova, 1–20.

———, ——— 1975. Tectonical and volcanological evolution of the western Afar margin (Ethiopia). *In:* Afar Depression of Ethiopia, ed. Pilger, A. & Rösler, A. Stuttgart, 300–9.

3

BASIL C. KING

Structural and volcanic evolution of the Gregory Rift Valley

The East African Rift System extends southwards from the Red Sea for a distance of about 3000 km. Within East Africa it is represented by the Western Rift and the Eastern Rift, of which the Gregory Rift Valley is that part lying within Kenya and northern Tanzania.

The Gregory Rift Valley was initiated in early Miocene times as a downwarp along the continental watershed on a land surface having considerable relief (up to 2000 m). Subsequent faulting produced a graben about 80 km wide and some 450 km in length, to the north and south of which the faults splay outwards over much broader zones.

Volcanic activity was not confined to the Rift, but developed both to the west in the Miocene and to the east from Plio-Pleistocene to Recent times. Within the Rift itself volcanism commenced in the north and extended to the south in the Pliocene, while both tectonic and volcanic activity tended later to restriction within a central zone, which from Lake Hannington northwards formed a narrow axial trough.

Episodes of volcanism and sedimentation largely or wholly infilled the downwarping trough, but only in the Pleistocene, by uplift of the shoulders and downfaulting of the floor, was the morphology of the present Rift Valley achieved. The altitude of the sub-volcanic surface averages about 1500 m along the shoulders of the Rift; its greatest depth below the floor may be southwards from near Lake Baringo, corresponding to an absolute displacement of some 4 km. About 8 km of crustal extension over about 80 km of rift width may have occurred, but there is no evidence for crustal separation.

Introduction

In 1896, J. W. Gregory defined a rift valley as one 'of subsidence with long steep parallel walls' and inferred that it had been let down between major normal faults, or by a series of step faults. Since, in central Kenya, the plateau rises on either side to the walls of the Rift, he likened its formation to the dropping of the keystone of an arch under tension. Gregory also recognised that the Rift Valley in Kenya was only a part of a system of rifts extending from south to north along most of the eastern side of Africa and thence via the Red Sea into the Levant.

Interest in this great continental rift system was greatly stimulated by discoveries made on the ocean floor during the 1950s and 1960s with the recognition of oceanic crustal spreading from a world-wide 'rift' system, some 60 000 km in extent. The apparent connection via the Gulf of Aden between oceanic and continental structures became a focus of attention for both geophysists and geologists. Such a connection would have been more difficult to postulate in the case of other continental rift systems, notably the Rhinegraben and the Baikal rift. Briefly it can be stated that the structures of continental rifts conform to Gregory's original description, whereas those of oceanic 'rifts' involve continuous crustal separation. It may not be too late to employ another term, such as 'crestal troughs', for the oceanic structures, in which at any given time the walls represent a parting along an original single fracture. It is sufficiently clear that the Red Sea and the Gulf of Aden imply crustal separation and it is likely that the former follows a sector of the East African rift system (Girdler 1965). Abnormal seismic velocities in the lower crust and upper mantle are dubious as a major criterion for identifying continental rift structures with the so-called oceanic 'rifts' for abnormalities of crustal and subcrustal structures are increasingly recognised within assumed stable continental areas. Moreover, crustal extension is implicit wherever there is a pattern of normal faulting, which is by no means confined to rift valleys. Igneous (volcanic) activity is the exception, rather than the rule, of rift systems, but the rift structures are everywhere entirely comparable. This is important to mention, since most geophysical work has been concentrated on volcanic sectors of the East African rift system, which have thus been regarded as typical.

The present account, while it obviously owes much to earlier work, including notably that of the Geological Surveys in East African countries, depends particularly on the largely unpublished results of the East African Geological Research Unit (directed by the writer), involving the detailed mapping of about 20 000 sq km over the past ten years (see Acknowledgements).

It is now possible to assess the evolution of the Gregory Rift Valley in terms of structure, volcanism and geomorphology.

Relationship of the Gregory Rift Valley to the East African Rift System

The Gregory Rift approximates to a fault-bounded graben for about 450 km between latitudes 2°S to 2°N, with a width of 70 to 90 km and a topographic relief of no more than 2 km, but to the north and the south the fault structures splay out over widths of 200 km or more, wherein the graben pattern is lost (Fig. 3:1). Indeed, over much of the rift system a simple graben structure is exceptional; for example, the pattern in Tanzania is largely of tilted blocks.

In general the Cenozoic rift system of eastern Africa shows a correspondence with older trends in the Precambrian basement, which, on a continental scale, are oriented approximately north-south (Dixey 1956). Earlier, post-Karroo troughs in southern Africa also conform with older grains, but these are mostly east-west or NE–SW. Different stress systems for Mesozoic and Cenozoic rifting seem likely, for only those older troughs which approach the Cenozoic north-south direction have been rejuvenated, such as the Malawi and Rukwa troughs.

In relation to the Cenozoic rift system the 'Tanganyika shield', with east-west trends, appears to have acted as a resistant block deflecting fault patterns to either side, thus forming the Western and Eastern Rifts. Even so, there are a number of subsidiary troughs that were directed into the shield, such as those of Kavirondo, Speke Gulf and the Lake Eyasi zone. It is also noteworthy that the abrupt termination of the Lake Albert trough with its relaying, en echelon faults, against the old Madi–Aswa zone shows quite clearly that the Western Rift is not a belt of crustal separation.

Pre-rift and early rift history

The older southern rifts and troughs were related to Karroo and post-Karroo faulting and downwarping within the Gondwana 'super-continent', but the Cenozoic rifts developed after continental separation.

An old, deeply weathered, well-planed and lateritised surface which must be of late Mesozoic age (Bishop & Trendall 1967, King *et al.* 1972) is recognisable extensively in Uganda at 1000 to 2500 m and as high level remnants in Kenya, from 2500 to 3500 m. The varying elevations reflect Cenozoic movements, in part associated with rift tectonics.

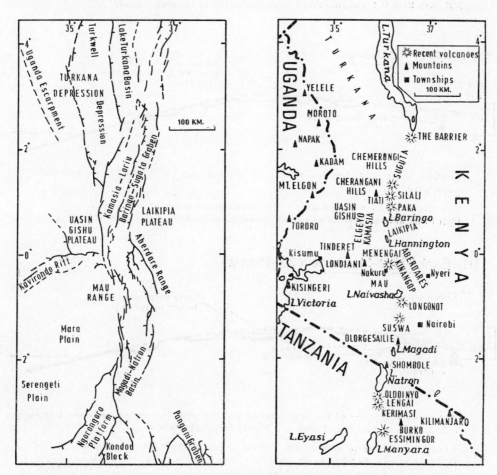

FIG. 3:1. General structural (A) and locality maps (B) of the Gregory Rift. The structural map is modified and corrected from Baker *et al.* 1972.

The post-Karroo, rising rim of Africa, developing intermittently over a long period of time, determined the major watershed between the newly forming Atlantic and Indian Oceans, and was accompanied by even greater downwarping in the coastal region of East Africa (Dixey 1956). The course of the Gregory Rift coincided rather closely, if fortuitously, with this uplift; the Western Rift developed

I. Mid-Mesozoic (Jurassic)

Old Gondwana Surface raised to form African continental watershed with major downwarp towards new Indian Ocean.

Not less than 0.5 km inherited from elevation of Gondwanaland.

II. Early Tertiary

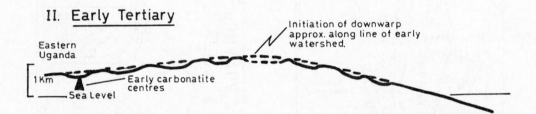

Further uplift caused active erosion of old surface, particularly towards the watershed,

The Eastern Rift was probably initiated as a downwarp at this time.
Uplift and active erosion resulted in sedimentation in early-formed basins.

III. Early-Mid-Miocene (25-15 m.y.)

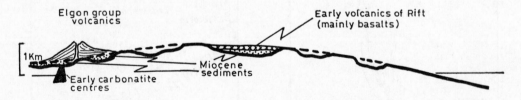

The large volcanoes of Eastern Uganda and the earliest volcanics in the Kenya Rift overlie sediments, infilling a highly irregular landscape.

Vertebrate faunas in sediments confirm overlying volcanics as about Mid-Miocene in age, but age determinations suggest that the earliest volcanics in the L. Rudolf area may be older.

FIG. 3:2. The early history of the Gregory Rift.

across the Atlantic drainage, which was not interrupted until the Pleistocene (Holmes 1965, Fig. 767). The subsequent histories of the rifts, the Gregory Rift with a series of shallow lakes formed by internal drainage, and the Western Rift with deep lakes (Tanganyika), or very thick sediments (Albert), reflect their different situations relative to the pattern of the ancestral drainage.

Erosion, following the marginal uplift and enhanced by gentle arching along the lines of the future rifts, produced a landscape which, in Miocene times, had considerable relief (up to 2000 m) and within which sediments accumulated in local basins (Fig. 3:2). In eastern Uganda and western Kenya these sediments are dateable by fossil vertebrates and underlie or are interbedded with the lower strata of the giant Miocene nephelinite volcanoes, such as Mt. Elgon. In eastern Kenya the corresponding surface is more planar and its form led to the notion of a sub-Miocene surface or bevel (Pulfrey 1960).

The Kenya Rift was initiated as a downwarp imposed upon the gentle arching roughly along the line of the old watershed, but much of the present elevation of dissected basement along its flanks has been inherited from its earlier history. This downwarped trough was wider than the present rift; the same is true of the Rhinegraben and the Baikal Rift. An early feature was the eastward downwarp along the northern part of the Kenya–Uganda border, which by subsequent erosion was carved into a scarp. The initial infilling of the subsiding trough was of sediments derived from the basement; these, known as the Turkana Grits, are thickest in the north (up to 300 m) and may be indirectly dated as early Miocene or older.

Outline of the structural and volcanic history of the Gregory Rift

Generalised accounts on this topic have already been presented (e.g. Baker *et al.* 1972; Logatchev *et al.* 1972, both with comprehensive lists of references). Table 1 gives a much simplified summary of the main volcanic events.

TABLE 1: *Generalised sequences of volcanics in the Gregory Rift Region*

Eastern Uganda		Rift Valley	East of Rift
	1 my	Central trachyte—basalt volcanoes	
		Nephelinite— carbonatite volcanoes in N. Tanzania	Multicentre basaltic lavas
	2–1 my	Plio-Pleistocene trachytes and tuffs.	Mt. Kilimanjaro—Trachyte, etc.
			Mt. Kenya —Phonolite, basalt, etc.
	7–2 my	Trachyte flood and central volcanoes. Basalts. Nephelinites in south.	
	10–7 my	Phonolites and Trachytes	
	16–11 my	Flood phonolites	
25–17 my Nephelinite volcanoes	20–15 my	Basalts (with basanites and phonolites)	
	? to 30 my	in north: Basalts	

Here I shall only indicate the most significant aspects and then elaborate on certain of these in later sections.

Sedimentation, followed by volcanism, apparently commenced in the northern part of the Gregory Rift, although the nephelinite volcanoes of eastern Uganda are of comparable age. Ages of up to 30 m.y. seem plausible in the north, whereas in the southern part of the rift valley there is not much evidence for volcanic rocks older than 7 to 8 m.y. Nevertheless, volcanic and tectonic activity has continued throughout the entire rift until the present day.

Single large faults have contributed little to the structure of the Rift; indeed, the greatest of these, the Elgeyo fault with a throw of about 3500 m is unique (see Figs. 3:5A and 16), but even so it developed by a succession of movements at widely varying rates. In many cases the displacements result from an aggregate of throws on a number of faults and, in some cases, merely by simple, major downwarps.

The rate of infilling by volcanics and sediments mostly kept pace with down-warping and faulting; thus the relief of the Rift throughout most of its history was generally subdued. In the later Miocene (13 to 11 m.y.) the early trough was completely overtopped by phonolitic flood lavas and similar infilling probably occurred after the eruption of the Plio-Pleistocene tuffs and lavas.

Both tectonic and volcanic activity show a trend towards concentration from the flanks towards the centre, so that in its latest stages the Rift is marked by a central 'graben' having very numerous faults with minor displacements and a series of late central volcanoes. It is to be recognised, however, that older formations and structures must lie concealed beneath younger in the central trough, so that information here is very incomplete, and that older marginal faults were rejuvenated into the Pleistocene.

The present spectacular morphology of the Rift was only achieved by such rejuvenation in later Pleistocene times. In general this resulted from fault movement and downwarping greatly exceeding the rate of accumulation within the Rift. A late Pliocene or early Pleistocene landscape, expressed over basement and Cenozoic volcanics alike, and marked by deep weathering resulting from periods of relative tectonic quiescence, was depressed beneath the Rift floor; nevertheless bore-holes show great thicknesses (up to 1000 m) of late Pleistocene volcanics within the central part of the rift, the floor of which, essentially the Plio-Pleistocene land surface, must have subsided faster than the deposits accumulated.

Thus, although earlier notions of well-defined episodes of faulting alternating with ones of volcanic activity cannot be sustained in detail, there is much evidence to support the idea that there were two important periods of movement, namely, at around 7 m.y. and from 2 to 1.5 m.y. The second was largely responsible for the present rift morphology, while much of the history of the southern part of the Rift post-dates the first. Each was preceded by a period of relative stability, so that thick chemically-weathered profiles were produced, and it is these, as well as the underlying volcanics upon which they were developed, that were involved in the movements. A still later period of faulting around, 0.5 m.y., is mostly marked by numerous 'grid' faults of small displacement along the axial graben or depression in the north and more widely in the central and southern sectors of the Rift. (Fig. 3:9).

The volcanics, including those which extend both to the west and the east, well beyond the confines of the Gregory Rift, have a great diversity of compositions, ranging from highly undersaturated alkaline melanephelinites to oversaturated peralkaline pantellerites and comendites, to the less alkaline associations of alkali basalts, mugearites and quartz trachytes. The distribution of the main rock types and their associations have already been well described (Williams 1972). To some extent this diversity has been explained by the assumption of two

TABLE 2: *Compositional relationships among rift volcanoes*

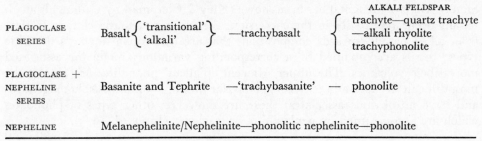

PLAGIOCLASE SERIES	Basalt { 'transitional' 'alkali' } —trachybasalt	{ ALKALI FELDSPAR trachyte—quartz trachyte —alkali rhyolite trachyphonolite
PLAGIOCLASE + NEPHELINE SERIES	Basanite and Tephrite —'trachybasanite' — phonolite	
NEPHELINE	Melanephelinite/Nephelinite—phonolitic nephelinite—phonolite	

Note: Analcite often takes the place of nepheline.

'fractionation series', the one strongly, the other mildly alkaline (King & Sutherland 1960; King 1965; Saggerson & Williams 1964), but intermediate compositional series exist. Table 2 summarises this, but it should be regarded as expressing petrographic rather than petrogenetic associations, although genetic sequences are generally also implied.

Basalts (sensu lato) have been erupted repeatedly throughout the history of

TABLE 3: *Volcanics of Region of Gregory Rift*

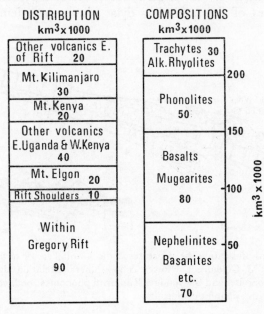

DISTRIBUTION
km³ × 1000

Other volcanics E. of Rift	20
Mt. Kilimanjaro	30
Mt. Kenya	20
Other volcanics E. Uganda & W. Kenya	40
Mt. Elgon	20
Rift Shoulders	10
Within Gregory Rift	90

COMPOSITIONS
km³ × 1000

Trachytes Alk. Rhyolites	30
Phonolites	50
Basalts Mugearites	80
Nephelinites Basanites etc.	70

km³ × 1000 — 200, 150, 100, 50

the Gregory Rift and in total volume have been estimated at not less than 70 000 km³ (Williams 1972), but this volume is certainly exceeded collectively by that of other rock types, notably phonolites and trachytes. Table 3 presents estimates of the volumes of volcanics in terms of geographic situation and of composition; a total of 230 000 km³ is here assumed. This is rather greater than earlier estimates (Baker *et al.* 1972), but the order of magnitude is similar.

For the Gregory Rift region as a whole there is no clear pattern of genetic evolution of volcanic sequences, but it is possible that evolutionary sequences may emerge from detailed studies within particular regions. Thus, in the Lake Baringo–Suguta sector of the Rift it has been shown (King & Chapman 1972) that, although basalts occur repeatedly in the succession, there is a systematic change in time from alkali basalts towards compositions that are transitional towards tholeiitic types; this is accompanied by a corresponding variation among the associated more silicic volcanics. The upper Miocene 'plateau' phonolites are among the most difficult to explain in terms of genesis, for they are voluminous, homogeneous and have no obvious associates; there are, however, other types of phonolites which are accompanied by nephelinites or trachytes (Lippard, 1973).

Volcanic structures and modes of emplacement

The convenient division that has been made between flood (fissure or multi-centre) eruptions and central volcanoes (Williams 1969) has understandably been shown by recent, more detailed, study to be an over-simplification.

The Miocene flood basalts were erupted from central zones of closely spaced fissures (marked by dykes). Great thicknesses of interdigitating flows were built up into a plateau structure on a regional scale; the actual positions of centres can be appreciated both from the localised occurrences of pyroclastics and the increasing thicknesses of flows. Similar dyke centres are demonstrable for the

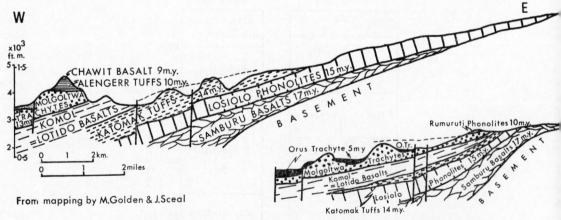

FIG. 3:3. Sections illustrating the structures of the Maralal sector of the eastern side of the Rift (after M. Golden). The lower section shows the relationships between the older Losiolo phonolites and the younger Rumuruti phonolites; both converge on the Rift shoulders with apparent conformity.

Pliocene Kaparaina basalts (Martyn 1969), which form coalescing very low angle volcanoes; nevertheless low angle basaltic cones of similar age in the Aberdare Mountains appear to have been erupted from typical central vents.

The late Miocene/early Pliocene phonolites are of particular interest; as 'plateau' phonolites they succeeded basalts, or older phonolites and basanites, in the central and north-central parts of the Rift and overflowed the shoulders of the early trough to distances of 100 km or more. From their wide distribution and uniform compositions they have been considered a useful time marker over a large part of the Rift region. Consistent dates of between about 12 and 13.5 m.y. from the Uasin Gishu and Kericho areas to the west of the Rift and from the Kapiti and Yatta phonolites to the south-east of Nairobi (Bishop *et al.* 1969) supported this view. More recent determinations confirm earlier findings, but place the Rumuruti phonolites in the Laikipia area on the eastern side of the Rift in the range 10 to 12 m.y. These show, moreover, a different succession of stratigraphic types from that of Kamasia and Uasin Gishu and the sequence thins towards the central trough, but owing to concealment under more recent formations the relationship between the eastern and western successions cannot be established. North of Laikipia, M. Golden (pers. comm.) has established that the Losiolo phonolites stratigraphically underlie the Rumuruti phonolites, although both form parts of the continuous plateau surface of the Rift shoulders, an observation that is confirmed by radiometric dates of around 15 m.y. (Fig. 3:3).

In the Kamasia range (Chapman *et al.* In press) much of the lower part of the sequence consists of phonolites in addition to basalts, ranging back to 15 m.y. and resting on basement. Some of the later flows can be directly correlated, petrographically and, more broadly, by radiometric dating, with those of the Elgeyo escarpment and Uasin Gishu plateau, whereas the higher, overlying flows of phonolite range up to about 7 m.y. The Kamasia section provides the clearest

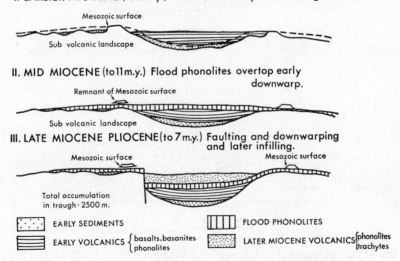

FIG. 3:4. Miocene history of the Gregory Rift.

evidence to be found within the Rift for the initial infilling of an early depression, its overtopping by later volcanics and further infilling after renewed subsidence, giving a total thickness of 2.5 km (Figs 3:4 and 5).

The volume of the 'plateau' phonolites, according to Lippard (1973) is of the order of 40 000 to 50 000 km³, which is almost a quarter of that estimated for the volcanics in the region of the Gregory Rift as a whole. Moreover, many of the individual flows are extremely voluminous (up to at least 500 km³), as compared with flows of other types of volcanics. Martyn, Chapman and Lippard (pers. comms.) and Lippard (1973) have remarked on the rarity of phonolite dykes and conclude that the prevalent idea of the phonolites having been erupted from fissures, extending over the area of their distribution (Logatchev *et al.* 1972) is

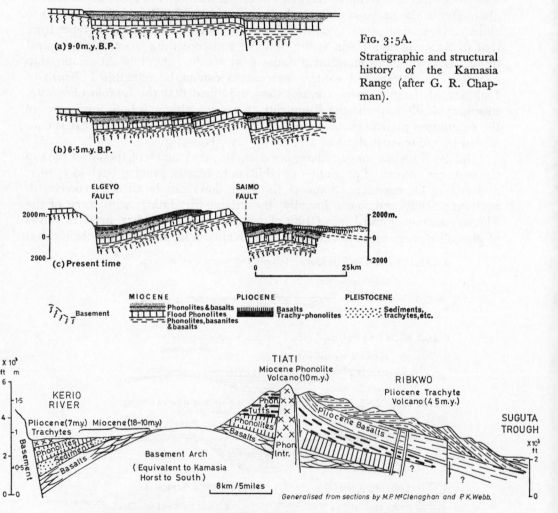

FIG. 3:5A.

Stratigraphic and structural history of the Kamasia Range (after G. R. Chapman).

FIG. 3:5B. The basement arch in south Turkana (after M. P. McClenaghan & P. K. Webb). Diagrammatic section across western side of Gregory Rift Valley at approximately 1°30′N.

without foundation. Lippard shows that the phonolite flows originated from a small number of sources, which gave rise to exceedingly low angle volcanoes with diameters of 70 to 80 km. The positions of centres can be inferred from the directions of thickening of the phonolite flows, and in a few instances actual centres can be identified, e.g. Sigatgat, on the southern flanks of Tiati (Figs. 5B and 8); the local occurrence of phonolitic pyroclastics supports this idea. All evidence suggests that the phonolites were erupted extremely rapidly and voluminously as unusually mobile flows.

On the Uasin Gishu plateau these extremely fluid lavas infilled initial depressions before they flowed thinly over the adjacent region. Lippard (1972) has shown, from the orientation of phenocrysts, that the lavas flowed just as freely towards the north and south as towards the west, so that the present declivity of 700 m over a distance of about 40 km, from the edge of the Elgeyo escarpment to Eldoret did not then exist. Moreover, the pattern of infilling of the older topography is only compatible with a gentler rise to the Rift than is now seen, and shows that later elevation occurred along the shoulders of the Rift.

The Yatta plateau, to the east of the Gregory Rift provides particularly convincing evidence for the rapid emplacement of extremely mobile phonolite lava. The flow has the same petrographic character throughout its length of around 300 km, with large phenocrysts of alkali feldspar, but has a thickness of only 10 to 15 m, and is now seen as a thin capping above a pediment of basement, 100 m or so in height, isolated by subsequent lateral erosion. It evidently followed a mature Miocene valley and had a probable original volume of about 20 km³, but again the present declivity of its course suggests enhancement by later elevation along the Rift shoulders.

It is also to be observed that the large Miocene volcanoes to the west of the Rift are dissected, low-angle cones of simple, regular construction, whether they are mainly pyroclastic (Elgon, Napak), or mainly composed of lava (Kisingiri).

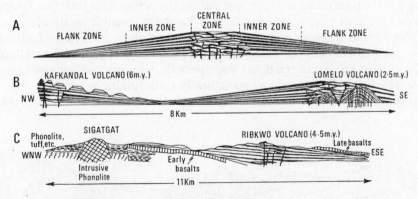

FIG. 3:6. Pliocene Trachyte (Ribkwo Type) volcanoes.

    A. Generalised section of 'Ribkwo Type' volcano.
    B. Relationships among trachyte volcanoes.
    C. Relationship between Ribkwo volcano and older Miocene volcanics.
    (B and C: vertical and horizontal scales approximately equal).

Here, too, there is evidence of build-up by voluminous volcanic units over relatively short periods of time (King *et al.* 1972), although the radiometric ages obtained from the various volcanoes range from about 22 to 16 m.y. (Bishop *et al.* 1969).

An important finding in the central and northern parts of the Rift is that, while 'flood' trachytes, such as the Kabarnet trachytes, exist, most of the formerly assumed Pliocene 'plateau' trachytes belong to very large low-angle central volcanoes of an unusual type (Fig. 3:6 and Webb & Weaver 1975). They range in age from about 6.5 to 2 m.y.; the later volcanoes, generally developed towards the central axis of the Rift, rest on the eroded flanks of the earlier. Basalts are commonly associated as initial and terminal phases, but the main structural feature of these volcanoes is that, although pyroclastic (including 'welded tuff') horizons occur, the trachyte lavas wedge out in an extraordinarily regular fashion from the centre to the flanks. The central zones are ill-defined and show not only intrusive and extrusive eruption, but also infillings by lake deposits, landslips, lahars and ladoes. Volcanic calderas also occur.

In the region to the west of the Suguta River a succession of widespread basalt extrusions (from Miocene to Recent in age) is punctuated by the formation and dissection of a series of trachytic centres (Fig. 3:7).

It should also be noted that, whereas the 'classic' examples of welded tuffs and ignimbrites relate to dacitic/rhyolitic members of the 'andesite suite', the majority of those in Kenya are of trachytic composition, sometimes ranging to pantelleritic, while examples of phonolitic welded tuffs have been described by Weaver (1974).

A general aspect that emerges from the above is that while lacustrine sedimentation (not discussed here in detail, but represented at many times and places during the succession of events) covered lengthy periods of time and often reveals fault growth, volcanic events were almost instantaneous, so that a stratigraphic sequence, although appearing to be continuous in the field, must be punctuated by long gaps, which make up the major part of the actual time scale (see Chapman, this volume). This partly explains how it has been possible to divide events so conveniently into volcanic and tectonic. Unconformities, marked by faulting and erosion, such as beneath the Kabarnet trachytes, may, as judged by radiometric ages, have been formed in relatively short periods, whereas much greater intervals must be represented within many structurally conformable sequences.

In the northern part of the Rift, although the successions on either side are

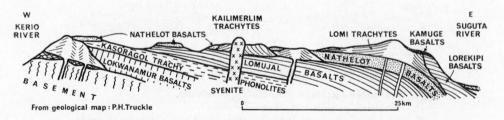

FIG. 3:7. Kerio–Suguta Section: South Turkana—south of Lokori (after P. Truckle).

well-established and are now effectively dated, a lithological correlation between them is still virtually impossible; in the central part of the Rift displacements within the general blanketing by Hannington phonolites (more correctly, trachy-phonolites) (c. 1 m.y.) show very clearly the latest movements within the Rift, but the phonolites conceal earlier events, while in the south the very thick later Pleistocene volcanics, shown by boreholes, could not have been predicted from evidence on the flanks. Projection of information both across and into the Rift floor is thus highly conjectural.

Observed structures of the Gregory Rift

A summary of the kinds of structures that characterise the margins of the Rift, namely, major boundary faults, step faults and downwarping is presented in Fig. 3:8. All of these structures result in a crustal depression of the original basement without involving discontinuity or separation; indeed, basement is extensively exposed or its presence is easily inferrable in the Rudolf and northern Tanzanian sectors; it outcrops within the Rift in the Kamasia range, and its occurrence in depth is shown by fragments brought up by volcanoes within the central graben. Moreover, successive members of the Cenozoic sequence show similar continuity across various sectors of the Rift.

The faults are all essentially of normal dip-slip type, although a minor strike-slip component is to be assumed where obliquely intersecting, contemporaneous faulting occurs. From the throws, (usually determinable) and the hades (rarely determinable accurately) of the faults an estimate of crustal extension may be made; a contribution is also to be inferred from dilation associated with the emplacement of parallel dyke swarms. An extension of not more than 1 in 10, i.e. 8 km over a Rift width of 80 km is possible.

Fault throws vary enormously in magnitude, from thousands of metres down to less than a metre. The latest grid faulting is marked by numerous, often sinuous faults, with small individual displacements and with throws in varying directions,

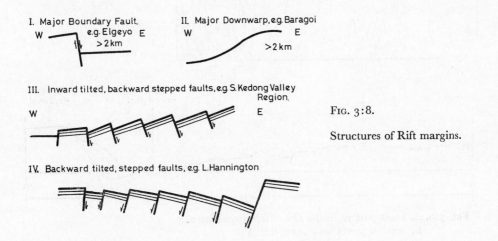

I. Major Boundary Fault.
e.g. Elgeyo
W   E
>2 km

II. Major Downwarp, e.g. Baragoi
W   E
>2 km

III. Inward tilted, backward stepped faults, e.g. S. Kedong Valley Region.
W   E

FIG. 3:8.

Structures of Rift margins.

IV. Backward tilted, stepped faults, e.g. L. Hannington

forming miniature graben and horsts (Fig. 3:9). Overall, however, the effect is subsidence towards the central zone of the Rift, so that it may represent a late

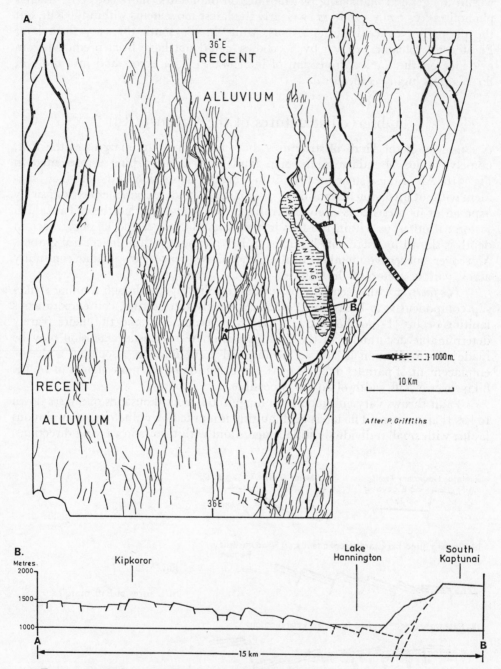

FIG. 3:9. A. Fault pattern in the Lake Hannington area.
B. Section across area along line A–B.

stage 'collapse' consequent upon stretching and attenuation at deeper levels. Such faulting has continued to the present day and affects even the latest volcanics and sediments.

Whereas topographic relief and morphology are fairly constant over the greater part of the Rift, the structural patterns by which the general depression has been achieved show rapid variations both along the flanks and across the Rift. Thus, displacements are relayed by 'en echelon' faults or are deployed along branch faults, so that the trend of the Rift Valley is in many places discordant with that of the fault trends; fault-bounded strips descend longitudinally into the Rift, forming 'ramp' or 'gang-plank' structures (Fig. 3:10); and throws along major faults may become dispersed among many movements along smaller faults, or even be replaced by monoclinal downwarps (Figs. 3:9 and 11).

As noted first by J. Martyn in 1965, in the Kamasia area, the hades of faults are commonly perpendicular to the dips of the formations. This is so whether the

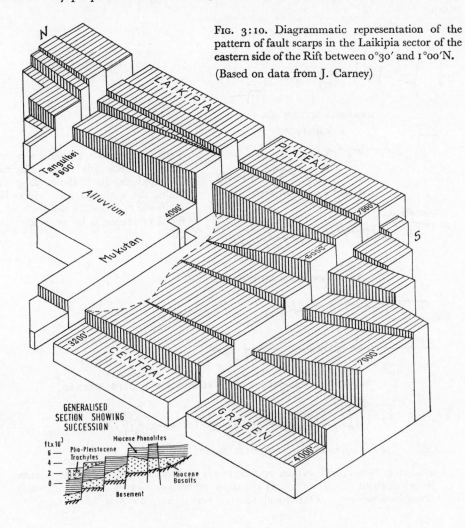

FIG. 3:10. Diagrammatic representation of the pattern of fault scarps in the Laikipia sector of the eastern side of the Rift between 0°30′ and 1°00′N.

(Based on data from J. Carney)

faults are synthetic or antithetic in relation to the rift margins and it applies even in the case of dipping Pleistocene formations. Variation from synthetic to antithetic relationships are seen along the Kamasia range (Fig. 3:11). An analogy may be made with the tilting of books on an incompletely filled bookshelf and in structural terms implies lateral extension and corresponding vertical thinning. Fault-bounded slices appear to have rotated in an extensional field and a plausible mechanism is that they rest on a major curving fault dislocation which gradually flattens in depth (Fig. 3:18A). In this connection the various types of land-slips are instructive and, since all faulting in the Rift involves dislocation of what is virtually or actually the present land surface it is not possible to make a firm distinction between land-slip and tectonic phenomena. Whereas all types of land-slip depend on a free 'out-fall', produced by either erosion or fault growth, a characteristic type involves the development of a series of back-tilted slices moving on a spoon-shaped slip surface, the entire train being driven forward as later slips occur.

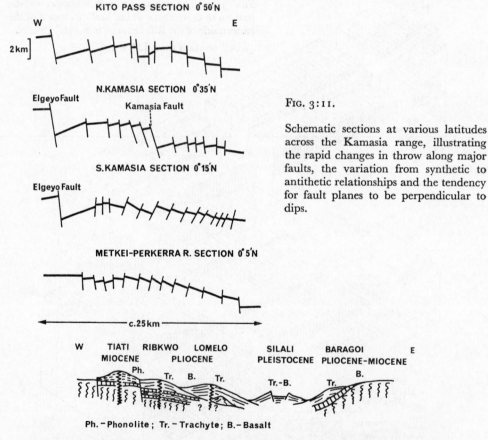

FIG. 3:11.

Schematic sections at various latitudes across the Kamasia range, illustrating the rapid changes in throw along major faults, the variation from synthetic to antithetic relationships and the tendency for fault planes to be perpendicular to dips.

FIG. 3:12. Generalised section, approximately 50 km in length across the Suguta trough at about 1°N. The Pliocene volcanoes rest on downfaulted Miocene volcanics; tilting increases towards the trough which is here a sharp downwarp.

Gradual warping towards the axis of the Rift is shown, for example, in south Turkana, where the trachyte volcanoes to the east and north-east of Tiati become progressively younger eastwards to the Suguta trough, while the older volcanoes are systematically more tilted than the younger. Farther to the north a succession of basalts ranging in age from Miocene to Pleistocene shows a similar increasing eastward tilt with age (Figs. 3:7 and 12).

An '*axial*' *trough* developed during the Pleistocene from Lake Hannington northwards via the Suguta to southern Lake Turkana and, indeed, is the structure which continues into the Wonji Rift of Ethiopia by way of Lake Stefanie. In places this is abruptly downfaulted, as to the east of Lake Baringo, but in the lower Suguta sector it is a sharp downwarp, with only incidental faulting (Figs. 3:7, 10 and 12). Other late troughs include the main Rudolf depression into which the rivers Kerio and Turkwell drain and, in the southern part of the Rift, the depressions containing lakes Magadi and Natron.

Evolution of the Rift landscapes and drainage systems

On the flanks of the Rift, the latest formations on the Uasin Gishu, Laikipia and Maralal plateaus are the upper Miocene phonolites (ranging from 16 to 11 m.y. in age). They were deeply weathered to form red lateritic 'soils' during the later Tertiary. Drainage that incised the weathered zone was always away from the Rift on the Uasin Gishu plateau, but initially towards the Rift in Laikipia and Maralal; it can be shown in Laikipia that later Pliocene faulting produced horst strips, which generally deflected the headwaters of the earlier rivers towards the east, but which were cut across as they formed by the more active rivers such as the Mukutan and Churo. These were favoured by structural depressions in the fault strips. At the same time a secondary drainage pattern followed the faults and was directed northwards by the slightly backward and northward tilted Tangulbei platform (Figs. 3:10 and 14).

Similar zones of deep lateritic weathering developed over the profiles of the Pliocene volcanoes and extended across the 'basement'. In the area to the west of Nairobi the dating of volcanics suggests that such weathering extended into the Pleistocene, but pre-dated the faulting that gave rise to the present Rift morphology. Not only is such surface weathering absent from the late tuffs of the Kinangop plateau and the Pleistocene–Recent infill of the central sector of the Rift, but also northwards from Thika to Rumuruti where the weathered zone has been progressively stripped from phonolites and basement alike by later erosion. Indeed, the surface can be traced almost continuously to the coast mostly across basement; it is dissected by recent erosion but underlies the Pleistocene–Recent volcanics of the Chyulu Hills, as well as the later flows of Mt. Kenya.

These observations have important implications regarding the age of uplift of the shoulders of the Rift, for not only have the soil profiles been dissected by recently rejuvenated drainage, but lateritic soils could not everywhere have been formed at present altitudes. Thus over the Aberdare Mountains such a profile is traceable to nearly 4000 m, but it has been modified superficially to a cool temperate forest soil.

The Gregory Rift has been essentially a closed drainage system for a long period of time, so that it should be possible theoretically to relate the volumes of sedimentary deposits to the volumes of formations eroded. Data for this are not currently available, but it does seem that extensive amounts of sedimentary formations are concealed by subsidence.

The earlier pattern of drainage in the Rift appears to have been directed inwards (Fig. 3:13) and, indeed, this is true of the eastern slopes of the northern Kamasia range, where rivers are incised within resistant arched blocks of trachyte. It is also supported by evidence from farther north in south Turkana, where mature valleys have pebbles of basement which must have been derived from a source, now completely cut off, to the west of the present deeply incised Kerio valley.

The present drainage pattern in the Rift has certainly depended largely on tectonic influences: a broad arching along the central sector, the backward tilting of the Kamasia block and the subsidence of the Lake Turkana and Suguta troughs

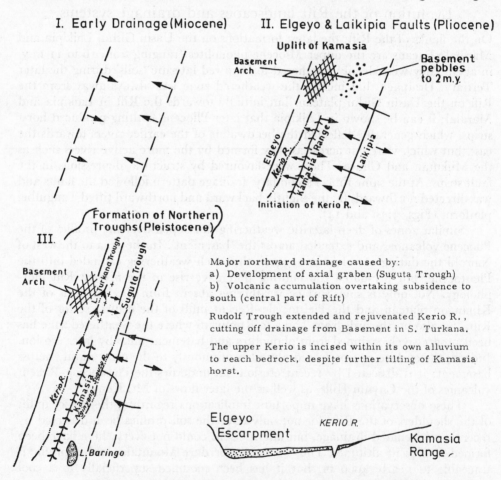

FIG. 3:13. Evolution of the drainage system in the northern Rift.

to the north and the Lake Natron trough to the south, but it also reflects volcanic accumulation overtaking subsidence in the arched region (Fig. 3:14). The course of the River Kerio, with its sources in the currently well-watered Timboroa–Metkei highlands, was clearly determined by the back-slope of the Kamasia block, but it evidently extended northwards in the later Pleistocene to behead the eastward drainage in south Turkana and with subsidence of the Lake Turkana trough cut rapidly into basement behind the volcanics, while its meandering upper reaches were incised through alluvium to reach Pliocene volcanics (Fig. 3:13). Continued tilting of the block has moved the lowest part of the valley towards the Elgeyo escarpment, but has failed to alter the course of the incised river. Somewhat similarly the Ewaso Ngiro was rejuvenated by the subsiding Natron trough and now flows in meanders incised within its own alluvium at some distance from the recently active Nguruman fault, across which rapidly eroding streams are cutting back towards the headwaters of the older, mature Pagasi River (Fig. 3:15).

The Perkerra River is structurally controlled in its upper reaches where it follows the boundary between the faulted and tilted Pliocene and older volcanics of the southern part of the Kamasia range and the younger, mainly Pleistocene flat-lying volcanics to the south and east. It has been directed northwards in its lower reaches to flow into the Lake Baringo basin, the subsidence of which has caused the incision of the older meandering course through Hannington phonolites (mid-Pleistocene) into older (late Pliocene/early Pleistocene) volcanics (Fig. 3:14). Further subsidence, related to the numerous grid faulting towards Lake Hannington, has lowered the axis of the Rift well below the level of the adjacent parts of the Perkerra River (Fig. 3:9).

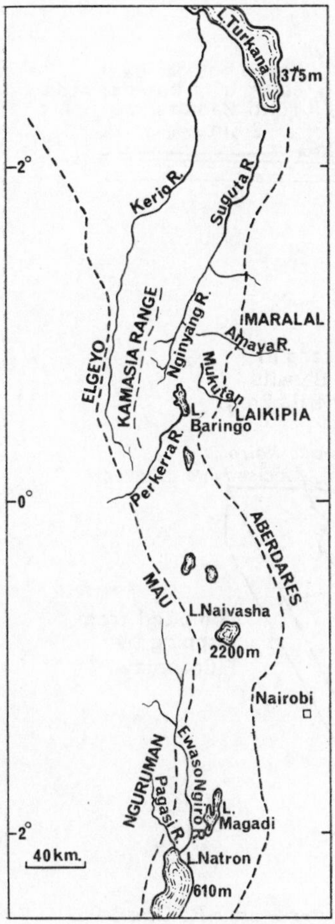

FIG. 3:14.

Main drainage pattern of the Gregory Rift.

Major structures of the Gregory Rift

Generalised structural maps of the Rift show a sinuous swing in trends between about 0° and 1°S latitude (Fig. 3:1). Detailed mapping of the Kamasia range and its eastern pediment by Martyn and Chapman revealed an interference pattern between two obliquely intersecting, but essentially contemporaneous fault trends, the one about north-south, the other more nearly NNE–SSW; the result is a system of sharp buckles or arches, suggesting a strike-slip component along the latter of these trends.

An analysis of ERTS photographic 'imagery', supported by geological field data, has proved very informative on major structures; such an analysis of a selected, but critical, area is presented in Fig. 3:16. This area covers both the 'sigmoid bend' and the junction with the Kavirondo trough. It reveals an inter-

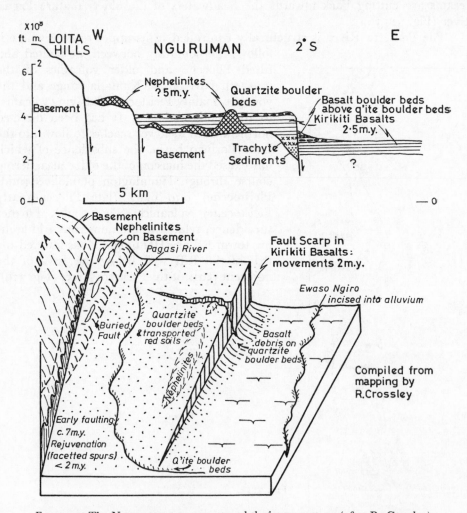

FIG. 3:15. The Nguruman escarpment and drainage pattern (after R. Crossley).

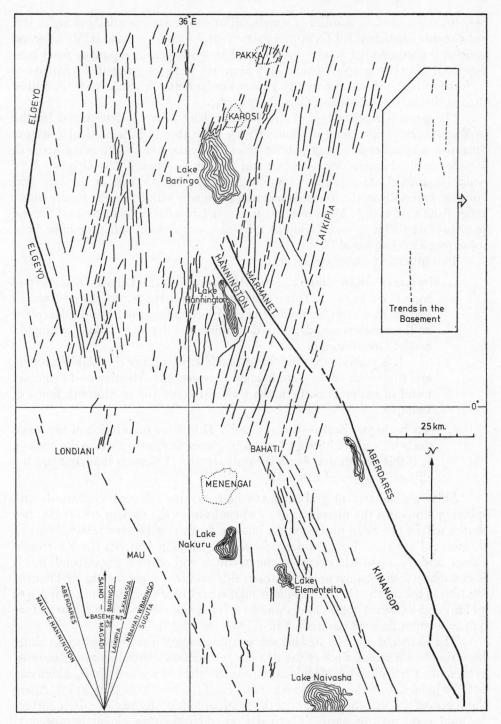

FIG. 3:16 Fault trends and lineaments of the central sector of the Rift as shown by ERTS 'imagery'.

play among a limited number of trends, all of which are essentially straight and not curved. Elsewhere ERTS imagery has been shown to provide information on structural patterns that cannot be obtained by geological mapping or from conventional air photography. Commonly seen are series of straight lineaments of regional extent, which apparently reflect fundamental fracture systems, rather than their surface expressions (Kronberg 1975).

The general morphology of the Gregory Rift valley is determined by the predominance of one trend of faulting with large throws in a particular sector; thus over a considerable length of the Rift a fairly constant width is maintained.

An approximately north-south trend is seen in much of the Gregory Rift, which is, indeed, its regional orientation and is characteristic of the late grid faulting, especially in the southern sector. This agrees with basement trends which range from 345° to 15°. Moreover, many major faults dating back to the Miocene (e.g. northern Elgeyo and Laikipia) have this trend, although they have been subsequently rejuvenated.

Two other important trends occur:

1. Marmanet—East Hannington—Aberdares—south Bahati—Kinangop—Mau (145° to 160°). This trend dominates a large part of the central Rift and produces the 'sigmoid bend'. It involves faults of large displacement and, beneath recent deposits, determines the shapes of Lake Hannington and of Lake Baringo.

   It is possible that the 'gang-plank' structures of the Laikipia region are the result of an interference between the Aberdares—Marmanet trend of major Pliocene faults (NW–SE) and the north-south faults of Laikipia.

2. West Baringo—Pakka—Suguta (20°). This is the orientation of the 'axial graben', which although faulted in places, is mostly a sharp downwarp. It is this trough that forms the only structural element that connects the rifts of Kenya and Ethiopia.

Although the typical graben passes both northwards and southwards into splaying structures the directions of these conform to the various trends that are shown within the main graben itself, but the faults mostly have relatively small throws and are spread over wide regions. The junction between the Kavirondo trough and the main Rift occurs at the northern end of the sigmoid bend, but it is essentially a downwarp on its southern side and is defined mostly by Pliocene faulting to the north. The Kavirondo trough is narrower than the main Rift (20 to 30 km) and shows no uplift of the shoulders; it may be considered a 'lag' structure on the western flank of the main Rift (Shackleton 1951).

The Kenya 'dome' is a striking topographic feature in which elevations along both the shoulders and floor of the Rift rise towards the central sector. It is, however, more a reflection of volcanic accumulation than of vertical uplift, which can only be judged from the displacement of levels of the basement (Fig. 3:17). Along the flanks of the Rift the altitudes of the residual hills of the late Mesozoic surface (around 3000 m) are partly inherited from their position along the ancestral watershed and partly the result of pre-rift movement together with uplift of the

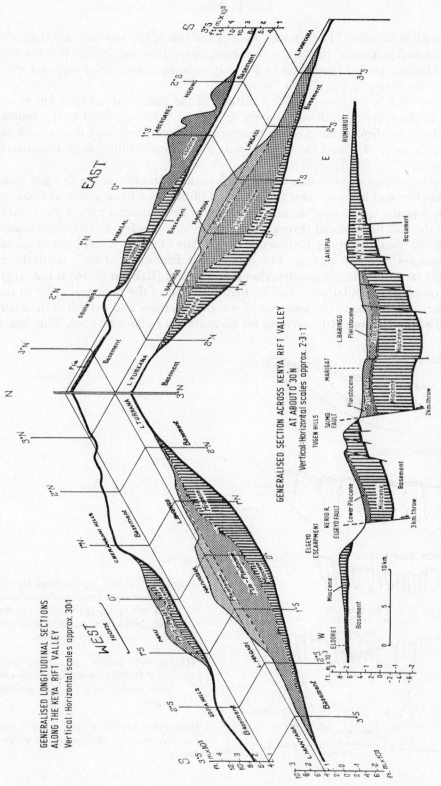

Fig. 3:17. Expanded longitudinal section along the Gregory Rift with transverse section at about 0°30′N.

rift shoulders in later, Pliocene and Pleistocene times, the last also affecting the superimposed volcanics. In general, however, the altitude of the basement is that of the Miocene eroded landscape and is from 1400 m to 2000 m along, and even beyond, the extent of the Kenya 'dome'.

Along the floor of the Rift the altitude of the basement ranges from about 300 m at the splaying ends to as much as 2500 m below sea-level in the central part. Here, the displacement of the basement between flanks and floor is at least 4000 m (4 km), of which the greater part must have resulted from downward movement.

Accumulation of volcanics above the basement has added up to 2500 m to the shoulders and has increased the elevation of the floor by as much as 1200 m; thereby forming the Kenya 'dome'. Boreholes in the Naivasha area of the central part of the Rift have passed through as much as 1000 m of volcanics (pryoclastics and lavas), mostly of upper Pleistocene age; a date of 0.4 m.y. has been obtained from samples at about 950 m. The even larger Ethiopian 'dome', similarly, is more the result of volcanic construction than of uplift (Baker *et al.* 1972, Fig. 3:5).

Volumetric calculations based on displacements of the floor and uplift of the shoulders, together with the amounts of volcanic overflow and overfill of the main graben of the Gregory Rift, account for no more than about 80 000³. This figure

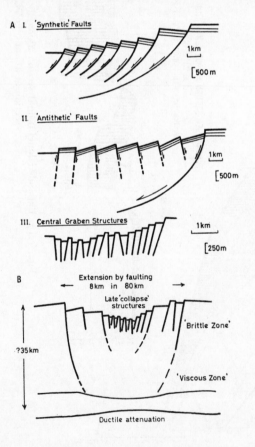

FIG. 3:18.

A. I and II: Possible mechanisms for the development of marginal 'synthetic' and 'antithetic' fault patterns in the Rift, largely by analogy with observed land-slip phenomena.

III. Central graben structures, with 'grid' faulting implying gravity collapse under extension at shallow depth.

B. Generalised crustal section relating surface structures to phenomena at greater depths.

is of the same order of magnitude as the volume of volcanics associated with the Rift zone itself (Table 3). It is tempting to suppose, therefore, that this accordance of volumes is related to the transfer of materials to upper levels, from within or below the crust. However, a consideration of rift systems in general shows that in those sectors, of limited extent, which approximate to simple graben, vertical displacements are greatest, with subsidence of the floors of 4 km or more and elevation of the shoulders of around 1 km, but the essential structures are similar, whether the graben are infilled (or overtopped) by volcanics (Gregory Rift), by sediments alone (Lake Albert Rift), or largely by water (Tanganyika Rift and Baikal Rift). Thus the fundamental control must be tectonic and the nature of the infill of minor significance.

Such rift graben are marked by negative gravity anomalies (Bouguer), which are partly, but incompletely, explained by infills of low density sediments. Crustal attenuation at deeper levels and the uprise of anomalously low density mantle material are reconcilable with geological data (Searle 1970 and Fig. 3:18B). Localised gravity 'highs' within rifts are confined to those unusual sectors with volcanics and are provisionally to be explained as reflecting the rise of primary magmatic reservoirs, by some 'replacive' process, rather than by the initiation of crustal spreading, for there is no regular correlation between 'highs' and the positions of central volcanoes at the surface. Seismic and gravity data are difficult to reconcile at present and even the evidence for anomalous mantle velocities and crustal depths extending across the entire rift system in eastern Africa (Baker *et al.* 1972, Fig. 3:19) has to be reviewed in the light of increasing knowledge about supposedly 'normal' continental crust.

In any case the major problem is to explain not so much the present pattern of anomalies, as how they originated and developed, for geological data suggest that the Gregory Rift is at the closing stages of its evolution, rather than marking the initiation of an episode of crustal spreading.

Acknowledgements

As indicated in the introduction, this short appraisal of important aspects of the history of the Gregory Rift would not have been possible without the opportunities afforded to the writer of making direct observations throughout the entire region, guided and stimulated by the detailed studies conducted over a period of 10 years by J. E. Martyn, G. R. Chapman, M. P. McClenaghan, P. K. Webb, J. Carney, J. S. C. Sceal, S. Rhemtulla, S. J. Lippard, M. Golden, Miss M. Reid, P. H. Truckle, W. B. Jones, P. S. Griffiths, R. Crossley and R. Knight and with the unfailing co-operative participation of Dr L. A. J. Williams. Close collaboration with Professor W. W. Bishop and his group is gratefully acknowledged, while Dr N. J. Snelling, Institute of Geological Sciences, has provided continuing support with K–Ar age determinations, notably by means of mutual co-operation in which Mrs M. Brook was the main worker. Geophysical investigators were also extremely helpful; among those particularly to be mentioned is Dr A. Khan.

Essential to the project has been the financial support by the Ministry of Overseas Development over the entire period, who are continuing currently to sustain the cost of the preparation of standard geological maps, as well as major contributions from the Government of Kenya and from the Natural Environment Research Council, the latter having made possible the employment of specialised staff and services.

Finally, I wish to acknowledge the important contribution made by Mr N. Sinclair-Jones not only for his skill in drawing the maps and diagrams, but also for his advice and inventiveness in improving their presentation.

References

BAKER, B. H. MOHR, P. A. & WILLIAMS, L. A. J. 1972. Geology of the Eastern Rift System of Africa. *Geol. Soc. Amer. Spec. Paper* **136**. 67 pp.

BISHOP, W. W. & TRENDALL, A. F. 1967. Erosion-surfaces, tectonics and volcanic activity in Uganda. *Q. Jl. geol. Soc. Lond.* **122** (for 1966), 385–420.

——, MILLER, J. A. & FITCH, F. J. 1969. New potassium-argon age determinations relevant to the Miocene fossil mammal sequence in East Africa. *Amer. Journ. Sci.*, **267**, 669–99.

CHAPMAN, G. R. MARTYN, J. E. & LIPPARD, S. J. Stratigraphy and structure of the Kamasia Range, Kenya rift valley (to be published).

DIXEY, F. 1956. The East African rift system. *Colonial Geol. & Mineral Resources Supp.* **1**. 71 pp.

GIRDLER, R. W. 1965. The formation of new oceanic crust: in 'A symposium on continental drift'. *Phil. Trans. R. Soc. London, Ser. A.* **258**, 123–36.

GREGORY, J. W. 1896. *The Great Rift Valley*. John Murray, London. 422 pp.

HOLMES, A. 1965. *Principles of Physical Geology*. Nelson, London and Edinburgh. 1288 pp.

KING, B. C. 1965. Petrogenesis of the alkaline igneous rock suites of the volcanic and intrusive centres of eastern Uganda. *Journ. Petrol.* **6**, 67–100.

——, & SUTHERLAND, D. S. 1960. Alkaline rocks of eastern and southern Africa. *Sci. Progress*, **48**, 298–321, 504–24, 702–20.

——, & CHAPMAN, G. R. 1972. Volcanism of the Kenya rift valley. *Phil. Trans. R. Soc. Lond.* **A271**, 185–208.

——, LE BAS, M. J. & SUTHERLAND, D. S. 1972. The history of the alkaline volcanoes and intrusive complexes of eastern Uganda and western Kenya. *Journ. Geol. Sci.*, **128**, 173–205.

KRONBERG, P. 1975. ERTS on Regional Fracture Patterns of Central Europe. *In 1975 Progress Rept., Geodynamics Project. Nat. Comm. Fed. Rep. of Germany and German Res. Soc.*, Bonn, 14–18.

LIPPARD, S. J. 1972. The stratigraphy and structure of the Elgeyo escarpment, southern Kamasia Hills and adjoining regions, Rift Valley Province, Kenya. Ph.D. thesis, Univ. London (unpublished).

—— 1973. The petrology of phonolites from the Kenya Rift. *Lithos*, **6**, 217–34.

LOGATCHEV, N. A., BELOUSSOV, V. V. & MILANOVSKY, E. E. 1972. East African rift development. *Tectonophysics*, **15**, 71–81.

MARTYN, J. E. 1969. The geological history of the country between Lake Baringo and the Kerio River, Baringo District, Kenya. Ph.D. thesis, Univ. London (unpublished).

PULFREY, W. P. 1960. Shape of the sub-Miocene erosion bevel in Kenya. *Bull. geol. Surv. Kenya*, **3**.

SAGGERSON, E. P. & WILLIAMS, L. A. J. 1964. Ngurumanite from southern Kenya and its bearing on the origin of rocks in the northern Tanganyika alkaline district. *Quart. Journ. Geol. Soc.*, **121**, 51–72.

SEARLE, R. C. 1970. Evidence from gravity anomalies for thinning of the lithosphere beneath the Rift Valley in Kenya. *Geophys. J. R. Astr. Soc.*, **21**, 13–31.

SHACKLETON, R. M. 1951. The Kavirondo Rift Valley. *Quart. Journ. Geol. Soc.*, **106**, 345–92.

WEAVER, S. D. 1974. Phonolitic ash-flow tuffs from northern Kenya. *Min. Mag.* **39**, 893–5.

WEBB, P. K. & WEAVER, S. D. 1975. Trachyte shield volcanoes: a new volcanic form from South Turkana, Kenya. *Bull. Volcanologique*, **39**-2, 1–19.

WILLIAMS, L. A. J. 1969. Volcanic associations in the Gregory rift valley, East Africa. *Nature, Lond.* **224**, 61–4.

—— 1972. The Kenya Rift volcanics: a note on volumes and composition. *Tectonophysics*, **15**, 83–96.

4

LAURENCE A. J. WILLIAMS

Character of Quaternary volcanism in the Gregory Rift Valley

The latest stages in the evolution of the rift in Kenya were marked by intense volcanic activity along the graben and massive eruptions across the plateau to the east. Within the rift, trachytic volcanoes south of the equator have calderas at least 10 km across. They erupted more abundant ash-fall and ash-flow deposits than the volcanoes in a northern group which are characterised by smaller calderas. Flood trachytes occupy much of the rift floor in central and southern Kenya, but fissure basalts become increasingly more prominent to the north where they are conspicuous in late eruptions across the caldera volcanoes. Variations in the nature of almost exclusively basaltic activity east of the rift produced impressive linear multicentre ranges in some areas, thin yet extensive flows in others, and one field dominated by maar formation. In northern Tanzania, strongly alkaline volcanism is superimposed on both the rift floor and plateau regimes mentioned above. Late caldera collapse on some trachytic shields was accompanied by build-up of large nephelinite cones and a complex volcano having a broad basaltic lava pile surmounted by phonolites and trachytes. Volcanic and tectonic processes combined to exert a profound influence on patterns of drainage and sedimentation throughout the region, thereby both creating and destroying environments favourable for occupation by early man.

The structural evolution, history of sedimentation and volcanological development of the Gregory Rift have become more fully documented in the last decade as a result of detailed investigations directed mainly at areas along the rift floor and margins. Contrasting styles of volcanism have been recognised or inferred in many of the Miocene and Pliocene formations, but the Quaternary volcanics (Fig. 4:1) provide a more rewarding field for studies of the character of eruptions both within and outside the rift.

An important aspect of Cenozoic volcanism in this region was a general eastward shift with time of the main areas of activity: Miocene eruptions took place west of and within the proto-rift, Pliocene volcanism tended to be more closely confined to a narrower zone along the rift, and Quaternary activity occurred along the rift floor and in areas well to the east (Williams 1970). Re-examination of evidence from regions east of the rift valley suggests, however, that some extensive but rather poorly known volcanics, traditionally accepted as Quaternary, may in fact represent more widespread episodes of Tertiary activity than previously envisaged.

Though it is generally true that Quaternary volcanism within the Kenya rift was predominantly trachytic whereas activity on the plateau to the east produced mainly basalts (Baker *et al.* 1971), this oversimplification conceals significant changes in the nature of eruptions in different parts of the rift floor and the increasingly important role of basalts to the north where rift and plateau regimes ultimately converge. Moreover, by considering only the Kenya rift, the generalisation conveniently excludes a complex province in northern Tanzania within which nephelinitic volcanism of distinctive character has been superimposed on trachytic and basaltic activity since the Pliocene.

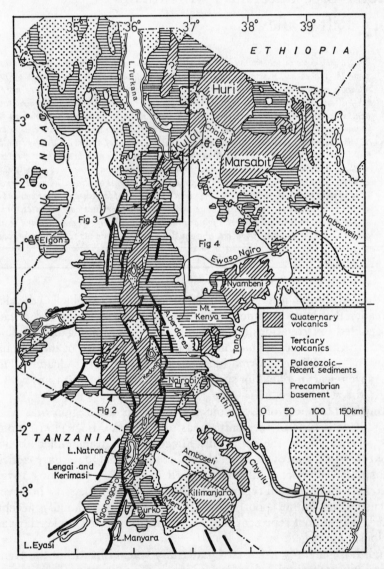

FIG. 4:1. The distribution of Quaternary and Tertiary volcanics associated with the Gregory Rift Valley, and the areas shown in more detail in Figs 4:2, 3 and 4.

Southern part of the Kenya rift

From the equator to latitude 1°30'S the rift is a 70 km-wide trough bounded by NNW–SSE faults, but the most recent volcanism was largely contained within an inner graben about half that width, separated from the eastern boundary scarps by the Bahati–Kinangop step-fault platform (Fig. 4:2). The rift floor attains its highest elevation (2000 m) between Naivasha and Nakuru, and declines southwards to 1500 m near Suswa and to a similar altitude just north of the equator where faults trend NNE–SSW. The rift margins rise to over 3000 m in the densely forested Mau range to the west, and to nearly 4000 m at peaks in the Aberdare range beyond the Kinangop–Bahati platform, much of which stands above 2500 m.

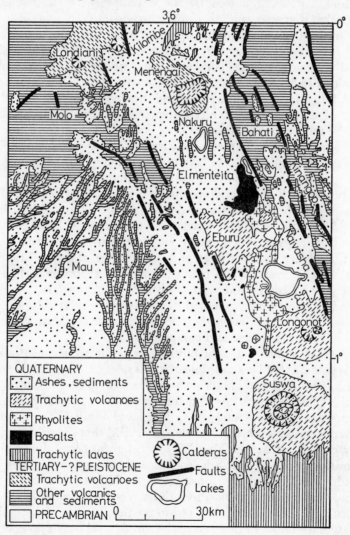

FIG. 4:2. Simplified map of part of the Kenya rift south of the equator.

This section of the rift is characterised by a chain of closely spaced trachytic caldera volcanoes. Some are situated on the western rift margin and are certainly late Tertiary or early Pleistocene in age (Londiani c. 3 m.y.; Kilombe c. 2 m.y.). All probably had complex early histories and they display significant structural, petrological and volcanological differences noticeably in the closing stages of activity. A feature of three well-known Quaternary calderas Menengai, Longonot and Suswa (Fig. 4:2) is similarity in size.

Late summit collapse and young effusive and explosive volcanism at the ignimbrite-trachyte shield of Menengai left a well-defined caldera 11 × 7 km across, floored with recent trachyte flows. The history of build-up of the shield could well date back to the Pliocene. Formation of an early caldera about 10 km across at Longonot was followed by the development of a smaller, perfectly preserved 1.5 km diameter summit caldera on a composite cone situated on the eastern side of the original shield. Glassy ignimbrites capping the south Kinangop platform were probably erupted from Longonot before subsidence of the inner graben, and the most recent flank eruptions produced trachytic block flows. Suswa is unique in having an annular graben within a caldera some 10 km in diameter. All the lavas are trachytes and trachyphonolites, and the state of preservation of flows together with widespread fumarolic activity suggest that Suswa was the most recently active caldera volcano in this part of the rift.

The significance of a marked semi-circular embayment in the fault scarps forming the eastern wall of the rift NW of Nairobi (Fig. 4:1) has largely escaped comment in previous accounts of the volcanism and structure of this region. The writer interprets this Kedong embayment as a rejuvenated portion of a gigantic caldera-like structure, now obliterated on the west by faulting, build-up of Longonot and Suswa shields, and by sedimentation in the rift floor. Plio-Pleistocene trachytes and ignimbrites which extend eastwards into the Nairobi area, similar formations underlying youngest flows of the Kinangop, and some trachytes in the rift south of Suswa, probably all constitute part of an original shield. Arcuate faults defining the Kedong embayment displace trachytes less than 2 m.y. old. If the above interpretation is accepted, these faults are related to a caldera some 35 km across. This is much larger than Ngorongoro in northern Tanzania which has a shield of comparable age and composition.

Eburu, which stands out as a buttress from the western (Mau) rift scarp (Fig. 4:2), is surmounted by a multiple crater system on the axial line of the inner graben but has no caldera. The craters are flanked to the west by a substantial portion of the volcano poorly exposed through a cover of young ashes and sediments which have yielded numerous artifacts and fossils. The western parts of Eburu perhaps represent remnants of an early shield that contributed ignimbrites sporadically exposed across the Mau. The volcano was certainly deeply dissected before deposition of the present ash mantle, and throughout its later history the ridge formed an effective southern margin of the Nakuru–Elmenteita lake basin.

Rhyolite domes related to north-south fissures extend across the eastern portion of Eburu, and the volcano is flanked to the north by the only significant outcrop of Quaternary basalt in this part of the rift. Several basalt flows are recognisable in an area known as the 'Badlands'. They are unfaulted, and surround and engulf older faulted basaltic cones, some of which have craters about 1 km across attributed to phreatic eruptions. The youngest basalts and associated cinder cones south of Lake Elmenteita rest on lake beds, but older flows are covered by and intercalated with sediments. Strand-lines preserved on some of the scoria cones provide further evidence of fluctuating lake levels in the Nakuru–Elmenteita basin.

Highly explosive episodes in the evolution of caldera volcanoes south of the equator led to deposition of trachytic ashes across much of the rift floor and the western fault scarps. These pyroclastics are up to 200 m thick in the Mau range. Interfluves still preserve a flat profile in north-western parts of the Mau, but farther east rapid incision of drainage accompanied by gentle tilt away from the rift carved a series of north-south valleys with asymmetric transverse profiles. Similar ashes draped across re-excavated valleys south of the Mau testify to the youthful character of the deposits in that area. On the opposite side of the rift, late Quaternary ashes form a thin veneer across the Kinangop and Bahati platforms.

Some of the most important archaeological sites in Kenya are located in areas covered by these ashes and associated sediments, but the sources of the pyroclastics can seldom be traced with certainty. Young deposits at Bahati and in northern parts of the Mau evidently came from Menengai. In earlier accounts, the writer favoured Longonot and Suswa as the most likely sources for the bulk of the Mau ash, despite difficulty in visualising eruptions on the scale required to account for the great thickness. Interpretation of the Kedong embayment as part of a very large caldera provides an alternative and more plausible origin for some of the earlier deposits.

The rift floor takes on a different character south of Suswa where the ash mantle thins rapidly to expose trachyte lavas that flooded the entire southern part of the rift. The detailed stratigraphy is under revision following recent investigations, but it is clear that the most extensive lavas are less than 1 m.y. old (Fairhead *et al.* 1972). Flows have not been traced convincingly to central volcanoes, and they were probably derived from fissure sources over a wide area. Scattered ash cones developed after the main eruptive episode, and all the volcanics were subsequently cut by numerous so-called 'grid' faults which produced minor horst and graben structures. Fault blocks are commonly tilted, and troughs are infilled with sediments and alluvium. In the Olorgesailie area, 45 km south of Suswa, diatomites and tuffaceous silts containing artifacts and fossils were deposited in a lake confined by graben faulting of Quaternary trachytes and basalts on the northern flanks of a late Tertiary volcano.

Northern section of the Kenya rift

North of the equator, closely spaced faults disrupt late Quaternary flood trachytes and trachyphonolites which occupy some 80 km of the rift floor between the group of caldera volcanoes described above and those of a NNE-trending chain in which all the centres display basalts as well as trachytes. Basalts outcrop between the southern volcanoes of Karosi, Paka and Silali which occur north of Lake Baringo and are about 25 km apart. With wider spacing of the centres, intervening basaltic fissure eruptions figure more prominently north of Silali despite a widespread cover of Holocene sediments and alluvium over the older flows. Silali has the most impressive caldera (8 × 6 km): the others are less than 5 km across.

A broad belt of Tertiary volcanism in the region south of Lake Turkana (Fig. 4:3) was dominated by extrusion of basalts and build-up of low-angle trachytic shields some of which were surmounted by calderas (e.g. Lomi), but by Quaternary times volcanic and tectonic activity had become more closely confined to a NNE-trending zone (the Suguta valley) about 25 km wide, bounded by faults and major downwarps. The floor of the trough slopes northwards to its lowest elevation of 255 m immediately south of the 'Barrier' caldera volcano which rises to about 1000 m and effectively cuts off the Suguta valley from Lake Turkana. The lake surface now stands at 375 m, but strand-lines indicate a previous level at least at 450 m. Reconnaissance mapping (Dodson 1963) showed that under-

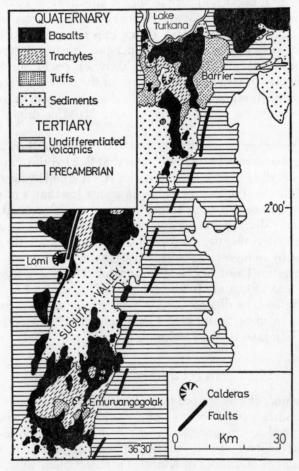

FIG. 4:3. Simplified map of the Kenya rift south of Lake Turkana.

saturated trachytic lavas forming the Barrier shield are accompanied by abundant pyroclastics that rest locally on lacustrine sediments now standing some 150 m above the level of Lake Turkana. Basalt eruptions, which followed development of trachytic satellite cones, were probably related to fissures concentrated mainly across the eastern flanks of the volcano and along the eastern margin of the Suguta valley. The same NNE–SSW zone was marked by faulting and tilting before the deposition of sediments from high-level Turkana and Suguta lakes. Summit collapse on the Barrier shield left a caldera 3.5 × 3 km across in which trachytes were erupted. The final phase of activity produced basaltic and trachy-basaltic flank eruptions north and south of the caldera, culminating in ash emission from one of the cones (Teleki's volcano) late last century.

Farther south, Emuruangogolak volcano (1340 m) restricts the Suguta river to a narrow channel close to the western margin of the trough. The early shield lavas are faulted trachytes which have yielded K–Ar ages in the range 0.5 to 1 m.y., but the precise eastern limits of the volcano are difficult to define in ground where shield trachytes are not readily distinguishable from similar Pliocene (c. 3.5 m.y.) volcanics of the Emuruagiring plateau. Exposed pyroclastics are trivial compared to the shield-forming lavas, and there are no significant accumulations of ash along the rift margins at this latitude. The rim of a shallow summit caldera 4.5 × 3 km across is largely concealed by trachytic lava cones, and more recent trachytes were erupted in an arcuate pattern across the caldera region. Fissure basalts appeared at late stages in the history of the volcano and are exposed over a large area (Fig. 4:3). Some of the earlier flows are intercalated with sediments deposited in a Suguta lake which left high strand-lines at an elevation of 450 m on trachytic cones on the north side of Emuruangogolak. Later basalts came from NNE-trending fissures which opened across the volcano and adjacent parts of the rift floor. They include a flow shown by C^{14} dating of charred wood to be only about 250 years old: and a trachytic flank eruption on the south side of the volcano is even younger. Phreatic eruptions produced numerous basaltic pseudocraters on low ground south of Emuruangogolak, and several maar-type explosion craters in the Suguta valley to the north.

Volcanism east of the rift

Though much of the volcanic terrain east of the Kenya rift has been covered in the course of reconnaissance mapping by the Geological Survey, vast tracts between Lake Turkana and Marsabit in northern Kenya remain largely uninvestigated. Comments on the general character of volcanism and its interaction with patterns of sedimentation must be viewed against this background.

Despite a likely Tertiary age, basalts which cap plateaus rising 20–130 m above the plains between the Nyambeni range and Marsabit (Fig. 4:4) merit some consideration here because of their importance in deciphering the subsequent history of the region. The surfaces of these plateaus decline steadily north-eastwards from nearly 1000 m close to the Ewaso Ngiro river at Archer's Post to approximately 500 m near the Kaisut desert. Recognition of similar lavas north of Kaisut (Randel 1970) suggests that the plateau surface there rises towards

Marsabit beneath younger volcanics. In some areas, basalts rest directly on a pediment of basement gneisses, but the more easterly plateaus expose extensive sub-volcanic sediments. These grits and conglomerates have not yielded fossils and, in the absence of isotopic dates, the age of the overlying lava has remained a matter of some speculation. Evidence suggesting a Pliocene age for basalts capping the Merti plateau has been presented elsewhere (Williams 1966). The plateau basalts were erupted from intersecting patterns of fissures: some feeder dykes closely follow the N to NNW basement grain, but many display E to NE trends.

On unpublished maps of the Geological Survey of Kenya, a Pleistocene age is assigned to basalts forming a barren and monotonous region north of Marsabit known as the Dida Galgalu. This vast boulder-strewn plain slopes southwards from about 1000 m near the Ethiopian border, where basalts rest on basement, to some 600 m at the fringes of the overlying Marsabit volcanics. Consequently, the Dida Galgalu lavas may well represent an extension of the plateau volcanics mapped farther south. Similarly, lavas in an unmapped region south-east of

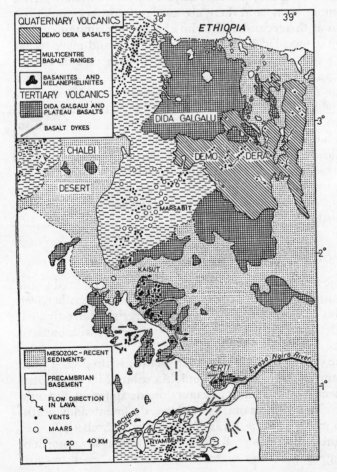

Fig. 4:4.

Outline geology of part of northern Kenya, east of the rift valley. Large areas remain unmapped so that distinctions between Tertiary and Quaternary volcanics are somewhat speculative.

Marsabit and some flows in the Chalbi desert are also tentatively correlated with the plateau basalts. Critical evidence will no doubt come from studies of relationships between volcanics and sediments in these areas.

Saggerson & Baker (1965) regarded the sub-volcanic surface in this region as a continuation of the end-Tertiary erosion surface that is well preserved in eastern Kenya, but it can be argued that lateritic soils and basement pebbles on plateau surfaces point to dissection of formerly more extensive basalt sheets during the end-Tertiary erosion cycle. Regardless of whether reconstruction of the end-Tertiary surface is based on elevations of plateau surfaces, the base of basalt sheets, or altitudes of the surrounding plains which are covered with thick lateritic soils, a marked linear depression must be invoked with an axis near Marsabit trending NW–SE: it was described as the Chalbi–Habaswein depression separating major domal uplifts in Kenya and Ethiopia (Saggerson & Baker 1965). The age of relative downwarping of the end-Tertiary surface in this region cannot be established with precision at present, but development of the broad depression had a profound influence on the character of later volcanism.

Numerous basanitic and basaltic flows and ridges up to 15 m thick break the regularity of most plateau surfaces between the Ewaso Ngiro and Marsabit. These lavas came from small vents and fissures, and eruptions probably coincided with activity at multicentre ranges described below. The most recent eruptions in this area produced NE to NNE chains of basaltic and melanephelinitic vents on the plains, and maar-type craters on or close to the plateaus where rising magma probably encountered aquifers in sediments.

A more impressive array of maars characterises the Marsabit region where a broad NE–SW belt of well-preserved explosion craters (local name: *gof*) can be traced for 70 km. The maars have low rims of basaltic scoria and range from 1 to 2.5 km across; some contain permanent lakes. They extend across an accumulation of basaltic lavas and pyroclastics forming the forested Marsabit mountain which rises to 1700 m, but the maars are more spectacular on the arid lava plains west of Marsabit where they are interspersed with cones almost to the edge of the Chalbi desert. Marsabit lies astride the Chalbi–Habaswein depression, and can be regarded as a unique type of multicentre range beneath which accumulation of groundwater led to development of maars rather than cinder cones.

Extensive basalts in the Demo Dera region north-east of Marsabit overlie both the Dida Galgalu flows and the Marsabit volcanics. The lavas flowed southwards and south-eastwards from NE–SW lines of vents and fissures, indicating eruption on an already significantly tilted end-Tertiary surface.

En echelon NNE-trending ranges composed of hundreds of basaltic cones rise to altitudes of over 1200 m in the Huri hills north-west of Marsabit, and above 2000 m at Kulal on the opposite side of the Chalbi desert. When these areas are mapped, it seems likely that it will be shown that the multicentre ranges built up on a foundation of older plateau basalts and sediments. The form and significance of a large crater or caldera structure on Kulal must similarly await investigation on the ground.

More accessible and better known is the Nyambeni multicentre range which extends 100 km NNE from the late Tertiary volcanic pile of Mt. Kenya. It

consists essentially of a broad apron of basalts between the Ewaso Ngiro and Tana
rivers, and a spine of later lavas surmounted by numerous cones. Much of the
central range stands above 1500 m, but rises to 2500 m near the southern end
where the volcanics are some 1500 m thick and include phonolites and tephrites
overlying basalts. In a synthesis of Nyambeni volcanicity, Rix (1967) considered
early basalts to have been derived from NE-trending fissures (probably more
correctly NNE) beneath the range, envisaging spread across a dissected end-
Tertiary surface to account for flows into the Ewaso Ngiro and Tana valleys. He
made no distinction, however, between these flows and other basalts derived from
eroded vents low on the northern slopes of Mt. Kenya, although the latter are
clearly older since they rest on a surface 100 m above the Ewaso Ngiro (Shackleton,
1946). Rix also inadvertently included in his Lower Olivine Basalts some very
recent flows, such as the one that dammed the Ewaso Ngiro at Archer's Post and
led to deposition of lacustrine sediments (Jennings 1967). Instances of earlier
ponding, sedimentation and diversion of the river are reported from observations
downstream (Williams 1966). Lack of interference with the more deeply incised
Tana river suggests that the bulk of the lava flowed northwards, perhaps on an
already tilted remnant of plateau basalt preserved on the interfluve. Minor flows
of melanephelinite came from a few scattered vents during eruption of the early
Nyambeni basalts. Explosive activity at the northern end of the range probably
occurred late in the apron-building phase and produced bedded pyroclastics
around a cluster of basaltic vents, including a single maar-type crater which
brought up massive basement blocks. The tuffs are overlain by basalts some 500
m thick which form a substantial part of the high range. Flows have well-preserved
fronts but can seldom be traced to cone sources, so a fissure origin is inferred.
Most of the basaltic scoria cones were built up on the lava units at a very late
stage. Lava emissions were rare though a trachybasaltic flow, perhaps the youngest
in the area, was erupted from a cone at the northern end of the chain. A continuation of the Nyambeni phase of volcanism across the flanks of Mt. Kenya resulted
in development of numerous satellite cones (Baker 1967). Basalt flows emanate
from some well-preserved vents on the northern side of the mountain, but much
more extensive lavas (Thiba basalts) flooded the southern slopes. Valley flows
occupied tributaries to the Tana river, and one of its main headstreams is incised
in the basalts.

Another Quaternary multicentre chain forms the Chyulu hills 300 km south
of the Nyambeni field (Fig. 4:1). This NW–SE range is composed entirely of
basaltic lavas and pyroclastics up to 1000 m thick. The bulk of the volcanics rest
on lateritic soils mantling a basement surface which stands at an altitude of about
1000 m, but some flows occupied river valleys cut into the 'end-Tertiary' surface.
Basalts and basanites at the NW end of the chain were attributed to Pleistocene
eruptions, whereas the more spectacular central and southern parts of the range
built up more recently (Saggerson 1963). Flows only a few centuries old at the
southern extremity show that this tendency for SE migration of volcanic activity
with time persisted through the latest episodes. Hundreds of cones on NW–SE
and N–S trends are related to faults, many of which still have surface expression
across cones and lavas. From time to time, mild earth tremors provide a reminder

of the youthful character of the main range. The Chyulu hills differ from multi-centre ranges farther north in several respects: there are no indications of under-lying Tertiary volcanics or widespread sediments; the orientations of feeder fissures are quite different; many of the exposed lava flows can be traced to sources at cones; maar-type explosion craters are absent; no phonolites or nephelinitic volcanics are reported; and the flanking basalts form a much narrower belt.

Northern Tanzania

The volcanics at the southern end of the Gregory Rift occupy a 300 km-wide zone in a complex region of diverging faults and downwarps (Fig. 4:1). Recent radio-metric dating projects (e.g. Macintyre *et al.* 1974) have been largely concerned with attempts to establish the age of a period of faulting which provides a con-venient basis for subdivision of the volcanics into older and younger groups. The older phase of activity resulted in build-up of Plio-Pleistocene trachyte-basalt volcanoes interspersed with mainly Pliocene nephelinite-phonolite centres. Trachytic volcanoes of the Crater Highlands, including the giant caldera at Ngorongoro, lie in a NE extension of the Eyasi trough: other trachyte-basalt centres form a chain trending SSE from Lake Natron. More relevant to the present discussion are several younger nephelinitic volcanoes (Lengai, Kerimasi, Burko) located close to major faults in the Natron–Manyara region, and two large central volcanoes farther east (Meru and Kilimanjaro) which largely infill an E–W depression.

Investigations of the late Quaternary Kerimasi and still-active Ol Doinyo Lengai centres have led to clearer understanding of many aspects of nephelinite-carbonatite volcanism, but have largely failed to stress the importance of using drainage patterns on young pyroclastics as sensitive indicators of late tectonic events in the region: yet the geomorphological evidence here is critical in establish-ing the full history of sedimentary basins.

The main activity at Kerimasi was followed by minor faulting and develop-ment of cones, craters and tuff rings; maar-type craters are attributed to eruptions in swamps or lakes (Dawson & Powell 1969). Subsequently, repeated explosive episodes at Lengai built up a steep cone 2000 m high near the foot of the Natron–Manyara fault scarp. The most recent phases of activity are well documented. Falls of carbonatitic ash were reported up to 130 km west of the volcano during the 1966 eruption (Dawson *et al.* 1968) and some nephelinite ashes draped across Olduvai Gorge, 60 km WSW of Lengai, may have been deposited during earlier eruptions (Hay 1963). Ash dunes near Olduvai show the effectiveness of wind action in the redistribution of pyroclastic material. Recognition within the Olduvai succession of older wind-worked nephelinite tuffs altered to zeolites and calcrete greatly assisted in clarification of the local stratigraphy (Hay 1967). The precise sources of these pyroclastics, and similar deposits bearing surface calcrete across large areas of the Serengeti Plains, are a matter of speculation. On the other hand, thick accumulations of pyroclastics at the northern end of the Crater Highlands clearly came from Lengai and Kerimasi. They provide exceptional opportunities for study of the initiation and development of drainage systems on ash-infilled topographies close to the edge of the Natron fault scarp. A striking feature is the alignment of valleys parallel to the scarp. Examples of north-flowing drainage recently impaled on a surface which now slopes westwards demonstrate convincingly the very recent age of substantial tilting in this area. Some deeper valleys have asymmetric transverse profiles and are reminiscent of those carved in the Mau ashes in central Kenya where similar late tilting away from the rift is envisaged.

The largest and in many respects the most complex Quaternary volcano in Tanzania is Kilimanjaro which rises more than 5000 m above a base 100 km long and 65 km across, with the long axis trending WNW–ESE. Early development of a multicentre basaltic shield, considerably thicker than any of the linear ranges in Kenya, was followed by concentration of activity at three main centres which initially contributed basalts but later erupted lavas of more diverse compositions, including rhomb porphyries and nephelinites (Downie & Wilkinson 1972). Collapse of part of the eastern cone (Mawenzi) during the closing stages of activity in early Quaternary times led to formation of a lahar which can be traced NE to the foothills of the Chyulu range. Another lahar south of the mountain resulted from crater rim collapse in the late Quaternary at the central vent (Kibo) where the most prolonged volcanism spanned several episodes of glaciation. Parasitic cones are grouped in distinct zones across the mountain, but controls over some linear patterns are not fully understood. Growth of the Kilimanjaro shield led to disruption of early drainage: one result was development of the Amboseli basin on the north side of the mountain (Williams 1972).

The full history of activity at Mt. Meru, west of Kilimanjaro, has yet to be described, but build-up of a large cone rising to over 4500 m culminated in recent eruptions of nephelinites. In the final stages of development of the volcano, collapse of the NE rim of a summit crater gave rise to a lahar and mud-flow field extending to the flanks of Kilimanjaro. The Meru lahar effectively completed the interruption of southward drainage from the Amboseli basin (Downie & Wilkinson 1972).

Discussion

The main areas of Quaternary volcanism in Kenya and northern Tanzania are indicated diagrammatically in Fig. 4:5. Ignimbrites in central Kenya (1) include many flows known to be pre-Quaternary but, until the precise relationships and ages are determined in all areas, there is some merit in regarding the field as broadly Plio-Pleistocene. At this stage, the rift depression was infilled periodically by volcanics and sediments so that trachytes and ignimbrites, perhaps related to large caldera structures in the floor of the trough, spread on to the flanking ground. Some ignimbrites west of the rift valley probably date back to about 5 m.y., and flows in the Nairobi area to the east are about 3 m.y. old. On the other hand, flat-lying ash-flow sheets capping the Kinangop platform along the eastern side of the rift are much younger and have no equivalents on the marginal plateaus.

Late Quaternary trachytes (2) and trachyphonolites (3) flooded sections of the rift floor after its present form had been largely determined by major faulting.

Lake basins developed in southern Kenya mainly after and as a result of disruption of the lavas by numerous faults, but volcanism played a more active role in central Kenya by diverting drainage and by ponding and damming to produce lakes.

Caldera volcanoes in the rift floor are divided into two groups. Thick deposits of ash draped across the floor and the scarps and plateau to the west testify to the highly explosive character of eruptions at the volcanoes in central and southern Kenya (4) where calderas are at least 10 km in diameter. Despite data from numerous archaeological sites, difficulties remain in correlating pyroclastics over large enough areas to establish specific sources: yet thorough understanding of the late tectonic and sedimentary history of the region depends heavily on correct interpretation of evidence preserved in the ash cover. The caldera volcanoes were largely responsible for confining lakes in the Nakuru–Elmenteita, Naivasha and Kedong basins. Overflow channels are preserved in volcanics and intercalated sediments near Longonot and Suswa. Lava flows from the latter volcano at times interrupted southward drainage, confirming the youthful character of this centre (Johnson 1969).

Strongly alkaline volcanism in northern Tanzania (6) continues to the present day and offers a unique opportunity for investigation of the role of nephelinitic and carbonatitic ashes in the development of nearby sedimentary basins. Pyroclastics intercalated with lacustrine deposits have proved invaluable in unravelling the

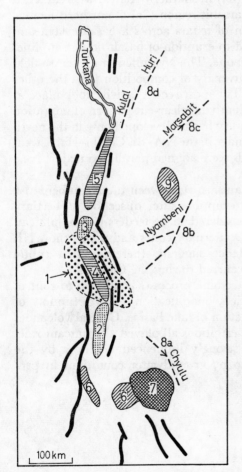

Fig. 4:5.

Diagrammatic representation of areas of Quaternary volcanism in Kenya and northern Tanzania. Key: 1, Plio-Pleistocene trachytic ignimbrites; 2, flood trachytes; 3, flood trachyphonolites; 4 & 5, caldera volcanoes; 6, nephelinite-carbonatite volcanoes; 7, Kilimanjaro; 8a to d basaltic multi-centre ranges (arrows indicate migration of activity); 9, nephelinite-basanite eruptions.

complex micro-stratigraphy in one area where late faulting led to reversal of
drainage. It is not generally appreciated, however, that the evolution of drainage
on the youngest ash surfaces provides critical evidence of recent tectonic events
close to the rift margin. As well as indicating the significance of tilting, studies of
this kind will perhaps throw further light on the question of the age of caldera
collapse on late Tertiary trachytic shields of the Crater Highlands and Ngorongoro.

The area of strongly alkaline volcanism extends eastwards to Kilimanjaro (7),
a broad basaltic lava pile surmounted by phonolitic and trachytic flows. Of
particular interest in this region are massive lahars which combined with volcanics
from Kilimanjaro and the neighbouring nephelinite cone of Meru to confine a
substantial lake, now represented by sediments of the Amboseli basin.

Predominantly basaltic activity east of the rift valley in Kenya was controlled
by fissure systems strongly oblique to the rift trends at the same latitudes. Volcanism culminated in construction of several linear ranges, each consisting of hundreds
of cinder cones on a thick foundation of lavas. A SE migration of volcanic activity
with time is reported in the Chyulu range (8a) in southern Kenya, whereas a NE
shift of eruptive centres occurred in the Marsabit region (8c) of northern Kenya.
Phreatic eruptions there produced a chain of maars across a basaltic lava and
pyroclastic pile, but subsequent activity led to eruption of basalt flows from lines
of cones in a continuation of the Marsabit zone. The Nyambeni multicentre field
(8b) in central Kenya displays a greater diversity of composition than the other
linear ranges that have been investigated. The most recent activity took place at
the NE end of the chain, but there was evidently no clear-cut pattern of migration
of volcanism with time. Significant sedimentary basins developed only in the north
where linear volcanic ranges cut across drainage in the NW–SE Chalbi–Habaswein
depression. Elsewhere, the multicentre fields were aligned parallel to major rivers
and caused, at most, merely local ponding.

Minor basanitic and nephelinitic volcanism (9) between the Nyambeni and
Marsabit fields took the form of scattered eruptions after dissection of Tertiary
plateau basalts. Many dykes which are considered to be feeders for the plateau
basalts trend E to NE, whereas chains of Quaternary cones and maars show NE
to NNE trends. Presently available evidence suggests, therefore, that subtle
changes in orientations of fissure systems occurred during the Cenozoic.

Throughout the region, volcanic and tectonic processes combined to control
patterns of sedimentation by the initiation, modification and disruption of
drainage, and by the formation and destruction of lake basins. Central volcanoes
and the products of fissure and multicentre eruptions all played a significant role.
Locally, the character of volcanism was strongly influenced not only by the
distribution of lakes and swamps, but also by groundwater conditions in pre-
existing sedimentary formations.

References

BAKER, B. H. 1967. Geology of the Mount Kenya area. *Geol. Surv. Kenya Rep.*, **79**.

——, WILLIAMS, L. A. J., MILLER, J. A. & FITCH, F. J. 1971. Sequence and geochronology of the Kenya rift volcanics. *Tectonophysics*, **11**, 191–215.

DAWSON, J. B. & POWELL, D. G. 1969. The Natron–Engaruka explosion crater area, northern Tanzania. *Bull. volcan.*, **33**, 791–817.

——, BOWDEN, P. & CLARKE, G. C. 1968. Activity of the carbonatite volcano Oldoinyo Lengai, 1966. *Geol. Rdsch.*, **57**, 865–79.

DODSON, R. G. 1963. Geology of the South Horr area. *Geol. Surv. Kenya Rep.*, **60**.

DOWNIE, C. & WILKINSON, P. 1972. *The geology of Kilimanjaro*. Univ. Sheffield.

FAIRHEAD, J. D., MITCHELL, J. G. & WILLIAMS, L. A. J. 1972. New K/Ar determinations on rift volcanics of S. Kenya and their bearing on age of rift faulting. *Nature phys. Sci.*, **238**, 66–9.

HAY, R. L. 1963. Zeolitic weathering in Olduvai Gorge, Tanganyika. *Geol. Soc. Am. Bull.*, **74**, 1281–6.

—— 1967. Revised stratigraphy of Olduvai Gorge. In: *Background to Evolution in Africa* (Eds W. W. Bishop & J. D. Clark), pp. 221–8. Univ. Chicago Press, Chicago.

JENNINGS, D. J. 1967. Geology of the Archer's Post area. *Geol. Surv. Kenya Rep.*, **77**.

JOHNSON, R. W. 1969. Volcanic geology of Mount Suswa, Kenya. *Phil. Trans. R. Soc. Lond.*, **A265**, 383–412.

MACINTYRE, R. M., MITCHELL, J. G. & DAWSON, J. B. 1974. Age of fault movements in Tanzanian sector of East African rift system. *Nature, Lond.*, **247**, 354–6.

RANDEL, R. P. 1970. Geology of the Laisamis area. *Geol. Surv. Kenya Rep.*, **84**.

RIX, P. 1967. Geology of the Kinna area. *Geol. Surv. Kenya Rep.*, **81**.

SAGGERSON, E. P. 1963. Geology of the Simba–Kibwezi area. *Geol. Surv. Kenya Rep.*, **58**.

—— & BAKER, B. H. 1965. Post-Jurassic erosion-surfaces in eastern Kenya and their deformation in relation to rift structure. *Q. Jl. geol. Soc. Lond.*, **121**, 51–68.

SHACKLETON, R. M. 1946. Geology of the country between Nanyuki and Maralal. *Geol. Surv. Kenya Rep.*, **11**.

WILLIAMS, L. A. J. 1966. Geology of the Chanler's Falls area. *Geol. Surv. Kenya Rep.*, **75**.

—— 1970. The volcanics of the Gregory Rift Valley, East Africa. *Bull. volcan.*, **34**, 439–65.

—— 1972. Geology of the Amboseli area. *Geol. Surv. Kenya Rep.*, **90**.

5

M. AFTAB KHAN and CHRISTOPHER J. SWAIN

Geophysical investigations and the Rift Valley geology of Kenya

During the last decade there has been considerable activity in geophysical investigations in the East African Rift under the auspices of the Upper Mantle Project. Studies in the fields of explosion and earthquake seismology, gravity, magnetism, heat flow and electrical conductivity have considerably increased our understanding of the deep structure and processes beneath the Rift. The gravity data also throw light on the near-surface geology. A gravity map of Kenya at 1 : 2 000 000 has just been compiled at the University of Leicester using data from over 6500 stations, most of which have been recently established in the Rift area. The regional field can be represented by a seventh order trend surface with a 'saddle' centred at Marsabit in the north. From the centre of this, the regional field rises north-westwards towards Lake Turkana and falls south-westwards towards a pronounced bowl-shaped minimum centred in the Narok–Naivasha region near the culmination of the Kenya dome. The trend surface correlates inversely with the topography and may be attributed to deep structure. The residual gravity field of the Rift Valley region contains a number of features which are clearly related to the visible geology.

The axial region is associated with an undulating high along which lie a number of prominent caldera volcanoes—Suswa, Menengai, Silali, Emuruangogolak and Lomi. Further north the most spectacular high occurs at the syenite plug, El Moiti, which is probably the remnant of a trachyte volcano. The flanking negative anomalies can usually be correlated with low density lavas and sediments—notably at Lake Turkana, near Lokichar, the Kerio Valley and further south at Londiani and Ol Kalou, but lows are notably absent over the Baringo and Suguta basins. Quantitative interpretation requires an estimate of the true regional, as opposed to the synthetic regional of trend surface analysis, and is often extremely difficult to make, due to lack of geologic data. Nevertheless, several east-west profiles have been modelled, yielding, amongst other parameters, estimates of the thickness of sediments and volcanics.

Introduction

In recent years there has been considerable interest and activity in geophysical investigation of the East African Rift following the suggestion that it may be an incipient divergent plate margin. This was suggested initially on the basis of the distribution of the shallow earthquakes which continued from the Carlsberg ridge into the Gulf of Aden, Red Sea and the East African Rift through a triple junction in the Afar triangle. The possibility was given further credence as work in the Red Sea and Gulf of Aden showed that they were young and oceanic in character. Their topography, volcanics, fracture zones, seismic velocities and linear magnetic anomalies are similar to those associated with the oceans. Palaeomagnetic measurements (Irving & Tarling 1961) on the Aden volcanics had suggested that the Red Sea was formed during the last 5 m.y. by the anticlockwise rotation of Arabia by 7°.

The geophysical work so far shows that the deep structure of the Rift is anomalous and similar to that of the mid-ocean ridges. Most of the information has come from seismic studies. The travel time corrections relative to Bulawayo on normal shield-type crust have been estimated for other stations in Africa. It is found that the stations close to the Rift at Nairobi and Addis Ababa have corrections of + 2.3 s and 2.7 s (Long *et al.* 1972), values which are more than twice that for any other part of Africa. This slowing down of the P waves may be accounted for by high temperatures or low density materials underlying the rift. The delay is comparable with the value of 2.5 s obtained for Iceland, astride the mid-Atlantic ridge. The limited data on the first motions (Fairhead & Girdler 1971) of earthquakes suggest tensional stresses normal to the length of the Red Sea in the north and to the Rift in the south and are clearly consistent with the idea of crustal extension across the system. Rayleigh wave dispersion studies show that most of Africa (Gumper & Pomeroy 1970; P. Maguire, personal communication) has a structure typical of normal shield areas, i.e. a P-wave velocity of about 5.8 km/s down to 26 km, 6.6 km/s from 26 to 42 km and then 8.0 km/s for the upper mantle. However, these studies suggest an anomalous structure underneath the Gregory Rift zone across which shear waves fail to propagate. A seismic refraction experiment using explosions in Lakes Hannington and Turkana in the Rift (Griffiths *et al.* 1971) shows a structure reminiscent of the mid-Atlantic ridge, i.e., a P-wave velocity of 6.4 km/s down to 20 km, and 7.5 km/s below. No higher velocity was observed. A similar structure has been obtained for the Afar triangle (Makris *et al.* 1970).

In the 1930s Bullard (1936) carried out a large-scale gravity survey in East Africa using pendulum apparatus. He established that the plateau as a whole is isostatically compensated, but that there are negative anomalies in the Rift areas. He had no seismic data available to him and he explained his observations in terms of a mass of light material underlying the Rifts and this also explained the plateau in terms of isostasy. More recent gravity surveys show that the gravity

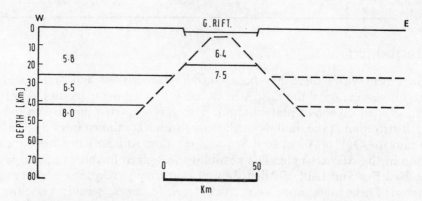

FIG. 5:1. The main features of the deep structure of the Gregory Rift as revealed by seismology. The P-wave velocities are in km/s.

field is much more complicated. In the Red Sea (Girdler 1963) and in Afar (Makris *et al.* 1972) the gravity anomalies are positive. Further south in Kenya, there is usually a narrow positive anomaly within the broad negative (McCall 1967a; Searle 1970; Khan & Mansfield 1971). The highest anomalies occur in the Red Sea where there is clear evidence of crustal separation and the intrusion of dense material. The high crustal velocities observed along the axis of the Gregory Rift suggest that the axial gravity high of the Rift is due to high density intrusive material within the crust. The association of the gravity high with major volcanics in the Rift supports this general conclusion. Recent observations (Williamson 1975) show that the heat flow in the vicinity of the rift is significantly higher than the region outside. Electrical conductivity studies using magnetic variations (Banks & Ottey 1974) show that there is a region of high temperatures beneath the Rift axis and a deeper more extensive zone in the vicinity of Mount Kenya to the east. The main features of the deep structure are summarised in Figure 5:1.

Despite this broad understanding, there is a need for detailed geophysical work to ascertain the structure of the Rift as well as the positions and extents of the anomalous regions underlying it, both within and below the crust. In particular, detailed explosion seismic and gravity studies are needed to refine the broad conclusions on the deep structure obtained from the gravity data. The confirmation of a zone of intrusion within the crust would provide critical evidence on whether or not crustal extension has occurred. The gravity data described in the remainder of this paper illustrate the complexities and possibilities.

The gravity field of Kenya

A. *The data*

The data stored on computer files at the University of Leicester, consist of observations made at 6600 stations throughout Kenya at spacings between 0.5 km and 10 km. The coverage is uneven. One of the principal gaps around Kavirondo has been covered but the data are still being reduced. The other between Marsabit and Wajir has been partly covered by an oil company who have not yet released the data and the remainder will be covered later this year. Some data from northern Tanzania (Darracott, 1972) and some from eastern Uganda (Geological Survey of Uganda 1962) have been added to make the area approximately rectangular to facilitate automatic contouring. The sources of the data for Kenya are as follows:

(i) The University of Newcastle Catalogue of gravity data (Searle & Darracott, 1971). 1620 stations from the catalogue have been used. The data were collected by workers from the University of Newcastle, J. H. McCall of the Kenya Mines and Geology Department, and the Overseas Geological Surveys. The Bouguer Anomalies were recalculated from the basic data given in the catalogue as some of the values were in error by up to 2 mgal due to inaccuracies in the calculation of the latitude correction. The terrain corrections were applied to most stations.

(ii) The British Petroleum Company for whom 7000 stations were occupied by Geoprosco (1956, 1957) and GSI (1960). The data from 1300 of these stations (every fifth on traverses) have been extracted for use in the compilation. The data for every tenth were included in the Newcastle catalogue but there were errors caused by using varying correction factors. There were also errors up to 2 km in the station co-ordinates given in the catalogue due to the scale of the location map used (1:500 000. These errors have now been removed by using the original (1:100 000) location maps which have now become available. Terrain corrections have not been made, but these are thought to be negligible for most stations.

(iii) Workers from the University of Leicester who have occupied 3326 stations. An interim catalogue (Khan *et al.* 1972) contains the data from 1900 stations south of 1°N and west of 38°E. The data from the remaining 1426 stations occupied by one of us (C. J. Swain) since then from parts of Kenya so far uncovered are also presented here. Terrain corrections are being estimated for these data and none have been included here. As some of these are about 10 mgal, there will be some slight revision of the map in due course.

B. *The Bouguer Anomaly map*

The Bouguer anomaly map shown in Fig. 5:2 has been obtained using automatic interpolation for estimating the values at regular grid points. The method is based on minimum total curvature (Briggs 1974) and is equivalent to fitting a flexible sheet to all the data points. It is suitable for potential field data as it produces the smoothest surface through the data points. The method has two of the limitations common to automatic contouring methods. At the edges of the area, e.g. near the Ethiopian border in the north, where there are no data, false anomalies occur. Secondly the number of grid points is limited by the size of the computer. In this case the rather wide spacing of 9 km had to be used so a lot of detailed information is lost—less than half the 6600 stations have been used in the contouring. The final corrected map will use a much smaller grid.

The most obvious feature of this map is the belt of high gradients round the north and east sides of the Kenya dome at the culmination of which the lowest values, 250 mgals, occur. The gravity field is related to the topography but is not a mirror image of it. This implies that the area is regionally, but not locally, isostatically compensated.

C. *The regional map*

To appreciate, at least qualitatively, the other features of the data, it is convenient to follow the normal procedure of fitting a regional surface which is generally attributed to the deep structure. This was done by fitting orthogonal polynomials to the gridded data (Grant 1957). The seventh order surface shown in Fig. 5:3 fits the data quite well. It contains a 'saddle' centred at Marsabit in the north from which the field rises north-westwards towards Lake Turkana and falls south-westwards towards a prominent bowl-shaped minimum centred in the Narok–Naivasha region at the culmination of the Kenya dome. This inverse correlation with the topography is due to deep structure. If we attribute the regional low in the south to an asthenolith penetrating the normal mantle, then the high in the north may be due to it penetrating the crust.

Fig. 5:4 shows a set of E–W profiles at 0.4° intervals with the regional and Bouguer anomalies superimposed as it shows how well the regional anomaly fits the data.

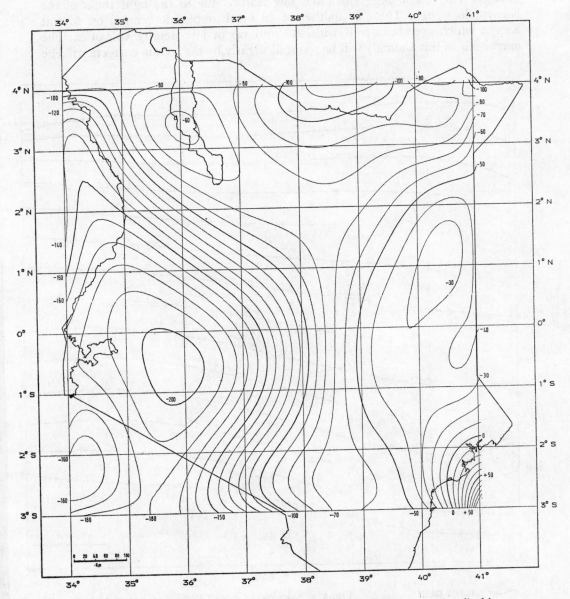

FIG. 5:3. The Regional Gravity Map of Kenya derived from Fig. 5:2 as described in the text.

D. *The Residual Anomaly map*

The difference between the observed Bouguer Anomaly and the regional surface defines the residuals and these are shown plotted and contoured on Fig. 5:5.

This map shows a number of features which clearly correlate with the visible geology. There is a large elongated low feature due to the light rocks of the Kavirondo trough. There is another of 50 mgal amplitude centred on Mount Kenya which correlates well with the outcrop of low density volcanics. The magnitude of this anomaly will be reduced slightly by the terrain corrections. The

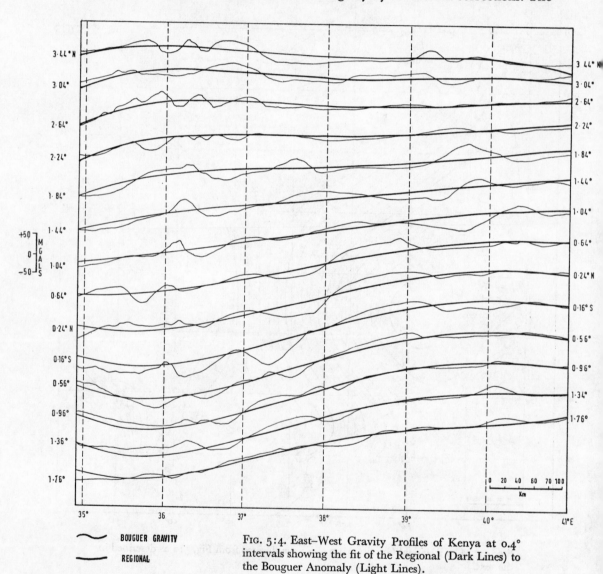

BOUGUER GRAVITY

REGIONAL

FIG. 5:4. East–West Gravity Profiles of Kenya at 0.4° intervals showing the fit of the Regional (Dark Lines) to the Bouguer Anomaly (Light Lines).

north-west trend of the anomalies in the north-east of Kenya does not correlate so clearly with the surface geology. An irregular high extends for 500 km from east of Garrissa to North Horr, following the eastern margin of the basement outcrop some of the way. This implies a considerable quantity of high density basement rocks. Further east near Wajir a remarkable step anomaly of 80 mgal occurs. This indicates a major fault with a throw of several kilometres.

The residual map also shows clearly the extent of the axial gravity high of the rift. South of the equator it is emphasised by flanking lows caused by low density volcanics on Londiani and Ol Kalou. It is prominent again between Lake Baringo

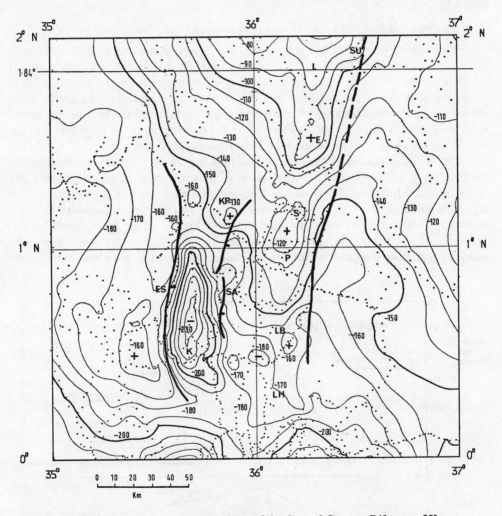

Fig. 5:6. Bouguer Gravity Anomaly Map of the Central Gregory Rift area of Kenya showing gravity stations, and major faults (after East African Geological Research Unit). Based on data digitised on a grid spacing of 0.02°, L—Lomi, E—Emuruango-golak, S—Silali, P—Pakka, LB—Lake Baringo, LH—Lake Hannington, SU—Suguta Basin, KP—Kito Pass Fault, ES—Elgeyo Escarpment, SA—Saimo Fault.

and Latitude 2°N. The calderas of four prominent trachyte volcanoes occur on the high in this region—Pakka, Silali, Emuruangogolak, and Lomi. A more detailed Bouguer anomaly map of this region is shown in Fig. 5:6 on which are also shown the positions of the Elgeyo Escarpment and the faults through the Kito Pass and the eastern margin of the rift. The fact that the Saimo Fault has no gravity expression indicates that the volcanic/sediment succession on its down-throw side has a similar mean density to the basement. The narrow axial high follows the eastern (Suguta) trough within the rift. This is superimposed on a much broader high associated with the rest of the rift as is shown more clearly on the profile of Fig. 5:7.

E. *Two-dimensional interpretation of profiles at Latitudes 1.84°, 2.7° and 3.3°N.*

The upper part of Fig. 5:7 shows the Bouguer anomaly profile through 1.84°N on Fig. 5:6. The two-dimensional model shown in the lower part of the figure assumes a 2-layer crust with densities of 2.7 and 2.9 overlying a mantle of density 3.3. Beneath the rift axis at a depth of 20 km, material of density 3.1 is introduced following Khan and Mansfield (1971) who used the seismic refraction data of Griffiths *et al.* (1970). The volcano Lomi appears to originate on the eastern slope of the wedge of intermediate material.

Fig. 5:8 shows an E–W profile at Latitude 3.3°N. It is shown on the detailed contour map of the Lake Turkana area (Fig. 5:9), and passes through Lake Turkana, El Moiti and North Horr. The regional curve in the upper part of the figure is taken from Fig. 5:3. It has not been interpreted quantitatively, but evidently the interesting structure postulated in Fig. 5:7 has disappeared by this latitude. The residual has been interpreted by two-dimension modelling. The model suggests a basin of sediments, perhaps interbedded with volcanics, 4000

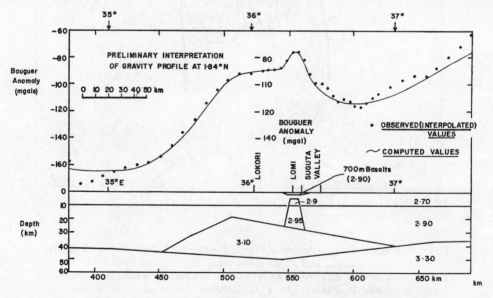

Fig. 5:7. Two-Dimensional Interpretation of Gravity Profile at Latitude 1.84°N.

metres deep in central Lake Turkana, faulted at its eastern margin against a dense basic intrusion underlying the syenite plug of El Moiti. This thickness of sediments is perhaps surprising. The figure is subject to two sources of uncertainty one in the regional anomaly and the other in the density. The regional anomaly is considered to be reliable to within 10 mgals. An error of this amount would result in an error of 900 m in the depth estimate. The density is the average saturated density of 4 samples of lake sediments (shelly limestones/grits) from the area. 2.4 is also a fairly typical value for Turkana grits, which probably underlie the lake beds. The standard deviation of these 4 measurements is 0.16. Even after making the maximum allowance for the two sources of error, a minimum thickness of 2200 m is indicated. However, a lower average density than 2.24 is possible. Further density data are clearly needed. The centre of the basin, as expressed by the gravity low, is displaced towards the west shore of the lake. Walsh & Dodson (1969) agree with Fuchs (1939) that 'the centre of the depression in which the Turkana grits were deposited lay somewhat to the west of the present lake'. However, they suggest a maximum thickness for these sediments of only 300 metres.

The intrusion postulated as the cause of the El Moiti anomaly may be similar to those thought to cause the 'axial highs' which are associated with the caldera volcanoes of the central and southern Kenya rift. Of these volcanoes, Silali and Lomi probably overlie syenite cupolas or plugs as evidenced by the existence of syenite enclaves in the lavas (McCall 1967b; P. Truckle, personal communication). El Moiti would therefore appear to be the eroded remnant of such a plug with an underlying basic root now much closer to the surface. The area is now the subject of further geological study.

The other features of this model are even more tentative. A dense intrusion is postulated under the Lothidok hills as the thickness of basalts is not thought sufficient to cause the positive anomaly. The low to the east of El Moiti is explained by a shallow trough of interbedded sediments and volcanics, with its base dipping to the north. The residual high to the east is again explained by a near

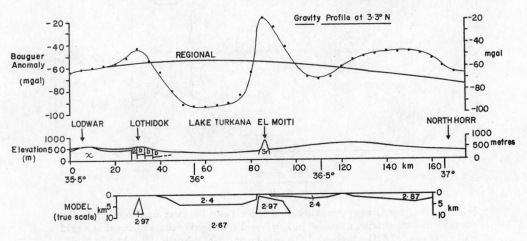

FIG. 5:8. Two-Dimensional Interpretation of Gravity Profile at Latitude 3.3°N.

surface density distribution. The width of the anomaly suggests that a deep origin is equally plausible. It is difficult to decide between these possibilities without further geological data.

The measurements on which the profile at Latitude 2.7°N (Fig. 5:10) are based are rather widely spaced, so that the anomalies are not very well defined. The regional anomaly as given by the trend surface analysis has again been used. The residual anomalies are here explained by (a) density contrasts within the basement, and (b) troughs containing 1 to 2 km of less dense rocks. It is suggested that between 35.5°E and 36°E the surface rocks are underlain by basement with a

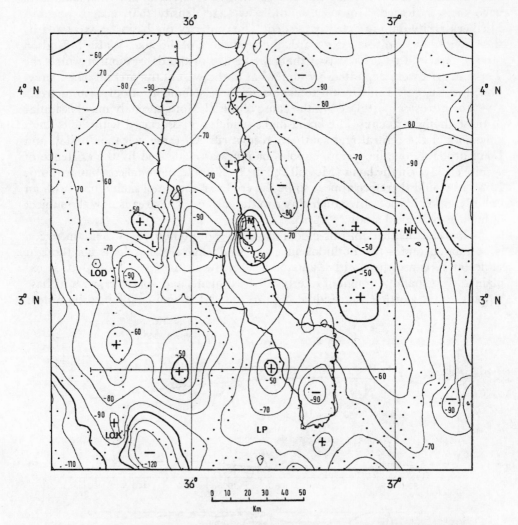

FIG. 5:9. Bouguer Gravity Anomaly Map of the Lake Turkana area showing station positions and the profiles modelled in Figs 8 and 10. Based on data digitised at a grid spacing of 0.02°. LOD—Lodwar, LOK—Lokichar, NH—North Horr, M—El Moiti, LP—Loriu Plateau, L—Lothidok.

rather higher than normal density (2.8). This is speculative as no basement rocks were collected in this area. The positive anomaly over the northern end of the Loriu plateau again seems to be associated with the basement rocks which are exposed along its eastern edge. The Loriu has been described briefly by Rhemtulla (1970) as an upfaulted and tilted basement block capped by basalts. The positive anomaly to the east of the lake is explained by a near surface density distribution. It is part of the same anomaly as that at the east end of the 3.3°N profile. The three intervening lows on this profile are all assumed to be due to troughs or basins of Pliocene to recent sediments with interbedded volcanics. These occupy the valleys of the Lokichar and Kerio rivers and the southern Lake Turkana basin. Rhemtulla (1970) considers that the Kerio trough was formed tectonically at a fairly early date, whereas the Suguta trough to the south is a younger feature in which sedimentation started in the mid-Pleistocene. It is worth noting that although it is a major tectonic feature, the Suguta trough has virtually no gravity expression. The south Lake Turkana basin, however, is marked by quite a large residual gravity low: the lowest values occur to the south of the 2.7°N profile, and imply a total sediment/volcanic thickness of 2000 metres or more.

Conclusions

A preliminary Bouguer anomaly map of Kenya based on 6600 stations is presented and shows a number of features of geological interest. The regional field, approximates a seventh order polynomial surface and shows the general inverse correlation with the topography to be expected from isostatic considerations— the field falls from the coast towards a minimum coincident with the culmination of the Kenya dome. The Turkana depression is associated with a gravity maximum which suggests that the 'anomalous mantle' there reaches high levels. The principal negative anomalies, revealed by the residual map are due to low density rocks in the Kavirondo trough, the Mount Kenya region, the Kerio valley and the Lake

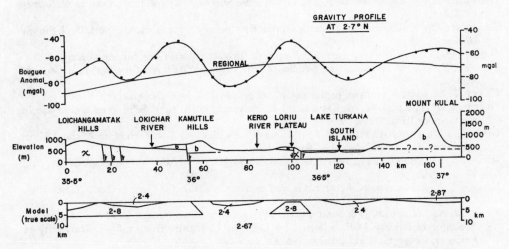

Fig. 5:10. Two-Dimensional Interpretation of Gravity Profile at Latitude 2.7°N.

Turkana basin. The principal positive anomalies are attributed to anomalous crust and mantle underlying the rift axis and to high density basement in the north-east of Kenya. There are a considerable number of smaller anomalies due to faulting, sedimentary basins, volcanic and basement structures. These will be better defined when the terrain corrections have been completed and further detailed field data already collected have been processed. A number of interesting anomalies occur in regions which have so far not been mapped geologically. A programme of detailed seismic work is being carried out to provide some control for the interpretation of the principal anomalies.

Acknowledgements

We thank the Kenya Department of Mines and Geological Survey and the Physics Department of the University College, Nairobi and the Geology Department of the University of Leicester, for their considerable help with this work which was carried out under NERC Grant GR/3/1486.

References

BANKS, R. J. & OTTEY, P., 1974. Geomagnetic deep sounding in and around the Kenya rift valley. *Geophys. J.*, **36**, 321–36.

BRIGGS, I. C., 1974. Machine contouring using minimum curvature. *Geophysics*, **39**, 39–48 (February, 1974).

BULLARD, E. C., 1936. Gravity Measurements in East Africa. *Phil. Trans. Roy. Soc., A.*, **235**, 445–531.

DARRACOTT, B. W., 1972. Ph.D. Thesis. University of Newcastle-upon-Tyne.

FAIRHEAD, J. D. & GIRDLER, R. W., 1971. The Seismicity of Africa. *Geophys. J.*, **24**, 271.

FUCHS, V. E., 1939. The Geological History of the Lake Rudolf Basin, Kenya Colony. *Phil. Trans. Roy. Soc., B.*, **229**, 219–74.

GEOLOGICAL SURVEY OF UGANDA, 1962. 1:250 000 sheets for Aloi, Kaabong, (Mbale, and Moroto showing gravity stations and Bouguer gravity contours.

GEOPROSCO, 1956. Report on the Kenya Gravimetric Survey, 1955 (private report, Geol. Survey of Kenya.)

GEOPROSCO, 1957. Report on the Kenya Gravimetric Survey, 1957 (private report, Geol. Survey of Kenya).

G.S.I., 1960. Gravity Survey, Garissa–Wajir Area (private report, Geol. Survey of Kenya).

GIRDLER, R. W., 1963. Geophysical Studies of Rift Valleys. *Physics and Chemistry of the Earth*, **5**, 122–54.

GRANT, F. S., 1957. A problem in the analysis of geophysical data. *Geophysics*, **22**, 309.

GRIFFITHS, D. H., KHAN, M. A., KING, R. F. & BLUNDELL, D. J., 1971. Seismic refraction line in the Gregory Rift. *Nature, Phys. Sci.*, **229**, 69–71.

GUMPER, F. & POMEROY, P., 1970. Seismic wave velocities and Earth structure of the African continent. *Bull. Seismol. Soc. Am.*, **60**, 651–68.

IRVING, E. & TARLING, D. H., 1961. The Palaeomagnetism of the Aden Volcanics. *J. Geophs. Res.*, **66**, 549–56.

KHAN, M. A. & MANSFIELD, J., 1971. Gravity measurements in the Gregory Rift. *Nature, Phys. Sci.*, **229**, 72–5.

KHAN, M. A., MANSFIELD, J. & SWAIN, C. J., 1972. Gravity Measurements in Kenya. An interim catalogue of Gravity Data collected by Leicester University from 1965–9. University of Leicester, Geophysical Publication, No. 1.

LONG, R. E., BACKHOUSE, R. W., MAGUIRE, P. K. H. & SUNDERALINGHAM, K., 1972. The Structure of East Africa Using Surface Wave Dispersion and Durham Seismic Array Data. *Tectonophysics*, **156** ($\frac{1}{2}$), 165–78.

MAKRIS, J., MENZEL, H., ZIMMERMANN, J., BONJER, K.-P., FUCHS, K. & WOHLENBERG, J. 1970. *Zeit. fur Geophys.*, **36**, 387–91.

MAKRIS, J., MENZEL, H. & ZIMMERMANN, J., 1972. A preliminary interpretation of the gravity field of Afar: North-east Ethopia. *Tectonophysics*, **15**, 31–9.

McCALL, G. J. H., 1967. Geology of the Nakuru–Thomson's Falls—Lake Hannington area. *Geol. Survey Kenya Report*, **78**.

McCALL, G. J. H., 1967. Silali—another major caldera volcano in the Rift Valley of Kenya. (Abstract). *Proceedings of the Geol. Soc. April, 1968*, 267.

RHEMTULLA, S., 1970. A Geological Reconnaissance of S. Turkana. *Geographical J.*, **136**, 61–73.

SEARLE, R. C., 1970. Evidence from gravity anomalies for thinning of the lithosphere beneath the Rift Valley in Kenya. *Geophys. J.*, **21**, 13–32.

SEARLE, R. E. & DARRACOTT, B. W., 1971. A Catalogue of Gravity Data from Kenya. University of Newcastle.

WALSH, J. & DODSON, R. G., 1969. Geology of Northern Turkana. *Report* **32**, *Geol. Surv. of Kenya*.

WILLIAMSON, K., 1975. 8th Colloquium on African Geology (Abstract) University of Leeds. In Press.

Part II

BACKGROUND:
Palaeontological and Archaeological Problems

6

ANDREW HILL

Taphonomical background to fossil man
–problems in palaeoecology

Taphonomy deals with the various processes affecting animals from their deaths to their possible fossilisation. The particular importance of work of this kind to the elucidation of features of fossil hominid behaviour and ecology is stressed. The subject of taphonomy is re-defined in terms of the differences existing between fossil assemblages and living communities of animals. A brief review is given of some of the factors to be considered in the interpretation of assemblages of terrestrial fossil vertebrates, particularly those that have some connection with hominids and hominid activity.

Introduction

A determinant of the character of modern ecosystems, and one which seems often to be ignored, is the evolutionary history of the organisms and environments concerned. Past ecosystems mark stages in the evolution of the present-day situation, and palaeoecology has the potential of providing a time perspective for modern ecological studies. A knowledge of the succession of fossil ecosystems offers insight into those factors that control their structure and cause change. Profound ecological and environmental modifications appear to have accompanied the evolution of man, and at present man's impact upon environments and animal communities is considerable. Thus, information regarding the dynamics of ecological change is particularly important in situations where man has been involved over the time period of his recent evolution.

The Rift System of eastern Africa preserves the best samples yet known of early man and other components of the ecosystems of which he was a part, and it is from an analysis of these fossil samples that most of our palaeoecological information must derive.

Taphonomy

When dealing with fossil collections it is frequently assumed that they are essentially equivalent to a once living community of animals. This is rarely if ever the case, as the individuals represented in a fossil assemblage have almost invariably been subjected to a variety of modifying processes between their death and fossilisation. Hence it is necessary to assess the extent of alteration of the material after death, before attempting palaeoecological reconstructions.

Taphonomy is a word proposed by Efremov (1940) to cover the study of the processes leading to fossilisation. It is merely a branch of actualism, encompassing the concerns of earlier neologisms such as biostratonomy (Weigelt 1927) and actuopalaeontology (Richter 1928). Taphonomy has been considered as the study of the processes involved in '. . . the transition from the biosphere to the litho-sphere . . .' (Efremov 1940), but there is obviously more than one means of transition, and many of the processes involved are not known in detail, either intuitively or from deliberate investigation. Taphonomy ultimately deals with the differences that exist between an assemblage, or collection, and the community or communities of animals from which it came. These differences constitute the bias in many fossil assemblages and are due to various taphonomic causes. Often, in practice, few relevant observations are made of the condition of a fossil accumu-lation. The conditions that are noted are often attributed to hypothetical causes without studying the effects of such causes or factors operating at the present day.

Statements concerning taphonomy can be placed in a number of categories. Palaeotaphonomy is concerned with observations of fossil assemblages. A second category refers to that more nebulous region of conjecture constituting hypo-thetical assertions about the causes of the observed bias in such assemblages. A third, neotaphonomy, involves relevant experimentation or observations of the condition of modern vertebrate remains in various closely defined environments. This is designed to test the ideas of the second category. In turn, it may suggest consequences for palaeoecological interpretation that may not be detectable from any features of the assemblage. For example, they may involve predictions con-cerning the absence of species from particular sorts of assemblages, or the absence of whole palaeocommunities from any representation in the fossil record.

Taphonomy and fossil hominids

This paper gives a very brief review of some of the factors to be considered in the palaeoecological analysis of fossil vertebrate assemblages; a fuller treatment will appear elsewhere (Hill *in MS(a)*). Most of the matters discussed are not limited to hominids but apply to vertebrates in general. However, because of the increasing involvement of hominids over the last few million years in the pattern of evolving ecosystems, coupled with the relative rarity of samples of fossil hominid populations, it is particularly important when investigating such cases as do exist that valuable information is not lost.

There is also another respect in which taphonomical work is especially important to hominid palaeoecology. Often the intention in taphonomy is to rectify the bias in fossil accumulations. But in the course of this it provides very positive information concerning the agents that caused this bias, and this in itself may be of great palaeoecological interest. When the agent of modification is believed to be a hominid the state of the assemblage potentially reflects aspects of human behaviour and relationship to surroundings. Collections of fossil bone which may represent hominid food débris are archaeologically interesting. A knowledge such as taphonomy provides of bones accumulating in situations where man has not been directly involved constitutes a valuable base against which the significance of possible hominid influenced accumulations might be better assessed. Similarly it is relevant to features of hominid behaviour involving food selectivity, butchery practices, domestication, use of bone for implements, and so on.

A descriptive framework

The way that taphonomy is formulated leads to its ideas and results being considered in terms of a sequence of events from death to fossilisation. However, this involves a certain amount of inference arising from the fact that it is usually only postulated causes that are fitted into the sequence, the details of which are often only poorly known. It is premature to attribute observed effects too rigidly to stages or processes in what are essentially hypothetical sequences. Such assumptions can either be eliminated from the descriptive framework, or at least be made more explicit, by emphasising the observed differences between assemblages of fossils and living communities, by separating them clearly from the suggestions advanced to account for them, and by describing observations relevant to testing and discriminating between these suggestions.

The state of a fossil assemblage can be described by a set of observations of its various features. The interpretation of some of these features as different from the original community relies upon a comparison with that community, and this is not possible as direct knowledge of it is lacking. The characteristics of a fossil assemblage can be separated into two groups. The interpretation of one set of characteristics as differences relies only on the assumption that individual animals contributing to the fossil record were similar in general to animals today. The second set cannot be interpreted as bias on so simple an assumption.

The first set come under the following headings:

1. the animals are dead
2. there are no soft parts
3. skeletons are often disarticulated
4. bones are often concentrated together
5. parts of the skeleton occur in different proportions to their occurrence in life
6. the remains are buried in sediment or other rock
7. bones are sometimes preferentially oriented within the rock
8. bones are altered chemically

The other group of features is only potentially different from the situation in life and involves questions of more immediate relevance to palaeoecology and palaeocommunity structure. They concern:

the association of different species

the numbers of individuals of different species (including the possible total absence of some species in the fossil assemblage)

the numbers of individuals of different age and sex groups within each species

This potential or cryptic bias is normally explicable in terms of the action of factors producing the more directly observable manifestations of bias that consti-tute the first set listed above. It is these that are discussed here.

1. *Death*

The obvious fact of death has significant consequences for the interpretation of fossil assemblages. Some of the causes of death in modern populations of animals that were no doubt equally effective in the past include:
– predation
– starvation and drought
– disease
– physical accident
– senility

Death affects several other directly observable characteristics. The mode of death may influence loss of soft parts, disarticulation, concentration of bones, damage to bones, the anomalous representation in terms of skeletal proportions, and burial in sediment. It may also affect the second category of features discussed above, that of potential or cryptic bias. Death may be selective of species, of sex, and of age groups. Assemblages that are the product of sudden mass mortality provide different palaeoecological information from those that are the result of attritional deaths.

The ratios of species represented will tend to differ according to their re-productive rates and the length of time involved in accumulation. This follows from the fact that, for example, in the time necessary for a pair of elephants to produce one young (i.e. a single potential skeleton) the number of potential and actual skeletons produced directly and indirectly by a pair of mice is considerably more; and it is not in linear proportion to their numbers in the original com-munity. These effects are significant if a fossil collection derives from a land surface accumulation of several years. Similarly, mortality patterns affect the age struc-ture of a species in a fossil assemblage as compared with the age structure of that species in life. The consequences of this for the interpretation of invertebrate fossil communities have been discussed by Olson (1957) and Craig and Oertel (1966, 1967) amongst others, but the situation in terrestrial vertebrates has received less attention.

Mass disasters might result in the death of a good representation of a whole community, but epidemics often affect some species more than others, and within a species affect each sex and age group differently. Predators are selective of their prey, and work on modern carnivores, such as that by Schaller (1972) on the lion, and Kruuk (1972) on the hyaena, is helpful to understanding the biases such animals produce. Natural traps are also selective, often being related to the behaviour of the animals trapped. The classic example is the Rancho la Brea tar pits, U.S.A. Here there is an unusually high proportion of carnivores, which presumably were tempted to their deaths by the attraction of large herbivores trapped in the tar. But similar effects are displayed by less spectacular accumula-tions.

Care must be taken before attributing to an assemblage any particular cause of death, such as catastrophic mass mortality or human predation, as the implications for palaeoecological interpretation of the various causes are considerably different.

2. *Absence of soft parts*

The vast majority of fossils possess no trace of soft parts. There are exceptions; amongst hominids the 'bog-bodies' of northern Europe are well known (Glob 1969), but they are geologically recent, and such specialised environments of preservation are probably transitory in terms of the fossil record.

The agents generally held to be responsible for the loss of soft parts include:
–large carnivores
–insects
–bacterial and chemical agents of decay

Animals that eat soft parts sometimes leave traces of their activities upon the bone. The speed and style of decomposition differs seasonally, and from one environment to another. The rate of decomposition affects the speed with which parts of the skeleton are exposed to damage, and to the possibility of disarticulation. In turn this influences the speed at which parts of the skeleton are made available for transportation by water and other agencies, and the probability of them being buried and preserved. Different styles of decomposition produce different results. The hydraulic behaviour of a mummified body is not the same as that of its constituent skeletal parts. The modes of decomposition prevalent in many environments probably result in the complete decomposition of the hard parts as well. This may lead to an under-representation of the whole faunas of such environments in the fossil record.

In the fossil situation fairly direct evidence of the causes of the loss of soft parts is sometimes found. Damage to bone attributable to carnivore action will be mentioned under the section concerning damage. Grooves are sometimes present on fossil bovid horn cores that were probably caused by the larvae of the moth *Ceratophaga sp.* whilst feeding on the horn, and such features can provide useful climatic information. Gautier and Schumann (1973) describe puparia of a blow-fly from the skull of a Pleistocene bison. Weigelt (1930, 1935) records other instances, and they are also known from fossils at East Lake Turkana, Kenya.

There has been some relevant work carried out on modern aspects of decomposition. Mégnin (1894) provides interesting information on the decay of human corpses. Payne (1965) confined his attention to pigs, and noted the sequence and rates of decomposition in different circumstances. This and later work (Payne *et al.* 1968) reveal large differences in sequences and rates of decay between pigs on the surface of a mesophytic pine wood, and those buried in the same environment. This was related to the influence of insects, that probably mediate to a large extent the effects of environment on decomposition, through the effects of the environment upon them.

Size is important too. Payne (1965) found the decay of species of small mammals such as rodents to be very rapid, and discrete stages of decay to be indiscernible. Dodson (1974) provides details of the subaqueous decay of small animals, which differs from that in terrestrial situations. Larger corpses in aquatic conditions have been dealt with by Schäfer (1962, 1972). His discussion of marine mammals in the North Sea provides basic information concerning the factors that affect decomposition, and also has a bearing upon the decomposition of terrestrial mammals in bodies of fresh water.

3. Disarticulation

Bones in fossil situations are mostly separated from those with which they were connected in life. This disarticulation is a consequence of the loss of soft parts, coupled with an amount of transport, but it is convenient to discuss it separately. The factors involved are those put forward to account for the absence of soft parts:
    – hominids
    – other large animals
    – insects
    – bacterial and chemical agents of decay
In general disarticulation is important because it exposes the ends of bones to damage and potential destruction. At the same time it reduces a skeleton into a number of smaller units which are more amenable to transport in certain conditions than is the whole animal. Such transport, by water or scavengers, may increase the likelihood of preservation. If the sequence of disarticulation in animals is variable then this variation may provide information about the agent involved or the general environment. This possibility is particularly interesting in situations where traces of hominids are found. Articulated remains of animals found there may suggest theories regarding hominid butchery and food practices. But before additional inferences are made concerning food consumption and economy it would be well to know how human butchery patterns differ from disarticulation caused by other factors. It is possible that the controls of the patterns are inherent in the anatomy of the dead animal itself, and thus disarticulation is independent of the agents whereby it is realised. Clark (1972) remarks how similar are human butchery practices over long periods of time, and cites examples from the prehistoric and historic record of bison butchery by American Indians. He also comments that variations in butchery technique can be related to the size and species of the animal involved. Perhaps this is really a comment about disarticulation in general.

Some work on disarticulation in modern environments has been carried out. Schäfer (1962, 1972) reports on the sequence found in marine mammals decomposing in water. Disarticulation is complete before the flesh has gone. My own observations suggest the same is true of *Hippopotamus* in East African lakes. Dodson (1974) looked at smaller animals breaking up in water. He believes the amount of tendon involved in different joints influences the sequence.

Disarticulation on land surfaces in some circumstances takes place after most of the soft parts have disappeared. Toots (1965a) describes the situation in the coyote in semi-arid grasslands in the U.S.A. The sequence differed from that described for cows by Weigelt (1927) and Müller (1950). Toots thought the pattern was determined by the type of joint and the relative amounts of easily decomposible and more resistant tissue involved in it.

My work on modern animals at East Lake Turkana, Kenya (Hill 1975, *in*

$MS(b)$) resulted in a statistical model of disarticulation in *Damaliscus korrigum* Ogilby, a large bovid, that accords well with known facts. It was also possible to produce a similar model from the data that Wheat (1972) provides concerning the Olsen-Chubbuck palaeoindian bison kill. The model reflects his conclusions based on independent criteria regarding the butchery technique at the site. It is also remarkably similar to the 'natural' disarticulation sequence in *Damaliscus*. In this case at least, differences due to human butchery, whilst significant are only slight.

4. *Concentration*

Most vertebrate fossil assemblages consist of relatively large numbers of bones concentrated together. Often this has to be so for fossil sites to be discovered or to be worth studying closely. It is normally inferred that one or more of the following agents has been operative in this accumulation:
- hominids for use as tools
- hominids in feeding
- other animals in feeding
- mass deaths
- natural traps
- moving water

All the agents other than natural traps and some mass deaths involve the removal of animals from their place of death. They may cause bias in the representation of particular species or of particular skeletal parts. In addition to mixing together remains of animals once separated in space, they may accumulate animals which in life were separated in time. It is thus important for palaeoecology that the agent of accumulation be identified and the features of its selectivity be known.

Distinguishing between agents of accumulation relies upon other aspects of the assemblage, particularly damage and anomalous proportions of skeletal parts, that the factor involved might affect.

There have been suggestions that some of these agents do not accumulate bone at all. Controversy has been particularly vigorous in hominid contexts. Dart (1957 *inter alia*) attempted to show that *Australopithecus* was responsible for the assemblage at Makapansgat Cave, South Africa, through having collected bones as tools. In doing this he rejected previous notions, such as those of Buckland (1822), that hyaenas accumulate bones in caves (Dart 1956). Hughes (1954, 1958) supported Dart by statements from authorities with long experience of South African hyaenas, and also from his own observations of hyaena lairs. Similar work by Sutcliffe (1969, 1970, 1973a) in East Africa, produced contrary results, illustrating that in some circumstances hyaenas do collect bones. The means of discriminating between these and various hominid accumulations is not yet well established, nor is it between these and collections produced by other animals such as porcupine and leopard. Simons (1966) records a modern assemblage that he believes has been formed by leopard, at least in part. Leopard activity has also been given as the origin of the fossil material, including the hominids, from Swartkrans, South Africa (Brain 1968, 1970).

5. *Damage*

Most fossil bones are broken and damaged. The main postulated causes of
this include:
– hominids in feeding
– hominids in tool manufacture and use
– other animals in feeding
– animal trampling
– climatic factors
– various agents operative after burial
– pathology

The type of damage may give clues as to the nature of the agent involved in
accumulation, or that has caused differential representation of parts of the skeleton
and perhaps the differential representation of species. Damage of some kinds may
totally destroy some bones, or all the bones of some species. It may also supply
information about environmental conditions and animals otherwise unrepresented
in the fossil record.

A great deal of the interest in damage comes from its relevance to anthropological considerations such as the nature of food débris, bone tools and palaeopathology. The recognition of damage, particularly distortion, is vital if measurements are taken of critical specimens to support taxonomic or functional
notions.

Bones found in archaeological contexts are usually thought of as being human
food refuse. Work on the food remains of modern peoples is being carried out by
Yellen (Smithsonian Institute, Washington), and in the East Lake Turkana area
of Kenya by Gifford (University of California, Berkeley). Brain (1967a) depicts
remains of goat left by Kuiseb River Hottentots. They show a great consistency in
the pattern of damage.

The consistency of pattern was one of the features that led Dart (1957) to
suggest that the assemblage of bones associated with *Australopithecus* at Makapansgat
had been collected by these hominids and modified by them for use as tools.
However, the pattern shown by them closely resembles that known from human
food remains. Mary D. Leakey (1971) sees no features in the Choukoutien bone
and antler industry (Breuil 1939) that could not be found in human food débris.
Leakey presents some features found in bones at Olduvai, such as animal limb
bones with large flake scars, that she sees as characteristic of human implemental
modification. None of the Olduvai suggested tools resemble those from Makapansgat.

At present, it seems impossible to distinguish the hypothesised Makapansgat
tools from the food remains of carnivorous animals. Simons (1966) describes
damage to baboon skeletons that may have been caused by leopard. Work at
East Lake Turkana, Kenya, and in Uganda National Parks (Hill 1975, 1976,
in press) shows that the pattern of damage suffered by bones in circumstances
where hominids are not involved is very consistent. It is also remarkably similar
to that attributed to osteodontokeratic cultures. The style of damage seems as
much determined by the nature of the bone itself as by the agent causing damage.

Damage is related to such factors as the age of the individual, the size of the species concerned, and its anatomy.

Sutcliffe (1970) records damage to bone caused specifically by hyaenas. More recently (Sutcliffe 1973b) he comments on bone damage caused by artiodactyls. Their chewing can produce very regular forms, which Sutcliffe compares with bone structures from the Pleistocene of Crete that Kuss (1969) believed to be Palaeolithic tools. Possibly a similar explanation is applicable to an antler fragment described by Kenyon and Churcher (1965) as being worked by man. Brain (1967b) has shown that polish resembling that on suggested bone tools can be produced on bone by sand abrasion.

Hill & Walker (1972) postulated trampling by large animals as being a significant factor in breakage. Behrensmeyer (pers. comm.) now has information on the extent of this in modern land surface assemblages.

Damage caused by climatic factors is discussed by Toots (1965a), Clark *et al.* (1967), Isaac (1967, 1968), Voorhies (1969), and Hill (1975, *in press*). Much seems to depend upon the micro-climate, but on East African grasslands bone can last in some circumstances for many years. Fossil assemblages derived from such land surfaces could easily represent the accumulation of ten years or more.

Breakage and damage to fossil bone is sometimes interpreted as pathological. Examples of palaeopathology in early man are given by Brothwell (1961), Wells (1964), Brothwell and Sandison (1967), and Goldstein (1969). Pathology is not strictly bias as it is a feature found in life, but it must be recognised to prevent an affected specimen being regarded as typical of the more normal healthy condition, and also to distinguish it from *post mortem* damage. Wells (1967) gives a valuable account of factors that can produce pseudopathological features.

6. *Anomalous proportions of skeletal parts*

The relative proportions of different skeletal parts found in fossil assemblages is rarely the same as that found in life. Agents connected with some other aspect of bias, such as damage or accumulation, are often invoked to account for this. They include:
- hominids, selection for use as tools
- hominids, damage for use as tools
- hominids, selection in feeding
- hominids, damage in feeding
- other animals, selection in feeding
- other animals, damage in feeding
- damage by climatic factors
- sorting by moving water
- removal from buried assemblages by burrowing animals
- differential chemical deterioration after burial
- differential non-collection

Some of these factors are believed to be affected by the following **controls**:
- size of element
- strength of element
- shape of element

Damage must completely and differentially destroy some parts of the skeleton, and it is suggested that in accumulation some selectivity is imposed upon the material collected. The nature of the selectivity may provide useful information about the behaviour of the accumulating agent, and where the agent is believed to be man such information is of considerable anthropological interest. Also the controls over the representation of skeletal parts that survive may relate to the controls over the representation of different species.

The main feature of the Makapansgat bones, other than damage, that persuaded Dart (1957) to believe they had been selected and modified as tools, was the anomalous proportion of skeletal parts. Brain (1967a, 1969) has subsequently shown that these proportions are characteristic of human food remains. An analysis of the food débris of Topnaar Hottentots showed a remarkable consistency of pattern with the Makapansgat assemblage. Brain correlated the representation of different ends of limb bones with their specific gravity and the time of epiphysis fusion. There is no need to postulate selection by man, as the observed proportions are simply a function of the anatomical construction of the animals themselves.

The interpretation of bone fragments as human food débris is common in the archaeological literature, and often with some justification. As yet, however, there are very few criteria by which such assemblages can be distinguished from the food remains of other animals. Frequently the mere association with lithic artefactual material is deemed sufficient. For instance, Clark (1972), whilst acknowledging the selective effects produced by other agents, comments, '. . . generally speaking, however, where culture exists there is little difficulty in distinguishing the agency of bone accumulation. . . .' Unfortunately he fails to elaborate upon this statement. On the basis of such food remains are propounded hypotheses regarding the economy of the groups concerned, and various features of their behaviour. But all such suggestions must be considered in the light of other factors that might produce similar effects.

Another interpretation of the Makapansgat assemblage is that it consists of the food remains of animals other than hominids. Occasionally the effects of animals on bone may be recognised from the type of damage present. Work on modern bone in East African environments (Hill 1975) shows that land surface assemblages unaffected by man reflect proportions of skeletal parts similar to those found by Brain in human food remains. Comparisons also reveal close resemblances with a range of fossil accumulations from both archaeological and non-archaeological situations.

Most fossil vertebrate accumulations have some connection with aquatic sedimentary environments. The effects of water sorting upon bones is therefore highly relevant to their interpretation and such matters are beginning to be looked at in detail. Voorhies (1969) conducted flume experiments to show the mode of dispersal of the disarticulated remains of sheep and coyote under the influence of moving water. This enabled him to explain apparent anomalies in a fossil assemblage in Nebraska. Dodson (1974) presented similar information concerning small animals. Behrensmeyer (1975) extended this work by further experiments and by investigating the fundamental properties of bones that influence their hydraulic behaviour. From this she was able to calculate for different bones the diameter of the quartz spheres to which they are hydro-dynamically equivalent. This information was applied to fossil assemblages at East Lake Turkana to test the hydraulic homogeneity of fossils with the sediments in which they were found, and to separate material that had been transported, from that which had remained more or less at the site of death. This enabled palaeoecological deductions to be made with some confidence. Later work (Behrensmeyer 1976a, 1976b and *in press*) has applied similar considerations to the sedimentary contexts of hominids from East Lake Turkana. Boaz and Behrens-meyer (1976) have investigated the hydraulic behaviour of modern human skeletal material when subjected to water flow in flume experiments. They relate this to features of the bones concerned, and comment on the occurrence of hominid remains at various East African sites.

Most of the factors discussed above concern anomalies in assemblages. However, it is only from collections of fossils that assemblages are known. Payne (1972) has clearly shown the poor nature of samples recovered by what are con-ventionally regarded as satisfactory excavation techniques. Bias occurs in repre-sentation of both skeletal parts and species.

7. *Burial in sediment*

For bones to become fossilised they must usually have been buried in sedi-ment. This condition has little effect on other features of bias, but assemblages may be influenced by those factors that lead to the formation of the rock in which they are incorporated. If bones are to be preserved they must be in a position whereby they can be buried in a sediment which itself will be preserved through time. Obviously not all bones are in such a position, and whole ecosystems may be so placed. This is important if generalisations are made about the ecology of whole regions on the basis of isolated and disparate collections of fossils. Bishop (1963, 1967, 1968) has remarked on the Tertiary sedimentary environments of East Africa and the nature of the associated fossil assemblages. And Butzer (1970) comments on the modern sedimentary environments of the Omo delta, Ethiopia, and their association or otherwise with modern bone remains.

8. *Orientation*

That bones may become oriented under the influence of moving water is well known, and Toots (1965b) emphasises that some preferred orientation pattern is characteristic of most sedimentary fossil assemblages, and that random orienta-tions are rare. Voorhies (1969) provides information on orientation of bones from his flume studies. Prevailing current directions can be detected from orientation information, but it also appears (Hill & Walker 1972; Hill 1975) that the total orientation pattern provides a useful summary of a number of features of an as-semblage and of its relationship to the sedimentary environment that produced it.

9. *Chemical alteration*

The word fossil carries with it the connotation of some form of chemical alteration. The constituents of bone react with their surroundings to achieve a higher degree of physico-chemical stability. Thus the chemical composition of fossil bone contains information about aspects of the environment at the time and place of its burial. Houston *et al.* (1966), for example, suggest that variations in the iron content of fossil bones from the Tertiary of Wyoming are influenced by climatic factors mediated by soil-forming processes.

Not all environments will favour a bone's ultimate preservation, and in this many factors are involved. Unless some bones or the bones of some species are differentially influenced by this it little affects the analysis of those bone assemblages that have survived. But it is important when considering what proportions and kinds of all palaeoenvironments have been preserved. Bishop (1963, 1968) points out that many Miocene hominoid localities in East Africa are associated with carbonatitic volcanics. Analyses show that the chemical composition of such rocks is similar to that of bone, producing a stable environment for fossilisation.

Similar work is needed to determine what chemical conditions are necessary for fossilisation, and which of the whole range of possible palaeoenvironments might have possessed them.

Conclusions

The interests lumped together under the heading taphonomy are various and at the moment rather diffuse, However, they all have the aim of understanding better those many factors and processes involved in the preservation of fossil assemblages. Such an understanding is vital if deductions concerning palaeo-ecology are to be made with any confidence. This is especially important where hominids are involved, both in monitoring the increasing influence of man on his surroundings, and in helping to discriminate between the multiplicity of theories to which fossil hominids give rise. Taphonomy potentially can provide a means of correcting the biases imposed upon fossil assemblages during their transition from living communities. By investigating the effects of hominids and other animals on bones it also supplies positive information on aspects of their behaviour. Although research in taphonomy may not immediately reveal the ultimate palaeoecological truth, it does provide very necessary knowledge about a great variety of relevant processes and conditions.

Acknowledgements

I thank Dorothy Dechant (University of California, Berkeley) for her comments on the manuscript.

References

BEHRENSMEYER A. K. 1975. The taphonomy and paleoecology of Plio-Pleistocene vertebrate assemblages east of Lake Rudolf, Kenya. *Bull. Mus. Comp. Zool.* **146** (10), 473–8.
—— 1976a. Fossil assemblages in relation to sedimentary environments in the East Rudolf succession *in* Coppens Y., Howell F. C., Isaac G. L. and Leakey R. E. (eds) *Earliest Man and Environments in the Lake Rudolf Basin* University of Chicago Press. 383–401.

—— 1976b. Taphonomy and palaeoecology in the hominid fossil record *in* Buettner-Janusch J. (ed) *Year Book of Physical Anthropology Am. Assoc. Phys. Anthrop.* 36–50.

—— *in press* The habitat of Plio-Pleistocene hominids in East Africa *in* Jolly C. (ed) *African Hominidae of the Plio-Pleistocene: Evidence, Problems and Strategies.* Butterworths, London.

BISHOP W. W. 1963. The later Tertiary and Pleistocene in eastern equatorial Africa *in* Howell F. C. and Bourliere F. (eds) *African Ecology and Human Evolution.* Methuen.

—— 1967. The later Tertiary in East Africa—Volcanics, Sediments and Faunal inventory *in* Bishop W. W. and Clark J. D. *Background to Evolution in Africa.* University of Chicago.

—— 1968. The evolution of fossil environments in East Africa *Trans. Leicester Lit. and Phil. Soc.* **62**, 22–44.

BOAZ N. & BEHRENSMEYER A. K. 1976. Hominid taphonomy: transport of human skeletal parts in an artificial fluviatile environment. *Am. J. Phys. Anthrop.* 53–60.

BRAIN C. K. 1967a. Hottentot food remains and their bearing on the interpretation of fossil bone assemblages. *Scient. Pap. Namib Desert Res. Stn.* **32**, 1–11.

—— 1967b. Bone weathering and the problem of bone pseudo-tools. *South African J. Sci.* **63**, 97–9.

—— 1968. Who killed the Swartkrans Ape-men? *SAMAB* **9**, 127–39.

—— 1969. The contribution of Namib Desert Hottentots to an understanding of Australopithecine bone accumulations. *Scient. Pap. Namib Desert Res. Stn.* **39**, 13–22.

—— 1970. New finds at the Swartkrans Australopithecine site. *Nature Lond.* **225** (5238), 1112–19.

BREUIL H. 1939. Bone and antler industry of the Choukoutien Sinanthropus site. *Palaeontol. Sinica N.S. D.* **6**, 40.

BROTHWELL D. R. 1961. The palaeopathology of early British man. *J. Roy. Anth. Inst.* **91**, 318–44.

—— & SANDISON A. T. (eds) 1967. *Diseases in Antiquity.* Springfield.

BUCKLAND W. 1822. Account of an assemblage of fossil teeth and bones . . . etc. *Phil. Trans. Roy. Soc. Lond.* **112**, 171–237.

BUTZER K. W. 1970. Contemporary depositional environments of the Omo delta. *Nature, Lond.* **226**, 425–30.

CLARK J., BEERBOWER J. R. & KIETZKE K. 1967. Oligocene sedimentation, stratigraphy, palaeoecology, and palaeoclimatology in the Big Badlands of South Dakota. *Fieldiana: Geol. Memoirs* **5**. 158 pp.

CLARK J. D. 1972. Palaeolithic butchery practices *in* Ucko P. J., Tringham R. and Dimbleby G .W. (eds) *Man, settlement and urbanism.* Duckworth. London. 149–56.

CRAIG G. Y. & OERTEL G. 1966. Deterministic models of living and fossil populations of animals. *Q. Jl geol. Soc. Lond.* **122**, 315–55.

—— 1967. The growth and death of computer populations. *Nature Lond.* **214**, 870–2.

DART R. A. 1956. The myth of the bone-accumulating hyaena. *Amer. Anthrop.* **58**, 40–62.

—— 1957. The osteodontokeratic culture of *Australopithecus prometheus. Mem. Transvaal Mus.* **10**. 105 pp.

DODSON P. 1974. The significance of small bones in palaeoecological interpretation. *Contrib. to Geol. Spec. Pap. no.* 2. Univ. Wyoming.

EFREMOV I. A. 1940. Taphonomy, a new branch of palaeontology. *Biol. Akad. Nauk. S.S.S.R.* (*Biol. Ser.*) **3**, 405–13 (in Russian).

GAUTIER A. & SCHUMANN H. 1973. Puparia of the subarctic or black blowfly *Protophormia terraenovae* (Robineau-Desvoidy, 1830), in a skull of a Late Eemian (?) bison at Zemst, Brabant, (Belgium). *Palaeogeogr. Palaeoclimatol. Palaeoecol.* **14**, 119–25.

GLOB P. V. 1969. *The Bog People.* Faber and Faber. London.

GOLDSTEIN M. S. 1969. The palaeopathology of human skeletal remains *in* Brothwell D. R. and Higgs E. (eds) *Science in Archaeology.* Thames and Hudson. 480–9.

HILL A. 1975. Taphonomy of contemporary and late Cenozoic East African vertebrates. Ph.D. thesis. University of London (unpublished). 331 pp.

—— 1976 On carnivore and weathering damage to bone. *Current Anthropology.* **17** (2) 335–6.

—— *in press* Early post-mortem damage to the remains of some East African mammals *in* Behrensmeyer A. K. and Hill A. (eds) *Taphonomy and Vertebrate Palaeoecology.* Proceedings of a Wenner-Gren Conference, July 1976.

—— *in MS(a) Vertebrate Taphonomy*. Academic Press. London and New York.

—— *in MS(b)* Butchery and natural disarticulation: an investigatory technique.

HILL A. & WALKER A. 1972. Procedures in vertebrate taphonomy; notes on a Uganda Miocene fossil locality. *Jl geol. Soc. Lond,* **128**, 399–406.

HOUSTON R. S., TOOTS H. & KELLEY J. C. 1966. Iron content of fossil bones of Tertiary age in Wyoming correlated with climatic change. *Contr. Geol. University of Wyoming.* **5**, 1–18.

HUGHES A. R. 1954. Hyaenas versus Australopithecines as agents of bone accumulation. *Am. J. Phys. Anthrop. n.s.* **12**, 467–86.

—— 1958. Some ancient and recent observations on Hyaenas. *Koedoe (J. Sci. Res. in Nat. Parks, S. Af.)* **1**, 1–10.

ISAAC G. L. 1967. Towards the interpretation of occupation debris: some experiments and observations. *Kroeber Anthrop. Soc. Papers.* **37**, Berkeley, Cal.

—— 1968. Traces of Pleistocene Hunters: an East African example *in* Lee R. B. and DeVore I. (eds) *Man the Hunter*. Aldine. Chicago. 253–61.

KENYON W. A. & CHURCHER C. S. 1965. A flake tool and a worked antler fragment from Late Lake Agassiz. *Canadian Journal of Earth Sciences.* **2**, 237–46.

KRUUK H. 1972. *The Spotted Hyaena.* University of Chicago Press. 335 pp.

KUSS S. E. 1969. Die palaolithische osteokerotische Kultur der Insel Krete (Griechenland) *Ber. Naturf. Ges. Freiburg. i. Br.* **59**, 137–68.

LEAKEY M. D. 1971. *Olduvai Gorge Vol. III. Excavations in Beds I and II 1960–1963.* Cambridge University Press. 306 pp.

MÉGNIN P. 1894. *La fauna des Cadavres: Applications de l'Entomologie a la médicine legale.* Gauthier. Paris. Villars et fils. 214 pp.

MÜLLER A. H. 1950. Grundlagen der Biostratonomie. *Abh. Deutsche Akad. Wiss. Berlin, Kl. Math, u. allg. Naturw. Jahrg.* 1950 nr 3.

OLSON E. C. 1957. Size frequency distributions in samples of extinct organisms. *J. Geology.* **65**, 309–33.

PAYNE J. A. 1965. A summer carrion study of the baby pig *Sus scrofa* Linnaeus. *Ecology.* **46**, 592–602.

PAYNE J. A., KING E. W. & BEINHART G. 1968. Arthropod succession and decomposition of buried pigs. *Nature. Lond.* **219**, 1180–1.

PAYNE S. 1972. Partial recovery and sample bias: the results of some sieving experiments *in* Higgs E. (ed) *Papers in Economic Prehistory*. Cambridge University Press. 49–64.

RICHTER R. 1928. Aktuopaläontologie und Paläobiologie: Eine Abrenzung. *Senckenbergiana.* **10**, 285–92.

SCHÄFER W. 1962. *Aktuo-Paläontologie nach Studien in der Nordsee.* Frankfurt am Main, Waldemar Kramer. 666 pp.

—— 1972. *Ecology and palaeoecology of marine environments.* Craig G. (ed) Oertel I. (transl.) University of Chicago Press. 568 pp.

SCHALLER G. B. 1972. *The Serengeti Lion.* University of Chicago Press. 480 pp.

SIMONS J. W. 1966. The presence of leopard and a study of the food debris in the leopard lairs of the Mt. Suswa caves, Kenya. *Bull. Cave Expln. Group E.A.* **1**, 51–69.

SUTCLIFFE A. J. 1969. Adaptions of Spotted Hyaenas to living in the British Isles. *Mammal Soc. Bull.* **31**.

—— 1970, Spotted hyaena: Crusher, gnawer, digester and collector of bones. *Nature Lond.* **227** (5263), 1110–13.

—— 1973a. Caves of the East African Rift Valley. *Trans. Cave Research Group of Great Britain.* **15** (1), 41–65.

—— 1973b. Similarity of bones and antlers gnawed by deer to human artefacts. *Nature Lond.* **246**, 428–30.

TOOTS H. 1965a. Sequence of disarticulation in mammalian skeletons. *Contr. Geol. University of Wyoming.* **4** (1), 37–9.

—— 1965b. Random orientation of fossils and its significance. *Contr. Geol. University of Wyoming.* **4** (2), 59–62.

VOORHIES M. R. 1969. Taphonomy and population dynamics of an early Pliocene vertebrate fauna, Knox County, Nebraska. *Contr. Geol. university of Wyoming. Special paper no. 1.* 69 pp.

WEIGELT J. 1927. *Rezente Wirbeltierleichen und ihre Paläologische Bedeutung.* Leipzig. Max Weg. 227 pp.

—— 1930. Vom Sterben der Wirbeltiere. Ein Nachtrag zu meinem Buch 'Rezente Wirbeltier-leichen und ihre Paläontologische Bedeutung'. *Leopoldina, Ber. Kais. Leopold. Dtsch. Akad. Naturforsch. Halle.* **6**, 281–314.

—— 1935. Was bezwecken die Hallenser Universitas—Grabungen in der Braunkohle des Geisel-tales? *Nat. Volk.* **65**, 347–56.

WELLS C. 1964. *Bones, bodies and disease.* Thames and Hudson. London.

—— 1967. Pseudopathology *in* Brothwell D. R. and Sandison A. T. (eds) *Diseases in Antiquity.* Springfield.

WHEAT J. B. 1972 The Olsen–Chubbuck Site; a palaeo-indian Bison kill. *Mem. Soc. Amer. Arch.* **26**, *Amer. Antiquity* **37**, 180 pp.

7

R. T. SHUEY, FRANK H. BROWN,
G. G. ECK and F. CLARK HOWELL

A statistical approach to temporal biostratigraphy

Various statistics (similarity coefficients) can be constructed for faunal assemblages which vary more or less regularly as these assemblages change. Methods are described for placing assemblages on an evolutionary scale, the units of which are equal amounts of evolutionary transformation in the species makeup of assemblages. The greatest single factor involved in the scaling appears to be time; the effects of zoogeography are also discussed. Characteristics of different similarity coefficients differ, and some of these characteristics are discussed.

Applying our method to faunal assemblages from East African hominid sites, reasonable correlations can be established between Olduvai Gorge, Tanzania, the Shungura Formation, Ethiopia, and the Koobi Fora Formation, Kenya.

Introduction

This paper reports computer-oriented research motivated by the problem of establishing a reliable chronology for the hominid fossils of East Africa. As the work progressed we became aware that our approach could have rather broad applications and was related to recently proposed statistical techniques in paleobiology, archaeology, and stratigraphy (Southam *et al.* 1975). The present paper emphasizes the philosophy of our method and its characteristics as revealed by our numerical experience. A further paper (Brown *et al.* this volume) focuses on some critical chronological questions we hoped to help resolve.

To keep perspective we begin by outlining the input and output of our system of computer programs. The input is a set of species lists. This data could be viewed as a matrix whose ij-th element is '1' when examples of the j-th species are found at the i-th collection station. Otherwise the matrix element is '0'. We refer exclusively to zoological species, although our approach could be tried at other taxonomic levels or with nonbiological categories such as pottery types. We have made no provision for incorporating information on relative or absolute abundances of species, only their presence or absence is considered. We do suppose that there is some stratigraphic control, that is, a certain collection may be known to predate or postdate others.

The output is a scaling of the collections, that is, each is assigned a value on a numerical scale. Equal intervals on this scale are supposed to be equal amounts of evolutionary transformation in the species composition of the faunal assemblages. Usually the ordering is constrained to be consistent with available stratigraphic information. The computer is to determine the rest of the ordering and also the relative sizes of the intervals between collections on this scale.

On the statistical nature of biological evolution

Underlying our approach is the philosophical view that biological evolution can fruitfully be modelled as a stochastic process. Recently others have actively developed this view (Schopf 1972; Kolata 1975). For a real-valued random time series a fundamental descriptive quantity is the autocorrelation function $A(t_1;t_2)$ expressing the statistical similarity between the behaviour at two times t_1 and t_2. This function has the value 1 when t_1 equals t_2, and in most cases it approaches zero as the difference between t_1 and t_2 becomes large. We postulate that a similar function can be defined for any mathematical model of evolution. A simple form for its dependence on time would be

$$A(t_1;t_2) = e^{-R|t_1-t_2|} \tag{1}$$

Although we have not yet considered how the 'autocorrelation' A is to be related to biological entities, it seems that equation (1) (Eq. 1) could not apply to any realistic model of biological evolution. It implies an equilibrium condition, whereas in the geological record evolution appears nonuniform. Although it would be an exaggeration to view the history of life as periods of stagnation punctuated by mass extinctions and originations, there is substantial evidence for inconstancy of the overall rate of evolution (Simpson 1953, p. 34; Kurten 1968, p. 261). But even so, Eq. 1 would still be applicable provided time were measured with a clock which speeded up and slowed down in synchronism with the rate of evolution. Let the reading of such an evolutionary clock at time t be $E(t)$. Then instead of Eq. 1 we could postulate

$$A(t_1;t_2) = e^{-k|E(t_1)-E(t_2)|} \tag{2}$$

The constant k is determined by the absolute magnitude of units on the E-scale. Reduction of the 'autocorrelation' A to $1/2$ occurs when E changes by the amount

$$E_{1/2} = 0.693/k$$

Systematic deviation from linearity in a plot of E versus time means a nonuniform rate of evolution.

Our computer programs determine a scale-value E for each species list. This is done by fitting the entire data matrix to the model given by Eq. 2. If the assemblages are from various levels of a single stratigraphic section, then our output gives a quantitative description of progressive evolutionary transformation. However, our goal is to place collections from different stratigraphic sections on the same scale, thereby making paleontological correlations between them.

Similarity coefficient

When we are dealing with real life rather than a stochastic model of it, we cannot rigorously define what we mean by the 'autocorrelation' function A. However, we can construct statistics having a numerical value for two specific comparable faunas and having properties we believe appropriate to an estimator of A. Cheetham & Hazel (1969) tabulate 22 distinct formulas giving a numerical measure of similarity between two species lists. To qualify as an estimator of a function A obeying Eq. 2, a formula should have the following four properties: 1) The numerical similarity of two assemblages must be zero when there are no species in common. This is not true for many coefficients used in numerical taxonomy (Sneath & Sokal, 1973, p. 129ff) and it holds for only 8 of the 22 formulas in the Cheetham–Hazel tabulation. (2) The similarity of A to B must be the same as the similarity of B to A, that is, the similarity coefficient must be symmetric. (3) The similarity should be 1 when the two assemblages are identical. (4) When the two assemblages are almost identical, the numerical similarity should decrease linearly (and not, for example, quadratically) with decrease in the number of species in common. Table 1 lists all those formulas from the Cheetham–Hazel list which have these four analytical properties.

Fitting the model

Suppose we have species lists for N different stratigraphic levels at a reference section. For each pair of assemblages we have a measure of similarity SIM_{ij}. The number of such similarity coefficients is

$$NP = N(N-1)/2 \qquad (4)$$

Is it possible to choose a 'scale constant' k and N 'scale values' E_i such that

$$SIM_{ij} = e^{-k|E_i - E_j|}? \qquad (5)$$

The answer in general is no. We have NP equations in $N + 1$ unknowns, so for N greater than 3 the number of equations is greater than the number of unknowns. Furthermore, as long as the number of species is greater than N the similarity coefficients will be mutually independent regardless of which formula in Table 1 is used. Thus in any real case there will be NP non-zero 'residuals'

$$R_{ij} = SIM_{ij} - e^{-k|E_i - E_j|} \qquad (6)$$

A standard mathematical remedy for this situation is to choose the values of k and E_i so as to minimize the 'objective function'.

$$\phi = \Sigma_{ij}(R_{ij})^2 \qquad (7)$$

Computational aspects of this problem are well documented (Kuester & Mise 1973; Bard 1974). But what is the meaning of these residuals R_{ij}? We can still postulate that there is some function A and some time-like scale $E(t)$ such that Eq. 2 is obeyed. Then each residual can be attributed to imperfection in the computed similarity SIM_{ij} as a measure of A. Errors of this kind would be due to non-temporal factors such as: paleoecology, paleoenvironment, taphonomic effects, incomplete collection, uncertainties in taxonomy, and arbitrariness of the similarity formula. If we had independent reliable information on the absolute age of each stratigraphic level sampled, and if we were willing to assume uniform rate of evolution over the span of the section, then we might display the residuals as indicators of nontemporal effects, notably paleoecology and paleoenvironment. However, our present interest is in developing a new approach to biostratigraphic correlation, independent of geophysical methods of absolute chronology. As indicated previously, this is to be done by generating values E_i on the same scale for collections from more than one stratigraphic section.

Defining the scale

The mathematical problem posed in the preceding section is to find a 'scale constant' k and 'scale values' E_i so as to minimize the sum of squares of residuals. But these values are not unique. The residuals would be unchanged if a constant were added to all E_i, or if all E_i were increased by some factor and k decreased by the same factor. This ambiguity is not an essential detriment for the purpose of biostratigraphic correlation. Indeed, Shaw (1964, p. 83) emphasizes that 'absolute time' has much the same ambiguity.

Two conditions must be arbitrarily imposed to make the minimization problem mathematically well defined. One possible condition would be to set $k = 0.693$. Then the unit of measurement on the E-scale would be $E_{1/2}$, that is, a difference of one unit on the scale would correspond to a faunal resemblance of $1/2$. A complementary condition would be to require that the sum of the E_1's be zero. This would be one way to make the problem mathematically well defined. A very similar scaling has been described by Kruskal (1964), although he had in mind a positioning of the assemblages in an abstract multidimensional space rather than along a single time-like axis.

We have chosen an alternative procedure. We assign numerical values to two of the E_i's, and then hold these fixed while adjusting k and the other $N-2$ values of E_i so as to minimize the sum of squares of residuals. The actual choice of two pre-assigned E_i values is arbitrary. Examples will be discussed later (Tables 2–4).

One argument for our way of removing the nonuniqueness is that scalings of the same stratigraphic section using different taxonomic orders can be directly compared. It seems that for any zoogeographic province and geologic epoch, different orders may differ greatly in their rate of evolution (Simpson 1953, Chap. 10). For the Pleistocene mammals of Europe, Kurten (1968, p. 264) finds a species half-life of 180 000 years for elephants and 1 600 000 for bats—a ratio of nearly an order of magnitude. Such differences may well be an artifact of taxonomy rather than something organic (Kolata 1975, p. 626), but that makes them no less real in the mathematical problem being posed. If we prescribed the scale constant k to have some value such as 0.693, then the E values for stratigraphic horizons would be much larger if we used rapidly evolving orders than if we used slowly evolving orders. But when we peg down the scale value at two stratigraphic horizons, we may hope that the values assigned by the computer to other horizons of the section will turn out to be rather independent of which taxonomic groups are being used to compute the coefficients of fauna resemblance between pairs of horizons.

Scaling the Shungura Formation

In the lower Omo Valley of south-western Ethiopia the Shungura Formation has been the object of intensive multidisciplinary study focused on early man. Along with hominid remains over 20 000 mammalian fossils have been collected and are under study by a team of specialists (Y. Coppens, F. C. Howell, D. A. Hooijer, S. Coryndon-Savage, and others). De Heinzelin, Haesaerts & Howell (1975) have divided the Shungura Formation into the 11 members A, B, C, D, E, F, G, H, J, K, and L each with a tuff at the base, plus an additional Basal Member partly exposed below Member A. Fossil collections are much more extensive for Members B through H than for the top or bottom of the section. By means of detailed lithostratigraphic profiles De Heinzelin & Haesaerts (1975) divided members into submembers. From a computerized data bank of specimen identifications, we tabulated the species reported in each submember. Many submembers are non-fossiliferous, while some submembers in the middle of the section have substantial species lists.

We have made dozens of alternative scalings of the Shungura Formation. In each case the computer was asked to find a scale constant k and scale values E_j each associated with a different stratigraphic horizon. The essential criterion was that the factors $\exp -k|E_i-E_j|$ agree as closely as possible with the numerical faunal similarity SIM_{ji}. The purpose of computing many alternative scales was to define empirically the dependence on procedural details. Many of the characteristics of our method thus revealed were confirmed with other types of fossil data (trilobites and nannofossils), although our computations with these were much less extensive.

Our input to the computer was a binary data matrix recording which of 109 species had been identified in each of 57 stratigraphic levels. We did not try to define scale values E_i for all 57 levels, primarily because it was early appreciated that subdivision of the Shungura Formation to this extent would lead to scale values E_i with a statistical uncertainty greater than the scale difference between adjacent stratigraphic levels. In addition, for many of these 57 levels the fossil record seemed clearly inadequate in that only a very few fossil species were reported, far fewer, we suppose, than the number of species extant at the time of deposition of the submember of the formation. Finally, in order to deal with more than about 25 unknowns E_i our computer programs would have to be redesigned.

The first step in each scaling was to combine the species lists from several consecutive fossiliferous submembers. Some of these lists would encompass two or even three members, while others would represent a single exceptionally rich locality in a single submember. Table 2 shows computer output for a typical grouping consisting of 15 species lists. It includes six different scales corresponding to the six different formulas in Table 1. They are all 'preliminary', and later in this paper we present alternatives we prefer. The purpose of Table 2 is to provide specific examples for the subsequent summary of our computational experience.

For the scales in Table 2 we exercised a computer program option to 'fill' the species lists. Each species was assumed present at all times between its first and last recognized occurrence. We fixed the scale values for the stratigraphically

TABLE 1: *Formulas for similarity coefficients*

N_A and N_B are the observed number of species in the two assemblages, and C is the number of species in common. N_1 is the smaller of (N_A, N_B) and N_2 is the larger.

1. Jaccard $\dfrac{C}{N_A + N_B - C}$

2. Burt–Pilot $\dfrac{2C}{N_A + N_B}$

3. Kulczynski $\dfrac{C(N_A + N_B)}{2N_A N_B}$

4. Otsuka $\dfrac{C}{\sqrt{N_A N_B}}$

5. Simpson $\dfrac{C}{N_1}$

6. Braun–Blaunquet $\dfrac{C}{N_2}$

TABLE 2: *Typical Preliminary Results, Shungura Formation*

Formula	1	2	3	4	5	6
Sigma (σ)	0.037	0.028	0.024	0.025	0.039	0.039
Half Life ($E_{1/2}$)	16	24	31	27	51	18

Level	Evolutionary Scale					
L	100	100	100	100	100	100
K	91	91	91	91	96	89
J	87	88	87	88	90	87
G27–H	80	81	80	81	86	79
G11–G24	73	74	71	73	73	73
G7–G10	73	74	71	73	72	73
G1–G6	66	68	63	66	63	67
F	62	64	58	61	54	64
E	58	61	53	58	46	62
D	57	60	53	57	45	61
C6–C9	55	59	51	55	42	60
C1–C5	52	56	48	53	39	57
B8–B12	47	51	41	47	31	52
B3	29	33	22	28	13	33
BASAL–A	0	0	0	0	0	0

lowest and highest groups at 0 and 100 respectively. The computer adjusted the scale constant k and the 13 intermediate E_i's so as to minimize the function in Eq. 7. The half-life $E_{1/2}$ is related to k by Eq. 3. The standard deviation of the numerical similarities from the exponential factors is given by

$$\sigma = (\phi_{min}/(NP{-}N + 1))^{1/2} \tag{8}$$

The denominator in this equation is the number of degrees of freedom in the multivariate adjustment, namely the number of pairs of levels minus the number of unknowns.

Dependence of scale on formula for faunal similarity

Table 2 shows that the output does depend on the formula used to express similarity between pairs of assemblages. Certain characteristic patterns were observed for all the Shungura work as well as the trilobite and nannofossil studies: (1) The best fit (lowest sigma) is for one of the formulas nos. 2, 3, 4. (2) These three formulas give quite similar values for $E_{1/2}$ and the E_i. Indeed the differences in Table 2 are unusually large. (3) Formula no. 1 gives the largest k (smallest $E_{1/2}$) while Formula no. 6 is next. The smallest k (largest $E_{1/2}$) is for formula no. 5. (4) The actual numerical similarities (not shown in Table 2) are smallest for Formula no. 1 and largest for Formula no. 5, that is, they differ in the same way as the estimated $E_{1/2}$.

Our conclusion is that Formulas nos. 2, 3, and 4 are practically interchangeable, despite their different appearance, and that when the fossil record is adequate they most closely estimate the 'autocorrelation' $A(t_1;t_2)$ postulated to obey Eq. 2. Is is interesting that Formula no. 4 is very similar to the well-known definition of autocorrelation for a numerical time series.

When the fossil record is far from ideal for our method, Formula no. 5 is often the best. This is based on instances (not specifically described here) of disagreement between the order of the output E_i and the independently known stratigraphic order. The species lists leading to such a wrong scaling are characterized by discontinuous range zones and wide variation in total number of species. In such a case Formula no. 5 often has the least tendency to give scale values out of order. Previously Simpson (1960) compared his formula (our no. 5) with nos. 1, 2, and 6. He argued that his was superior because it was least liable to give spuriously low similarity estimates when the fossil record was inadequate. This is just what our experience shows. However, in compensation we find the Simpson formula gives similarities on the high side (for our purposes) when the data quality is good.

Table 2 illustrates the response of the different formulas to marginal data. For the middle of the Shungura Formation the total number of species in each list is of the order of 50. By extrapolating a plot of species versus specimens we believe the actual number of species of large mammals extant in this region at any time to be about 60. However the total number of species actually found in the Basal Member and in Member A is only 9. This is a poor sample. Most of the difference between the six different scales in Table 2 is due to a fixed value $E = 0$ being assigned to this poorly sampled stratigraphic level. In addition the upper members J, K, and L are not well sampled. If each of the six scales were to be shifted and expanded or contracted so that they matched at the levels B3 and G27–H, they would be nearly identical at the 9 intermediate stratigraphic levels. In other words, the scaling is rather independent of formula when the data are good.

Table 2 illustrates well the distinctive character of scalings by Formula no. 5 where the fossil data is sparse. At the extreme top and bottom of the section in column no. 5 the scale values E_i are relatively close together. This is because when compared to the others the Simpson formula maintains high similarities even if the total number of species listed is unusually small.

Accuracy of the scale

In addition to finding the k (or equivalently $E_{1/2}$) and the E_j which minimize expression 7, our computer program also gives standard deviations for each of these parameters and correlations between them. The statistical theory is standard and will not be given here in detail (Bard 1974, p. 1976). In brief this information describes how the output values $E_{1/2}$ and E_i would vary if the input SIM_{ij} were independently perturbed by amounts of about sigma (Eq. 8).

The standard deviations of $E_{1/2}$ and the E_i are sometimes smaller than the systematic changes described in the previous section which are due to change of similarity formula. For all six examples in Table 2 the standard deviation of $E_{1/2}$ is about 1 and the deviations of the E_i (excluding those two which are prescribed) are 1 to 2. Thus the computed statistical uncertainty in each scaling is not a reliable measure of the total uncertainty. For this reason we do not give specifically this part of our computer output.

The statistical uncertainties in the E_i are useful for deciding when two species lists are not significantly different. On all size scales of Table 2 the difference between Members D and E is less than the standard error in either scale value. Thus in our method they are indistinguishable and the two could be combined into one. With the species lists 'filled' both members have 55 species and there are 53 in common.

The estimates of the scale values E_i are not statistically independent. Almost invariably we found a positive correlation between the values inferred for adjacent stratigraphic levels. Possibly this is a 'closure' effect due to the scale being fixed at the top and bottom of the section. Almost always the correlation coefficient is less than 0.5 in absolute value, so for most purposes it could be ignored. The correlation does have the consequence that the standard deviations of the several E_i's are usually about the same.

There is also a correlation between $E_{1/2}$ and individual E_i's. The correlation coefficient is positive for large E_i and negative for small E_i, which reflects the fact that the similarity coefficients fit just as well if the $E_{1/2}$ is increased (k decreased) and the scale expanded. These correlations are greater when the two fixed levels chosen to define the E-scale are not strongly tied (through the similarity coefficients) to numerous other levels.

Revision of the Shungura scale

Examination of the preliminary scaling of the Shungura Formation suggested several ways in which it could be improved. The improvements were as follows: First, submember B3 was taken as the lower fixed point, rather than the sparsely sampled members Basal and A. The scale value for B3 was arbitrarily chosen as 50. Some aesthetic considerations in this choice are that it avoids negative scale values and keeps the scale unit roughly equal to the statistical standard deviation. These same considerations were also used by Shaw (1964) in his scaling procedure.

As a second step statistically indistinguishable assemblages were merged. From Table 2 this applies to members D and E and to the submember groups G7–G10 and G11–G24. The systematic difference of 3 units between groups C1–C5 and C6–C9 is of marginal significance, and we decided to merge these too. The benefits to be gained by such merging are a more complete sampling of the extant species and possibly some averaging of faunal fluctuations due to environmental and ecological factors.

Finally, we elected not to fill the species lists. It is a moot question whether or not gaps in range zones should be filled for our method of temporal biostratigraphy. Certainly the biozone (all rocks deposited anywhere during the entire time the species existed) must be free of gaps, since extinction is permanent. Filling a composite section should tend to generate biozones. However, the filling process would amplify error due to reworking of fossils from older deposits or leaking of fossils from younger deposits. If these possibilities are taken seriously, perhaps it is wiser to leave gaps even when only temporal information is sought. Another consideration is that the filling cannot be done from the Shungura data alone for those stratigraphic levels where it would make the most difference. Some species recorded from middle members are absent from upper members, and yet are still extant. Should they be added to the species lists for the upper members? Conversely, some species recorded for middle members and not for the Basal member or Member A are represented in the Mursi Formation, which is older than the Shungura Formation. Should these be added to the species list for Basal + A? Most importantly we would like to include assemblages from different sites in East Africa with no lithostratigraphic connection on the same scale. For such a data set filling could not be completely accomplished without assuming the relative temporal order of all assemblages, which would be begging the question.

Effect of filling the species lists

Table 3 shows the scalings computed with the three modifications just described. To help analyse the differences between these and the preliminary scales (Table 2), we made some supplementary computations using only formula no. 3 (Table 4). Column 3A gives the entries from Table 2, averaged for the pairs of levels being merged, then scaled to 50 units at level 'B3'. Alternatively (Column 3B) we first merged pairs of levels, then estimated scale values holding 'Basal + A' at 0 units, and finally made a linear transformation to 50 units at B3. The only difference is at level 'C'. Comparison of columns 3A and 3B confirms that merging statistically

TABLE 3: *Evolutionary Scale for the Shungura Formation (unfilled)*

Formula	1	2	3	4	5	6
Sigma (σ)	0.126	0.119	0.122	0.120	0.126	0.128
Half Life ($E_{1/2}$)	7	13	14	13	18	10

Level	Evolutionary Scale					
L	100	100	100	100	100	100
K	95	95	95	95	95	95
J	91	93	92	93	93	92
G27–H	83	84	84	84	83	84
G7–G24	77	77	76	77	75	78
G1–G6	72	72	72	72	73	73
F	69	70	69	70	70	70
D–E	66	67	66	66	65	67
C	64	65	64	64	63	65
B8–B12	59	60	60	60	60	60
B3	50	50	50	50	50	50
BASAL–A	38	35	41	37	47	32

TABLE 4: *Influence of Grouping and Filling on Evolutionary Scale Values.*
(Methods 3A–3D explained in text)

Method	3A	3B	3C	3D
Sigma (σ)	0.024	0.025	0.025	0.122
Half Life ($E_{1/2}$)	20	20	20	14

Level	Evolutionary Scale			
L	100	100	100	100
K	94	94	94	95
J	91	91	92	92
G27–H	87	87	87	84
G7–G24	81	81	82	76
G1–G6	76	76	76	72
F	73	73	73	69
D–E	70	70	70	66
C	68	67	68	64
B8–B12	62	62	62	60
B3	50	50	50	50
BASAL–A	36	36	36	41

indistinguishable assemblages gives a new scale value which is the mean of the old ones. The comparison also reminds us that the upper and lower parts of Member C do have a marginal difference in evolutionary stage.

For a third variation (Column 3C) we merged pairs of levels and then generated scale values holding 'B3' fixed at 50 units. The comparison with column 3B shows that for a given similarity formula the scale depends very little on which two points are held fixed.

Column 3D of Table 4 is the same as column 3 in Table 3. The only change from the computational procedure used for column 3C is that gaps in the range zone are not filled. Is is clear that filling of the species lists accounts almost entirely for the difference between Tables 2 and 3 in the distribution of the Shungura members along the evolutionary scale. The principal difference is that filling brings G27–H and G7–G24 closer to J. The distribution of B12 through G6 is essentially unaffected. As discussed later, it seems that the present faunal sampling of member J in particular is inadequate, and the evolutionary scale in Table 3 may contain a systematic distortion due to inclusion of the actual species list for Member J.

Filling also has several other well-defined effects on the output of our scaling program. When lists are filled, the similarity coefficients fit our model much better, as expressed by a smaller standard deviation σ. The statistical standard errors in the E_i are proportional to σ and hence are also decreased by filling. These computed deviations (not shown) for the scales in Table 3 are in the range of 2 to 5 units, and are larger than the systematic differences caused by differences between formulas. Finally, filling increases the computed similarity between any pair of assemblages and increases $E_{1/2}$.

Numerical tests with trilobites and nannofossils

While our first interest has been in the Pliocene/Pleistocene mammals of East Africa, we have also applied our ideas to two other sets of paleontological data. The purpose was to acquire broader computational experience, and thus a deeper and more reliable understanding of the possibilities and limitations of our statistical approach. One set of data is for Late Cambrian trilobites from the Riley Formation in central Texas (Palmer 1954). In his book on temporal biostratigraphy, Shaw (1964) devotes over 100 pages to construction of a continuous scale for the Riley Formation, using the same species lists from Palmer (1954). Comparison of his approach with ours brought out a practical limitation of each. Shaw's method uses measured sedimentary thickness to help smooth and edit the range zones in the several measured and sampled sections. It is applicable only if rate of deposition is uniform within each section, excepting only a few discrete horizons. Thus it appears restricted to layer-cake marine formations. In contrast our scaling does not use thickness information at all but rather uses the faunal similarity coefficients to define an orderly progression of species. It is applicable only if the assemblages are spaced in time more closely than a species half-life. An extreme example of what happens when this is not the case is the Morgan Creek section of the Riley Formation. Palmer (1954) collected at 26 stratigraphic levels, but both the lowest level (222' above base) and the highest level (574' above base) have no species in common with other levels. Thus it is impossible to place these two assemblages on the same scale as the other 24 solely on the basis of their similarity coefficients.

The continuous scales produced for the Riley Formation by our method and by Shaw's are not linearly related. However, our scale is closely related to the conventional zoning given by Palmer (1954). Four zones were defined by Palmer for the middle of the Riley Formation. The lower two zones, *Cedarina-Cedaria* and *Coosella*, span the same length on our *E*-scale as do the upper two zones, *Maryvilla* and *Aphelaspis*. However, on Shaw's scale the lower two span over twice the length of the upper two. This is because Shaw's scale is based on sedimentary thickness rather than faunal resemblance, and in all the sections reported by Palmer the lower zones are over twice as thick as the upper zones. Thus there is a distinct disparity between rate of evolution and rate of sedimentation. Which is more uniform with respect to absolute time we do not know in this case.

There is a difference between Palmer's zoning and our scaling of the same species lists, namely that the scaling is continuous. At Morgan Creek the *Coosella* zone was sampled at 10 stratigraphic levels, and no two of these have the same species list. Consequently each has a different scale value E_i, even though all are in the *Coosella* zone. On our scale the interval $E_{1/2}$ (Eq. 3) was somewhat less than the size of a zone.

As a second test case we considered the Cenozoic calcareous nannofossils of Deep Sea Drilling Project (DSDP) Leg 29 (Edwards & Perch-Nielsen 1974). We generated scalings for half-a-dozen cores, in each case using only the nannofossil species lists for that core. Then we compared our results with the chronological interpretation given in the DSDP initial report. This chronology was based primarily on the shipboard paleontological work, notably on foraminifera in addition to the calcareous nannofossils. Site 281 was of particular interest in that it has an unconformity between late Eocene and early Miocene, with substantial nannofossil collections on both sides. The unconformity was apparent in our results as a large gap in the scale values E_i.

It seems that between late Paleocene and early Pliocene our scaling has a fairly linear relation to absolute time, with $E_{1/2}$ of the order of 10 m.y. Although Edwards & Perch-Nielsen (1974) reported a distinctly greater diversity below mid-Oligocene than above, from the data at sites 277 and 282 we could not detect a corresponding break in slope of E versus absolute time. This encourages us in regarding our scaling as primarily temporal.

For the nannofossil data of DSDP site 278 we did notice a break at the early Pliocene in the relation of our scale to absolute time. There seemed to be a corresponding break in the zoning used on DSDP Leg 29. For the Pleistocene and late Pliocene, a period of about 4 m.y., there are two zones (*Coccolithus pelagicus* and *Pseudoemiliania lacunosa*). From mid-Oligocene to early Pliocene, a period of about 32 m.y., there are four zones (*Reticulofenestra pseudoumbilica*, *Cyclicargolithus neogammation*, *Discoaster deflandrei*, and *Reticulofenestra bisecta*). The number of zones per unit of absolute time is distinctly different, but the number of zones per scale interval $E_{1/2}$ is about the same. This observation reinforces the indication from the

Riley Formation that a conventional zoning approach tends to gives zones of equal value on our scale.

Rate of faunal evolution in the Shungura Formation

We have previously published a scale of absolute time for the Shungura Formation based on magnetic polarity measurements and K-Ar dates (Shuey, Brown & Croes 1974). Figure 7:1 is a plot of the evolutionary scale of Table 3 against this absolute time scale. The slope of such a plot is the rate of evolution in absolute time. Evolution was relatively slow during the time of members C, D, E, and F. We already noted this from the indistinguishibility of Members D and E in evolutionary stage. Conversely evolution was relatively fast during the time of members G and H. These two features persist through all variations of our computational technique. Similar variations have previously been noticed in the rate of change of molar dimensions in the single species *Mesochoerus limnetes* (Cooke & Maglio, 1972; H. B. S. Cooke, pers. comm.). The interval of rapid evolution is centered on a thick lacustrine sequence unique in the Shungura Formation, and it is tempting to speculate on a causal relation.

Geography and enviroment in East Africa

It would be important to the study of early man if the scale established for the Shungura Formation could be extended to other East African hominid localities such as East Lake Turkana and Olduvai Gorge. Of first concern is the extent to which paleoenvironmental diversity and paleogeographic barriers introduce error.

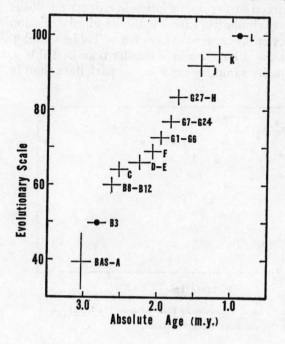

FIG. 7:1.

Evolutionary scale (Table 3) versus absolute age (Shuey *et al.* 1974) for the Shungura Formation. Vertical and horizontal bars indicate uncertainties. The two fixed points on the evolutionary scale are 50 and 100 units. Letters designate members of the Shungura Formation.

To address this question we tabulated the large mammals reported in each of the present National Parks of East Africa. The number of species in a park was not correlated with the area of the park and seemed to have about the same range as the number of species in a member of the Shungura Formation. We suspect that some of the smaller species lists are incomplete. To reduce sampling error we considered only 16 parks which had more than 38 species (about the mode).

Figure 7:2 shows the similarity coefficient computed from formula no. 5 for pairs of parks, plotted against the distance between parks. Table 5 gives the parameters of a linear fit to this plot and similar plots using the other similarity formulas. Since the slope of the linear fit is small, the influence of distance alone is insignificant. Within present East Africa, a separation of 1000 km does not inherently imply a lower faunal similarity (for large mammals) than does a separation of 50 km.

TABLE 5: *Linear regression of faunal similarity against distance for East African National Parks*

Similarity formula	1	2	3	4	5	6
Reciprocal slope (km)	3800	4500	4300	4400	3700	5200
Intercept	0.70	0.82	0.84	0.83	0.92	0.75
Standard deviation	0.101	0.082	0.082	0.082	0.103	0.087

The intercept of the linear fit is always less than unity, even though all formulas give unity for identical faunas. We interpret this shift as the response of each formula to environmental diversity (and any sampling error). The ranking of formulas by intercept in Table 5 is essentially the same as the ranking by half-life in Tables 2 and 3, for as we remarked earlier the variation in estimated half-life is due to variation in average value of similarity. Likewise the standard deviation of similarity from the linear fit varies in the same way as sigma in Tables 2 and 3. The magnitude of the deviation for the present parks is smaller than in Table 3, which we interpret to indicate a smaller sampling error in the park data than in

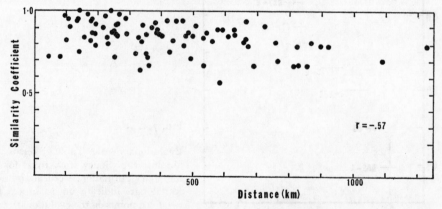

FIG. 7:2. Faunal similarity by formula no. 5 for pairs of East African National parks plotted against the distance between parks.

the Shungura data. Table 2 shows by far the smallest deviation (sigma). We believe this is because the 'filling' greatly suppresses the environmental factor(s).

Geographic extension of the temporal theory

As explained near the beginning of this paper, our objective is to make biostratigraphic correlations by placing faunal assemblages from different localities on the same scale. To do so, we must slightly extend our stochastic theory. We conceive of a 'cross-correlation', C, between the two evolving biological communities. A simple law for its temporal behaviour would be

$$C(t_1;t_2) = Ge^{-k|E(t_1)-E(t_2)|} \qquad (9)$$

which is a generalization of Eq. 2.

The numerical value of $C(t_1;t_2)$ is supposed to express the correlation between the community in the first location at time t_1, and the community in the second location at time t_2. Thus Eq. 9 allows statistically for both temporal and geographic factors. The geographic coefficient 'G' is a measure of the decoupling of the two evolutionary sequences. For any specific paleontological correlation problem it could be argued that the model indicated in Eq. 9 is inappropriate. As an extreme example it could be hypothesized that species originated only at one location and diffused, or migrated to the other. In this case, if a fossil assemblage from the second site were compared with a chronological sequence of assemblages from the first site, then the best match would be with the fossils of an earlier absolute date. Fitting the data to Eq. 6 would still be appropriate, but the relation between time and scale value would depend on location. It would be a case of biostratigraphy being well-defined but diachronous.

The study of present-day East African National Parks (Table 5) suggests that the coefficient G should have a value of 0.7 to 0.9, depending on which formula is being used for numerical estimate of faunal resemblance. This supposes, however, that the two locations are in the same zoogeographic province.

The distribution maps of Dorst & Dandelot (1970) were used to compile a species list for southern Libya; similarity coefficients computed between this locality and East African Parks are about 0.2 to 0.3. They are a function mainly of the size of the East African species list, since the number of animals in common is nearly constant, being restricted to species whose distribution is nearly pan-African. Because of this, the statistical resolution of our method might be quite poor when dealing with assemblages from different zoogeographic provinces, since the numerical similarity would not be much larger than sigma (the standard error in measuring similarity). Zoogeographic provinces known for trilobites (Cook & Taylor 1975) and nannofossils (Edwards & Perch-Nielsen 1974) appear to be similar in the sharpness of their boundaries and the magnitude of the numerical similarity between contemporaneous assemblages from different provinces.

Glynn Isaac (personal communication, May 1975) raised the following intriguing question: What if a zoogeographic boundary moved across one or both of the locations during the time of fossil deposition? Certainly in such a case Eq. 9

would be inappropriate because it has no time-dependence on the coefficient G, which would be lower for locations in different provinces than for locations in the same province. We will not comment here on the possibility of identifying or excluding such a province crossing by analysis independent of our procedure.

Correlation techniques

We have developed two computer programs for biostratigraphic correlation. In either case species lists and a scaling (*E*-values and the scale constant *k*) are required for a reference section. The first program works with a single assemblage from another location. It computes the numerical similarity between this new assemblage and each of those in the reference section. The residuals

$$R_{in} + SIM_{in} - Ge^{-k|E-E_n|}$$

measure the misfit between the computed similarity and the theoretical Eq. 9. The subscript 'n' denotes the new assemblage, and the subscript 'i' denotes one of the N assemblages of the reference data. The program adjusts the geographic coefficient G and the new scale value E_n so as to minimize the function

$$\phi' = \Sigma_i (R_{in})^2 \tag{11}$$

The scale constant '*k*' and the N values E_1 are held fixed.

At the optimum values of G and E_n the function ϕ' in Eq. 11 has its minimum value, ϕ'_0. The ratio $(\phi'-\phi'_0)/\phi'_0$ obeys the variance-ratio or *F*-distribution. By evaluating this ratio for near-optimum values of G and E_n, statistical confidence limits can be established for these two quantities (Draper & Smith 1966, p. 274).

Our second computer program for biostratigraphic correlation works with a sequence of assemblages from the new section. It determines a scale value E_n for each of them, and also a single coefficient G for the entire section. The determination is made by minimizing the function

$$\phi'' = \Sigma_{nm}(R_{nm})^2 + \Sigma_{in}(R_{in})^2. \tag{12}$$

The first sum is over all pairs within the new section, while the second sum is over all pairings between the new section and the reference section. Eq. 6 is used for the *intrasectional* residuals R_{nm}. Eq. 10 is used for the *intersectional* residuals R_{in}. As in the first program the scale constant *k* and the scale values for the reference section are held at prescribed values. The correlations by the second method have a greater statistical precision. This is because the scale values for the new section are constrained not only by the relative faunal similarities to the reference section, but also by the faunal changes within the new section. In Tables 6 and 7 below we give examples of results from this second correlation method.

Correlation of the Usno and Shungura Formations

The Usno Formation outcrops in a small area just 30 km north-east of the northernmost Shungura exposures. Some 37 species of large mammals have been identified (Coppens & Howell, 1975). Similarity coefficients computed between this assemblage and the Shungura Formation are highest for Member C and the upper part of Member B. This is true for all formulas and with the Shungura species lists filled or not. Using the first of the computer programs just described, the Usno assemblage is placed on the Shungura scale between upper Member B and Member C. Estimates of the coefficient 'G' range from 0.75 to 0.81.

TABLE 6: *Correlation of Olduvai Gorge with Shungura Formation*

Formula	Unfilled			Filled		
	1	3	5	1	3	5
Sigma (σ)	0.11	0.11	0.11	0.08	0.07	0.06
Coefficient (G)	0.42	0.61	0.66	0.44	0.59	0.65
Level	Evolutionary Scale					
Upper Bed II	90	90	92	91	91	106
Lower Bed II	84	85	86	84	85	89
Bed I	78	79	82	79	80	87

By magnetostratigraphy and radiometric dates we have independently correlated the members of the Usno and Shungura Formations (Brown & Shuey 1975). The main Usno fossil site (White Sands) was correlated with upper B2 and a second Usno site (Flat Sands) with upper Member B. Thus there is a perceptible difference between the faunal and geophysical time lines. If the geophysical correlations are correct, the Usno fauna is somewhat 'ahead of its time' on the Shungura evolutionary scale. This discrepancy is of marginal statistical significance, since the lower limit of the Usno standard error range is usually about midway between the B3 and B8–B12 Shungura levels.

TABLE 7: *Preliminary faunal correlation of East Lake Turkana with Shungura Formation*

Formula	Unfilled			Filled		
	1	3	5	1	3	5
Sigma (σ)	0.11	0.10	0.10	0.08	0.07	0.07
Coefficient (G)	0.51	0.71	0.77	0.51	0.68	0.72
Level	Evolutionary Scale					
Chari	82	86	87	88	90	90
Koobi Fora	78	80	82	81	84	87
KBS	72	74	75	75	77	78

Correlation of Olduvai Gorge with the
Shungura Formation

As a second test we took the fauna reported for the lower strata at Olduvai Gorge, Tanzania, 500 km south of the Lower Omo Valley. Species lists were prepared for Bed I and for Bed II below and above the internal unconformity. We explored as follows the correlations made by our technique. We took six alternative scalings for the Shungura Formation as the reference section, namely with species lists filled or unfilled and with similarity formulas nos. 1, 3, and 5. (It is unnecessary to consider other formulas, since it has already been shown that the results are always in the same order with that from no. 3 near the middle and the results from nos. 1 and 5 at the two extremes.) In all cases we grouped the Shungura strata as in Table 3, and we prescribed the scale values at 'B3' and 'L' to be 50 and 100 respectively.

Table 6 gives some results from our second computer program for biostratigraphic correlation. As previously observed in Table 4, filling the species lists decreases sigma, the standard deviation of the numerical similarities from the best-fitting exponential model. The estimates of geographical coefficient 'G' show the usual dependence on formula. The actual G values are somewhat lower than those found for present-day National Parks of comparable geographic separation (Table 5).

In comparing the six alternative sets of scale values for the Olduvai assemblages, keep in mind that the scale values of the Shungura strata (except B3 and L) are also slightly different in each of the six cases. (The Shungura scale values for unfilled species lists are in Table 3, while column 3C of Table 4 gives the scale values for filled species lists and Formula no. 3.)

For all six variations, Bed I at Olduvai Gorge is assigned a time between the Shungura levels G7–G24 and G27–H. The standard deviation is about 2 units, so that generally both G7–G24 and G27–H are within a standard deviation of Olduvai Bed I. This is consistent with the radiometric and magnetostratigraphic ties between the two sections. We have previously reported (Shuey, Brown & Croes 1974) that the Olduvai magnetic polarity event, for which Olduvai Bed I is the type location, is represented in the Shungura Formation by G27 throuhg Member H.

In Table 6 results for Formula no. 5 with filling seem to be anomalous. In part this is because the Shungura levels are crowded together at the top for this definition of similarity (cf. Table 2). The root of the anomaly is the inadequacy of the faunal data for Shungura Member J. The similarity between an Olduvai level and the Shungura Formation is generally bimodal, with values for J less than for higher or lower Shungura levels. In all columns of Table 6 except the last, Upper Bed II is placed about one unit below J, while the last column shows a strong distortion with Lower Bed II actually below G27–H and Upper Bed II above L. Formula no. 5 with filling gives the highest similarities and probably this is the reason the distortion occurs particularly in the last column of Table 6. The computed standard deviations of Olduvai scale values are not at all diagnostic, as they are always below 2 units. If level J is deleted from the Shungura species lists, computations show both Lower and Upper Bed II with a standard error range from below G27–H to above L, while the range for Bed I is still between G7–G24 and G27–H. Fig. 7:3 below displays a similar situation on another data set.

Paleomagnetic work and K-Ar dating at Olduvai Gorge fixes the age of Bed II between 1.0 and 1.7 m.y. A reasonable estimate for the age of the lower part of Bed II is 1.5–1.7 m.y., and for the upper part a reasonable estimate is 1.0–1.4 m.y. (Hay, 1973).

Correlation of East Lake Turkana

As a final test example of biostratigraphic correlation in East Africa, we considered East Lake Turkana, Kenya. Faunal lists were prepared for three major faunal zones which were originally defined by Maglio (1972). These faunal zones (*Loxodonta africana*, *Metridiochoerus andrewsi*, and *Mesochoerus limnetes*) correspond to strata exposed below the Chari, Koobi Fora, and KBS tuffs respectively, and in the ensuing discussion we refer to them by the name of the overlying tuff in each instance. The Chari exposures are only about 75 km from outcrops of the roughly contemporaneous Members K and L of the Shungura Formation, but the Koobi Fora and KBS sites are some 120 km from the main Shungura outcrops. For the computations described below we used the published data of Harris (1975a, b, c), Cooke (1975), M. Leakey (1975a, b), Beden (1975), Eisenmann (1975), and Savage (1975). R. E. Leakey (written communication, July, 1975) has emphasized that the East Lake Turkana faunal lists are being significantly revised and expanded by current study, so that any inferences from the published lists can only be very tentative.

Table 7 shows the results of a set of computer runs identical in design to those used with the Olduvai faunal lists. The values of σ and G are much the same as in Table 6, except that the geographical coefficient G is systematically higher for East Lake Turkana than for Olduvai. The statistical standard deviation of each

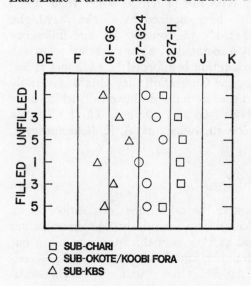

□ SUB-CHARI
○ SUB-OKOTE/KOOBI FORA
△ SUB-KBS

Fig. 7:3.

Preliminary faunal correlation of East Lake Turkana assemblages and assemblages from the Shungura Formation. The symbols were placed by comparing the East Lake Turkana scale-values (Table 7 with the Shungura scale-values (Table 3).

scale value in Table 7 is about 1 scale unit. Differences among the six alternative scale values for each individual assemblage (Chari, Koobi Fora, and KBS) are much larger than this. Again it must be remembered that this is also true for the Shungura assemblages (except the two designated B3 and L). Fig. 7:3 shows where the East Lake Turkana assemblages are placed relative to the Shungura assemblages. It appears that the KBS fauna is being correlated with lowest G, the Koobi Fora fauna with middle or upper G, and the Chari fauna with H. However, the similarity coefficients for the Chari assemblage show the bimodal behaviour previously discussed for Olduvai Bed II, that is, the similarity with the J fauna is less than the similarity with stratigraphically higher or lower Shungura faunas. When J is deleted from the Shungura species lists, similar competitions lead to a faunal correlation of Chari with K and Koobi Fora with uppermost G or H. No comparable effect has been found for the KBS assemblage. Over a period of a year and a half we have made dozens of computerized comparisons between the reported KBS fauna and the Shungura species lists. In most of these the poorly sampled BASAL–A was deleted from the Shungura section. In every case KBS was correlated with Member F or lower Member G of the Shungura Formation.

Radiometric dates for the East Lake Turkana tuffs have been published by Fitch et al. (1974) and compatible paleomagnetic measurements were reported by Brock & Isaac (1974). The dates on the Chari and Koobi Fora Tuffs are about 1.2 and 1.6 m.y. respectively. From Fig. 7:1 the geophysical evidence makes the Chari tuff roughly time-equivalent to Member K and the Koobi Fora Tuff roughly time-equivalent to Member H. The estimates in Fig. 7:3 are distinctly lower stratigraphically. In some part this is because the East Lake Turkana faunas underlie the respective tuffs, but the bias described earlier is probably more important.

The KBS fauna correlates with Shungura faunas which are substantially younger (according to the chronology of Fig. 7:1) than the published radiometric date of 2.6. m.y. on the overlying KBS tuff (Fitch & Miller 1970; Fitch et al. 1974). This discrepancy has already been noticed by Cooke & Maglio (1972, p. 327), and by Cooke (1975). Frankly we consider the age difference greater than any demonstrated uncertainty associated with our statistical approach to temporal biostratigraphy. A possible resolution is afforded by radiometric age determinations by Curtis et al. (1975). For the Chari Tuff they obtain essentially the same age as did Fitch et al. (1974), but they actually find two tuffs which have been called the KBS tuff, with ages of about 1.60 and 1.82 m.y. The KBS fauna is derived from levels stratigraphically below the older tuff (A. K. Behrensmeyer, pers. comm., 1975).

Conclusion

We have started with a simple idea, namely the adjustment of position on an arbitrary scale so that similarity coefficients vary regularly. We have rather extensively explored the ramifications and variations of this idea. Not all of our investigations could be summarized in this paper, and development is still proceeding in some areas. We feel our approach has merit in a situation where biostratigraphic data are abundant but inconsistent. Considerable effort has been expended in designing and debugging the computer subroutines for data management and multivariate optimization. Indeed our software has evolved. We would welcome correspondence from biostratigraphers interested in applying our methods to their problems.

Acknowledgements

H. B. S. Cooke & A. A. Ekdale read the manuscript in its original form, and we deeply appreciate their comments and suggestions which resulted in major changes in the organizations of the paper. Noel T. Boaz is thanked for his efforts in compiling the faunal lists of the Shungura Formation.

References

BARD, Y. 1974. *Nonlinear Parameter Estimation.* Academic Press, New York. 341 pp.

BEDEN, M. 1975. Proboscideans from Omo Group Formations. In: *Earliest Man and Environments in the Lake Rudolf Basin: Stratigraphy, Paleoecology and Evolution.* Y. Coppens, F. C. Howell, G. Ll. Isaac, and R. E. F. Leakey, eds, University of Chicago Press. Chicago.

BROCK, A. & ISAAC, G. Ll. 1974. Paleomagnetic stratigraphy and chronology of hominid-bearing sediments east of Lake Rudolf, Kenya. *Nature Lond.* **247**, 344–8.

BROWN, F. H., HOWELL, F. C. & ECK, G. G. 1976. Observations on problems of correlation of Late Cenozoic hominid-bearing formations in the North Rudolf Basin. In: *Geological Background to Fossil Man.* W. W. Bishop, *ed.* Geological Society of London, London.

——, & SHUEY, R. T. 1975. Magnetostratigraphy of the Shungura and Usno Formations, lower Omo valley, Ethiopia. In: *Earliest Man and Environments in the Lake Rudolf Basin: Stratigraphy, Paleoecology, and Evolution.* Y. Coppens, F. C. Howell, G. Ll. Isaac, and R. E. F. Leakey, *eds.* University of Chicago Press, Chicago.

CHEETHAM, A. H. & HAZEL, J. E. 1969. Binary (presence–absence) similarity coefficients. *Jour. Paleo.* **43** (5), 1130–6.

COOK, H. E. & TAYLOR, M. E. 1975. Early Paleozoic continental margin sedimentation, trilobite biofacies, and the thermocline, western United States, *Geology.* **3** (10), 559–62.

COOKE, H. B. S. 1975. Suidae from Pliocene/Pleistocene successions of the Rudolf Basin. *In: Earliest Man and Environments in the Lake Rudolf Basin: Stratigraphy, Paleoecology, and Evolution.* Y. Coppens F. C. Howell, G. Ll. Isaac, and R. E. F. Leakey, *eds.* University of Chicago Press, Chicago.

——, & MAGLIO, V. J., 1972. Plio-Pleistocene stratigraphy in East Africa in relation to proboscidean and suid evolution. *In: Calibration of Hominoid Evolution.* W. W. Bishop and J. A. Miller *eds.* Scottish Academic Press, Edinburgh.

COPPENS, Y. & HOWELL, F. C. 1975. Vue general des faunes mammaliennes des depots des groupe de l'Omo. In: *Earliest Man and Environments in the Lake Rudolf Basin: Stratigraphy, Paleoecology, and Evolution.* Y. Coppens, F. C. Howell, G. Ll. Isaac, and R. E. F. Leakey, *eds.* University of Chicago Press, Chicago.

CURTIS, G. H., DRAKE, R., CERLING, T. & HAMPEL, J. 1975. Age of the KBS tuff in the Koobi Fora Formation, East Rudolf, Kenya. *Nature Lond.* 258, 395–8.

DRAPER, N. R. & SMITH, H. 1966. *Applied Regression Analysis.* John Wiley and Sons, New York.

DORST, J. & DANDELOT, P. 1970. *A Field Guide to the Larger Mammals of Africa.* Collins, London.

EDWARDS, A. B. & PERCH-NIELSEN, K. 1974. Calcareous nannofossils from the southern southwest Pacific, Deep Sea Drilling Project, Leg. 29. In: Kennett, J. P., Houtz, R. E. *et al.*, *Initial Reports of the Deep Sea Drilling Project*, **29**: U.S. Gov't. Printing Office, Washington, D.C. 496–539.

EISENMANN, V. 1975. Equidae from the Shungura Formation. *In: Earliest Man and Environments in the Lake Rudolf Basin: Stratigraphy, Paleoecology and Evolution.* Y. Coppens, F. C. Howell, G. Ll. Isaac and R. E. F. Leakey, *eds.* University of Chicago Press, Chicago.

FITCH, F. J. & MILLER, J. A. 1970. Radioisotopic age determinations of Lake Rudolf Artefact Site. *Nature Lond.* **226**, 226–8.

—— FINDLATER, I. C., WATKINS, R. T. & MILLER, J. A. 1974. Dating of the rock succession containing fossil hominids at East Rudolf, Kenya. *Nature Lond.* **251**, 213–15.

HARRIS, J. M. 1975a. Rhinocerotidae from the East Rudolf succession. *In: Earliest Man and Environments in the Lake Rudolf Basin: Stratigraphy, Paleoecology and Evolution.* Y. Coppens, F. C. Howell, G. L. Isaac, and R. E. F. Leakey, *eds.* University of Chicago Press, Chicago.

—— 1975b. Giraffidae from the East Rudolf succession. *In: Earliest Man and Environments in the Lake Rudolf Basin: Stratigraphy, Paleoecology, and Evolution.* Y. Coppens, F. C. Howell, G. Ll. Isaac, and R. E. F. Leakey, *eds.* University of Chicago Press, Chicago.

—— 1975c. Bovidae from the East Rudolf succession. *In: Earliest Man and Environments in the Lake Rudolf Basin: Stratigraphy, Paleoecology and Evolution.* Y. Coppens, F. C. Howell, G. Ll. Isaac, and R. E. F. Leakey, *eds.* University of Chicago Press, Chicago.

HAY, R. L. 1973. Lithofacies and environments of Bed I, Olduvai Gorge, Tanzania. *Quaternary Research,* **3**, 541–60.

HEINZELIN, J. DE, HAESAERTS, P. & HOWELL, F. C. 1975. Plio-Pleistocene formations of the lower Omo basin, with particular reference to the Shungura Formation. *In: Earliest Man and Environments in the Lake Rudolf Basin: Stratigraphy Paleoecology, and Evolution.* Y. Coppens, F. C. Howell, G. L1. Isaac, and R. E. F. Leakey, *eds.* University of Chicago Press, Chicago.

KOLATA, G. B., 1975. Paleobiology; random events over geologic time. *Science,* **189**, 625–6, 660.

KRUSKAL, J. B. 1964. Multidimensional scaling. *Psychometrika,* **29**, 1–42.

KUESTER, J. L. & MISE, J. H. 1973. *Optimization techniques with Fortran.* McGraw-Hill Book Co., Inc., New York. 500 pp.

KURTEN, B. 1968. *Pleistocene mammals of Europe.* Aldine Publishing Co., Chicago. 317 pp.

LEAKEY, M. 1975. Carnivora of the East Rudolf succession. *In: Earliest Man and Environments in the Lake Rudolf Basin: Stratigraphy, Paleoecology and Evolution.* Y. Coppens, F. C. Howell, G. Ll. Isaac, and R. E. F. Leakey, eds. University of Chicago Press, Chicago.

—— 1975b. Cercopithecoidea of the East Rudolf Succession. *In: Earliest Man and Environments in the Lake Rudolf Basin: Stratigraphy, Paleoecology, and Evolution.* Y. Coppens, F. C. Howell, G. Ll. Isaac, and R. E. F. Leakey, *eds.* University of Chicago, Chicago.

MAGLIO, V. J. 1972. Vertebrate fauna and chronology of hominid bearing sediments east of Lake Rudolf, Kenya. *Nature Lond.* **239**, 379–85.

PALMER, A. R. 1954. The faunas of the Riley Formation in central Texas. *Jour. Paleont.* **28**, 709–86.

SAVAGE, S. C. 1975. Fossil Hippopotamidae from Pliocene/Pleistocene successions of the Rudolf Basin. *In: Earliest Man and Environments in the Lake Rudolf Basin: Stratigraphy, Paleoecology, and Evolution.* Y. Coppens, F. C. Howell, G. Ll. Isaac, and R. E. F. Leakey, eds. University of Chicago Press, Chicago.

SCHOPF, T. J. M. (ed.) 1972. *Models in paleobiology.* Freeman, Cooper & Co. San Francisco.

SHAW, A. B. 1964. *Time in stratigraphy.* McGraw-Hill Book Co., New York, 365 p.

SHUEY, R. T., BROWN, F. H. & CROES, M. K. 1974. Magnetostratigraphy of the Shungura Formation, southwestern Ethiopia: Fine structure of the lower Matuyama polarity epoch: *Earth and Planet. Sci. Letters,* **23**, 249–60.

SIMPSON, G. G. 1953. *The Major Features of Evolution:* Columbia University Press, New York. 434 pp.

—— 1960. Notes on the measurement of faunal resemblance: *Amer. Jour. Science.* **258A**, 300–11.

SNEATH, P. A. & SOKAL, R. H. 1973. *Numerical Taxonomy.* W. H. Freeman, San Francisco. 573 pp.

SOUTHAM, J. R., HAY, W. W. & WORSLEY, T. R. 1975. Quantitative formulation of reliability in stratigraphic correlation: *Science.* **188**, 357–9.

8

BERNARD A. WOOD

Allometry and hominid studies

When the shape of a bone changes during growth the relationship between its shape and its size can be expressed mathematically and the phenomenon is known as allometric or relative growth. The same concept applies to samples of adult bones of different mean overall size. It is likely that some of the morphological distinctions between early hominids are due to size related allometric features. However, even if such differences can be credited with an allometric basis their significance for hominid taxonomy is uncertain.

This paper reviews the development of allometric theory and deals with technical advances which have improved our understanding of the theory and application of allometry.

Three relevant applications of allometry are considered. These are its use in attempts to compensate for allometrically based intra-group differences which distort attempts to differentiate between taxonomic groups; its role in taxonomy and its use as a tool to increase our understanding of the relationships between function and morphology in hominids.

Introduction

Allometry has been succinctly defined as 'the study of size and its consequences' (Gould 1966). There is now a growing realization that body size discrepancies between early hominid populations may be the cause of important morphological differences (Brace 1972, Wolpoff 1973, Pilbeam & Gould 1974, and Wood 1975a). In order to distinguish which differences these are and whether this makes them more or less important for an understanding of early hominid evolution and taxonomy some knowledge of allometric theory and application are necessary. This paper is offered as a guide to allometry and it also reviews three areas where allometric principles can be applied in the study of hominid palaeontology

Historical development of allometric theory

Allometric theory developed from the observation that if various parts of an organism grew at different rates its overall shape would depend on the stage of growth that had been reached. It was then realized that this relative growth concept also applied to populations of animals of different size. Thus it was apparent that relative growth differences of parts could lead to differences in shape between samples of animals. These shape differences were predictable and were solely consequent on overall size discrepancies and not on any differences that necessarily required taxonomic recognition.

The relationship between size and shape was apparently recognized as early as 1638 by Galileo (Gould 1975). However, it was not until the end of the last century that the first attempts were made to characterize the relationship by Snell (1891). Snell found that when he plotted log brain weight, y, against log body weight, x, of a sample of adult animals a straight line resulted, and he devised the notation $y = bx^{\alpha}$ where b is the intercept on the y axis and α is the slope. Klatt (1919) studied the behaviour of two variables in a growth study and also found a straight line relationship in a log plot indicating an unchanging ratio between the growth rates of the variables. Pezard (1918) termed such growth 'heterogonic'. Champy (1924) introduced the term 'dysharmonic' and Huxley (1924) introduced a third term, 'constant differential growth'.

D'Arcy Thompson (1917) utilized some of the concepts of relative growth but only in a static way. Its subsequent analysis and refinement as a dynamic concept was due to Julian Huxley (1924, 1927, 1931, 1932). He elaborated his earlier ideas in a treatise on relative growth (1932) and in it he included the first application of the formula $y = bx^{\alpha}$ to primate data when he used material collected by Zuckerman (1926) on baboon skulls. He demonstrated a constant ratio between the growth rates of skull length and muzzle length. Huxley's derivation of the formula was based on the three assumptions. First, that growth was essentially multiplicative, second, that it slowed with increasing size or age, and third, that changes in the rate of self-multiplication affected all parts equally. More complicated formulae, with additional constants, have been derived but to quote Reeve and Huxley (1945) 'the introduction of further constants is likely to make the formula impossible to apply in practice and at the same time no easier to justify theoretically than its simpler prototype'. Huxley never claimed that the formula represented a fundamental law of growth and its use can still only be justified by its utility for a wide variety of biological data.

The term 'allometry' was introduced by Huxley and Teissier (1936). They also agreed on the now widely adopted notation

$$y = bx^{\alpha}$$

where α, the ratio of growth rates or the relationship between variables in a population, was called the 'allometry coefficient'. It was realized that when the geometric growth rates were equal, and therefore shape did not change with size, $\alpha = 1$. This was called 'isometry' or 'isometric growth'. If α had values greater or less than unity shape would alter with size and such growth was described as 'allometric', either positive or negative according to the value of α.

Examples of the results of allometric relationships between variables are legion and they have been excellently reviewed by Gould (1966). For example, it is a manifestation of an allometric relationship that large animals have relatively small brains. This is because the expected 'scale' relationship between the brain, a volume, and body size, a length, is not followed. The value of $\alpha = 1$ for isometry applies only to comparable measurements. If the reference variable is the length of an animal then allometric coefficients of volume, surface area and diameter dependents measurements are isometric at $\alpha = 3$, 2 and 1.5 respectively. Distinction should be made between the effects of 'scaling', for example the tendency for longer bones to be more robust, and those of allometry, which are due to departures from the predicted scale relationships.

Huxley, Needham & Lerner (1941) introduced terms for the two main applications of allometry. The first application, to cover the relationship between the growth rates of variables in a single developing organism (a longitudinal growth study) or between organisms of different ages (a cross-sectional growth study) they termed 'heterauxesis'. The second application, to cover the relationships between variables in a group of individuals all at the same stage of growth, usually adult, was called 'allomorphosis'. The relationship between the two main applications is such that in allomorphic studies adults are studied whose final shape is the result of heterauxetic growth relationships during ontogeny. Kavanagh & Richards (1942) quite correctly thought of allomorphic studies as investigations of relative 'size' rather than relative 'growth'.

The terminology is cumbersome but some classification is imperative in order to communicate which type of allometric relationship is being studied and on what basis allometric coefficients are being calculated. The curve that results from a heterauxetic study is known as an ontogenetic curve. Allomorphic studies can be undertaken either using data from within a species, an intraspecific curve, from a group of very different taxa, an interspecific curve, or from a sample that represents an evolutionary sequence, a phylogenetic curve. In the case of the widely studied relationship between brain and body weight each of these types of study produces curves of different slope (Gould 1975), and a section of this paper is devoted to the importance of using the correct curves for comparison.

Significant recent advances

Allometry, despite the prominence given to it by Huxley, has always been an interest of only a minority of biologists. The recent work and excellent reviews by Gould (1966, 1971, 1975) and others have gone some way to correct this neglect. Developments have therefore been few but those discussed below have been selected because of their possible importance in the application of allometric theory to hominid studies.

A. *Reference parameters*

There have been attempts to replace the variable that is chosen *a priori* to represent overall body size by a reference parameter that represents simultaneously all the various measurements that have been taken. Teissier (1955) proposed using the factors derived from a matrix of variables for this purpose but in a later paper, Teissier (1960), he suggested using the first principal component of a correlation matrix as a multivariate size factor. Matsuda and Rohlf (1961) used the centroid solution to extract the first factor from a correlation matrix of standardized variables for use as their 'size' factor. Jolicoeur (1963a) supported Teissier's second suggestion but preferred the first principal component of a covariance matrix for its ease of manipulation. A disadvantage that underlies the use of the first linear function, be it either a factor or a principal component, is that they do not always represent size alone but they also include shape information. The details of why this is are outlined below.

B. *Multivariate allometry*

In an attempt to extend the allometry concept to more than two dimensions simultaneously, Jolicoeur (1963a) used the equation for the first principal component of a covariance matrix as a multivariate generalization of the allometry equation. He then compared the angles made by the co-ordinate axes of two original variables with the first principal component axis. If their cosines were equal then the dimensions were isometric with respect to one another. He extended his concept (Jolicoeur 1963b), and devised a method to test the isometry hypothesis simultaneously for all the variables.

Indirect analyses of growth have also been made by using multivariate methods to analyse intra-group structure. Jolicoeur & Mosimann (1960) used principal component analysis alone in an analysis of size and shape dimorphisms in turtles. Reyment (1969) combined distance and principal component analysis in a study of sexual dimorphism. The D^2 generalized distance of Mahalanobis was used as a measure of overall morphological difference and the orientation of the major axes of the multidimensional figures and their relative inflation were used to compare group structure. Principal axes angulated with respect to one another were taken to indicate significant growth pattern differences, and the variables which contributed in a major way to the non-parallel axes were identified.

It is debatable whether these techniques add sophistication or just complication to allometric studies. Many different hypotheses are capable of being tested using allometric methods, and while some may benefit from being treated on a multivariate basis the more specific questions are probably still more appropriately dealt with using one of the bivariate methods. Multivariate generalizations of the allometry equation for their own sake are as likely to be as confusing as they are enlightening. However, for those interested in exploring these techniques but who are unfamiliar with the multivariate methods themselves, they are clearly explained by Oxnard (1973).

C. *Importance of b*

Greater understanding has been gained of what is represented by the constant, b, which is the intercept on the y axis of a log plot. Huxley (1932, 1950), and others, claimed that the constant b 'had no biological or general significance'. Other workers claimed it was a size independent factor. However, White & Gould (1965) and Gould (1971) showed quite convincingly that when the allometry coefficients, α, are the same for two sets of data the values of the constant b are neither unimportant nor size independent. In these circumstances they claim that b is a scale factor and in order for such animals of different size to maintain the same proportions the value of b should increase or decrease depending on whether α for both sets of data is greater or less than unity. Fig. 8:1 shows a variable representing overall size plotted against a variable of similar scale. In such a case $\alpha = 1$ represents isometry and the dashed line will join specimens of the same shape. For values of $\alpha < 1$ (Fig. 8:1 (a)), then a proportionate increase in size necessitates an increase in b. If $\alpha > 1$ (Fig. 8:1 (b)) then proportionate size increase must be accompanied by a reduction in constant b. Gould (1971) devised a formula to calculate the amount of change in b necessary to maintain shape.

By using this method, the analysis of situations where size changes, but α is invariant, can now be taken a further stage and combined with estimates of shape differences, Wood (1975a). If it is determined whether or not b is adjusted to maintain geometric similarity in a new size range it may provide evidence about whether the maintenance of a constant shape is taxonomically significant or not.

D. *Estimation of shape differences*

An aspect closely related to allometry which has seen relatively recent development is the estimation of shape differences. Unless the extent of shape differences can be quantified in some way the contribution, if any, of allometric effect cannot easily be assessed. There are two main groups of method, transformation grids and linear functions.

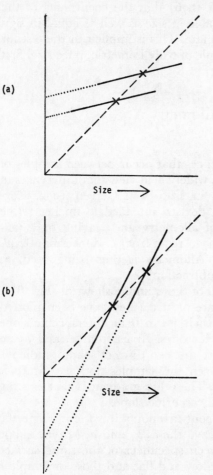

D'Arcy Thompson (1917) pioneered the use of transformation grids but did not devise any means of measuring the overall deformation that occurred in the grids. Medawar (1945) and Richards & Kavanagh (1945) amplified Thompson's concept but a comprehensive numerical estimate of shape eluded them. The nearest to fulfilling Medawar's criteria of an analytically based numerical method has been the trend-surface analysis of transformation grids presented by Sneath (1967). This has not received the attention it deserves because of the complexities of

FIG. 8:1.

Graphs to show relationship between shape, slope and the value of 'b', the y intercept. (after Gould (1971)).

The dashed line joins specimens of the same shape.

trend-surface analysis. These have largely been resolved and Sneath's work is worthy of re-examination and adaptation to complement allometric studies.

Two of the most widely used linear functions to estimate shape differences are Penrose's size and shape distance and principal component analysis. Penrose's statistic of mean square distance, C_H^2, takes no account of the intercorrelations between variables, but nevertheless it can be usefully analysed into size and shape components. Details can be found in Penrose (1954).

Principal components are simply series of uncorrelated axes through points scattered in a space that has as many dimensions as there are measurements. Each axis represents progressively less variance and the original variables have a co-efficient or correlation with each component. Jolicoeur & Mosimann (1960) suggested that if the coefficients that related to the first (usually the dominant component) principal component were all the same sign then it represented 'size'. Subsequent components, usually of mixed sign, were then designated 'shape' components. However, it is apparent (Wood 1976) that the coefficients of the first principal component need to be comparable in size as well as equal in sign before it can be accepted as a pure 'size' estimate. This is implicit in the test for isometry used by Jolicoeur (1963b). An example of truly isometric, pure size, first principal component is given by Gould (1975).

Applications of allometry

A. *Correcting for allometric effect*

The identification and analysis of differences that occur between samples of fossils is at the heart of hominid studies. It is a widely held view that differences in shape are more significant than differences in size (Van Gerven 1972, Corruccini 1973). Methods have therefore been devised intended to further refine shape differences by removing the effects of allometry and leaving only tax-onomically valent differences, Corrucini (1972, 1973, 1975), Zuckerman et al. (1973) and McHenry & Corrucini (1975). Allometry is thus being used, as Gould (1975) aptly terms it, as 'criteria for subtraction'.

This can best be explained with the help of the example shown in Fig. 8:2. Two hypothetical species samples, a and b, are plotted (they have been plotted with equal slopes for convenience). Joining their mean is an interspecific slope which, as is usual, differs from the intraspecific ones. Apart from a_1 and b_3 no other specimens are the same shape but there are two distinct shape gradients. One is represented by the differences between the samples and its nature is dictated by the slope of the interspecific line. The other gradient is that existing within each sample and is governed by the intraspecific slope.

If *a priori* judgements have been made about taxonomy it is the intraspecific shape differences that need eliminating to leave the a_2/b_1 and a_1/b_2 type shape differences as taxonomic discriminants between specimens of the same surface area. At each size specimens of sample a are low and flat and those of sample b tall and thin. In order to remove intraspecific shape differences allometric co-efficients for each character need to be calculated and then used as exponents to correct each measurement. Only if the allometry coefficients are the same for all samples can a general correction be applied. Corrections based on a regression of the sample means (i.e. an interspecific slope) would achieve the opposite effect and remove shape differences corresponding to the alleged taxonomic classification.

This scheme has been adopted by Zuckerman *et al.* (1973) except that they corrected the shape of each specimen to a standard size. Corruccini (1972, 1975), however, uses different methods to calculate regression coefficients for corrections. In his earlier paper he apparently uses the mean of all the regression coefficients and in the later paper the regression coefficient of only one genus is used to correct all the other samples. This is proper if the intraspecific slopes are equal for each sample but there are many examples where this is not so (Wood 1975a), in which case any correction made may well be inappropriate. Corruccini (1972) used the subjective criterion of taxonomic relevance to judge the efficacy of his correction technique. This is a notoriously fickle yardstick when fossils are concerned and it may well be wiser to adopt another test to judge the efficacy of any developments of the method.

A full discussion of the statement that shape variables are superior to size ones is inappropriate. It depends of course whether shape can be satisfactorily divided into its intraspecific and interspecific components. It must also be recognized that many biological relationships are allometric, and as Giles (1956) stated, 'in such a situation size and shape are correlated and a change in one should be as relevant as a change in the other'.

B. *Allometry as a taxonomic indicator*

Huxley (1932) speculated that genetic difference between samples would be characterized by different intraspecific slopes. Hersh (1934) propounded the idea and using Osborn's data on Titanotheres showed that intraspecific values for α and b were distinct between genera and were equal in congeneric species. Allo-

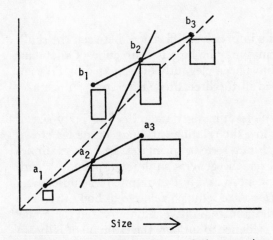

FIG. 8:2.

Allometric and taxonomic shape differences.

Size ⟶

The dashed line joins specimens of the same shape

metry coefficients have been considered for taxonomic assessment in primates by Lumer (1939), Lumer & Schultz (1941) Freedman (1957, 1963), Frick (1959), Bibus (1967) & Hemmer (1964, 1967a and 1967b) with varying degrees of enthusiasm and success. Gould (1971, 1975) has reviewed the considerable literature dealing with brain/body weight data. The application of allometry to hominid taxonomy is limited to Kinzey (1972) and Pilbeam & Gould (1974).

The simple Huxley–Hersh hypothesis is clearly not a general rule, for example the brain/body weight intraspecific slopes are the same for dogs, rats and monkeys. There has also been little examination of significant causes of intra-group variation, such as sexual dimorphism, to see whether they could be excluded as causes of differences in α. There were indications in Lumer (1939) and Kurten (1964) that sexual differences in allometry coefficients may occur. This hypothesis was examined in detail by Wood (1975a, 1976) who found that in primates there were no significant sex differences in α but there were sex differences in the values of *b*.

Gould (1975) claims that interspecific slopes say nothing about evolutionary mechanisms. It is true that they cannot be depended on for phylogenetic information but as is demonstrated in the next section displacement of taxa from the general interspecific slope may be indicative of a profound adaptive shift. The progress of such a displacement can be traced if enough of a lineage is preserved and a 'phylogenetic' slope can be calculated. However, lineage plots and interspecific plots of extant or fossil animal groups are not comparable. Pilbeam & Gould (1974) plot slopes for a putative *Homo* lineage and compare it with interspecific slopes of extant apes and of fossil hominids. Unless it can be shown that the subhuman primate lineage or a hominid lineage other than that leading to man did not follow a similar process of displacement disparities between lineage and interspecific coefficients have little meaning. Likewise, the fact that 'gracile' and 'robust' australopithecines are capable of being linked in a straight line plot, Pilbeam & Gould (1974), is not particularly relevant to the problem of whether they are subspecifically, specifically or generically distinct.

C. *Allometry and functional interpretation*

It is now being recognized that the impressive differences between the skulls of 'robust' and 'gracile' australopithecines may be largely a function of allometry and scaling. Some of these differences have been elegantly investigated by Pilbeam & Gould (1974) and are being debated in the literature (Kay 1975, Pilbeam & Gould 1975).

The importance of relative size plots utilizing data reflecting body size is illustrated by Figs. 8:3 and 4 which show the relationship between molar crown area, canine base area and femur length in a series of primates. In both examples the *Homo* sample is clearly displaced from the general subhuman primate trend. The molar crown area in *Homo* is equivalent to that in *Pan* despite large differences in body size, and the canine teeth in *Homo* are reduced relative to body size. The inter and intraspecific slope values both indicate positive allometry and the molar crown value, α = 2.36, is independent evidence to support the estimate of Pilbeam & Gould (1974).

Relative size plots can also be used to investigate more specific problems such as the relationship between mandibular morphology and tooth size. In *Gorilla*, *Papio* and *Colobus* there is considerable sexual dimorphism in the cross-sectional area of the horizontal ramus of the mandible yet there is little dimorphism in the molar teeth that are embedded in the ramus. How much mandibular dimorphism is determined by molar crown area as opposed to the size of the anterior teeth and can the two effects be separated?

Relative size plots used molar crown area, Fig. 8:5 (a) and (b), and canine base area, Fig. 8:6 (a) and (b) as the reference variable. The interspecific slopes for molar crown area are isometric but the canine plots are negatively allometric, particularly the plot relating it to the horizontal ramus (Fig. 8:6 (a)). Thus

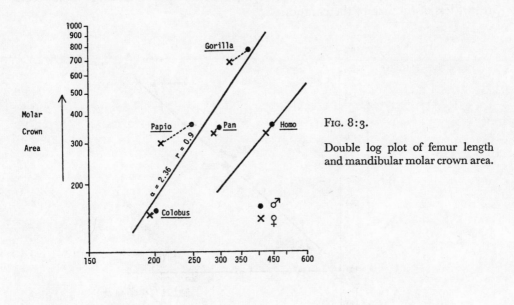

FIG. 8:3.

Double log plot of femur length and mandibular molar crown area.

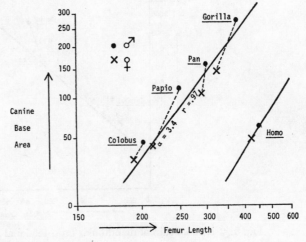

FIG. 8:4.

Double log plot of femur length and mandibular canine base area.

animals with large canines have relatively slender rami but a proportional symphysis. Inspection of the molar crown area intraspecific slopes shows much steeper slopes for forms with extremely dimorphic canines than for *Pan* and *Homo*. Thus though generally molar crown area and mandibular thickness increase *pari passu* disproportionately large canines impose mechanical strains on the mandible which overide this relationship. The relationship between molar size and the mandible can be tested for independence from canine size by including data from two nearly complete 'robust' australopithecine mandibles taken from Wood (1975b). They follow the general interspecific molar size/mandible slope and their massiveness is simply a function of their larger posterior teeth. However, the 'robust' mandibles are displaced well away from the canine/mandible inter-specific slope in Fig. 8:6. Thus a thick symphysis is required for large molar teeth regardless of the size of the canines.

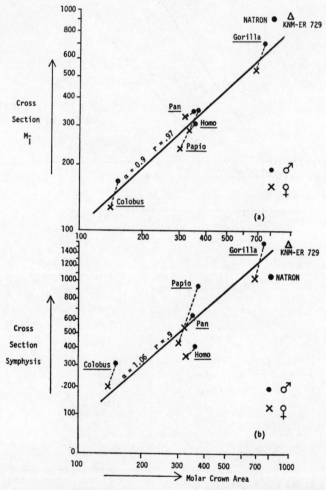

FIG. 8:5. Relationship of mandibular cross-sectional area to molar crown area.

Conclusions

Allometry has been demonstrated as a powerful tool for the investigation of the functional anatomy of early hominids, but its role in taxonomic studies has been shown to be much less well defined.

Now that 'difference' is under much more rigorous examination than it was even a decade ago, it is more important than ever to test analytical techniques in situations where there is a good deal of data available. Only after techniques have been refined, and the less useful and confusing discarded, should the more difficult problem of analysing differences in fossil hominid samples be attempted. Fossil material, by virtue of its scarcity and incompleteness, introduces its own special

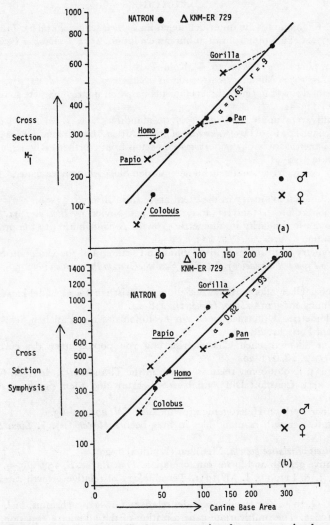

FIG. 8:6. Relationship of mandibular cross-sectional area to canine base area.

difficulties into what are, even under the best circumstances, a series of most complex analytical problems.

Acknowledgement

Research which contributed to this review has been generously supported by the Central Research Fund of the University of London, the Boise Fund of Oxford University, the Royal Society and the Natural Environment Research Council. I thank the Trustees of the National Museums of Kenya for access to the fossil material cited in the paper.

References

BIBUS, E. 1967. Kraniometrische untersuchungen an *Colobus* schäldeln. *Folia. Primatol.* **6**, 92–130.

BRACE, C. L. 1972. Sexual dimorphism in human evolution. *1972 Yearbook of Physical Anthropology*, **16**, 31–49.

CHAMPY, C. 1924. *Sexualite et Hormones*. G. Doin, Paris.

CORRUCCINI, R. S. 1972. Allometry correction in taximetrics. *Syst. Zool.* **21**, 375–83.

—— 1973. Size and shape in similarity coefficients based on metric characters. 1973. *Am. J. phys. Anthrop.* **38**, 743–54.

—— 1975. Multivariate analysis of *Gigantopithecus* mandibles. *Am. J. phys. Anthrop.* **42**, 167–70.

FREEDMAN, L. 1957. The fossil *Cercopithecoidea* of South Africa. *Ann. Transv. Mus.* **23**, Pt. 2, 122–259.

—— 1963. A biometric study of *Papio cynocephalus* skulls from Northern Rhodesia and Nyasaland. *J. Mammal.* **44**, 24–43.

FRICK, H. 1959. Allometrische untersuchungen an den Schädeln von Pavianen. *Anthrop. Anz.* **23**, 64–71.

GILES, E. 1956. Cranial allometry in the great apes. *Hum. Biol.* **28**, 43–58.

GOULD, S. J. 1966. Allometry and size in ontogeny and phylogeny *Biol. Rev.* **41**, 587–640.

—— 1971. Geometric similarity in allometric growth: A contribution to the problem of scaling in the evolution of size. *Am. Nat.* **105**, 113–36.

—— 1975. Allometry in primates, with emphasis on scaling and the evolution of the brain. *In: Approaches to Primate Paleobiology*. Ed. F. S. Szalay. *Cont. to Primatology*: No. **5** S. Karger (Basel), 244–92.

HEMMER, H. 1964. Über allometrische Beziehungen zwischen Hirnschädel-kapazitat und Hirnschädelwölbung im genus *Homo*. *Homo*, **15**, 218–24.

—— 1967a. Allometrie—Untersuchungen zur evolution des menschlichen Schädels und seiner Rassentypen. *Fortschr. Evol.* **3**.

—— 1967b. Zur allometrischen Charakterisierung von populationen des europäischen Neolithikums. *Homo.* **18**, 211–20.

HERSH, A. H. 1934. Evolutionary relative growth in the Titanotheres. *Am. Nat.* **68**, 537–61.

HUXLEY, J. S. 1924. Constant Differential growth ratios and their significance. *Nature, Lond.* **114**, 895–6.

—— 1927. Further work on Heterogonic growth. *Biol. Zbl.* **47**, 151–63.

—— 1931. Relative growth of mandibles in Stag-beetles (*Lucanidae*). *J. Linn. Soc. (Zool).* **37**, 675–703.

—— 1932. *Problems of relative growth*. Methuen (London) 1–276.

—— 1950. Relative growth and form transformation. *Proc. R. Soc. B*, **137**, 465–9.

—— NEEDHAM, J. & LERNER, I. M. 1941. Terminology of relative growth rates. *Nature, Lond.* **148**, 225.

—— & TEISSIER, G. 1936. Zur terminologie des relativen Grössenwachstums. *Biol. Zbl.* **56**, 381–3.

JOLICOEUR, P. 1963a. The multivariate generalization of the allometry equation. *Biometrics*, **19**, 497–9.

—— 1963b. The degree of generality of robustness in *Martes americana. Growth,* **27**, 1–27.

—— & MOSIMANN, J. E. 1960. Size and shape variation in the painted turtle. A principal component analysis. *Growth,* **24**, 339–54.

KAVANAGH, A. J. & RICHARDS, O. W. 1942. Mathematical analysis of the relative growth of organisms. *Proc. Rochester Acad. Sci.* **8** (4), 150–74.

KAY, R. F. 1975. Allometry and early hominids. *Science,* **189**, 63.

KINZEY, W. G. 1972. Allometric relationships of brain/body size relationships in hominid evolution. *Am. J. phys. Anthrop.* **37**, 442–3.

KLATT, B. 1919. Zur Methodik vergleichender metrischer Untersuchungen, besonders des Herzgewichtes. *Biol. Zb.* **39**, 406–20.

KURTEN, B. 1964. The evolution of the polar bear, *Ursus maritimus* Phipps, *Acta zool. fenn.* **108**, 1–26.

LUMER, H. 1939. Relative growth of the limb bones in the anthropoid apes. *Hum. Biol.* **11**, 379–92.

—— & SCHULTZ, A. H. 1941. Relative growth of the limb segments and tail in macaques. *Hum. Biol.* **13**, 283–305.

McHENRY, H. M. & CORRUCCINI, R. S. 1975. Distal humerus in hominoid evolution. *Folia primatol.* **23**, 227–44.

MATSUDA, R. & ROHLF, F. J. 1961. Studies of the relative growth in *Gerridae* (5): comparison of two populations, (Heteroptera: Insecta). *Growth,* **25**, 211–17.

MEDAWAR, P. B. 1945. Size, Shape and Age. *In: Essays on Growth and Form.* Ed. W. E. Le Gros Clark and P. B. Medawar, Oxford Univ. Press (Oxford and London), 157–87.

OXNARD, C. E. 1973. *Form and pattern in human evolution. Some mathematical, physical and engineering approaches.* Univ. of Chicago Press (Chicago) 1–218.

PENROSE, L. S. 1954. Distance, size and shape. *Ann. Eugen.* **18**, 337–43.

PEZARD, A. 1918. Le conditionnement physiologique des caracteres sexuels secondaires chez les oiseaux. *Bull. biol. Fr. Belg.* **52** 1–176.

PILBEAM, D. & GOULD, S. J. 1974. Size and scaling in human evolution. *Science,* **186**, 892–901.

—— & GOULD, S. J. 1975. Allometry and early hominids—a reply. *Science,* **189**, 64.

REEVE, E. C. R. & HUXLEY, J. S. 1945. Some problems in the study of allometric growth. *In: Essays on growth and form.* Ed. W. E. Le Gros Clark and P. B. Medawar, Oxford Univ. Press (Oxford and London), 121–56.

REYMENT, R. A. 1969. Some case studies of the statistical analysis of sexual dimorphism. *Bull. geol. Instn. Univ. Upsala N.S.* **1** (4), 97–119.

RICHARDS, O. W. & KAVANAGH, A. J. 1945. The analysis of growing form. *In: Essays on growth and form.* Ed. W. E. Le Gros Clark and P. B. Medawar, Oxford Univ. Press. (Oxford and London) 188–230.

SNEATH, P. H. A. 1967. Trend surface analysis of transformation grids. *J. Zool.* **151**, 65–122.

SNELL, O. 1891. Das Gewicht des Gehirnes und des Himmantels der Säugetiere in Beziehung zu deren geistigen Fähigkeiten. *Sber. Ges. Morph. Physiol. Münch.* **7**, 90–4.

TEISSIER, G. 1955. Allométrie de taille et variabilité chez *Maia squinado. Arch. Zool. exp. gen.* **92**, 221–64.

—— 1960. Relative growth. *In: The Physiology of the Crustacea.* **1**, Ed. T. H. Waterman, Academic Press (New York and London) 537–60.

THOMPSON, W. D'ARCY 1917. *On growth and form.* Cambridge Univ. Press (Cambridge and London) 1–793.

VAN GERVEN, D. P. 1972. The contribution of size and shape variation to patterns of sexual dimorphism of the human femur. *Am. J. phys. Anthrop.* **37**, 49–60.

WOLPOFF, M. 1973. Posterior tooth size, body size, and diet in South African gracile Australopithecines. *Am. J. phys. Anthrop.* **39** (3), 375–94.

WOOD, B. A. 1975a. An analysis of sexual dimorphism in primates. Ph.D. Thesis, Univ. of London 1–296.

—— 1975b. Remains attributable to *Homo* in the East Rudolf succession. *In: Earliest man and environments in The Lake Rudolf basin: Stratigraphy, paleoecology and evolution.* Eds. Y, Coppens, F. C. Howell, G. L. Isaac and R. E. Leakey. Univ. of Chicago Press, (Chicago)

—— 1976. The nature and basis of sexual dimorphism in the primate skeleton. *J. Zool. Soc. (Lond.)* **180**: 15–34.

WHITE, J. F. & GOULD, S. J. 1965. Interpretation of the coefficient in the allometric equation. *Am. Nat.* **99**, 5–18.

ZUCKERMAN, S. 1926. Growth in the skull of the baboon, *Papio porcarius. Proc. zool. Soc. Lond.* 843–873.

—— ASHTON, E. H., FLINN, R. M., OXNARD, C. E. & SPENCE, T. F. 1973. Some locomotor features of the pelvic girdle in primates. *Symp. zool. Soc. Lond.* (33) 71–165.

9

GLYNN LL. ISAAC

The first geologists—the archaeology of the original rock breakers

What good is a broken stone—then or now? How and when did this venerable practice begin? Did the Stone Age start with a big bang or a series of little taps? Aspects of East Africa's long record of stone tools are considered from a geological and an evolutionary perspective.

Introduction

What has research in the Gregory Rift Valley taught us concerning the long-term relationships between men and rocks?

This question has to be approached with great caution because the archaeological record establishes that geologists are a very primitive group. There are important traits that they share with ape-men; traits that are not part of normal modern human behaviour.

I use the word 'primitive' deliberately and in the technical biological sense which refers to the retention of an ancestral condition that has been superseded in the evolution of other related organisms. As geologists walk about the landscape with hammer in hand they frequently, with obvious gratification, hammer at outcrops and smash rocks and other stones. They collect a small proportion of the debitage and carry this off to accumulate in central depots of various kinds.

More than a hundred years of archaeological research has demonstrated that this is extraordinarily similar to the behaviour of early man. We can follow back through the Pleistocene a trail of smashed stones, and can find central depots where broken stone has been accumulated. The faint beginnings of this trail are best known from the Gregory Rift Valley and I propose to discuss archaeological research aimed at deciphering the long if patchy record so well preserved there.

Two questions are fundamental:

How, when, and why did ancestral human involvement with stone-breaking begin?

What can the study of the trail of broken stone contribute to our overall understanding of human evolution?

Tool use in animals

There are, however, other questions to be examined as a preliminary to the main issue:

Are geologists and fossil men unique? Can living apes do it too?

Various animals are involved in using stones as tools. Two of the most famous examples are the California sea otter which lies on its back in the waves with a pebble on its chest and smashes shellfish against the stone, and Egyptian vulture which picks up stones to break ostrich eggs (Goodall 1970). What of man's closest relatives, the apes? From studies carried out in the wild, we now know that apes use tools to a small extent but in a highly significant way, for example, the famous termite 'fishing' discovered by Jane Goodall (1968). Clearly if the last common ancestor of men and chimps had behavioural proclivities of this kind, evolutionary selection pressure could have taken them and intensified them until tool-making became a crucial and habitual part of adaptation.

However, non-human animals do not break stones in order to use the sharp edges that result. The question remains: could the apes do it? This problem was taken up by a prehistorian, Richard Wright, who set up an experiment at Bristol Zoo to test whether apes could be taught to make a stone tool (Wright 1972). There were difficulties with the chimpanzees and Wright experimented with an orang-utan. He put an edible prize in a box, the lid of which could only be opened by severing a lashing. He provided a suitable block of flint, demonstrated the striking of a sharp-edged flake and its use for opening the box. The orang was then left with the materials and the problem. Eventually it struck flakes and used them to open the box.

This neat experiment tends to confirm that apes are perfectly capable of making stone implements. However, in practice they do not do so, except under artificial situations that have been 'rigged' by human brains. In conversation, Professor Sherwood Washburn, of the University of California, Berkeley, has pointed out the parallel between the orang experiment and recent investigations into the linguistic capabilities of chimps. These also suggest that when guided by a human brain, apes can acquire signalling and communication abilities that far exceed their own natural levels.

Wright's experiment helps to show that the evolution of early man's stone-flaking activities involved subtle interplay of changes in a number of different components of normal anthropoid systems, most notably the hand-brain system.

Tools as an adaptive threshold

Why did some ancestral ape-like creature start to bother with these things? The answer probably lies in the fact that there are major classes of high reward foods that an ape cannot obtain without equipment.

During field work east of Lake Turkana a graphic demonstration was seen of the difference that even the most minimal stone tools can make. A topi carcass was abandoned by lions. A local Shangilla boy finding the kill but lacking a knife banged off a minute stone flake and skinned a leg of the topi with it. The 'tool' looked trivial and would have been termed 'waste' if found in an archaeological context and yet it made a critical difference in the quest for food. This is simply one illustration of the functional potential of equipment. If one stops to think of the usefulness of even such simple items as a knife, a club, a bag and a length of cord, then the general adaptive importance of tools is apparent. (cf. Isaac in press).

How and when did it all begin? Archaeology has had its own version of the cosmic debate. There are two rival models which one can caricature as the 'big bump' hypothesis versus the 'succession of little knocks' hypothesis.

For many years it was usual to think of hominid involvement in stone tool-making as having originated through a long gradual succession of stages. One version postulates a first stage in which stone tools are made by the removal of a few flakes only in one direction. The second stage was reached when eventually the hominids realised that in addition it was possible to knock flakes off in another direction. Later there followed the realisation that multidirectional flaking was possible.

This system, which the above description parodies rather unfairly, has the attraction of suggesting an orderly, logical progression. However, archaeological science received a sharp jolt when Dr Mary Leakey excavated and examined the first substantial samples of very early stone tools from a stratified context (M. D. Leakey 1966). She quickly noted that these tools showed clearly that the range of forms was far more diverse than the conventional model might have predicted. Thus another possibility needed to be considered. This is that the beginnings of stone tool involvement might have been a threshold phenomenon. Stone-breaking caused by percussion that is so controlled as to produce conchoidal fracture yields two series of objects. These are relatively large lumps with jagged edges, which one can term generically core-tools, together with smaller flakes and splinters of stone with sharp edges and pointed angles. Perhaps the discovery of the empirical facts of conchoidal fracture led immediately to the making of a fairly wide, opportunistic range of sharp and jagged edged core-tools and flakes?

It still remains to be shown which of these models comes closest to being correct or whether the true origin of tool-making lies somewhere between these two extremes.

Evidence for the early stages

To turn to the evidence from the Gregory Rift: there are several sites yielding relevant evidence for an age group of about two or more million years. Of these, Bed I at Olduvai Gorge, which has been intensively studied and published by Dr Mary Leakey (1971) has yielded the largest volume of definite evidence from strata dated between 1.8 and 1.6 million years.

More recently other sites have been discovered but these have been less extensively explored than Olduvai. At the Omo localities, artifacts recovered from Member F of the Shungura Formation, have an age of about 2.0 million years (Merrick *et al.* 1973; Chavaillon 1976). In the top part of the Lower Member at East Lake Turkana several artifact-bearing sites have been discovered that certainly belong with these other early occurrences (Fitch & Miller this volume; Curtis this volume).

In addition there are different sets of artifacts just slightly younger at about one and a half million years. At Olduvai Gorge these are the Developed Oldowan and the Acheulian of Bed II; at East Lake Turkana the 'Karari Industry'. (Harris & Herbich this volume) The early set of Chesowanja artifacts surely belong here though they have not yet been dated firmly (Bishop *et al.* 1975 and this volume). The lowest levels at Melka Kunture may also prove to be of similar antiquity.

In addition there are sites such as Kanam and Kanyatsi, discovered some years ago, that would now repay further investigations. (cf. S. Cole 1963).

I will restrict my attention to the early man-stone relationships evident at Olduvai (Tanzania), Omo (Ethiopia) and east of Lake Turkana (Kenya).

At Olduvai Gorge in Bed I, the work of Mary Leakey (1971) has shown that artifacts of fairly diversified morphology were being made. She and R. L. Hay have shown that the hominids responsible were perceptive geologists who selected with care different kinds of rock to break for different purposes. Several series of forms can be recognised: (For good general accounts of stone artefacts and the terminology used to describe them see Bordes (1968), Howell (1965) or Oakley (1956)).

> *Core-tools* mainly for 'heavy duty' functions (e.g. choppers, discoids, proto-
> bifaces, polyhedrons).
> *Small tools* mainly for 'light duty' functions (e.g. scrapers, awls, burins,
> chisels, etc.).
> *Flakes and splinters*—some of these show signs of use, but many more which
> may have been used as knives, do not show such traces.
> Plus, battered hammerstones, anvils and hominid transported but unmodified
> stones (manuports).

The choppers, etc., were made usually but not exclusively of lava, apparently carefully selected for the purpose. Lava produces a tough edge and many specimens show signs of battering. At most sites flakes, flake-based scrapers and other tools, were made preferentially of quartz. Dr Richard Hay and Dr Mary Leakey have worked out an intriguing pattern of changing raw material preferences through time at Olduvai that involves, for the later Oldowan tools, distances of transportation by man of over 10 miles.

In member F of the Shungura Formation, Dr Harry Merrick (Merrick *et al.* 1973) and Dr Jean Chavaillon (1976) have found traces of a different system. Two million years ago at the Omo, the early hominids seem mainly to have been smashing small quartz pebbles to generate a myriad of little jagged flakes and splinters. This is not very attractive material for archaeologists with a bent towards art history, but in functional-adaptive terms this sharp-edged debris may have been perfect for the task on hand. However, there does not seem to be much point in classifying most of this material according to an elaborate typological system.

The areas of the Shungura Formation that are exposed, do not contain pebbles larger than pigeon eggs and these are mainly of quartz. Thus we may be looking at the oldest evidence for hominid frustration. This perhaps helps to explain the vigour with which these unsatisfactory materials have been smashed!

To the east of Lake Turkana we have recovered artifacts from the KBS Tuff that are predominantly made of lava. These more nearly resemble those of Olduvai Bed I than those of Shungura F. Within the assemblages there are two clear sets: *Core-tools* that include choppers, discoids and polyhedrons; and *flakes and splinters*. (Isaac 1976 (b)).

However, there are some subtle differences between the KBS assemblages and the Oldowan samples from Bed I as follows:

the edges of the core-tools are less battered,

we do not as yet have definite 'scrapers' from the sites of KBS Tuff age.

the modal size of all classes seems to be smaller,

the maximum observed density of artifacts on the sites is much lower than at Olduvai. The material is much more sparsely distributed.

The KBS industry has been recovered mainly from the top of the Lower Member in Area 105. The beds which are of delta flood plain facies have been traced and studied over distances of several kilometres around the sites and it appears that there are no stone particles in them larger than about the size of a pea. Thus, the early hominids were carrying stones into the sites. They used mainly pebbles and small cobbles of a partly glassy basaltic lava. Some glassy, green welded tuff and a few splinters of quartz and chert have also been found. Presumably the materials were obtained from stream conglomerates in the vicinity of the basin margin.

Are any generalisations possible from these data?

1. Clearly by the Plio-Pleistocene (2.5 to 1.5 million years ago) the acquisition of sharp-edged tools was of considerable importance for at least some hominid forms. These early artifact assemblages could be regarded as the outcome of deft but opportunistic percussion applied to local stones yielding the variable outcome that one might expect. This is a partly subjective view and not the only way of interpreting the data.

2. Stone might be transported up to several miles to be used as tool material. However, the Omo case makes it look as though some early hominids tried to make do with what seems unsatisfactory local material in preference to going further afield. Perhaps some early hominids managed almost entirely without stone when it was not readily available. Certainly some of the areas east of Lake Turkana that are richest in hominid fossils have not yet yielded many artifacts.

3. Is any developmental sequence evident amongst this set of traces that seem to span about a million years commencing about two and a half million years ago? It is probably premature to ask this question but one can offer some comments. The distinctive features of the Omo material seem best regarded as due to the form of the only available stone rather than as being representative of a developmental stage. The apparent addition by Olduvai Bed I times of flake-based 'scraper' forms to the KBS type of repertoire may not turn out to be a real development.

What do we know about the adaptive significance of these early man-rock relationships? At both Olduvai and east of Lake Turkana we have evidence that seems to suggest that the early tool-making hominids were involved in behaviour patterns that constituted a departure for primates. These formed the evolutionary basis on which the subsequent structure of human evolution was built. Stone tools were part of a novel adaptive complex that involved the following components: life in relatively open habitats; bipedal locomotion; eating meat from the carcasses of animals as large or larger than the hominids themselves; and the carrying of at least some of the meat back to a central base or 'camp' Very probably the pattern also involved meat-sharing in some degree and perhaps the gathering and sharing of plant foods.

We know mainly about the stone artifacts, many of which would have made handy butchery knives and which also include items suitable for hacking and whittling branches and staves. In spite of their absence from the record owing to their perishable nature, digging sticks, spears and clubs made of wood, also at some stage, bags, pouches and baskets would be simple but crucial items of equipment. The earliest tools that survive are to be understood as much as tools for making tools as direct agents of adaptation.

Stone tools are not only fascinating objects in their own right, they are also invaluable indicators of foci of proto-human presence and activity.

Where does this put us with regard to the alternative models that I contrasted with each other? The question remains open. As we attempt to trace the record back beyond two and a half million years we may find successively simpler assemblages of fractured stone. Or, if the 'big bump' mode is more nearly true, at a certain time-horizon, which corresponds to the 'threshold', we may find almost the full Oldowan kind of assemblage appearing suddenly, while in earlier layers there is virtually nothing. Only time and persistent exploration can settle this question.

The next stage

It is only possible to discuss very briefly the further development of man-stone relationships, as the later record in the Rift Valley and elsewhere becomes more and more complex.

The second stage in stone tool-making, which can be taken as beginning about 1.5 million years ago, involves a number of changes:
  the addition of strikingly different new tool forms, notably handaxes and
    cleavers. These, however, are not present at all sites;
  a pronounced increase in the size range of artifacts so that the largest speci-
    mens come to weigh 2 or 3 kg or more;
  an increased diversity of shaped tool forms;
  an increased tendency for local, individual archaeological samples to differ
    from other samples.

The Gregory Rift has the world's best-documented and dated examples of this second stage as well as having virtually all those of previous series. Olduvai and the area east of Lake Turkana have yielded assemblages showing these new features stratified above the early Oldowan assemblages. Other occurrences that belong in this time-range are the Peninj Acheulian sites near Lake Natron that seem to date between 1.3 and 1.6 million years ago (Isaac 1967; Isaac & Curtis 1974), and perhaps Chesowanja east of Lake Baringo in Kenya (Bishop *et al.* 1975) and the base of the Melka Kunture sequence in Ethiopia. (Chavaillon 1973).

Those assemblages from this time range that are dominated by handaxes and cleavers are termed *Acheulian*. The term, derived from the town of St Acheul in France, was imported to Africa but it has recently been established that the earliest

Acheulian in East Africa may be 3 times as old as the oldest examples in Europe (cf. contributions to Butzer & Isaac 1975). The Acheulian form of assemblage may have been ushered in by a quantum jump: the discovery of techniques for knocking off big flakes suitable as blanks for handaxes and cleavers. (Isaac 1969).

There is a lively discussion going on amongst archaeologists about how best to explain the increased differentiation between groups of artifact assemblages. However, of more evolutionary significance than the details of artifact form is the appearance of spectacular concentrations of artifacts. These involve the accumulation 'by hand' of more than a ton of stone at some localities. Natural processes may have exaggerated these concentrations at some sites. Whatever the detailed socio-economic meaning of such striking masses of humanly fractured stone, it seems clear that by this time one of the most basic human contraptions, the bag or basket, must have been in use!

We also have clear-cut cases of extremely proficient hunting. For example the broken up remains of more than 50 *Simopithecus* baboons have been found at one of the sites at Olorgesailie in Kenya (Isaac this volume). There are also good examples from Olduvai Gorge.

During the vast span of time in which Acheulian stone tool kits were being made, there was a distinct rise in the maximum level of refinement of these tools. Yet, sets of more or less crude artifacts also continued to be made.

It is important for geologists to note that artifact *assemblages cannot be dated from their morphology*.

The increase in maximum finesse surely indicates increasing mental and manual ability on the part of the evolving hominids, but it is the subsequent set of stone industries that really shows what had been accomplished by evolution in the long time reaches of the Middle Pleistocene.

The end of the Stone-Age

The face of Africa, including the Gregory Rift is littered with discarded stone tools belonging to the last 100 000 years or so, which is less than 5 per cent of archaeological time. To those of us who work primarily with early materials, these latter-day relics seem rather fussy, hide-bound and unadventurous. There is characteristically a marked reduction in the size of the most prominent tool forms. Each industry tends to be segmented into more distinct, standardised, replicating types. Clearly, the hafting of pointed stones onto spears, javelins and other implements had been invented and had become an established practice. (See Clark 1970 for a more complete account).

Many of us would guess that the new fussy, regionally differentiated, rule-bound stone artifacts owe their distinctness from earlier material to crucial developments in aspects of human behaviour that have nothing to do with stone tools. It seems possible that they reflect dramatic increases in the information capacity of human communication systems, notably language. If this is so, then the vast and archaeologically uneventful Middle Pleistocene period was the crucial formative period that saw the transformation of a non-human early hominid adaptive pattern to a fully human one.

Little, pernickety artifacts of the Later Stone Age form particularly conspicuous litter. There are thousands, indeed millions, of tiny, spikey geometric forms lying about. Many archaeologists think these are arrow armitures and that their manufacture in bulk was associated with the spread of knowledge of the bow. They are now known to go back at least 20 thousand years and perhaps more. They ceased to be made only with the spread of iron technology two or three thousand years ago.

The overwhelming interest in the earliest cultural materials in East Africa has been such that more thorough and intensive interdisciplinary studies have been made on the early material than on the more abundant set of traces from the Upper Pleistocene and Holocene. It is now clear that in order to understand fully what happened in human evolution, we also need to study these later sites in such a way that they can be closely compared with earlier ones and with ethnographic records. Young research workers are tackling this with imagination and vigour.

Ironically, although archaeologists can give an exhaustive account of the fine details of change in artifact morphology they know distressingly little about the real functions of particular tools. What was the function of a chopper or of a handaxe? We have to confess that we don't yet know.

References

BISHOP, W. W., PICKFORD, M. & HILL, A. 1975. New evidence regarding the Quaternary geology, archaeology and hominids of Chesowanja, Kenya. *Nature, Lond.* **258**, 204–8.

BORDES, F. 1968. *The Old Stone Age.* London, Thames and Hudson.

BUTZER, K. W. & ISAAC G. L. (Eds). 1975. *After the Australopithecines: Stratigraphy, ecology and cultural development during the Middle Pleistocene.* Mouton; the Hague.

CHAVAILLON, J. 1973. Chronologie des niveaux paléolithiques de Melka Kontouré (Ethiopie). *C. R. Acad. Sci.* **276**, 1533–6.

—— 1976. Evidence for the technical practices of early Pleistocene hominids: Shungura Formation, Lower Valley of the Omo Ethiopia. In: *Earliest Man and Environments in the Lake Rudolf Basin: Stratigraphy, ecology and evolution*, Y. Coppens, F. C. Howell, G. L. Isaac and R. E. F. Leakey. (Eds), University of Chicago Press, Chicago.

CLARK, J. D. 1970. *The Prehistory of Africa.* Thames and Hudson, London.

COLE, S. 1963. *The Prehistory of East Africa.* New York, Macmillan.

GOODALL, J. van Lawick, 1968. The Behaviour of Free-living Chimpanzees in the Gombe Stream Area. *Animal Behaviour Monographs* **1** (3), 161–311.

GOODALL, J. van Lawick 1970. Tool using in primates and other vertebrates. In: *Advances in the Study of Behaviour* edited by D. S. Lehrman, R. A. Hinde and E. Shaw. **3**, 195–299 New York, Academic Press.

HOWELL, F. C. 1965. *Early Man*, New York, Time-Life Inc.

ISAAC, G. LL. 1967. *The stratigraphy of the Peninj Group—early Middle Pleistocene formation west of Lake Natron, Tanzania*, In: *Background to Evolution in Africa*, W. W. Bishop and J. D. Clark (Eds) pp. 229–57. Chicago: University of Chicago Press.

—— 1969. Studies of Early Culture in East Africa, *World Archaeology* **1** : 1–28.

—— In press (a) Stone tools: an adaptive threshold? In: *Essays in honour of Grahame Clark*, edited by G. de G. Sieveking.

—— 1976. Plio-Pleistocene artefact assemblages from East Rudolf, Kenya. In: *Earliest Man and Environments in the Lake Rudolf Basin* edited by Y. Coppens, F. C. Howell, G. L. Isaac and R. E. F. Leakey. University of Chicago Press, Chicago.

ISAAC, G. LL. & CURTIS, G. H. 1974. The age of early Acheulian industries in East Africa—new evidence from the Peninj Group, Tanzania, *Nature, Lond.* **249**, 624–7.

LEAKEY, M. D. 1966. *A review of the Oldowan Culture from Olduvai Gorge, Tanzania. Nature, Lond.* **210**, 462–6.

—— 1971. *Olduvai Gorge Volume 3 Excavations in Beds I & II, 1960–1963.* Cambridge at the University Press.

MERRICK, H. V., DE HEINZELIN, J., HAESAERTS, P. & HOWELL, F. C. 1973. Archaeological occurrences of Early Pleistocene Age from the Shungura Formation, Lower Omo Valley, Ethiopia. *Nature, Lond.* **242**, 572–5.

OAKLEY, K. P. 1956. *Man the toolmaker.* London: British Museum of Natural History.

WRIGHT, R. V. S. 1972. Imitative learning of a Flaked Stone technology—the case of an ourang-utan. *Mankind* **8**, 296–306.

Part III

Regional Studies in the Gregory Rift
Valley

A. Olduvai Gorge and Laetolil, Tanzania;
   at Olorgesailie, Kenya

B. The Lake Baringo Basin, Kenya

C. The Lake Turkana (Rudolf) Basin, Kenya
   and Ethiopia

D. The Afar area, Ethiopia

Part III.

Regional Studies in the Oregon Mill
Valley

10

MARY D. LEAKEY

Olduvai Gorge 1911–75: a history of the investigations

Olduvai Gorge lies in the eastern Serengeti plains of Northern Tanzania and is at an altitude of 4400 feet above sea level. It drains eastwards into the fault trough of the Olbalbal Depression from two lakes at the head of the gorge, known as Ndutu and Masek. These occasionally overflow into the Olduvai river during particularly wet years. The river is seasonal and flows through a shallow valley in its upper reaches, but where the basement rocks give place to sediments the valley changes dramatically to a steep-sided gorge. A southern branch, known as the Side Gorge, has its headwaters on Lemagrut mountain, to the south-west, and it is in this gorge that the Olduvai deposits can be seen to overlie the earlier Laetolil Beds.

Olduvai first became known to the scientific world in 1911 when a German entomologist, Professor Kattwinkel, who was collecting butterflies, came across the gorge accidentally. He picked up a number of fossils including teeth of *Hipparion* and took them to Berlin, where they aroused great interest. As a result, a German expedition under the leadership of a geologist, Professor Hans Reck, spent three months at Olduvai during 1913. The results were published the following year (Reck 1914a, 1914b). Reck set up the system of numbering the Olduvai lithological units I to V, in ascending order, a system that is still in use, although some alterations have been made for Beds IV and V.

A great deal of faunal material was collected during this first expedition. Regrettably, most of it was destroyed during the last war, to the extent that only a single bovid horn core has survived, from among the many that were collected. This material was described in papers by W. O. Dietrich (cf. 1916, 1925, 1926, 1928, 1937a, 1937b) while descriptions of the fauna found subsequently were published by L. S. B. Leakey (1967) in volume 1 of the Olduvai monographs.

The most controversial discovery made by Reck was a human skeleton which he claimed to be of Middle Pleistocene age and contemporary with Bed II. It was clearly *Homo sapiens* and lay in the crouched position typical of late Stone Age burials in East Africa and elsewhere. Reck's claim that *Homo sapiens* of entirely modern type was contemporary with Bed II not unnaturally aroused scepticism. The controversy over this skeleton continued for a number of years (Leakey, L. S. B. 1928), although it seems inconceivable that the mineralisation of the bone could have been the same as in *bona fide* Bed II fossils. Equally, the antenatal position of the skeleton, which indicated a burial, seems not to have been given due importance. Finally, after heavy mineral analysis of Bed II and of the deposit near the skeleton had been carried out by P. G. H. Boswell (1932), L. S. B. Leakey and others (1933) came to the conclusion that the skeleton was a burial and intrusive to Bed II. The matter has now been brought to a close by the publication of a carbon 14 date of 1690 ± 920 B.P. obtained by Reiner Protsch (1974) on some of the post-cranial bones.

The 1914–18 war put a stop to any further work at Olduvai since the gorge was in German East Africa, where fierce fighting took place between British and German forces.

The next expedition, under the leadership of L. S. B. Leakey, reached Olduvai in 1931. He was accompanied by Professor Reck, Dr A. T. Hopwood, palaeontologist from the British Museum of Natural History, D. McInnes, V. E. Fuchs (now Sir Vivian) and others.

L. S. B. Leakey had visited Germany in 1928 and had been delighted to find a number of handaxes and other artefacts classed as 'foreign stones' among the geological specimens. Reck, however, remained unconvinced about the presence of stone tools at Olduvai until the 1931 expedition reached the gorge and found a number of well-made handaxes in Bed IV.

In 1932 L. S. B. Leakey returned briefly to Olduvai with E. J. Wayland, then Director of the Geological Survey of Uganda, who wrote a report on the geology of the Gorge. It is regrettable that this was never published and only circulated privately since it expressed views much in accordance with the final interpretation of the stratigraphy as determined more than thirty years later by R. L. Hay (1963, 1967, 1975).

The results of the 1931–2 expeditions were not published until 1951 owing to the war and to difficulties in meeting the cost of printing (Leakey, L. S. B. 1951). A preliminary report, however, was presented at a meeting in Cambridge convened by the Royal Anthropological Institute and later published (Leakey, L. S. B. 1933).

Reck had died during 1934, but the 1951 volume included a chapter summarising his conclusions on the geology of the gorge, (Reck 1951). Unfortunately, no diagrams or sections could be included since these had been lost or destroyed during the bombing of Berlin.

In the 1951 volume, the stone industries were interpreted as a single evolutionary sequence beginning with the Oldowan of Bed I, or Stage I of the Chelles-Acheul culture, through ten subsequent stages. We now know that this was an over-simplification, but at the time there was a strong tendency in favour of gradually evolving cultural lineages. The fauna was described by Hopwood (1951) who maintained that the majority of the fossil mammals were the same as the species living in the area today. Furthermore, he did not differentiate between the fauna from the different beds, lumping the entire collection together as a single faunal unit. In the concluding paragraph of his report he wrote: 'there is no faunistic evidence to suggest that the lower part of the Olduvai series is of Lower Pleistocene age'.

L. S. B. Leakey spent several months at Olduvai in 1935 in company with P. E. Kent (now Sir Peter), Stanhope White, Peter Bell and the writer. From then until 1960 only brief visits were possible on account of lack of funds, although the

Boise Fund contributed as far as it was able and L. S. B. Leakey raised a certain amount of money elsewhere. However, the financial support and consequently the amount of time it was possible to spend in the field were wholly inadequate for research on the scale that Olduvai required. The discovery of the cranium of *Australopithecus boisei* in 1959 (Leakey, L. S. B. 1960) and the potassium argon dates obtained for Bed I at the Berkeley laboratory (Leakey *et al.* 1961), (Evernden & Curtis 1965), radically altered the position. The Research Committee of the National Geographic Society, Washington D.C. decided to support the work at Olduvai and have continued to do so most generously ever since. With their backing it was possible to embark on a long-term exploration of the Olduvai sequence with particular emphasis on detailed excavations of living sites and other localities where *in situ* material could be obtained. The excavations were initiated in the lowest horizon, at the base of Bed I, and continued upwards through the deposits, sampling hominid living sites at successive levels. Thus, Beds I and II were explored first and the results published in 1971, in the third Olduvai monograph (Leakey, M. D. 1971). Excavations in Beds III, IV and the Masek Beds were begun in 1968 and continued until 1973. These results are now being prepared for publication. The geological study of the Olduvai sequence by R. L. Hay was carried out during the same period and is the basis upon which the archaeological succession has been established (Hay 1975).

The more important results of the work at Olduvai during this period may be briefly summarised as follows. Hominid remains are known from lower Bed I and from various levels up to the Masek Beds (formerly known as Bed IVB) and extend over a time span from about 1.8 to 0.3 m.y. *Australopithecus boisei* (Tobias 1967), has been found only in Beds I and II; the latest occurrences being teeth from two localities in upper Bed II. The remains referred to *Homo habilis* have been found only in Bed I and the lower part of Bed II, the most recent in time being the skull H. 13, which occurred just above the Lemuta Member of Bed II. There is, however, some doubt now as to whether all the fossils formerly attributed to *Homo habilis* belong to a single taxon. It seems possible that H. 16 and H. 39, both from the base of Bed II, may belong to the same lineage as the cranium of '1470' from East Lake Turkana and to some specimens from the Hadar in Ethiopia, as well as those from the Laetolil Beds, 20 miles to the south of Olduvai.

Homo erectus is found in Bed II and upper Bed IV (Leakey, M. D. & Day 1971). The sole hominid fossil known from the Masek Beds is a small piece of mandible which is too abraded for positive taxonomic identification.

Definitive reports on the larger faunal groups such as the Equidae, Suidae and Bovidae are still not published, but studies of the small mammal population of Bed I have given interesting indications of wet and then progressively drier conditions during the time the bed was being laid down.

Besides fossil hominids and fauna Olduvai has provided the longest known record of stone industries, spanning a period of about $1\frac{1}{2}$ million years (Leakey, M. D. 1976). In Bed I there are abundant tools belonging to the Oldowan industry which is characterised mainly by various forms of choppers made on waterworn pebbles and cobbles as well as small cutting and scraping tools, (Leakey, M. D. 1966). Living sites of this period are mainly along the shore of the former Olduvai lake. This industry continues throughout the Olduvai sequence into upper Bed IV. In the later stages the tool-kit is expanded and includes a greater variety of implements than in Bed I. This more advanced industry has been termed the Developed Oldowan.

The Acheulian or handaxe culture first appears in middle Bed II at about 1.4 m.y. and continues into the Masek Beds. There is no discernable pattern of technical improvement in the Acheulean from Olduvai and, indeed, the bifacial tools from Bed II resemble most closely those from a site in upper Bed IV, rather than from sites in lower Bed IV. Associated small tools, with the exception of scrapers, are less varied and less abundant in the Acheulean than the Developed Oldowan.

References

BOSWELL, P. G. H. 1932. The Oldoway human skeleton, *Nature, Lond.* 237–8.

DIETRICH, W. D. 1916. Elephas antiquus recki n.f. aus dem Diluvium Deutsch-Ostafrikas I. *Arch. Biontol.* **4**, 80.

—— 1925. Elephas antiquus recki n.f. aus dem Diluvium Deutsch-Ostafrikas II. *Wiss. Eregbn. Oldoway-Exped. 1913*, (N.F.) **2**, 1–38.

—— 1926. Fortschritte der Sangetierpaläontologie Afrikas. *Forsch. Fortschr. dtsch. Wiss*, Berlin **2**, 121–2.

—— 1928. Pleistocene deutschostafricanische Hippopotamus-Reste. *Wiss, Ergebm. Oldoway-Exped. 1913*, (NF) **3**, 3–41.

—— 1937a. Pleistozäne Suiden-Reste aus Oldoway, Deutsch-Ostafrika. *Wiss. Ergebn. Oldoway-Expedn. 1913*, (NF) **4**, 91–104.

—— 1937b. Pleistozäne Giraffinen und Boisden aus Oldoway, Deutsch-Ostafrika. *Wiss. Ergebn. Oldoway-Expedn.* 1913, (NF) **4**, 105–10.

HAY, R. L. 1963. Stratigraphy of Bed I through IV, Olduvai Gorge, Tanganyika, *Science*, **139**, 829–33.

—— 1967. Revised stratigraphy of Olduvai Gorge, in *Background to Evolution in Africa*, (ed. W. W. Bishop and J. D. Clark). 221–8, Chicago University Press, Chicago.

—— 1975. *Geology of the Olduvai Gorge*, University of California Press, Berkeley.

—— In press. Olduvai Fossil Hominids: their stratigraphic positions and associations.

HOPWOOD, A. T. 1951. The Olduvai Fauna, In: *Olduvai Gorge* by L. S. B. Leakey: 20–4.

LEAKEY, L. S. B. 1928. The Oldoway Skull, *Nature, Lond.* **121**, 499.

—— 1933. Report on a conference convened by the R.A.I. *Man*, **33**, 66–8.

—— 1951. *Olduvai Gorge*, Cambridge University Press.

—— 1960. *The discovery of Zinjanthropus boisei, Current Anthropology*, **I**: 76–7.

—— 1967. *Olduvai Gorge 1951–1961*, **1**, Cambridge University Press, Cambridge.

——, BOSWELL, P. G. H. et al. 1933. Letter in *Nature, Lond.* **131**, 397.

——, EVERNDEN, J. F. & CURTIS, G. H. 1961. Age of Bed I, Olduvai, *Nature, Lond.* **191**, 478–9.

LEAKEY, M. D. 1966. A review of the Oldowan culture from Olduvai Gorge, Tanzania, *Nature, Lond.* 462–6.

—— 1971. *Olduvai Gorge, Excavations in Beds I and II*, Cambridge University Press, Cambridge.

—— 1976. Cultural patterns in the Olduvai sequence, In *After the Australopithecines*, eds. G. Ll. Isaac & K. Butzer World Anthropology, Mouton & Co., The Hague.

—— & DAY, M. H. 1971. Discovery of postcranial remains of *Homo erectus* and associated artefacts in Bed IV at Olduvai Gorge, Tanzania, *Nature, Lond.* **232**, 380–7.

PROTSCH, R. 1974. The age and stratigraphic position of Olduvai hominid I, *Journ. of Human Evolution*, **3**, 379–85.

RECK, H. 1914a. Erste vorläufige Mittelilung über den Fund eines fossilen Menschenskelets aus Zentralafrika. *Sond. aus den Sitzungsberichten der Gesellschaft naturforschendes Freunde*, **3**, 81–95, Berlin.

—— 1914b. Zweite vorläufige Mitteilung über fossile Tier-und Menschenfunde aus Oldoway in Zentralafrika. *Sond. aus den Sitzungsberichten der Gesellschaft naturforschenden Freunde*, **7**, 305–18, Berlin.

—— 1951. A preliminary survey of the tectonics and stratigraphy of Olduvai, in *Olduvai Gorge* by L. S. B. Leakey, 5–19.

TOBIAS, P. V. 1967. *The cranium and maxillary dentition of Australopithecus (Zinjanthropus) boisei*, Cambirdge University Press, Cambridge.

11

MARY D. LEAKEY, R. L. HAY, G. H. CURTIS,
R. E. DRAKE, M. K. JACKES
& T. D. WHITE

Fossil hominids from the Laetolil Beds, Tanzania[1]

Remains of 13 early hominids have been found in the Laetolil Beds in northern Tanzania, 30 miles south of Olduvai Gorge. Potassium–argon dating of the fossiliferous deposits gives an upper limit averaging 3.59 Myr and a lower limit of 3.77 Myr. An extensive mammalian fauna is associated. The fossils occur in the upper 30 m of ash-fall and aeolian tuffs whose total measured thickness is 130 m.

The fossil-bearing deposits referred to variously as the Laetolil Beds, Garusi or Vogel River series lie on the southern Serengeti Plains, in northern Tanzania. 20–30 miles from the camp site at Olduvai Gorge.

Fossils have been collected from the area on several occasions, the largest collection being made by L. Kohl-Larsen in 1938–9, who also found a small fragment of hominid maxilla which was named *Meganthropus africanus* by Weinert. L. S. B. Leakey and M. D. L. visited the area in 1935 and in 1959, while a day trip was made in 1964 in company with R. L. H. The faunal material recovered on these occasions was all collected before the advent of isotopic dating and the age of the fossils remained uncertain until potassium-argon dating was carried out on samples of biotite obtained during the 1975 field season. In 1974, however, lava flows which unconformably overlie the Laetolil Beds had been dated by G. H. C. at 2.4 Myr.

We now report evidence that fossiliferous deposits of several different ages exist in the Laetolil area and that specimens found on the surface are not necessarily derived from the same beds. There are, however, noticeable variations in the colour and physical condition of the surface fossils which provide indications of their origin.

Discrepancies in the fauna were noted by Dietrich and Maglio who both postulated faunal assemblages of two different ages. In view of this, it was proposed for a time that the name Laetolil should be abandoned in favour of the more generalised term Vogel River Series, based on the colloquial German name for the Garusi river, which abounds in bird life. As the early fossiliferous deposits here referred to as the Laetolil Beds are not confined to the Garusi valley, this change of name seems unnecessary. Furthermore, the name Laetolil embraces a larger

[1] Reprinted from *Nature, Lond.* **262**, 460–5, August 5 1976, by kind permission of the Editor.

area, because it is the anglicised version of the Masai name (laetoli) for *Haemanthus*, a red liliaceae that is abundant in the locality. The Laetolil Beds, *sensu stricto*, form a discrete unit, distinguishable from later deposits, and M. D. L. considers that the original name proposed by Kent in 1941 should be retained for this part of the sequence.

The relationship of the Laetolil to the Olduvai Beds had been under discussion for some years, but in 1969 R. L. H. noted that they underlay Bed I at the Kelogi inselberg in the Side Gorge and established that they antedated the Olduvai sequence. This has been further confirmed when tuffs correlatable with Bed I as well as an earlier, fossil-bearing series of tuffs were found to lie unconformably on the Laetolil Beds.

Interest in the area was renewed in 1974 after the discovery by George Dove of fossil equid and bovid teeth in the bed of the Gagjingero river, which drains into Lake Masek at the head of Olduvai Gorge. These fossils were found to be eroding from relatively recent deposits, probably the beds named Ngaloba by Kent. Exposures of the Laetolil Beds, not hitherto seen by M. D. L., were found to the east of the Gagjingero river, at the headwaters of the Garusi river, and of the Olduvai Side Gorge (referred to as Marambu by Kohl-Larsen). Several fossils, including a hominid premolar, were found at these localities and subsequent visits yielded further hominid remains.

The possibility of establishing the age of the hominid fossils from the Laetolil Beds by radiometric dating and of clarifying the discrepancies in the faunal material led to a 2-month field season during July and August 1975. Samples from the fossiliferous horizons, collected by R. L. H., have now been dated. On the basis of these results the hominid remains and associated fauna can be bracketed in time between 3.59 and 3.77 Myr.

No trace of stone tools or even of utilised bone or stone was observed in the material from the Laetolil Beds, although handaxes and other artifacts occur in conglomerates which are present in certain areas and which are unconformable to the Laetolil Beds.

Stratigraphy of Laetolil area

The bulk of the faunal remains and all of the hominid remains were found within an area of about 30 km² at the northern margin of the Eyasie plateau and in the divide between the Olduvai and Eyasie drainage systems (Fig. 11:1).

Kent studied this area as a member of L. S. B. Leakey's 1934-5 expedition, and his short paper is the only published description of the stratigraphy. He recognised three main subdivisions of the stratigraphic sequence, which overlies the metamorphic complex of Precambrian age. The lower unit he named the Laetolil Beds and the upper the Ngaloba Beds. The middle unit consists of olivine-rich lava flows and agglomerate, which are much closer in age to the Laetolil Beds than to the Ngaloba Beds. Kent briefly noted the local occurrence of tuffs younger than the lavas and older than the Ngaloba Beds. He described the Laetolil Beds as subaerially deposited tuffs, and he gave 30 m as an aggregate thickness in the vicinity of Laetolil. The Ngaloba Beds he described as tuffaceous clays, and he gave a thickness of about 5 m at the type locality. Pickering mapped this area as part of a 1:125 000 quarter degree sheet and extended the known occurrence of the Laetolil Beds. He also recognised that at least some of the tuffs are of nephelinite composition.

This picture was modified considerably by stratigraphic work in 1974 and 1975. The Laetolil Beds proved to be far thicker and more extensive than previously recognised. The thickest section, 130 m, was measured in a valley in the northern part of the area (geological localities *A* to *C*, Fig. 11:2). The base is not exposed and the full thickness of the Laetolil Beds here is unknown. Sections 100–120 m thick and representing only part of the Laetolil Beds were measured about 10 km south-east of Laetolil. The Laetolil Beds are 15–20 m thick at a distance of 25–30 km to the south-west and 10–15 m thick at Lakes Masek and Ndutu, 30 km to the north-west. The Laetolil Beds are tuffaceous sediments, dominantly of nephelinite composition.

The lavas and agglomerates noted by Kent overlie an irregular surface deeply eroded into the Laetolil Beds. The lavas were erupted from numerous small vents to the south, south-west and west of Lemagrut volcano. Although designated as nephelinite by Kent, the flows proved to be vogesite, a highly mafic lava with interstitial alkali feldspar and phlogopitic biotite. The lavas have reversed polarity as determined by field measurements (personal communication from A. Cox). A sample of lava from geological locality *D* (Fig. 11:2) gave K–Ar dates of 2.38 ± 0.5 Myr and 2.43 ± 0.7 Myr (Table 1).

The 130-m section of the Laetolil Beds (Fig. 11:3) is divisible into an upper

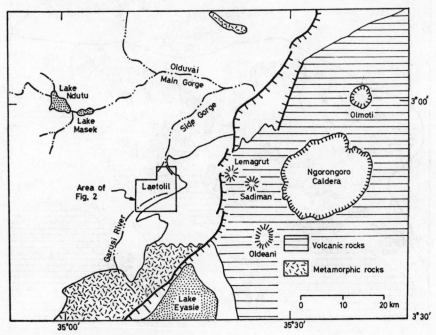

FIG. 11:1. Map of southern Serengeti and volcanic highlands.

half consisting largely of wind-worked, or aeolian, tuff and a lower half consisting of interbedded ash-fall and aeolian tuff with minor conglomerate and breccia. Tephra in the lower half are nephelinite, whereas nephelinite and melilitite are subequal in the upper half. Between these two divisions is a distinctive biotite-bearing coarse lithic-crystal tuff 60 cm thick. Hominid remains and nearly all of the other vertebrate remains are confined to the uppermost 30 m of aeolian tuffs beneath a widespread pale yellow vitric tuff 8 m thick (tuff *d* of Fig. 11:3). Several other marker tuffs, between 1 m and 30 cm thick, can be used for correlating within the fossiliferous 30-m thickness of sediments. Three prominent marker tuffs, designated *a*, *b*, and *c* (Fig. 11:3, column 2), can be recognised throughout the area shown in Fig. 11:2. Widespread horizons of ijolite and lava (mostly neph-

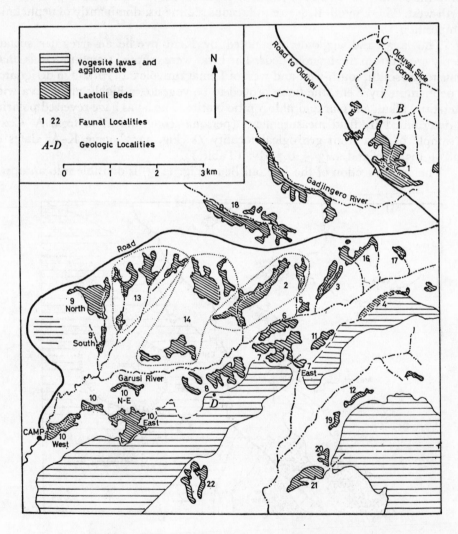

FIG. 11:2. Map of the Laetolil area showing the fossil beds.

elinite) xenoliths are at several levels in the fossiliferous part of the section and assist in correlating. Biotite is common in some of the ijolite.

Additional fossiliferous deposits, of mineralogical affinity to the Laetolil Beds and 10–15 m thick lie between tuff *d* and the Bed I (?) tuffs at several places. They are locally separated from tuff *d* and the underlying aeolian tuffs of the Laetolil Beds by an erosional surface with a relief of 8 m. These sediments comprise water-worked tuffs, aeolian tuffs, clay-pellet aggregates of aeolian origin and limestone. The tephra are of phonolite and nephelinite composition. It is not yet clear whether or not these sediments should be regarded as part of the Laetolil Beds.

Beds I and II are represented by sedimentary deposits that lie stratigraphically between the lava and the Ngaloba Beds. The Bed II deposits locally contain fossils and artifacts.

Sadiman volcano, about 15 km east of the fossiliferous exposures, seems to have been the eruptive source of the Laetolil Beds. The one K–Ar date, of 3.73 Myr, obtained from Sadiman lava, fits with the K–Ar dates on the Laetolil Beds presented here. This date was published previously as K–Ar 2238[8], where it was incorrectly assigned to Ngorongoro because of a mistake in listing the sample numbers.

TABLE 1 : *K-Ar dates from the Laetolil area.*

SAMPLE	KA	DATED MATERIAL	SAMPLE WT.	% K	moles/gm ^{40}Ar rad x 10^{-11}	% ^{40}Ar Atmos	AGE x 10^6 YEARS	REMARKS
Vogesite lava, Loc. D	2835	Whole Rock	10.06588	1.013	0.419	74.0	2.38 ± .05	Unconformably overlying Laetolil Beds
	2835R	"	11.27408	1.013	0.427	84.4	2.43 ± .07	
tuff c. Loc. A	2929	Biotite	1.98629	6.96 ± .03	4.118	77.5	3.41 ± .08	Treated with dilute HCl
	2977	"	2.544	6.98 ± .04	4.462	78.5	3.68 ± .09	Treated with warm dilute acetic acid
	2979	"	1.94298	6.95 ± .04	4.454	79.1	3.69 ± .10	
xenolithic horizon , Loc.A	2930	"	3.07730	7.58 ± .02	4.771	50.8	3.62 ± .09	Treated with dilute HCl
	2930R	"	1.83909	7.58 ± .02	4.733	39.7	3.59 ± .05	
ash-fall tuff Loc. B	2932	"	1.05978	6.00 ± .1	3.996	75.0	3.82 ± .16	Crushed, hand picked, treated with dilute HCl
	2938	"	1.42092	6.49 ± .04	4.182	78.8	3.71 ± .12	

^{40}K/K = 1.18 x 10^{-4}; ^{40}K$_\lambda$ = 5.480 x 10^{-10}yr^{-1}; ^{40}K$_{\lambda\beta}$ = 4.905 x 10^{-10}yr^{-1}; ^{40}K$_{\lambda\epsilon}$ = 0.575 x 10^{-10}yr^{-1}

Potassium-argon dating

The vogesite lava is composed of approximately 85–90 per cent olivine and augite. The remainder is principally anorthoclase in the groundmass together with a very small amount of phlogopitic biotite. These two minerals proved too fine-grained and sparse for effective separation, and whole-rock samples were used for dating.

In addition to the vogesite lava unconformably overlying the Laetolil Beds, three of the tuffaceous layers within the Laetolil Beds have been dated by the conventional, total degassing K–Ar method (Table 1), and one of these tuffaceous layers has also been dated by the $^{40}Ar/^{39}Ar$ method, using incremental heating.

Tuff *c* is the uppermost of the dated horizons, lying near the top of the fossiliferous deposits. It is a widespread crystal-lithic airfall tuff cemented with calcite and generally 10 to 15 cm thick. Abundant biotite crystals 1–2 cm in diameter occur in the upper part of the tuff, which is composed of nepheline and melilitite. The dated crystals were hand-picked from two outcrops of the tuff at locality *A* (Figs 11:2 and 3). The cementing calcite adhering to and interleaving the biotite books was removed by treatment with dilute HCl for a few minutes on one sample and with warm dilute acetic acid on two other samples. This treatment

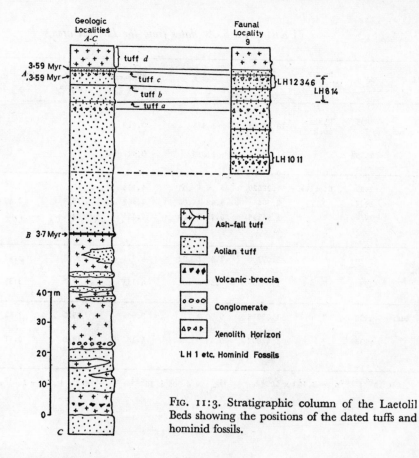

FIG. 11:3. Stratigraphic column of the Laetolil Beds showing the positions of the dated tuffs and hominid fossils.

was found to have negligible deleterious effects on biotite standard samples. The three dates obtained for tuff *c* (3.41 ± 0.08 Myr, 3.69 ± 0.09 Myr, and 3.69 ± 0.10 Myr) have about equal precision so were averaged to give a date of 3.59 Myr.

Two conventional K–Ar dates and one ⁴⁹Ar/³⁹Ar were obtained from a single biotite crystal from a xenolithic horizon at locality A (Figs 11:2 and 3) approximately 1–2 below the youngest tuff dated (tuff *c*). This horizon lies within the upper part of the fossiliferous beds and is distinguished by its ejecta of ijolite xenoliths together with nephelinite lava xenoliths. Although the biotite occurs in some of the ijolite clasts, a large single free crystal picked from the tuff was used for dating. The good agreement between the two conventional K–Ar dates for this sample (3.62 ± 0.09 Myr and 3.59 ± 0.05 Myr) and the ⁴⁰Ar/³⁹Ar date which yielded an isochron age of 3.55 Myr (Fig. 11:4) indicates that initial excess ⁴⁰Ar is not a problem with this sample, and that these dates give an average age for this horizon of 3.59 Myr.

Two dates were obtained from a biotite-bearing crystal-lithic tuff 60 cm thick lying approximately 50 m below tuff *c* near the middle of the thickest section of the Laetolil Beds at location B (Fig. 11:2). The tuff is of nephelinite composition, containing abundant augite and altered nepheline and 2–3 per cent of biotite, some crystals of which are as much as 1 cm in diameter. Calcite cements the crystals and fragments together. Biotite was separated by crushing, screening and hand picking and was cleaned of calcite with dilute HCl. Two dates, 3.82 ± 0.16 Myr and 3.71 ± 0.12 Myr, average 3.77 and give a lower limit to the age of the hominid remains (which occur below but close to the dated xenolithic horizon higher in the section) in the 30 m section at locality A (Fig. 11:2) whose base projects approximately 20 m above this tuff at locality B (Fig. 11:3).

Fauna

The fossiliferous deposits in the Laetolil area have been subdivided, for purposes of collecting, into 26 localities. These subdivisions are based on existing topographic features such as grassy ridges, lines of trees, stream channels and so on. This has provided a means of dividing the fossiliferous area into units of restricted size, but does not relate to the former topography.

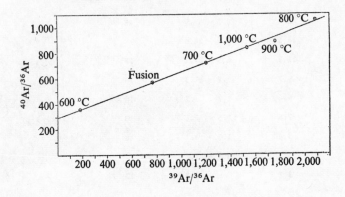

FIG. 11:4. ⁴⁰Ar/³⁹Ar incremental heating of biotite from xenolithic horizon, locality C. Isochron age, 3.55 Myr; [⁴⁰Ar/³⁹Ar]₀ = 294; ⁴⁰Ar/³⁹Ar = 0.35318; J = 0.005204.

Eighteen of these localities are in the Garusi valley, one in the valley at the head of the Olduvai Side Gorge, one in the Gadjingero valley, five in a valley to the south of the Garusi river and one in an isolated position to the west (Fig. 11:2).

Identifiable fossils noted on the exposures were either collected and registered or listed on the sites. Specimens from the Laetolil Beds were distinctively cream coloured or white and sometimes chalky in texture, but the surface material also included brown, grey or black specimens, often rolled. These have been excluded from the material under review, together with fossils which have adhering matrix clearly dissimilar from the tuffs of the Laetolil Beds. Among these are all remains of *Hippopotamus*, *Equus*, *Theropithecus*, *Phacochoerus* and *Tragelaphus*, formerly included in the Laetolil fauna, with the exception of *Hippopotamus*.

The fossil material from the Laetolil Beds is dispersed and fragmentary and it is not possible to assess the number of individuals represented. In this article, 'numbers of fossils' refer to individual bones and teeth, except in the case of clear association, confined almost entirely to remains of *Serengetilagus* and *Pedetes*, some of which were associated and even articulated.

Table 2 shows the mean percentage frequency of the more common vertebrate groups at several of the richer localities. A total of 6288 fossil specimens was identified from the localities considered here.

Reptiles are represented by snake vertebrae at three localities and by tortoises at all localities. The latter have an average frequency of 2.2 per cent and include several giant specimens. Avian remains occur widely and at several localities birds' eggs were completely preserved. There is one example of a shattered clutch of at least eight eggs, rather smaller in size than eggs of domestic fowl. Primates were found at 15 localities, and in one area they constitute 3.8 per cent of the fauna. Both cercopithecines and colobines are present (personal communication from M. G. Leakey).

Rodents are fairly well represented, although not abundant. Of the specimens identified by J. J. Jaeger, the most common are *Pedetes*, *Saccostomus* and *Hystrix*.

TABLE 2 : *Mean percentage frequency of bones of more common vertebrate groups.*

Group	Mean percentage	Range	Numbers of Localities
Bovidae	43.0	29.5 – 57.9	18
Lagomorpha	14.4	5.6 – 24.0	18
Giraffidae	11.2	6.8 – 23.2	18
Rhinocerotidae	9.7	5.7 – 17.3	18
Equidae	4.4	1.7 – 7.1	18
Suidae	3.6	1.0 – 7.1	18
Proboscidea	3.4	1.0 – 5.1	18
Rodentia	3.3	0.9 – 7.2	17
Carnivora	3.1	0.8 – 8.2	17

The carnivore fauna is characterised by a high percentage of viverrids, constituting 32 per cent of all the carnivore specimens. Large carnivores are represented by hyaenids, of which there are several genera, by felids and a machairodont.

Proboscidea include *Deinotherium* and *Loxodonta sp.* (M. Beden, personal communication). There is no evidence that the equid material (other than that derived from later deposits) includes any genus except *Hipparion*. The suids consist only of two genera, *Potamochoerus* and *Notochoerus* (J. Harris, personal communication). The presence of *Ancylotherium* and *Orycteropus*, noted in previous collections, is confirmed and the existence of two rhinocerotids has been established by the discovery of skulls of both *Ceratotherium* and *Diceros*, although only the former was listed previously.

Among the giraffids, *Sivatherium* and a small form of giraffe are equally common. *Giraffe jumae* is also present but is much less well represented. The bovid fauna is chiefly characterised by the very high percentage of *Madoqua* (dikdik). In 18 localities dikdik range from 1.5 to 37.7 per cent of all bovid specimens, with a mean percentage of 15.1 per cent.

The 1975 field season, although mainly confined to surface collection, has established that previous collecting had sampled faunas of several time periods. It is now possible to exclude some genera from the published lists of fauna from the Laetolil Beds[2, 9] such as *Theropithecus*, *Tragelaphus*, *Equus* and *Phacochoerus*.

Fossil hominids

Thirteen new fossil hominid specimens were recovered from the Laetolil site during 1974 and 1975. The remains include a maxilla, mandibles and teeth. This sample displays a complex of characters seemingly demonstrative of phylogenetic affinity to the genus *Homo*, but also features some primitive traits concordant with its great age.

The hominid specimens are listed in Table 3. Provisional stratigraphic correlation and dating have placed the hominid remains as shown in Fig. 11:3. With the exception of Laetolil hominids (LH) 7 and 8, all specimens retain the matrix characteristic of the Laetolil Beds from which they have been weathered or excavated. There is no reason to doubt that all the specimens derive from the Laetolil Beds as reported.

The Laetolil hominid sample consists of teeth and mandibles. Important features of these specimens are described here, followed by a brief preliminary discussion regarding the phylogenetic status of the fossils.

Remains of deciduous and permanent dentitions have been recovered. The dentitions consist of two maxillary and four mandibular partial tooth rows. Compared with the rest of the East African Pliocene/early Pleistocene hominid sample, the Laetolil anterior teeth are large and the post-canine teeth of small to moderate size.

Deciduous dentition

Canines (LH 2, 3) Single upper and lower deciduous canines are known. The lower is a slightly projecting, sharp conical tooth in its damaged state. It is smaller but its overall morphology is similar to its permanent counterpart.

First molars (LH 2, 3) The upper first deciduous molar displays spatial domin-ance of the protocone and a well-marked mesiobuccal accessory cusp defined by a strong anterior fovea. The lower first deciduous molar is molarised, with four or five main cusps depending on hypoconulid expression. There is a spatially dominant protoconid with a large flat buccal face, a lingually facing anterior fovea, and an inferiorly projecting mesiobuccal enamel line. Vertical dominance of the protoconid and metaconid is marked in lateral view.

Second molars (LH 2, 3, 6) Upper and lower deciduous second molars take the basic form of the analogous permanent first molars but are smaller in overall size. The dM_2 of LH 2 is the only lower molar in the Laetolil sample bearing any indication of a tuberculum sextum.

TABLE 3 : *List of hominid remains recovered in 1974–75.*

Laetolil Hominid (L.H.-)	Locality	Specimen Consists of	Discovered by
1	1	RP^4 fragment.	M. Muoka
2	3*[1]	Immature mandibular corpus with deciduous and perma-nent teeth.	M. Muluila
3 (a-t)	7*	Isolated deciduous and permanent teeth, upper and lower.	M. Muoka
4	7	Adult mandibular corpus with dentition.	M.Muluila
5	8	Adult maxillary row: I^2 to M^1.	M.Muluila
6 (a-e)	7*[1]	Isolated deciduous and permanent teeth, upper.	M. Muoka
7	5	RM^1 or ^{2} fragment.	M. Muoka
8 + 10	11	RM^2, RM^3.	E. Kandindi
10	10W	Fragment left mandibular corpus with broken roots.	E. Kandindi
11	10W	LM^1 or ^{2}.	E. Kandindi
12	5	LM^2 or ^{3} fragment.	E. Kandindi
13	8	Fragment right mandibular corpus with broken roots.	M. Jackes
14	19	Isolated permanent teeth, lower.	E. Kandindi

* In situ

[1] L.H.-3 and 6 associated in mixed state

+ L.H. 9 not valid.

Permanent dentition

Incisors (LH 2, 3, 5, 6, 14) The single upper central incisor is very large and bears pronounced lingual relief. The upper lateral incisors are smaller, with variable lingual relief. The lower incisors are narrow and very tall, with minimal relief.

Canines (LH 2, 3, 4, 5, 6, 14) The incompletely developed upper canine LH 3 is large, stout and pointed, bearing pronounced lingual relief. LH 6 is slightly smaller, with less lingual relief and a tall, pointed crown. LH 5 bears an elongate dentine exposure on its distal occlusal edge but does not project beyond the occlusal row in its worn state. The lower canines are similar to the uppers in their great size, height and lingual relief.

Premolars (LH 1, 2, 3, 4, 5, 6, 14) The upper premolars tend to be buccolingually elongate and bicuspid, with the lingual cusp placed mesially and an indented mesial crown face. The two lower third premolars each have a dominant buccal cusp with mesial and distal occlusal ridges, and a weak lingual cusp placed mesial of the major crown axis. The long axis of the oval crown crosses the dental arcade contour from mesiobuccal to distolingual, donating a 'skewed' occlusal profile to the tooth. The lower fourth premolars are fairly square in shape with major buccal and lingual cusps and moderate talonids.

Molars (LH 2, 3, 4, 5, 6, 7, 8, 11, 12, 14) The upper molars are of moderate size their relative proportions unknown. They show a basic four-cusp pattern with spatial dominance of the protocone and typical expression of a pit-like Carabelli feature. The lower molars display progressive size increase from first to third in LH 4. They have a fairly square occlusal outline with hypoconulida pressed anteriorly between hypoconid and entoconid and no trace of a tuberculum sextum. The Y-5 pattern of primary fissuration is constant.

Mandibles

Juvenile mandible (LH 2) This specimen (Figs 11:5 and 6) is incompletely fused at the midline, and the developing crowns of the permanent canines and pre-molars are exposed in the broken corpus. The first molars have just reached the occlusal plane. Only the posterior aspect of the symphysis shows fairly intact contours, with a concave post-incisive planum and incipient superior transverse torus. The genioglossal fossa is obscured by midline breakage. Associated distortion has artificially increased the bimolar distances.

Adult mandible (LH 4) The adult mandibular corpus is well preserved, with the rami missing (Figs 11:5 and 7). The dental arcade is essentially undistorted, and presents fairly straight sides which converge anteriorly. The anterior dentition has suffered post-mortem damage and loss, except for the right lateral incisor which seems to have been lost in life. Largely resorptive alveolar pathology has obliterated its alveolus and has affected the adjacent teeth. There is development of wide interproximal facets for the canine teeth but no C/P$_3$ contact facet. This combines with observation of extensive wear on the buccal P$_3$ face to suggest that

C/P$_3$ interlock has prevented mesial drift from eliminating the C/P$_3$ diastema. Judging from the preserved posterior incisor alveoli, these teeth were set in an evenly rounded arcade, projecting moderately anterior to the bicanine axis. The internal mandibular contour is a very narrow parabola in contrast to the wider basal contour which displays great lateral eversion posteriorly. There are weak to moderate superior and inferior transverse tori, the latter bearing strong mental spines.

The anterior root of the ramus is broken at its origin, lateral to M$_2$. Occlusal and basal margins diverge strongly anteriorly, resulting in a deep symphysis. The symphysis is angled sharply posteriorly and the anterior symphyseal contour is rounded and bulbous. The lateral aspects of the corpus have very flat posterior portions and distinctive hollowing in the areas above the mental foramina at the P$_3$ to P$_4$ position. The corpus is tall and fairly narrow, especially in its anterior portion.

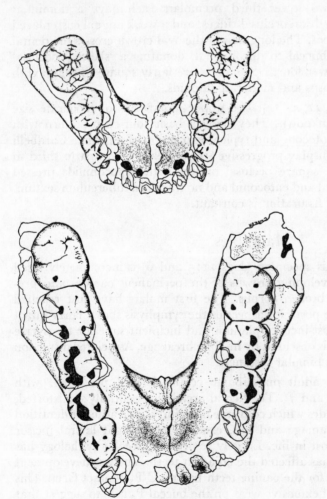

FIG. 11:5

Occlusal views of juvenile and adult mandibles (LH 2 and 4).

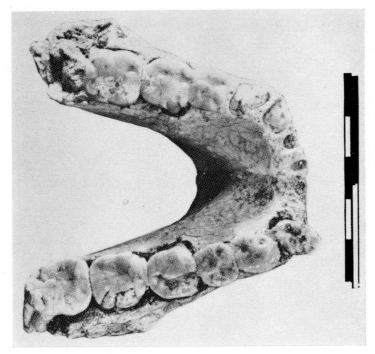

Fig. 11:7. Adult mandible from the Laetolil Beds (LH 4) occlusal view.

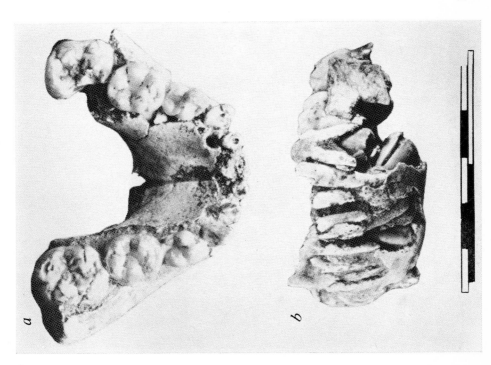

Fig. 11:6. Juvenile mandible from the Laetolil Beds (LH 2) occlusal and frontal views.

Implications of the specimens

The Laetolil fossil hominid sample, including the original Garusi maxillary fragment[5], seems to be representative of only one phylogenetic entity or lineage. The variations observed in the material seem to be primarily size-based and stem from individual and sexual factors.

The deciduous teeth, particularly the lower deciduous first molars, display remarkable similarity to hominid specimens from South Africa (Taung; STS 24)[10] as well as to individuals tentatively assigned to *Homo* in East Africa (KNM ER 820, 1507)[11, 12]. They depart strongly from the pattern of molarisation displayed in the South African 'robust' specimens (TM 1601; SK 61, 64)[10] as well as from East African specimens generally assigned to the same hominid lineage (KNM ER 1477)[13].

The Laetolil permanent canines and incisors are relatively and absolutely large and bear a great deal of lingual relief. They ally themselves similarly to earlier hominid specimens such as STS 3, 50, 51, 52; MLD 11; OH 7, 16; KNM ER 803, 1590[10, 15, 18-21] as well as to the younger African and Asian specimens usually assigned to *Homo erectus*. These features set the Laetolil specimens apart from the sample including SK 23, 48, 876 and so on; Peninj; KNM CH 1; OH 5, 38; KNM ER 729, 1171[10, 15, 18-21]. The Laetolil permanent premolars show none of the molarisation seen in the latter specimens and bear particularly strong resemblances to South and East African material (STS 51, 52, 55; OH 7, 16, 24; KNM ER 808, 992[10, 11, 14, 12].

The Laetolil permanent molars are consistent in aligning with the South African 'gracile' australopithecine and the East African *Homo* material in both size and morphology. The molars do not display the increased size, extra cusps or bulging, expanded, 'puffy' development of the individual cusps seen in high frequency among South and East African 'robust' forms (SK 6, 13, 48, 52; TM 1517; Peninj; KNM CH 1; KNM ER 729, 801, 802[10, 17-20]. The adult mandible has resemblances to certain East African specimens such as KNM ER 1802, with similar corpus section and basal eversion.

The Laetolil fossil hominids have several features possibly consistent with their radiometric age. These traits include the large crown size and lingual morphology of the permanent canines; the morphology and wear of the C/P_3 complex; the buccolingually elongate upper premolars; the overall square occlusal aspect of the lower permanent molars; the low symphyseal angle; the bulbous anterior symphysis; the relatively straight posterior tooth rows; the low placement of the mental foramina, and the distinctive lateral corpus contours including small, superiorly placed extramolar sulci.

Preliminary assessment indicates strong resemblance between the Laetolil hominids and later radiometrically dated specimens assigned to the genus *Homo* in East Africa. Such assessment suggests placement of the Laetolil specimens among the earliest firmly dated members of this genus. It should come as no surprise that the earlier members of the genus *Homo* display an increasing frequency of features generally interpreted as 'primitive' or 'pongid like', which indicate derivation from as yet largely hypothetical ancestors.

Much of the recently discovered comparable fossil hominid material from the Hadar region of Ethiopia shows strong similarity to the Laetolil specimens[23], and further collection combined with detailed comparative analysis of material from both localities is essential for the further understanding of human origins. The Laetolil collection adds to the developing phylogenetic perspective of the early Hominidae and emphasises the need for taxonomic schemes reflective of and consistent with the evolutionary processes involved in the origin and radiation of this family.

Acknowledgements

We thank the following for facilities, financial support, permission to examine originals and for discussion: the United Republic of Tanzania, the trustees of the National Museums of Kenya, the NSF Graduate Fellowship Program, the Scott Turner Fund, the Rackham Dissertation Fund, the National Geographic Society, G. Brent Dalrymple, C. K. Brain, T. Gray, D. C. Johanson, K. Kimeu, P. Leakey, R. E. F. Leakey, P. V. Tobias, E. Vrba, A. Walker, C. Weiler and B. Wood.

References

1. WEINERT, H. 1950. *Z. Morph. Anthrop.*, **42**, 138–48.
2. DIETRICH, W. O. 1942. *Palaeontographica*, **94A**, 43–133.
3. MAGLIO, V. J. 1969. *Breviora*, **336**.
4. KENT, P. E. 1941. *Geol. Mag.* London, **78**, 173–84.
5. KENT, *op. cit.*
6. KOHL-LARSEN, L. 1943. *Auf des Spuren des Vormenschen*, 2, 379–81.
7. KENT, *op. cit.*
8. PICKERING, R. 1964. *Endulen, Quarter degree sheet 52*, Geol. Surv. Tanz.
9. HAY, R. L. 1963. *Bull. Geol. Soc. Am.*, 1281–6.
10. HAY, R. L. 1976. *Geology of the Olduvai Gorge*, Univ. of Calif. Press.
11. DIETRICH, W. O. *op. cit.*
12. HOPWOOD, A. T., 1951. In *Olduvai Gorge*, by L. S. B. Leakey, Cambridge Univ. Press.
13. KOHL-LARSEN, L. *op. cit.*
14. ROBINSON, J. T. 1956. *Transv. Museum Mem.*, 9, 1.
15. LEAKEY, R. E. F. & WOOD, B. A. 1973. *Am. Jour. Phys. Anth.*, **39**, 355.
16. LEAKEY, R. E. F. & WOOD, B. A. 1974. *Am. Jour. Phys. Anth.*, **39**, 355.
17. LEAKEY, R. E. F. 1972. *Nature*, **242**, 170.
18. LEAKEY, L. S. B. 1960. *Nature*, **188**, 1050.
19. LEAKEY, L. S. B. & LEAKEY, M. D. 1964. *Nature*, **202**, 3.
20. DAY, M. H. & LEAKEY, R. E. F. 1974. *Am. Jour. Phys. Anth.*, **41**, 367.
21. LEAKEY, R. E. F. 1974. *Nature*, **248**, 653.
22. CARNEY, J., HILL, A., MILLER, J. & WALKER, A. 1971. *Nature*, **230**, 509.
23. LEAKEY, R. E. F., MUNGAI, J. M. & WALKER, A. C. 1972. *Am. Jour. Phys. Anth.*, **36**, 235.
24. LEAKEY, R. E. F. & WALKER, A. C. 1973. *Am. Jour. Phys. Anth.*, **39**, 205.
25. TOBIAS, P. V. 1967. *Olduvai Gorge*, 2, Cambridge Univ. Press.
26. LEAKEY, M. D., CLARKE, R. J. & LEAKEY, L. S. B. 1971. *Nature*, **232**, 308.
27. JOHANSON, D. C. & TAIEB, M. 1976. *Nature, Lond.* **260**, 293–7.

12

ROBERT M. SHACKLETON

Geological Map of the Olorgesailie Area, Kenya

This 1:10 000 coloured geological map of the area around the Olorgesailie archaeological reserve, as well as a geological map of the area between Olorgesailie and Ngong on a scale of 1:50 000, was prepared in 1946 for a conference in Nairobi organised by Dr L. S. B. Leakey. Funds were not then available to publish these maps.

A reduced and simplified version of a 1:600 geological map of the Olorgesailie archaeological reserve, prepared in 1944 was published by the Geological Survey of Kenya (Bull. 42, Geology of the Magadi Area, B. H. Baker, 1958). The three maps were based on theodolite triangulation and made on a plane table using a rangefinder for the 1:10 000 and 1:50 000 maps and tacheometer for the 1:600 map.

On this published version of the 1:10 000 map, the names of the units have been changed to those used by Baker (1958 and 1976; also see Isaac, this volume paper 13 for further details of the mapped lithological units of the Olorgesailie Formation and a full list of references). The map was kindly redrawn, for colour reproduction, by Mrs Barbara Isaac. Grants towards the cost of publication, from the Royal Society, London and the L. S. B. Leakey Foundation, are gratefully acknowledged.

12

ROBERT M. SHACKLETON

Geological Map of the Ol'orgesaille Area, Kenya

13

GLYNN LL. ISAAC

The Olorgesailie Formation: Stratigraphy, tectonics and the palaeogeographic context of the Middle Pleistocene archaeological sites

This report accompanies the 1:10,000
geological map surveyed by ROBERT M. SHACKLETON

The Olorgesailie Formation consists of sedimentary deposits which were laid down during the Middle Pleistocene in a small basin of internal drainage on the floor of the Gregory Rift Valley 65 kms south-west of Nairobi. The particular interest of the beds stems firstly from the fact that they preserve a rich and abundant archaeological record of the life patterns and handicrafts of early man; and secondly from the fact that the excellent exposure of the stratigraphy makes it possible to work out the history of sedimentation and to reconstruct the conditions under which the beds were deposited.

For various reasons, Olorgesailie has come to be one of the best-known sites in the Rift Valley; this is partly due to the fact that it was discovered at a relatively early stage in the development of Quaternary studies in East Africa, and partly owing to the fact that access to Olorgesailie from Nairobi is easy by comparison with many other East African sites. However, the fame of Olorgesailie is mostly attributable to the spectacular abundance of well-preserved early stone age artefacts that occur there.

The site has been known for more than 30 years (see below) and yet for various reasons no comprehensive account of either the archaeology or the geology has hitherto appeared. A monograph on the prehistory is now in press (Isaac, in press) and this volume provides a welcome opportunity to present an outline of the stratigraphy and a palaeoenvironemental interpretation. This accompanies a contoured 1:10 000 geological map of the Olorgesailie area which was surveyed by R. M. Shackleton in 1946 but which has not previously been published.

Historical background

In 1919, while on a foot safari from Nairobi to Magadi, the pioneer geologist J. W. Gregory pitched his camp at the foot of Mount Olorgesailie, which name he transcribed as Ol Gasalik. He observed the diatomites and other pale sediments which he took to belong within the series of beds attributed to 'Lake Kamasia' and which he believed to be of Miocene age. On some of the sedimentary outcrops at Olorgesailie, Gregory found 'stone picks', but given his view of the age of the beds, it is not surprising that he concluded that the picks represented diatomite mining tools of stone age men, rather than archaeological relics that had been stratified within the Formation and re-exposed by erosion (Gregory 1921: 221). It was left to that other great pioneer of East African studies, Louis Leakey, to correct this misunderstanding.

The effective discovery of the great wealth of palaeoanthropological evidence that is preserved at Olorgesailie, took place in 1943. Louis and Mary Leakey were at that time tied by petrol rationing and by war-time duties to the vicinity of Nairobi, so they began a systematic archaeological survey in the sector of the Rift Valley adjoining Nairobi; because of Gregory's reports, the Olorgesailie area was recognised as being particularly promising. Louis Leakey (1974:159) has described the discovery as follows:

> Having left the car under a tree for the day, we carried food and water with us and moved off on a series of quick traverses, as far as possible keeping parallel with each other and in touch by occasional shouts. Suddenly, at almost exactly the same moment, Mary, Menengetti and I each found exposures with quantities of handaxes and some fossil bones. Mary's was the most prolific site and she kept shouting to me to come over and see it. I, on the other hand, had found a site not nearly as rich as hers, but with plenty of fossils and some handaxes, and was calling her to come over to me! Meanwhile in the area between us, we heard Menengetti calling excitedly to both of us that he had found handaxes.
>
> I abandoned my site, having first tied a handkerchief to a thornbush in the middle of it, so that I could relocate it later on, and went across to Mary. When I saw her site I could scarcely believe my eyes. In an area of about fifty by sixty feet there were literally hundreds upon hundreds of perfect, very large handaxes and cleavers.

In the years following this dramatic discovery, Mary and Louis carried out fairly extensive excavations with the help of Italian prisoners of war. Their trenches exposed large segments of ancient living floors and in hindsight, we can recognise that this work was the start of a new research movement. Thus the development of the Olorgesailie style of excavations was continued by J. D. Clark at Kalambo Falls and by F. Clark Howell at Isimila, while the Leakeys themselves went on to extend this kind of research method at Olduvai Gorge. The approach is now standard.

In 1947 the scientists attending the first Pan African Congress on Prehistory, visited the excavations and shortly afterwards, through the generosity of the Masai inhabitants of the area, the prime archaeological sites were set aside as an enclosed National Park. The spectacular surface concentrations and excavated parts of stone age camp sites became public exhibits. Hundreds of thousands of visitors have been afforded a glimpse of early prehistory through the Olorgesailie museum on the spot. The reserve is now under the care of the National Museum, Nairobi, and continues to fulfil a valuable role in public education.

In 1944 at the invitation of Louis and Mary Leakey, R. M. Shackleton began to make a detailed geological survey of the area, and the mapping was completed during a return visit in 1946. An outline of these findings appeared in *Geologische Rundschau* (Shackleton 1955), and later B. H. Baker incorporated part of the results of this work into his account of the geology of the Magadi area (Shackleton in Baker 1958). This geological survey report included Shackleton's

1:600 map of the Prehistoric Site, but the 1:10 000 map has remained unpublished until the opportunity provided by this volume.

There has been a succession of wardens at the Prehistoric Site: F. Andrews, G. della Giustina, M. Posnansky, R. Wright and G. Ll. Isaac, most of whom have carried out some archaeological research operations. By 1961, a few brief general reports had been published (e.g. L. S. B. Leakey 1952; Sonia Cole 1954, 1963) but only one site report (Posnansky 1959). Because no comprehensive account had yet been compiled, Louis Leakey appointed Glynn Isaac to the post of warden and specifically charged that the research be completed and published. A full report was presented as a Ph.D. thesis at Cambridge University (Isaac 1968a) and a copy lodged in the National Museum Library. As already mentioned a condensed version of this has recently been accepted for publication by the University of Chicago Press (Isaac in press).

As part of the second round of research, the author took the maps previously made by Shackleton and the stratigraphic subdivisions that he had established, and set out to reconstruct the palaeogeographic circumstances under which deposition had occurred. This work was inspired by the methods of R. L. Hay at Olduvai, and it involved recording numerous closely spaced stratigraphic columns and using these to work out the geometry of the beds. From the spatial configuration, lithological variation and the differing thicknesses of the strata it has been possible to reach preliminary interpretations regarding the processes controlling sedimentation.

Detailed work has not been done on the petrology and the micropalaeontology. However, the studies reported here provide at least an outline of the palaeoenvironmental and stratigraphic setting of the archaeological sites. This account offers an outline of the geological history and of the processes of sedimentation. The results can be checked, revised and refined by more detailed researches which we hope will be carried out in the future.

This paper presents a formal definition of the Olorgesailie Formation, and summarises the main features of its stratigraphy. Based on the primary evidence, discussion is largely concerned with the following aspects of interpretation:

1. Tectonic influence on sedimentation.
2. Reconstruction of hydrological regimes.
3. Reconstruction of the palaeoenvironmental setting of the archaeological sites.
4. The correlation and dating of the Formation.

Various features of the Olorgesailie sediments have been used in the past as indicators of a pluvial episode during the Middle Pleistocene (Leakey 1955; S. Cole 1954) and the evidence from the site has also been discussed in various critiques of climate-stratigraphy in East Africa (e.g. Cooke 1958; Flint 1959a, b). All of the interpretations and counter-interpretations in these papers were formulated without access to the results of a comprehensive study of the sedimentary strata and their geometry. In this report I am making available a summary of the basic data that have been missing from previous discussions. I also seek to show that it is essential for studies of structure, stratigraphy and micropalaeontology to precede palaeoclimatic interpretations. In the Rift Valley, tectonically induced changes in topography and hydrology can produce effects on sedimentation that are particularly readily mistaken for climatic changes.

The geology of the region surrounding Olorgesailie has been mapped and studied by B. H. Baker (1958, 1963, 1967), F. J. Matheson (1966), R. M. Shackleton (1955 and unpublished), L. A. J. Williams (1967), followed by B. H. Baker & J. G. Mitchell (1976). The results of these studies constitute the broader context within which the detailed stratigraphic research on the Olorgesailie Formation can be understood.

Table 1 summarises the regional sequence for Olorgesailie and immediately adjacent areas. The stratigraphic units are predominantly volcanic rocks which have erupted and accumulated in and alongsde a Rift trough which was undergoing progressive deformation and subsidence. The Olorgesailie Formation was deposited on the floor of the Rift Valley and it conforms to the regional pattern by being largely composed of volcanic silts (Table 2) and by showing abundant evidence of tectonic control over sedimentation patterns.

The structure of the Olorgesailie–Magadi segment of the Gregory Rift has developed gradually since the late Miocene or early Pliocene. The Rift in this region is assymetric with the largest displacements occurring in the west along the Nguruman Fault, where an impressive boundary escarpment occurs. From the eastern margin, the PreCambrian rocks have been progressively flexed down towards the Nguruman Fault, thereby gradually forming the rift. Concurrently with flexure the limb of the monoclinal fold has been broken up by a very large number of sub-parallel faults arranged *en echelon*.

Lavas, pyroclastic deposits and sediments have accumulated throughout the deformation and have been progressively affected by displacements on the faults. As a consequence, the eastern margin and the present floor of the Rift is broken

---

TABLE I: *The regional sequence in the Olorgesailie area*
(after Baker & Mitchell, in press)

Ages are given in millions of years and are based on K/Ar and palaeomagnetic determinations.

Alluvium, travertines and swamp deposits	–
OLORGESAILIE FORMATION (= Legemunge Beds)	0.42 and 0.48
Ol Doinyo Nyegi volcanics	0.66
(Magadi) "Plateau trachyte series"	0.66 – 1.25
Ol Tepesi basalts	1.4 – 1.6
Ol Keju Nero basalts	
Limuru trachytes	1.9
Singaraini basalts	2.3
Olorgesailie volcanics	2.2 – 2.8
Ol Esayeti volcanics	3.6 – 6.7

up by the surface expression of numerous small horst and graben structures. Most of the graben are sediment traps and many are also depressions which lack surface outlets and hence which contain lakes or seasonal pools and swamps. It is common for these smaller graben to resemble the larger structure in having relatively high east-facing fault scarps, and minor opposing margins. The system of sub-parallel horst and graben structures which is very conspicuous in the present landscape, has come to be known as 'grid faulting' (Gregory 1921; Baker 1958).

There has been some tendency to treat the faulting in the region as being divisible into a series of major episodes; however, it seems more probable that movements have occurred continually. The interaction of varying rates of stratigraphic accumulation and intermittent movement can readily give rise to an illusion of pronounced episodicity. Differential deformation within the Olorgesailie Formation and subsequent faulting of the entire Formation certainly seems to attest the continual intermittent character of 'grid faulting'. (See below.) For this reason the extent of fault deformation makes a very poor basis for inter-area correlation, though within any area there is a tendency for cumulative increase in the magnitude of fault displacement of formations, proportional to age.

The former controversy regarding the Tertiary or Quaternary 'age' of the Rift has proved to involve false propositions. A complex feature of this kind does not have an 'age' but an evolutionary history. Shackleton (1955) has shown that in general displacements affecting 'Middle Pleistocene' formations such as those at Olorgesailie, Munya wa Gicheru and Kariandusi are very much smaller than the sum of preceding displacements. However, in the Natron graben it has been shown that the largest visible faults postdate the early Middle Pleistocene and loc-

TABLE 2: *Estimates of the proportional representation of the dominant lithologies in various parts of the Olorgesailie Formation*

	Western outcrops D - G	Central outcrops Z	East central: Main Site area b,d,T
Diatomites	31%	29%	19½%*
Pale tuffaceous siltstones	28	23	31½
Vitric tuff and volcanic sands	4	10	21½
Brown clays	29	30	20½*
Other	9	8	6½

* These lithologies are under-represented in this area because Member 13 has been removed by erosion.

ally involved displacements of the graben centre by 300–500 metres. (Isaac 1965, 1967). Demonstration of post Lower Pleistocene movements of this magnitude in the southern portion of the Rift should perhaps lead to a fresh review of Shackleton's (1955:262) suggestion that the floor of the Rift may have been tilted by 0°30′ to the south during the Middle Pleistocene (see also Baker & Mitchell 1976). As is shown below, it may be easiest to explain the history of the Olorgesailie lake basin by postulating the occurrence of such a tilt.

The Olorgesailie Formation

Definition

The formal name Olorgesailie Formation is proposed for a series of well-stratified sediments consisting principally of pale, volcanic siltstones, volcanic sandstones, diatomites and brown non-volcanic siltstones. Lesser quantities of claystones, pyroclastic deposits, and limestones also occur. Fig. 13:3 designates and describes a type section

This stratigraphic unit was designated as 'Olorgesailie Lake Beds' in Baker 1958, and the informal term 'Legemunge Beds' has been proposed in Baker & Mitchell (1976). However, I feel that the formal definition of an entity of formation rank is justified, and that the name Olorgesailie is so widely known in association that any change will merely cause confusion.

Distribution

In the Olorgesailie area, the known outcrops of sediments with these lithological characters are restricted to a comparatively small area, but it is inferred that the Formation underlies more recent soils and alluvium in large parts of the graben complex to the north of Mt. Olorgesailie. The known and inferred distribution of the Formation is shown in Fig. 13:1. Fig. 13:2 shows the location of section measurement traverses.

Thickness: upper and lower limits

Wherever exposed the lowest strata showing the characteristic lithology of the Formation rest unconformably on faulted and eroded surfaces of various older volcanic formations, including the following units as defined by Baker (1958): 'Mount Olorgesailie Volcanic Series', 'Ol Keju Nyiro Basalts' (and basaltic tuffs), and the 'Plateau Trachyte Series'. The surface underlying the Formation clearly shows considerable relief and the thickness of the Formation varies accordingly (Figs 13:5 and 6) attaining its maximum measured thickness of c. 60 m at locality Z. The sum of the maximum observed thicknesses of the individual members is greater than 80 m and it seems probable that the Formation attained actual thicknesses of this order in localities where exposures are now incomplete or lacking.

The upper limit of strata with the lithological characteristics of the Formation is in all available exposures an erosional disconformity and the Formation is only discontinuously overlain by a variety of deposits with partly contrasting lithologies. (See below.)

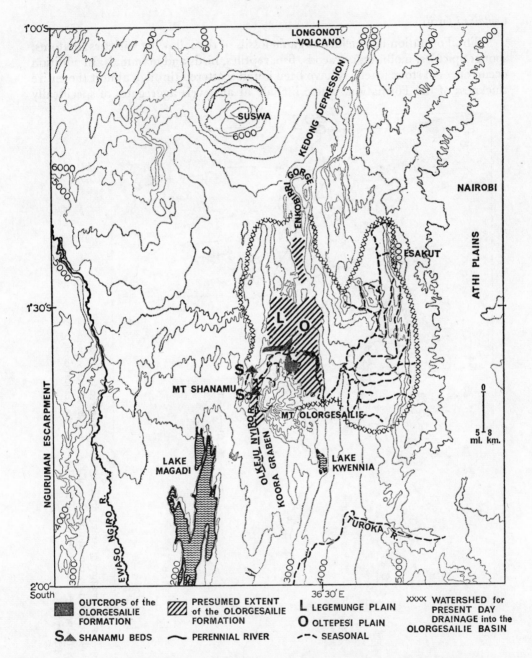

FIG. 13:1. Topographic map of the sector of the Rift Valley within which the Olorgesailie sedimentary basin is situated. The watershed for the drainage system that terminates in the Koora graben is marked. Note that if the Kedong depression to the north had a positive water balance it would overflow into the Olorgesailie basin.

S = Shanamu Beds; So = Sonorua Beds; both are isolated outcrops tentatively grouped with the Olorgesailie Formation.

Contained fossils

The Formation has yielded organic fossils of the following varieties: diatoms, sponge spicules, molluscs, ostracods, fish, reptiles, birds and mammals. In addition occurrences of stone artefacts have been found scattered through almost the entire thickness of the Formation. These include at all levels aggregates of specifically

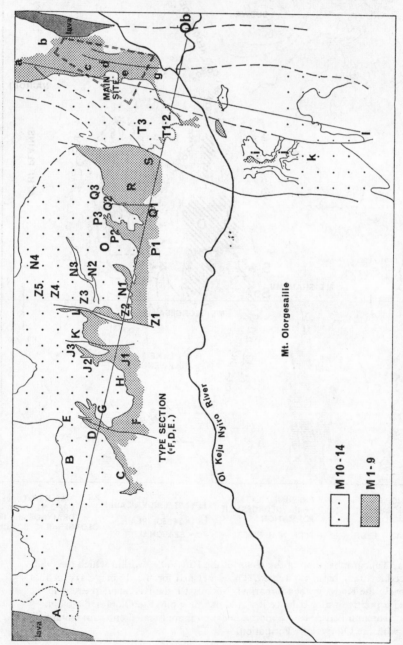

FIG. 13:2. Plan showing the location of stratigraphic sections measured by Isaac, and of the two section planes used in the compilation of Figs 13:4–8. Dense stipple = Members 1–9; Open stipple = Members 10–14. For detail see the coloured map by R. M. Shackleton (this volume).

Acheulian character as well as aggregates lacking the large bifacial tools specifically diagnostic of the Acheulian. Further details concerning the fossils and the artefacts are given in subsequent sections of the paper.

Comments

The lithological and stratigraphic unity of the proposed Formation is believed to result from quasi-continuous deposition of the strata within a portion of a drainage basin near where drainage terminated or was ponded. The subsequent sedimentary units in the area show predominant lithologies which contrast with the Formation and which are partially attributable to deposition under conditions in which more vigorous drainage had been established through this portion of the basin.

At present, recognition of the Formation poses few problems: most of the outcrops can be seen to be attached to a coherent, continuous body of sediments which include the strata represented in the type section. There are, however, three isolated groups of outcrops showing similar lithology, but having uncertain stratigraphic relations. These are provisionally referred to the Formation, but have been given the following informal designations for ease of reference:

1. *Oltepesi beds*, (Map loc. Ob in Fig. 13:2) comprising a down thrown fault block mapped by Shackleton (unpublished) as a post Olorgesailie Formation unit (Gamblian 1) and regarded by the author as a probable lateral equivalent of the uppermost Members of the Olorgesailie Formation.
2. *Sonorua beds*, (Map loc. So in Fig. 13:1) comprised of various small remnants along the western margin of the Koora graben.
3. *Shanamu beds*, (Map loc. S in Fig. 13:1) comprising poorly exposed sediments in a small depression upon the crest of the horst which forms the western boundary of the Legemunge and north Koora graben.

Problems of definition and recognition may well become more acute if studies of subsurface borehole samples are undertaken. It is very probable that both before and after the deposition of the known lower and upper members, units of comparable lithology were deposited in parts of the basin where surface exposures do not now occur.

The Internal Stratigraphy of the Olorgesailie Formation

Knowledge of the Olorgesailie Formation is derived solely from observations of the surface exposures which have resulted from dissection of the sediments by the Ol Keju Nyiro river and its tributary gullies. Three groups of exposures can conveniently be recognised: (See Map Fig. 13:1).

1. A broad band of exposures along the southern and south-eastern margins of the Legemunge plain.
2. Exposures around the northern foot hills of Mount Olorgesailie.
3. A narrow line of exposures along the gorge by which the Ol Keju Nyiro traverses the Oltepesi plain. This third group has been less intensely studied than the other two. It appears to be laterally equivalent to Members 1–9, but detailed relationships have not been worked out and hence this portion of the Formation is not discussed further.

During detailed geological survey and mapping done in 1944 and 1946,

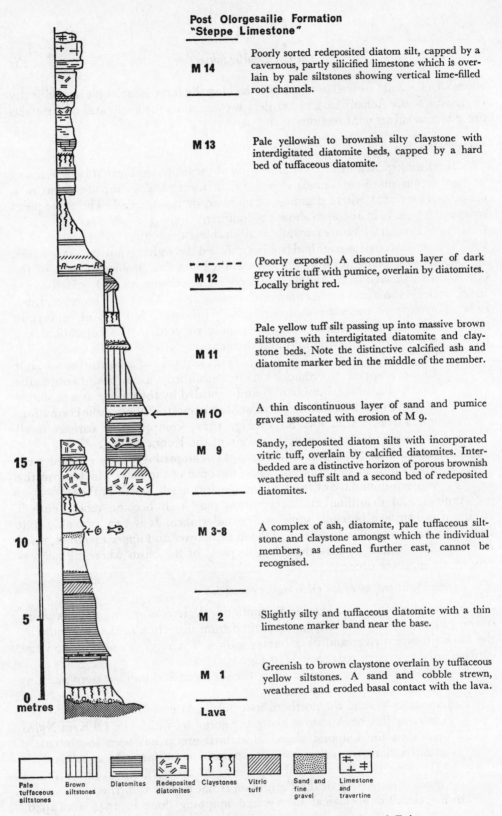

Post Olorgesailie Formation "Steppe Limestone"

M 14 Poorly sorted redeposited diatom silt, capped by a cavernous, partly silicified limestone which is overlain by pale siltstones showing vertical lime-filled root channels.

M 13 Pale yellowish to brownish silty claystone with interdigitated diatomite beds, capped by a hard bed of tuffaceous diatomite.

M 12 (Poorly exposed) A discontinuous layer of dark grey vitric tuff with pumice, overlain by diatomites. Locally bright red.

M 11 Pale yellow tuff silt passing up into massive brown siltstones with interdigitated diatomite and claystone beds. Note the distinctive calcified ash and diatomite marker bed in the middle of the member.

M 10 A thin discontinuous layer of sand and pumice gravel associated with erosion of M 9.

M 9 Sandy, redeposited diatom silts with incorporated vitric tuff, overlain by calcified diatomites. Interbedded are a distinctive horizon of porous brownish weathered tuff silt and a second bed of redeposited diatomites.

M 3-8 A complex of ash, diatomite, pale tuffaceous siltstone and claystone amongst which the individual members, as defined further east, cannot be recognised.

M 2 Slightly silty and tuffaceous diatomite with a thin limestone marker band near the base.

M 1 Greenish to brown claystone overlain by tuffaceous yellow siltstones. A sand and cobble strewn, weathered and eroded basal contact with the lava.

Lava

| Pale tuffaceous siltstones | Brown siltstones | Diatomites | Redeposited diatomites | Claystones | Vitric tuff | Sand and fine gravel | Limestone and travertine |

FIG. 13:3. The composite section comprised of sections F, D and E is designated as a type section for the Olorgesailie Formation.

Shackleton found it convenient to group the strata which crop out in these exposures into 14 numbered members. These 14 divisions were accepted by Baker and published as layers L1–L14 (1958:34). They have been used with minor modifications by Isaac in field work carried out between 1961 and 1965. An alternative scheme was used by L. S. B. Leakey (1952) and Posnansky (1959) who employed a stratigraphic nomenclature involving 12 or 13 numbered 'land surfaces', believed to represent widespread disconformities. Subsequent work has shown that many of these are highly localised and therefore unsuitable as a basis either for correlation or stratigraphic designation.

The fold-out, coloured map shows the distribution of the outcrops of each member at a scale of 1 : 10 000. Fig. 13:4 is a section compiled by R. M. Shackleton to show the present disposition of units.*

Detailed palaeogeographic and palaeoecological reconstruction was desired as a prerequisite for the interpretation of the artefact occurrences. Isaac made records and measurements of stratigraphy at all well-exposed sections; lateral relations were checked where possible by tracing units between sections. The data have been used to compile a series of correlation diagrams drawn as transects, with the relevant measured sections projected onto the section planes shown on figure 13:2 as A—Ob and a—l (Also as A—A', B—B' on the large map.

Fig. 13:5 shows the west-east composite section, A—A', drawn with the base of Member 12 as a horizontal section datum. As can be clearly seen, it is not possible to plot the members as continuous bands across the section. Field studies and analysis of the record showed conclusively that faulting and deformation took place during deposition. The section shows the dislocation of beds by faults and the thickening of beds in localities where subsidence was actively taking place. In order to reduce the visual confusion of such a diagram, the section has been divided into upper and lower halves, and these were replotted with different correlation planes (Figs 13:6 and 7). It is clear that the principal deformation took place during the deposition of Members 10 and 11.

Fig. 13:8 shows a similar composite section based on the north-south alignment of exposure. This transect is parallel to the regional trend of fault lines and consequently does not show the same degree of deformation. The members overlapped one another as they buried the volcanic rocks forming the lower slopes of Mt. Olorgesailie.

Several members show significant facies differentiation, and these provide important palaeogeographical evidence. In general, where a difference is observed the finer grained and more diatomaceous facies occur in the west around the mouth of the Koora graben, and the coarser or less diatomaceous facies occur in the easterly exposures away from the Koora (see for example Members 4, 9 and 10). This suggests that at times the segment of the basin at the mouth of the Koora was flooded while the area immediately north-east was exposed. The evidence of drainage channels and crossbedding corroborates this inference: observed alignments of channels and current bedding are shown as arrows on Fig. 13:9 and it

* Space precludes publication of detailed descriptions of the Members. Copies of a typescript giving additional information have been lodged in the libraries of the Geological Society (London), National Museum (Kenya) and the Geological Survey (Kenya).

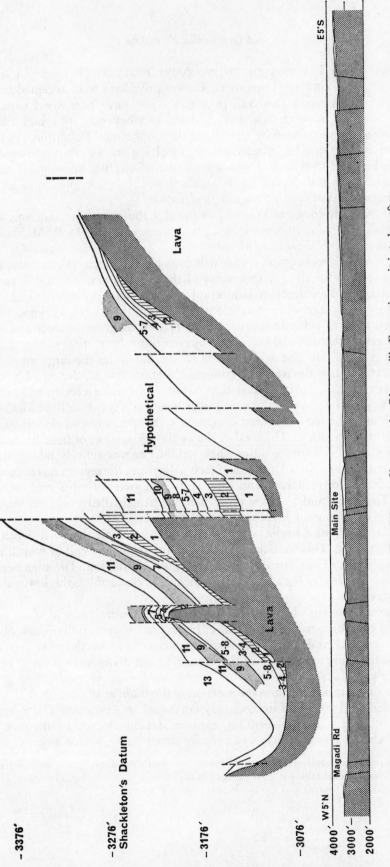

FIG. 13:4 An East-West profile across the Olorgesailie Formation as it is today after fault dislocation. Surveyed and compiled by R. M. Shackleton.

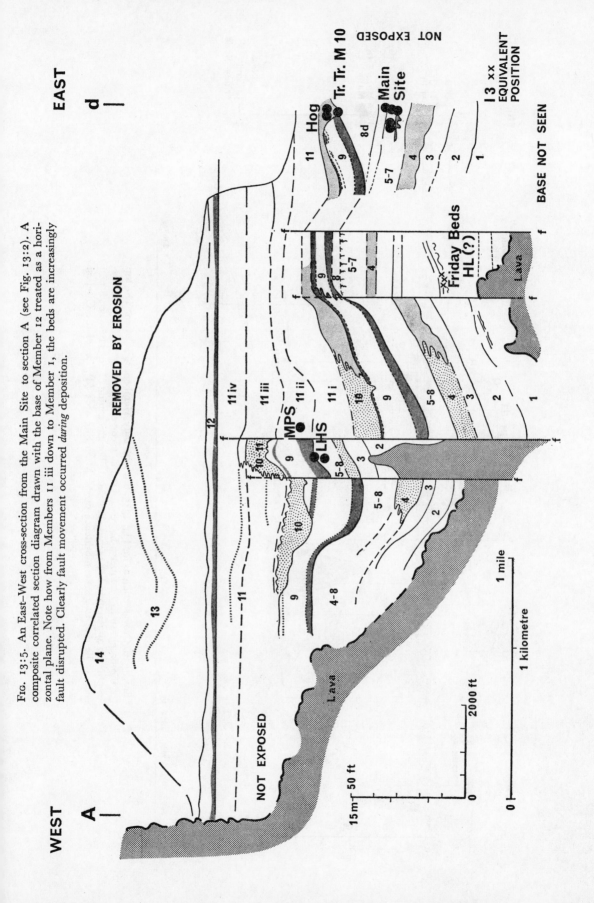

FIG. 13:5. An East–West cross-section from the Main Site to section A (see Fig. 13:2). A composite correlated section diagram drawn with the base of Member 12 treated as a horizontal plane. Note how from Members 11 iii down to Member 1, the beds are increasingly fault disrupted. Clearly fault movement occurred *during* deposition.

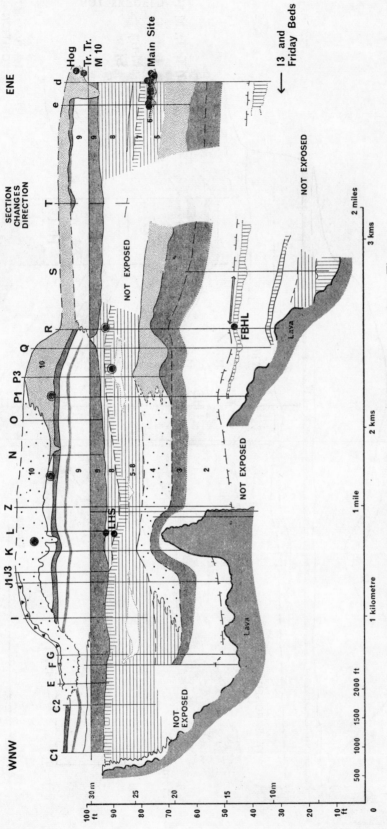

FIG. 13:6. Cross-section as for Fig. 13:5, but with only Members 1–9 shown. The top of the basal ash in Member 9 has been used as a correlation horizon. The fault disruption so apparent in Fig 13:5 is eliminated. The thickening of units between S and d may be due to gentle subsidence of the Main Site area (d), i.e. of the Oltepesi graben.

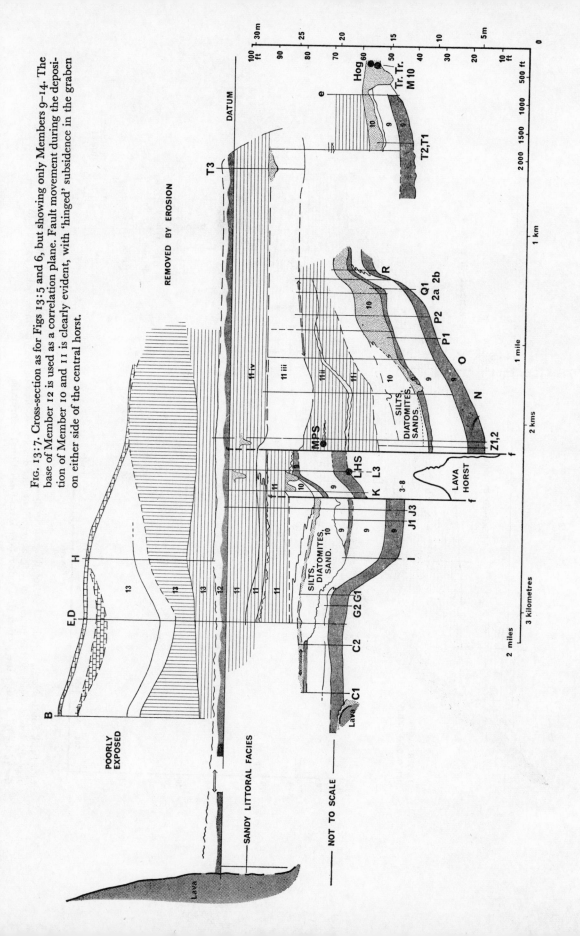

FIG. 13:7. Cross-section as for Figs 13:5 and 6, but showing only Members 9–14. The base of Member 12 is used as a correlation plane. Fault movement during the deposition of Member 10 and 11 is clearly evident, with 'hinged' subsidence in the graben on either side of the central horst.

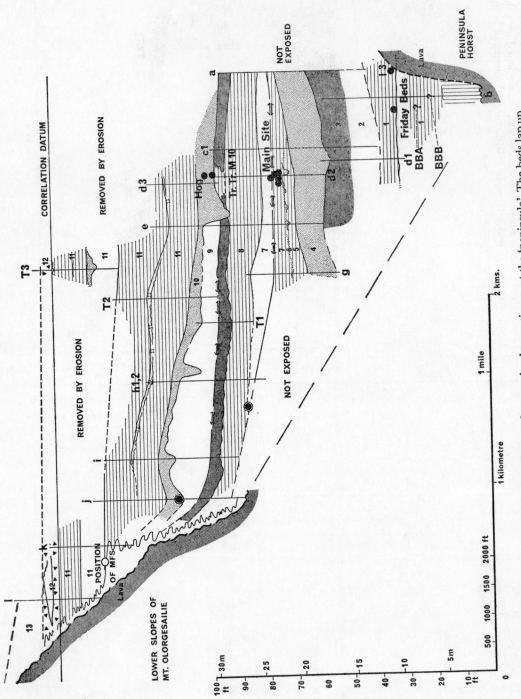

FIG. 13:8. South–North correlated section diagram from section l to section a, at the 'peninsula'. The beds lap up over the lower slopes of the mountain with a gravelly facies developed close to the interface. Being parallel to the plane of faulting the section does not reflect intra-formational deformation.

can be seen that most point in the direction of the opening into the Koora graben.

Detailed petrographic and sedimentological studies have not been undertaken so that the lithological designations given in these sections and the tables are based largely on the field characteristics and the features of hand specimens. The *diatomites* are gleaming white, porous rocks with low specific gravity (0.5–1.0). They grade into diatomaceous silts. The *volcanic-siltstones* are generally pale yellowish to greenish deposits with a fine-grained, clay-rich matrix. Microscopic examination of samples of these sediments reveals a sufficient number of sherds of volcanic glass to indicate that they are composed of fine pyroclastic material which has undergone a relatively long history of transport and alteration. Tests by R. L. Hay (pers. comm.) show that alteration has *not* resulted in the formation of the suite of zeolitic minerals so characteristic of sediments in adjoining basins such as the Magadi, Natron and Olduvai basins. (See R. L. Hay 1966.)

Claystones may be under-represented in the tables and sections since this designation was ordinarily restricted to sediments showing signs of plasticity when wet.

Calcium carbonate concretions are abundant through the Formation, but since these are generally in the form of large macroscopic nodules, the term marl which was previously used, has been dropped.

The primary diatomites and diatomaceous silts clearly represent deposition under aquatic or marshy conditions and the preliminary report by Dr J. L. Richardson (n.d.) gives some tentative palaeolimnological indications. Out of a column of 40 samples taken along a traverse at the type section, Richardson found some 14 to be effectively non-diatomaceous or to contain only broken, reworked diatoms. Reworked diatomites sometimes appear as a distinctive lithology with particles of consolidated material included in a silty matrix. The remaining 26 showed varied diatom assemblages in which planktonic (i.e. free floating) forms predominate only in some 6 or 7 samples. At least in the vicinity of the type section, diatomaceous silts were mainly laid down under fluctuating shallow water conditions of a lake margin or swamp, with abundant reed beds and emergent vegetation. According to Richardson, the water chemistry also seems to have fluctuated. There are no indications of extreme salinity, but the diatom assemblages suggest that frequently the water was comparable in its concentration and alkalinity to Lake Rukwa. About 9 samples are said to indicate more or less fresh water conditions, while about 14 show signs of moderate alkalinity. (Additional details on the diatoms are given in the supplementary archive.) The geometry of the two most widespread diatomite units, M 2, and the middle subdivision of M 11, shows that at the time of deposition of these units, at least the entire southern end of Oltepesi and Legemunge graben were flooded.

Changes in lithology do *not* involve very marked persistent developmental trends, but rather appear as oscillations. Normal deposition of silts, clays and diatomaceous beds was interrupted at least twice by a comparatively sudden influx of volcanic sands and fine gravels. These formed sheets or tongues that rest on a weakly eroded surface of earlier beds. The clearest examples of this are provided by Members 4 and 10, but Member 6 and the basal, sandy, volcanic

ash of Member 9 represent minor expressions of the same phenomenon. The intermittent ability of the incoming drainage to deliver quantities of sand to areas of
the basin where silt and clay deposition normally prevailed may have been caused
in a number of ways:

1. Climatic, vegetational volcanic, or topographic changes in the catchment
   area, affecting flood velocities and sediment load.
2. A drop in base level due to partial drying-up of the lake waters.
3. A drop in effective base level due to tectonic subsidence of the Koora
   graben.

The Source of the Sediments

Table 2 shows the approximate percentage thickness of the various principal
sediment types as identified in the field. In three out of four of the selected sections,
diatomite, a dominantly biogenic deposit of silica, is present in the highest proportion; however, it forms only a third or less of the total thickness. Of the remainder, the brown silts and some of the clays resemble the beds which have been
accumulating in the area in recent times, (e.g. the extensive alluvium beds at the
north end of the Koora graben). These brown beds, like their recent equivalents,
are presumed to be derived from the erosion of soils which formed on the volcanic
rocks of the area. These make up less than twenty per cent of the total. The remaining fifty to sixty per cent are distinctive sediments without counterparts

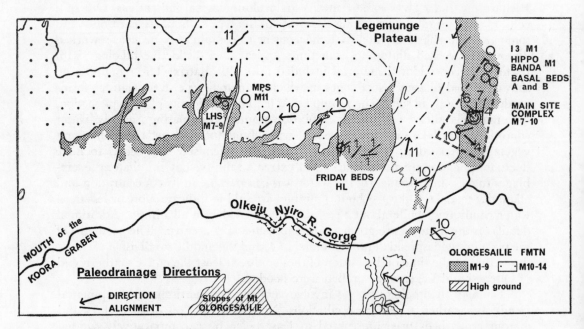

FIG. 13:9. Plan showing obervations of channel alignment (lines without arrow marking) and of current direction as gauged from cross-beddings (arrows). Numbers
adjacent to the markings refer to the Member in which the observation was made.
The location of archaeological sites is also shown.

amongst the material at present entering the basin. The distinctive characters appear to be due to the presence of large quantities of pyroclastic particles. Vitric tuffs, pumice and volcanic sands probably reached the basin by aerial transport or by stream transport, involving little chemical alteration. The finer grained volcanic silts are probably accumulations of volcanic dust which was being deposited subaerially over the whole catchment area. Some of the volcanic sediments undoubtedly derive from the reworking of extensive tuffs on the eastern Rift escarpment (e.g. Kerichwa Valley Tuffs), but since these continue to be eroded without giving rise to deposits with distinctive lithology, it seems likely that the bulk of the fresh and altered volcanic material must derive from volcanic vents which were active during the deposition of the Legemunge Formation. The nearest vent of any size is Ol Doinyo Nyegi (Mt. Shelford) twelve miles to the south, (Fig. 13:1). However, it is a comparatively small volcano and is in a direction opposite to the drainage evidenced in the Formation. The large central volcanoes of Suswa and Longonot (see Thompson & Dodson 1963; Johnson 1969) were actively being built up in the Middle and early Upper Pleistocene, and are the most probable source of the vast quantities of pyroclastic material involved. The channel linking the Kedong Valley and Olorgesailie (see below) may well have carried pumice and tuff southward into the Legemunge basin.

Tectonic Control over Sedimentation

The distribution and field relations of the Olorgesailie Formation make it clear that the sediments were deposited within a series of interconnected depressions which had formed as a result of 'grid faulting'. The establishment of a sediment trap and a lake in the area can thus be regarded largely as the outcome of tectonic changes (Shackleton 1955:259; Baker 1958; Flint 1959; Baker & Mitchell in press).

Shackleton (1955:262 and Fig. 13:4 of this paper) has shown that individual fault displacements amounting to at least 50 metres, affect the whole Olorgesailie Formation and are thus later in time. More recent stratigraphic studies by the author demonstrate also that appreciable movement of the fault blocks occurred during deposition. The maximum measured displacement being circa 15 m (50 feet) during the deposition of Member 11. (See Fig. 13:8). Where relationships can be determined, it generally proves that the same fault lines were involved in the pre-Formation, intra-Formation and post-Formation movements.

The cessation of deposition in the study area of beds with the characteristic Olorgesailie Formation lithology was certainly determined by tectonic changes. Consideration of the present-day topography makes it clear that post-Formation changes have occurred which are of a much larger order of magnitude than the 50-metre fault displacements documented in Fig. 13:5. Olorgesailie Formation diatomites occur around the north end of the Koora graben at elevations between circa 940 m and 1040 m above sea level. The sediment covered graben floor now slopes gently south for 50 km to reach elevations of circa 650 m at the point where the convergence of fault lines terminates the structure. The rim of the depression and the 'floor' of the Rift Valley slope in a similar fashion. The present southward sloping trough could not retain lake waters at the latitude of the Olorgesailie

Formation outcrops unless the southern rim of the Koora graben was raised by approximately 300 m. The necessity to invoke regional tectonic changes of this magnitude might be avoided by postulating that the bounding faults of the Koora graben had not yet formed a deep trench through the Olorgesailie–Shanamu volcanic massif. (See Baker 1958). However, this hypothesis is virtually untenable because the evidence of facies changes and drainage orientation within the Formation show clearly that surfaces at the north end of the Koora graben were already at a slightly lower altitude than surfaces in the Legemunge and Oltepesi graben. The reverse tilt which would be necessary to permit the damming of drainage at Olorgesailie is approximately 1 in 160, or 0° 20′. As mentioned above, a regional tilt of this amount was suggested by Shackleton on the basis of other evidence.

Hydrography and Palaeo-Geography

The tectonic depression within which the Olorgesailie Formation was deposited remains a closed basin. The map in Fig. 13:1 shows the form of the basin. For the most part the bounding watershed runs close to the floor of the depression and the Ol Keju Nyiro is the only major river course now entering the Oltepesi, Legemunge and Koora graben. This channel collects run off from circa 360 km² of high ground along the west-facing rift escarpment. By means of a substantial gorge, this river traverses the southern end of the Oltepesi and Legemunge plains and enters the Koora graben down which it flows within a channel of diminishing depth. Eventually the channel disappears by distributing its flow over the Koora alluvial plain, parts of which are seasonally inundated. Rainfall averages between 26 and 20 inches (700–500 mm) *per annum*, over much of the catchment area (Thompson & Sansom 1967) but diminishes to 15 inches (384 mm) at Magadi in the south-west. Precipitation and flow are strongly seasonal; prodigious spates may occur during the two rainy seasons, while for most of the year flow ceases or is negligible. During flooding, some water is lost to a side lobe of the basin, namely the Kwennia depression where a shallow lake is intermittently formed.

At the present time, evaporation and seepage readily dispose of the water delivered by the Ol Keju Nyiro.

From the present negative water balance in the basin it may appear that a more favourable overall precipitation—evaporation balance must have been involved in the prolonged existence of the lake which deposited the Olorgesailie Formation diatomites. This is very probably true, but unfortunately other hydrographic variables may also have been involved, so that the importance of climatic changes relative to differences in drainage regime cannot be assessed.

Two adjoining catchment areas may formerly have drained into the Olorgesailie graben complex:

1. The Kedong depression which lies 30 km to the north is still linked to the Oltepesi plain by a dry channel which appears to be in part a narrow graben and in part a water-cut gorge, (Map Fig. 13:1) (Sikes 1926). A high terrace of alluvium lines part of its course (Isaac unpublished; Baker & Mitchell 1976).

2. The Turoka river to the south-east of Mount Olorgesailie now turns south to feed the ephemeral 'Lake' Kabongo; however, prior to regional tilting, if such has occurred, drainage from this catchment may have joined the Ol Keju Nyiro.

Several closed lake basins exist along the Gregory Rift, and are discharged by evaporation. In consequence they are mostly strongly saline and highly alkaline. Examples include Lakes Manyara, Natron, Magadi, Elmenteita, Nakuru and Hannington. Lakes Baringo and Turkana have lower salinities, but Lake Turkana is known to have overflowed intermittently into the Nile drainage in times as recent as the Mid-Holocene (K. Butzer 1971). Lake Naivasha is a fresh water lake allegedly with discharge due to seepage. R. L. Hay (1965) and H. Eugster (1970) have shown that the sediments of the saline-alkaline lakes tend to have a distinctive set of authigenic minerals including various zeolites, feldspars and unstable sodium silicates which alter to cherts. The Olorgesailie Formation lacks any extensive development of this suite and their absence is almost certainly attributable to relatively low salinity during deposition.

The continued existence of comparatively fresh water conditions throughout the deposition of the Olorgesailie Formation suggests that an outlet mechanism must have existed. Whether this outlet was a perennial overflow channel such as exist for the fresh water lakes of the western Rift Valley, an intermittent overflow as is believed to have discharged Lake Rudolf periodically throughout the Quaternary, or a subterranean 'leak' as is claimed to be the case of the modern Lakes Naivasha and Baringo, cannot at present be determined.

The development cycle of freshwater lakes with a surface outlet commonly involves progressive reduction in average water depth because sedimentation raises the floor level and channel erosion lowers the overflow. Since the Olorgesailie Formation gives evidence of oscillations between deep water, shallow water and flood plain deposition through the accumulations of 55 m or more of sediments, it seems unlikely that a surface outlet could have occupied a fixed altimetric relationship to the rock floor of the basin. However, it is known that certain fault blocks within the basin were subsiding relative to fault blocks forming the margins, and this mechanism may have served to maintain a more or less constant height difference between the outlet and the floor. However, an hypothesis involving an underground exit would explain the observed situation just as well and perhaps rather more simply.

Whatever the outlet mechanism, the maintenance of fresh water conditions for a prolonged period appears to be consistent with an hypothesis of more humid conditions but does not constitute definitive proof.

Taken together, the sedimentary evidence and the hydrographic indicators permit a reconstruction of aspects of the palaeoenvironment.

1. The subsidence of a group of interconnecting graben created a localised basin of internal drainage. Sediments began to accumulate and mantled the floor of the depression.

2. A comparatively small fluctuating lake formed and occupied the topographically lowest portion of the basin. Sediments of lacustrine facies (*sensu stricto*) do not occur continuously through available sections, but since these have not

necessarily sampled the lowest portions of the basin at all times, it is not known whether lake waters were in permanent existence.

3. It is possible that during the deposition of Members 1 and 2 the lowest point in the basin was within the Oltepesi or Legemunge graben and an extensive stable lake in this portion of the basin is certainly evidenced by the diatomites of Member 2.

4. Thereafter it is certain that the lowest portion of the floor of the depression lay within the Koora graben south of all surface exposures. With alternate expansions and contractions, the lake waters extended out of the narrow Koora trough into the wider Legemunge and Oltepesi graben, and then withdrew again into the Koora where they may or may not have persisted Fig. 13:10. During phases of extension, diatomites and littoral sediments were deposited in the areas where exposures of the Formation are available (e.g. Members 9, 11 b–d, 12, part of 13), while during phases of contraction, alluvial deposition occurred in these areas: (e.g. Members 3–8, 11 a and 11 e). The evidence indicates the existence of a flat flood plain traversed by braided ephemeral stream courses. Roor markings indicate a dense grass and shrub vegetation, perhaps with trees along water courses. A series of concentrations of Acheulian artefacts are associated with the channels crossing this flood plain and were generally at least 3 km distant from actual lake waters (Isaac 1966b, 1968a).

5. The intra-formational fault deformation evidence in Members 9–11 and the development of the hypothetical regional tilt probably caused the lowest part of the basin, and hence the area of most stable lacustrine deposition, to shift southward down the Koora, until drainage was permanently established through the area north of Mount Olorgesailie with consequent erosion of the portions of the Olorgesailie Formation situated in that area.

The Age of the Olorgesailie Formation

Assessment of the age of the Formation is of importance both for the compilation of prehistoric sequences and for understanding the development of this part of the Gregory Rift Valley. There are two principal lines of evidence with relevance to this problem:

(i) Isotopic age measurements.

(ii) Correlation of contained fossils: (a) fauna,
(b) artefacts.

None of these methods yet provide an age estimate with entirely satisfactory confidence limits.

Potassium-argon measurements have been made on volcanic materials from several formations in the Magadi regional sequence, and from two members of the Olorgesailie Formation. The results are shown in Tables 1 and 3. It can be seen that the Olorgesailie Formation unconformably overlies the Magadi Plateau trachytes which have given a range of dates from 1.25 to 0.63 m.y. This age range provides a *terminus post quem* for the Formation.

Pumice from Members 4 and 10 has been dated by Evernden and Curtis (1965) and pumice from Member 10 by J. Miller (1967). The scatter of values suggests that the material dated may have derived in part from reworked deposits of various ages, but there is a good chance that the two lowest values of 0.425 and 0.486 million potassium-argon years may be valid estimates of age. However, one or two potassium-argon dates, especially when selected from a scatter of values cannot be relied upon in situations where there are no other checks on their reliability. When the measurements were first reported they were greeted with widespread scepticism. However, the establishment of a chronometric time scale for East African stratigraphy shows that these age values are consistent with many other dates. (See Bishop & Miller 1972; Isaac 1972a; Isaac & Curtis 1974; M. D. Leakey 1975.)

Palaeontological Correlations

The taxonomic composition of the Olorgesailie sample of mammalian fossils is listed in Table 4.

The biostratigraphic implications of the Olorgesailie fossil mammal assemblage have been considered by MacInnes & Leakey (in L. S. B. Leakey 1951) and by Cooke (1963). Both these authors agree that the sample is most like the assemblages from Olduvai Bed IV, and from Kanjera. All of these contain a fairly high proportion of extinct species and can be confidently classified as Middle Pleistocene (i.e. between 0.7 and 0.1 million years ago). (See Bishop & Miller 1972; Butzer & Isaac 1975). Greater precision does not seem possible at present, but a fresh review would be welcome. Meave & Richard Leakey (1973) have studied the *Theropithecus (Simopithecus) oswaldi* fossils from the Olorgesailie Formation and find them to be broadly comparable with a series of fossils from Olduvai Bed II. Unfortunately the Bed IV material is not sufficiently complete to allow of firm conclusions regarding its resemblance to the Olorgesailie material.

TABLE 3: *Potassium argon dates for volcanic materials from the Olorgesailie Formation*

Stratigraphic Unit	Laboratory and Number	Age Determination $(\times 10^6 \text{ years})$	Notes
Member 10	Cambridge 834	$0.425 \pm .009$	Pumice
Member 4	Berkeley 413	0.486	Anorthoclase
Member 10	Berkeley 925	1.45	Anorthoclase
Member 4	Berkeley 923	1.64	Anorthoclase
Member 4	Berkeley 435	2.9	Rerun of 413

Sedimentation rates and the time span represented by the Formation

Consideration of data for diatomaceous ooze from Lake Naivasha (Richardson & Richardson pers. comm. and 1972), for Lake Tanganyika (Livingstone 1965) and for Val del Inferno (Bonnadonna 1965) suggest that 15 000 years is an

TABLE 4. *A list of zoological taxa identified from the Olorgesailie Formation*

Taxon	Synonyms	Authority (See key below)
Mammal fossils		
Theropithecus oswaldi mariae	*Simopithecus*	3, 5
Elephas recki (advanced stage)	*Archidiskodon recki*	1, 6
Hipparion albertense	*Stylohipparion albertense*	4
Equus aff. grevi		4
Equus oldowayensis		1, 4
Ceratotherium simum		1, 4
Metridiochoerus meadowsi	*Tapinochoerus*	2, 4
Mesochoerus cf. *major*		8
Hippopotamus gorgops		1, 4
Giraffa camelopardus		4
Giraffa gracilis (?)		7
Sivatherium maurusium	*Libytherium oldowaiensis*	1, 4, 8
Tragelaphus sp.	*Strepsiceros* sp.	7
Taurotragus sp.		7
Homoioceros sp. ?		7
Kobus sp.		7
Redunca sp.		7
Connochaetes sp.	*Gorgon* sp.	7, 8
Antidorcas sp. ?	*Phenacotragus* sp. ?	7, 8
Aepyceros sp.		

* Denotes an extinct species.

Key to authorities:
1. MacInnes in L. S. B. Leakey 1951.
2. L. S. B. Leakey 1958.
3. Leakey and Whitworth 1958.
4. Cooke 1963.
5. M. G. Leakey and R. E. F. Leakey 1973.
6. Coppens personal communication.
7. Gentry personal communication.
8. J. M. Harris personal communication.

Molluscan fossils (identified by B. Verdcourt).
Bithynia neumanni Van Marten (? equivalent to *Gabbia* cf. *subbadiella*)
        found in Member 1 section a.
                Members 5–8 section F, K–L.
                Member 9 section 1, d, J.
Biomphalaria sp. found in Member 9 section 1, J.

(In addition *Melanoides tuberculata*, *Biomphalaria* cf. *sudanica* and *Lymnaea* sp. ? have been identified in the post Olorgesailie Formation travertines near section A.)

absolute minimum estimate for the time of deposition of a Formation which contains 20 m of diatomites in addition to c. 36 m of other fine-grained deposits. On the basis of these data 50 000 years would appear to be a better estimate. The data for Olduvai Gorge (Hay 1976) and for various lakes in the more arid parts of North America (e.g. Flint & Gale 1958:706) suggest that a time span of one or two hundred thousand years might have been involved.

Post Olorgesailie Formation Sediments and Events

The deposition of sediments with the characteristics of the Olorgesailie Formation in the area north of Mt. Olorgesailie presumably ceased when subsidence in the Koora Graben and regional tilting shifted the terminus of drainage away to the south. Very probably there are sediments buried in the Koora that are younger than M 14 and yet which, according to lithological criteria, would belong to the Formation. The Oltepesi Beds are also probably best included in the Formation, though they may well have been deposited after Member 14. However this may be, as a result of subsidence and/or southerly tilt, vigorous drainage was established by a proto-Ol Keju Nyiro River which has traversed the area north of Mt. Olorgesailie, and has cut a deep gorge in the lavas at the foot of the mountain. The Olorgesailie Formation has been uplifted as a horst in the area now termed the Legemunge Plain. The edges of this elevated block are being dissected by gullies that are tributary to the Ol Keju Nyiro. As the incision proceeded the Ol Keju Nyiro dumped beds of conglomerates, gravel, sand and brown alluvium in the developing drainage way. There are thus widespread relict patches of fluvial deposits in the area. Some of them are marked as Ps on the coloured map (Post Olorgesailie Sediments). These beds, which have not been studied in detail, include tuffs and pumice gravels. Huge boulders weighing many tons occur on the outcrops as stranded residuals of the river's former bed load. They provide impressive evidence of the power of peak floods during the period of erosion. Baker & Mitchell (1976) suggest that a single catastrophic flood was responsible for many of these deposits. I am not aware of specific evidence for or against this view.

In the Koora graben, large piedmont fans have developed along the foot of Mt. Olorgesailie. These consist of sandy brown silts, silty sands and conglomerate beds. Richard Wright discovered a site with Levallois technique artefacts in these beds. The centre of the graben is filled to a considerable depth with brown clayey-silt alluvium. At the northern end, the river is incised within the alluvium, but further south it discharges onto the surface of a flood plain which is periodically flooded and which is still actively aggrading. The bounding faults of the Koora graben converge at a point some 50 km south of the outcrops of the Olorgesailie Formation. In the angle thus formed is a small closed basin with precipitous margins. It is currently dry, but superficial diatomite deposits indicate comparatively recent lacustrine episodes, and there is a water-cut gorge leading out of the graben into the Magadi trough. Presumably during the late Pleistocene, or early Holocene, a small lake formed in the southern extremity of the graben. This can be regarded as a linear successor of the lake that laid down the Olorgesailie Formation diatomites, displaced 50 km by progressive regional tilting.

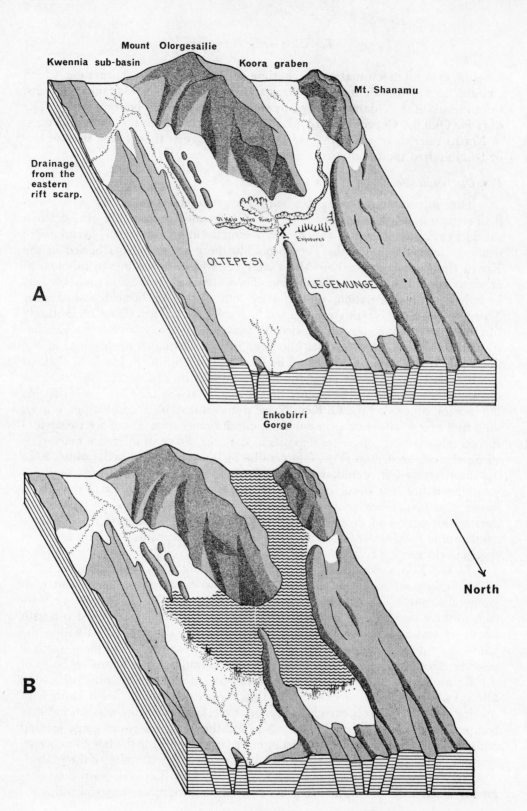

A

Kwennia sub-basin

Mount Olorgesailie

Koora graben

Mt. Shanamu

Drainage from the eastern rift scarp.

Ol Keju Nyiro River

Exposures

OLTEPESI

LEGEMUNGE

Enkobirri Gorge

B

North

The archaeological contents of the Formation

Archaeological evidence in the Olorgesailie Formation includes the following different kinds and configurations of materials:

1. A general *low-density scatter* of discarded artefacts in most of the sedimentary units that were not laid down under stable lake waters. Isaac (1968a) has given a rough estimate of 1 piece per m³ for this general background scatter. More precise measurements should be made in future research.

FIG. 13:10. Schematic block diagrams of the Olorgesailie sedimentary basin.

A. Present-day physiography with through drainage established into the Koora graben and the headwards erosion of a gorge across the foot of Mt. Olorgesailie and across the Oltepesi plain. X marks the Prehistoric Sites enclosure.

B. During the deposition of the Olorgesailie Formation at a stage of maximum extent of lake waters (e.g. Member 2). Note that the placement of the shoreline is speculative since it lay north of all natural outcrops.

C. As for B, but at a time when the lake had contracted into the confines of the Koora graben (e.g. Member 7). In the absence of exposures in the Koora we do not know if the lake was reduced to a small swamp or not. It was under these palaeogeographic circumstances that the Acheulian sites on display in the Cat Walk area were occupied (indicated by o).

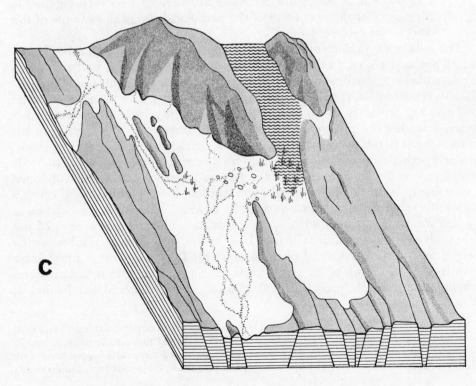

2. *Concentrated patches of artefacts and broken-up animal bones.* These are interpreted as being the result of the repeated use, by the prehistoric inhabitants of the particular locality as a 'camp' or 'home base'. The bone is presumed to be food refuse accumulated as a result of the transport of food back to the area for social sharing. The stone artefacts seem to represent discarded tools[1] and implements, plus the debris formed by stone-knapping undertaken in order to make tools and implements or to resharpen them. Most of the known patches of artefacts in the Olorgesailie Formation are distant by a mile or more from outcrops of the raw materials and occur in beds that are virtually devoid of stone particles larger than a few millimetres in diameter. The main transport agency in these cases was clearly the hominids themselves. However, in a number of cases, localised patches of artefacts appear to have been further concentrated by fluviatile processes (see below and Isaac 1968a; Isaac in press).

3. *Concentrated patches of artefacts* without large amounts of broken-up bones. These may either represent sites where relatively little bone refuse accumulated, or sites where bone dispersal and bone decay has destroyed the food refuse (see Isaac 1967b). Other such sites, like MFS near the foot of Mt. Olorgesailie, where new material crops out, may represent workshop sites.

4. Partial remains of the *carcass of a large animal plus a few artefacts.* These are generally interpreted as butchery sites. The locality HBS where parts of a hippopotamus skeleton and various artefacts are preserved together in Member 1 at the north end of the park enclosure is an example of this category of evidence.

The palaeogeographic conditions indicated by the work that has been summarised in this paper enabled the author to examine site location patterns and to offer hypotheses regarding aspects of land-use by the ancient prehistoric inhabitants (Isaac 1968a, c, 1972b, 1975 and in press). Briefly it emerges that camps show a strong tendency to be located along the sandy beds of seasonal water courses. Probably the preference for these kinds of situation stemmed from the fact that groves of relatively large trees grew along the channels, providing shade, fruits and perhaps at times, a refuge from predators. The camps were commonly a few kilometres distant from the lake margins, and water may have been obtained by scooping holes in the sand. In addition to these considerations, sand offers a comfortable substratum on which to sit or lie down. The recurrent association of sites and water courses within the East African lake basins was first pointed out on the basis of the Olorgesailie evidence (Isaac 1966b, 1968c), but has subsequently emerged as a fairly widespread phenomenon (M. D. Leakey 1971, 1975; Isaac 1972b, 1975). As a result of the predilection for camping in stream beds, it seems probable that fluviatile processes have rearranged the material left behind at

---

[1] All pieces of stone that are broken or shaped by man are *artefacts* but palaeolithic archaeologists reserve the term *tool* for objects which have been deliberately shaped to a certain pattern, usually by secondary trimming. Many other objects such as sharp edged flakes and jagged cores were probably used; these we have covered by the vague and technically undefined term, 'implements'.

many of the sites. In particular, it seems likely that some of the spectacular concentrations of material at Olorgesailie and elsewhere in East Africa have been exaggerated by the operation of the kinematic wave effect (Isaac 1967b).

The Middle Pleistocene humans seem also to have been attracted to camp repeatedly on or near rocky ridges that projected out into the alluvial flats that surrounded the lake itself, as at site I 3 and site LHS. However, sites also occur in varied situations other than the two just mentioned. In Member 10 there is a scattering of artefacts eroding out of sands that may represent beach or delta front conditions. This locality has not yet been excavated. A few patches of artefacts also occur in flood plain deposits away from any distinguishing topographic features. Densities are commonly much lower at such sites, perhaps because there was less persistent reason for relocating successive occupation in exactly the same spot.

Those Olorgesailie sites that contain fossil bone food refuse contribute valuable evidence towards the reconstruction of Middle Pleistocene dietary and economic patterns. (Isaac 1968a, c, 1971, 1975). A varied series of species are represented in differing combinations and proportions at different sites. These include medium-sized animals such as bovids (antelope), equids and suids, as well as rarer remains of large animals like giraffes, hippopotami and elephants. At almost all sites there are a few scattered catfish, frog and rodent bones, but since these occur throughout the beds, it is not clear whether or not they can be positively identified as a part of the human diet. The most impressive hunting pattern yet discovered at Olorgesailie is that which is documented by the remains of more than 40 individual *Simopithecus* baboons, at the site of DE/89 Horizon B.

The known archaeological sites in the Olorgesailie Formation are so distributed in relation to stratigraphy that they divide naturally into three sets (Fig. 13:11). A cluster in Member 1 has been designated the Lower Stratigraphic Set (LSS), another in Member 7 comprises the Middle Stratigraphic Set (MSS) while a more scattered series in Members 10 and 11 comprise the Upper Stratigraphic Set (USS). The stone artefacts recovered from the Formation vary considerably from site to site. There are differences both in the *percentage composition* of each assemblage as gauged by classification according to a standard typological scheme (Kleindienst 1961), and in the morphological norms of each sample of a given tool category. This is particularly true of the handaxe tool category.

Setting up these divisions has allowed comparisons to be made of variation between sites *within* a set and of variation *between* sets. In general, variation within a set is so great that between set differences should only be interpreted with great caution (Isaac 1972d). Similar conclusions with regard to the apparently erratic, non-cumulative patterns of change in Middle Pleistocene artefacts have also been reached at Isimila (Howell *et al.* 1962), Olduvai (M. D. Leakey 1975) and elsewhere. Given this situation, it would clearly be extremely unwise to use the artefacts as a basis for correlation between sedimentary basins, or as a crucial part of any process of chronological estimation. All that can be said at present is that the features of the varied assemblages in the Olorgesailie Formation do not seem to contradict the K/Ar age estimate of $0.4 - 0.5 \times 10^6$ years.

The interpretation of the differences between assemblages at Olorgesailie and other comparable sites has become a keenly debated question amongst archaeologists. (L. S. B. Leakey 1954; Posnansky 1959; Kleindienst 1961; Isaac 1969, 1972d; Clark 1970; M. D. Leakey 1971; Binford 1972). The discussion turns principally on a divergence of opinion over whether or not two or more markedly contrasting cultural systems coexisted and persisted within East Africa during most of the Middle Pleistocene. An alternative hypothesis involves complex assemblage differentiation in relation to different mixtures of activities being performed at different sites. This debate affects the archaeological taxonomy and nomenclature of the material. If one judges that all the material derives from one varied cultural system then it can all be included within the Acheulian Industrial Complex, perhaps with assemblages dominated by numerous bifaces being called Acheulian Type A, and those dominated by small scrapers being called Acheulian Type C (Kleindienst 1961). Alternatively, if one believes that each of these contrasting kinds of assemblage represents a separate culture, they will be termed Acheulian (*sensu stricto*) and Developed Oldowan respectively. This last name was

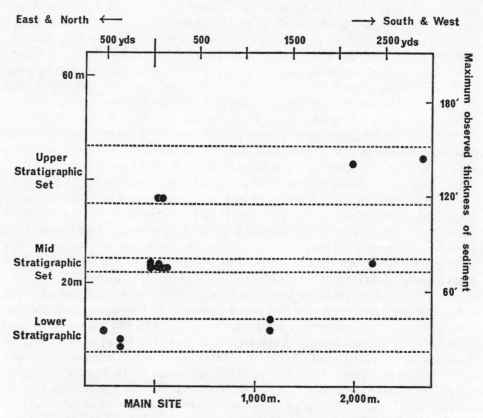

Fig. 13:11. Diagram showing the horizontal and vertical (stratigraphic) dispersal of excavated archaeological sites within the Olorgesailie Formation. The Lower Stratigraphic Set are in Member 1, the Middle in Member 7, and the Upper in Members 10 and 11.

introduced by Mary Leakey (1971) and supersedes unsatisfactory previous terms such as African Tayacian and Hope Fountain. The issue cannot be resolved at present, and detailed discussion of the pros and cons fall outside the scope of this paper.

Conclusion

In summary, the Olorgesailie Formation is a body of sedimentary rocks that represents deposition in a complex of interleading graben on the floor of the Gregory Rift Valley. By comparison with many other Rift Valley sedimentary formations, it is on a fairly small scale, both as regards area ($\sim$ 100 km²) and as regards thickness ($\sim$ 60 m). Furthermore, the sedimentary regime was rather different from those that seem to have prevailed in such other Pleistocene basins as Olduvai (Hay 1971 and in press), Peninj (Isaac 1967), Kapthurin (Bishop this volume), Lake Turkana (Coppens *et al.* 1975). Diatomites and pale diatomaceous sediments predominate in the Olorgesailie Formation in a degree that is not characteristic of most other well-known formations. A number of factors may all have contributed to this:

1. The Olorgesailie basin was a graben complex surrounded by fault dislocated lava terrain. Rivers delivering water to the lake flowed over appreciable distances with low gradients and consequently dumped the coarser load before reaching the sump itself. Fine tuffaceous silt, clay and biogenic sediments thus predominate in the central depression.
2. Although often somewhat alkaline, the lake does not ever seem to have been highly saline and it thus did not deposit sediments like those of the salty lakes that existed at Olduvai, Peninj and elsewhere.
3. Shallow, relatively fresh water conditions very probably encouraged dense growth of marsh and fringing aquatic vegetation. This in turn may have affected deposition characteristics.

The Olorgesailie Formation may well represent deposition under 'pluvial' conditions (Flint 1959a, b) but as has been pointed out, given the possibility of extensive tectonic modifications to the catchment area and drainage pattern, the geology of the Formation cannot stand on its own as conclusive evidence of a pluvial. Clearly at times the surrounds of the small lake and swamp were extensively used by stone age men, and a rich archaeological record of their activities is preserved. The Formation is richly fossiliferous, and it provides valuable opportunities for further work. Detailed micropalaeontological studies could elucidate aspects of the hydrology and vegetation, and in combination with the geology this would make possible an unusually well-rounded reconstruction of environmental conditions.

This report represents only a preliminary study of the geology of the Olorgesailie Formation. I hope that by the publication of a first approximation of the complex stratigraphy, plus R. M. Shackleton's excellent map, further research will be facilitated.

Acknowledgements

The late Dr L. S. B. Leakey gave me the opportunity to work at Olorgesailie from 1961–5, and he and Dr Mary Leakey gave me help and encouragement through-out that time. Professor Robert Shackleton was extremely generous in making his maps and some section data available to me. I used his map and stratigraphic framework as the starting point for this study.

Barbara Isaac has helped in the field, in the lab and by innumerable hours of drawing and redrawing sections, maps and diagrams. It has been through her efforts that it has been possible to reproduce Shackleton's detailed 1 : 10 000 map alongside this paper, published with generous assistance from the L. S. B. Leakey Foundation, Pasadena, and of the Royal Society of London.

Many scientists visited me in the field and gave valuable assistance and advice amongst them were B. H. Baker, W. W. Bishop, R. L. Hay, D. A. Living-stone, J. Richardson & R. M. Shackleton.

The National Parks and the National Museum helped support the work, as did the Wenner-Gren Foundation, the British Institute in Eastern Africa and the Boise Fund. The final revision of this paper was undertaken while benefiting from the hospitality of Peterhouse, Cambridge, during the tenure of a J. S. Guggenheim Memorial Fellowship.

References

BAKER, B. H. 1958. *Geology of the Magadi Area*. Report No. **42**, Geological Survey of Kenya. Nairobi: Government Printer.

—— 1963. *Geology of the Area South of Magadi*. Report No. **61**, Geological Survey of Kenya. Nairobi: Government Printer.

—— & MITCHELL, J. G. Volcanic stratigraphy and geochronology of the Kedong–Olorgesailie area and the evolution of the South Kenya rift valley. *Jl geol. Soc. Lond.* **132**, 467–84.

BINFORD, L. R. 1972. Contemporary model building: paradigms and the current state of Palaeo-lithic research. In: Clarke, D. L. (ed.) *Models in Archaeology*, pp. 109–66. London: Methuen.

BISHOP, W. W. & MILLER, J. A. (eds). 1972. *Calibration of Hominoid Evolution*. Edinburgh: Scottish Academic Press.

BONADONNA, F. P. 1965. Further information on the research in the Middle Pleistocene diatomite quarry of Valle dell'Inferno (Riano, Rome). *Quaternaria*, **7**, 279–99.

BUTZER, K. W. 1971. *Recent History of an Ethiopian Delta*. Research Papers No. 136. Department of Geography: University of Chicago.

—— & ISAAC, G. Ll. (eds). 1975. *After the Australopithecines: Stratigraphy, Ecology and Culture Change in the Middle Pleistocene*. World Anthropology Series. The Hague: Mouton.

CLARK, J. D. 1970. *The Prehistory of Africa*. London: Thames and Hudson.

COLE, S. 1954. *The Prehistory of East Africa*. London: Pelican Books.

—— 1963. *The Prehistory of East Africa*. New York: Macmillan.

COOKE, H. B. S. 1958. Observations relating to Quaternary environments in East and southern Africa. *Geological Society of South Africa Bulletin* (Annexure to), **60**, 1–73.

—— 1963. Pleistocene mammal faunas of Africa with particular reference to southern Africa. In: Howell, F. C. and Bourlière, F. (eds). *African Ecology and Human Evolution*, pp. 65–116. Chicago: Aldine.

COPPENS, Y., HOWELL, F. C., ISAAC, G. Ll. & LEAKEY, R. E. F. (eds). 1976. *Earliest Man and Environments in the Lake Rudolf Basin*. Chicago: University of Chicago Press.

EUGSTER, H. P. 1970. Chemistry and origin of the brines of Lake Magadi, Kenya. *Mineralogical Society of America*, Special Publication No. 3 (50th Anniversary Symposia) pp. 215–35.

EVERNDEN, J. F. & CURTIS, G. H. 1965. Potassium-argon dating of Late Cenozoic rocks in East Africa and Italy. *Current Anthropology*, **6**, 343–85.

FLINT, R. F. 1959a. On the basis of Pleistocene correlation in East Africa. *Geological Magazine*, **96**, 265–84.

—— 1959b. Pleistocene climates in East and southern Africa. *Bulletin of the Geological Society of America*, **70**, 343–74.

—— & GALE, W. A. 1958. Stratigraphy and radiocarbon dates at Searles Lake, California. *American Journal of Science*, **256**, 689–14.

GREGORY, J. W. 1921. *The Rift Valley and Geology of East Africa*. London: Seeley, Service and Co., Ltd.

HAY, R. L. 1966. Zeolites and zeolitic reactions in sedimentary rock. *Geological Society of America*, Special Paper No. 85. New York.

—— 1971. Geological background of Beds I and II. Stratigraphic Summary. In: Leakey, M. D. *Olduvai Gorge Volume Three*, pp. 9–18. Cambridge at the University Press.

—— 1976. *The Geology of Olduvai Gorge*. Berkeley: University of California Press.

HOWELL, F. C., COLE, G. H. & KLEINDIENST, M. R. 1962. Isimila, an Acheulian occupation site in the Iringa Highlands. *Actes du IVe. Congrès Panafricain de Préhistoire et de l'Etude du Quaternaire.* pp. 43–80. Tervuren: Musée Royale de l'Afrique Centrale.

ISAAC, G. Ll. 1965. The stratigraphy of the Peninj Beds and the provenance of the Natron Australopithecine mandible. *Quaternaria*, **7**, 101–30.

—— 1966a. The geological history of the Olorgesailie area. *Actas del V Congreso Panafricano y de Estudio del Cuaternario*, **2**, 125–33. Tenerife: Museo Arqueologico.

—— 1966b. New evidence from Olorgesailie relating to the character of Acheulian occupation sites. *Actas del V Congreso Panafricano de Prehistoria y de Estudio del Cuaternario*, **2**, 135–45. Tenerife: Museo Arqueologico.

—— 1967a. The stratigraphy of the Peninj Group—early Middle Pleistocene formations west of Lake Natron, Tanzania. In: Bishop, W. W. and Clark, J. D. *Background to African Evolution*, pp. 229–57. Chicago: University of Chicago Press.

—— 1967b. Towards the interpretation of occupation debris—some experiments and observations. *Kroeber Anthropological Society Papers*, **37**, 31–57.

—— 1968a. The Acheulian site complex at Olorgesailie, Kenya: a contribution to the interpretation of Middle Pleistocene culture in East Africa. Cambridge, doctoral thesis.

—— 1968b. Traces of Pleistocene hunters: an East African example. In: Lee, R. B. and DeVore, I. (eds). *Man the Hunter*, pp. 253–61. Chicago: Aldine.

—— 1969. Studies of early culture in East Africa. *World Archaeology*, **1**, 1–28.

—— 1971. The diet of early man: aspects of archaeological evidence from Lower and Middle Pleistocene sites in Africa, *World Archaeology*, **2**, 278–98.

—— 1972a. Chronology and the tempo of cultural change during the Pleistocene. In: Bishop, W. W. and Miller, J. A. (eds). *Calibration of Hominoid Evolution*, pp. 381–430. Edinburgh: Scottish Academic Press.

—— 1972b. Comparative studies of Pleistocene site locations in East Africa. In: Ucko, P. J., Tringham, R. and Dimbleby, G. W. (eds). *Man. Settlement and Urbanism*, pp. 165–76. London: Duckworth.

—— 1972c. Early phases of human behaviour: models in Lower Palaeolithic archaeology. In: Clarke, D. L. (ed.). *Models in Archaeology*, pp. 167–99. London: Methuen.

—— 1972d. Identification of cultural entities in the Middle Pleistocene. *Actes du VI Congrès Panafricain de Préhistoire et de l'Etude du Quaternaire, Dakar 1967*, pp. 566–00. Paris: Chambéry.

—— 1975. Stratigraphy and cultural patterns in East Africa during the middle ranges of Pleistocene time. In: Butzer, K. W. and Isaac, G. Ll. (eds). *After the Australopithecines*, pp. 495–542. The Hague: Mouton.

—— In press. *Olorgesailie: Archaeological Studies of a Middle Pleistocene Lake Basin in Kenya*. Chicago: University of Chicago Press.

—— & CURTIS, G. H. 1974. Age of early Acheulian industries from the Peninj Group, Tanzania. *Nature, Lond.* **249**, 624–7.

JOHNSON, R. W. 1969. Volcanic geology of Mount Suswa, Kenya. *Philosophical Transactions of the Royal Society of London* Series A, **265**, 383–412.

KLEINDIENST, M. R. 1961. Variability within the Late Acheulian assemblage in eastern Africa. *South African Archaeological Bulletin*, **16**, 35–52.

LEAKEY, L. S. B. 1951. *Olduvai Gorge*. Cambridge at the University Press.

—— 1954. Fourth edition. *Adam's Ancestors*. London: Methuen.

—— 1955. The climatic sequence of the Pleistocene in Africa. *Congrès Panafricain de Préhistoire Actes de la IIe. Session, Alger, 1952*, pp. 293–4. Paris: Arts et Metiers Graphiques.

—— 1974. *By the Evidence: Memoir 1932–1951*. New York and London: Harcourt, Brace, Jovanovich.

LEAKEY, M. D. 1971. *Olduvai Gorge Volume 3. Excavations in Beds I and II, 1960–1963*. Cambridge at the University Press.

—— 1975. Cultural patterns in the Olduvai sequence. In: Butzer, K. W. and Isaac, G. Ll. (eds). *After the Australopithecines*, pp. 477–93. The Hague: Mouton.

LEAKEY, M. G. & LEAKEY, R. E. F. 1973. Further evidence of *Simopithecus* (Mammalia, Primates) from Olduvai and Olorgesailie. In: Leakey, L. S. B., Savage, R. J. G. and Coryndon, S. C. (eds). *Fossil Vertebrates of Africa*, pp. 101–20. London and New York: Academic Press.

LIVINGSTONE, D. A. 1965. Sedimentation and the history of water level change in Lake Tanganyika. *Limnology and Oceanography*, **10**, 607–10.

McCALL, G. J. H., BAKER, B. H. & WALSH, J. 1967. Late Tertiary and Quaternary sediments in the Kenya Rift Valley. In: Bishop, W. W. and Clark, J. D. (eds). *Background to Evolution in Africa*, pp. 191–220. Chicago: Chicago University Press.

MATHESON, F. J. 1966. *Geology of the Kajiado Area*. Report No. 70, Geological Survey of Kenya. Nairobi: Government Printer.

MILLER, J. A. 1967. Problems of dating East African Tertiary and Quaternary volcanics by the potassium-argon method. In: Bishop, W. W. and Clark, J. D. (eds). *Background to Evolution in Africa*, pp. 259–72. Chicago: Chicago University Press.

POSNANSKY, M. 1959. The Hope Fountain site at Olorgesailie, Kenya Colony. *South African Archaeological Bulletin*, **16**, 83–9.

RICHARDSON, J. L. & RICHARDSON, A. E. 1972. History of an African Rift lake and its climatic implications. *Ecological Monographs*, **42**, 499–534.

SHACKLETON, R. M. 1955. Pleistocene movements in the Gregory Rift Valley. *Geologische Rundschau*. **43**, 257–63.

—— Unpublished. Various maps and geological profiles made available to G. Ll. Isaac.

SIKES, H. L. 1926. The structure of the eastern flank of the Rift Valley near Nairobi. *Geographical Journal*, **68**, 385–402.

THOMPSON, A. O. & DODSON, R. G. 1963. *Geology of the Naivasha Area*. Report No. 55, Geological Survey of Kenya. Nairobi: Government Printer.

THOMPSON, B. W. & SANSOM, H. W. 1967. Climate. In: Morgan, W. T. W. (ed.). *Nairobi: City and Region*, pp. 20–38. Nairobi: Oxford University Press.

WILLIAMS, L. A. J. 1967. Geology. In: Morgan, W. T. W. (ed.). *Nairobi: City and Region*, pp. 1–13 Nairobi: Oxford University Press.

14

GREGORY R. CHAPMAN & MAUREEN BROOK

Chronostratigraphy of the Baringo Basin, Kenya.

The results of a potassium-argon radiometric dating programme, based on field-work over 14 000 km² of the northern Kenya Rift Valley, are described. The Cenozoic volcanic rocks of the area, chiefly phonolites, trachytes and basalts, and their pyroclastic equivalents, cover a time-span from about 20 m.y. to immediately prehistoric times, and the age determinations on the volcanics have greatly assisted temporal correlation of the volcanic and tectonic events in the stratigraphic record. In any one section the validity of the dates obtained can be checked against the rigorous constraints imposed by the geological mapping. Despite the limited definition of the K–Ar method in this time range it has been possible to draw tentative conclusions regarding the rates of geological processes in the Rift and it is suggested that the sedimentary units and non-sequences record the greater part of the time-span. In the context of the present volume, the K–Ar programme has made possible the calibration of a remarkable series of fossiliferous sedimentary units which have yielded rich assemblages of fossil vertebrates.

The 'Baringo Basin' is taken to mean, in this account, that portion of the Gregory Rift lying between the Equator and latitude 1°N., that is, roughly between Lake Hannington in the south, Silali volcano in the north and the Elgeyo and Laikipia escarpments in the west and east respectively (Fig. 14:1). The total area is about 14 000 km².

Only the south-eastern part of the area so defined drains into Lake Baringo since drainage from the Elgeyo Escarpment and the western slopes of the Kamasia Range flows to the Kerio River and hence ultimately into Lake Turkana (Rudolf). All drainage north of Lake Baringo and, it is conjectured, the subterranean out-flow of the lake itself joins the Suguta system. Lake Baringo can be viewed as the logical successor to a series of 'wet' sedimentary basins which have existed in the general area throughout the last 15 million years, and has been the nodal point for a number of geological and palaeontological investigations over the past decade.

These include the regional mapping programme, on a scale of 1:50 000, of the East African Geological Research Unit (EAGRU) directed by Professor B. C. King, based at Bedford College (University of London) and Nairobi University and the Kenya Department of Mines and Geological Survey. The majority of the radiometric age determinations on the Cenozoic sequence in this area have been on samples collected by EAGRU research students in the course of mapping, for the purpose of erecting a time scale for the geological events recorded by the stratigraphic sequence. A smaller number of determinations has been made for the specific purpose of bracketing fossiliferous sedimentary formations at the Cambridge University Geochronology Laboratory on samples collected by palaeontological follow-up teams. The most recent determinations have been made at Liverpool University (Sub-department of Geophysics) in connection with palaeomagnetic investigations.

Sample selection and analytical details[1]

More than 60 age determinations have been made at the Institute of Geological

[1] Institute of Geological Sciences analyses only.

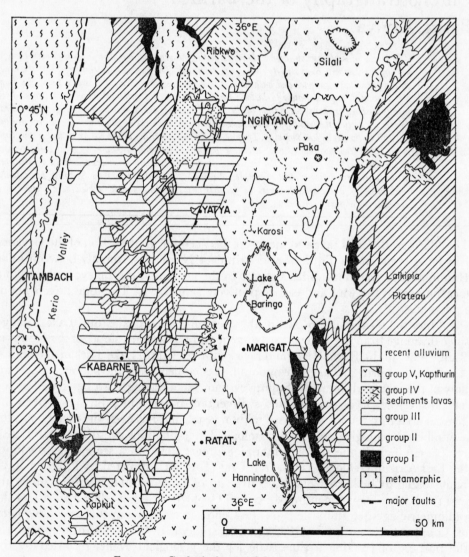

Fig. 14:1. Geological map of the Baringo Basin.

Sciences Geochemistry Division (Isotope Geology Unit) on samples collected from the Baringo area in the course of EAGRU mapping. The main objective has always been the calibration of established rock sequences. This is emphasised since this mode of operation has placed useful stratigraphic constraints on the interpretation of the ages obtained and has thus guided field selection of suitable material. Wherever it was available the most favoured rock-type for sampling has been trachyte lava which, as well as having the advantage of consisting largely of sanidine feldspar and having an average K_2O content (whole rock) of $>$ 5 per cent, is the co-dominant crystalline rock in the upper part of the Cenozoic succession. Basalt, which is almost equally abundant in the younger part of the sequence, and phonolite have not, as a rule, given such consistent or stratigraphically sensible results.

Not all the determinations made are given in this paper since many relate either to local details of successions, particularly in the Quaternary lavas, or were made to assist resolution of particular problems such as the tentative identification of isolated faulted inliers. However, no determinations have been omitted from descriptions of the particular sequences described.

After passing through a jaw crusher and roller mill, the -60 $+120$ sieve fraction was taken and split for both potassium and argon analysis. Potassium was determined in triplicate by flame photometry with a lithium internal standard. Samples were loaded for argon analysis in a vacuum system and baked overnight at 180°C, to reduce the concentration of adsorbed atmospheric argon.

An ^{38}Ar tracer (or spike) was introduced into the system before the start of each fusion. Conventional clean-up procedures were used, i.e. titanium gettering and liquid nitrogen cold trap. The isotopic ratios $^{36}Ar:^{38}Ar$ and $^{40}Ar:^{38}Ar$ were measured in a 5 cm radius, 180° sector, gas source mass spectrometer with permanent magnet operating in static mode.

Errors on the potassium–argon analyses are quoted at the two sigma level and take into account uncertainties in the potassium analysis, the argon spike calibration, isotopic ratio determinations and the error enhancement involved in correcting for atmospheric argon contamination.

Stratigraphic sequence

The general sequence of volcanic and structural events is as follows ('Group' numbers follow King & Chapman 1972):

Quaternary flood lavas and central volcanoes of the Rift centre	Group V
—2nd major faulting—	
Trachyte strato-volcanoes and associated sedimentary units	Group IV
Basalts (Kaparaina basalts) and flood trachytes	Group III
—1st major faulting—	

Plateau phonolites of Rift shoulders and Kamasia Range	Group II
Basalts	Group I

(Metamorphic basement)

Sedimentary units occur throughout the succession and are a significant factor in the time scale.

The plateau phonolites and associated sediments comprise about two-thirds of the whole succession in the Kamasia Range where they total a maximum 2 km in thickness.

Structural regions

The chief structural feature of this part of the Rift Valley is the Kamasia Range (or Tugen Hills), a north-south trending range of hills equal in height in many places to the Rift shoulders; it is essentially a faulted anticline or tilted block complex between the main rift boundary faults (Gregory 1921), or may be thought of as forming, with the Elgeyo Escarpment, the inner component of a double western boundary of the rift (Fig. 14:2).

Although the Kamasia has been an active tectonic region throughout most of the time recorded by the volcanic sequence, the present topography is largely the result of major fault movements within the early Quaternary. As a result the 3000 m thickness of lavas and sediments preserved in the Kamasia is effectively a 'within-rift' sequence, in contrast to the much thinner successions seen on the escarpments of the rift-margin. This fact has led to the Kamasia sequence being used to a large extent as a type-section for the entire Baringo basin.

The structural regions which serve in the following sections as a basis for the description of the chronostratigraphy are as follows (Fig. 14:2):

Elgeyo Escarpment: This is a 1200-m high escarpment in basement metamorphic rocks capped by plateau phonolites (group II) which extend
  from the scarp westwards across the Uasin Gishu plateau.

Kamasia main-range: Faulted and tilted group II phonolites overlain by
  extensive sheets of trachyte make up the main range. The trachytes form

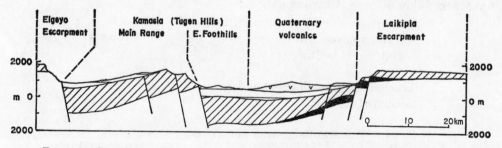

FIG. 14:2. Generalised cross-section of the Baringo basin, showing structural regions. (Key as for Fig. 14:1.)

a gentle dip-slope on the western flanks which decline into the Kerio Valley; the eastern slopes are defined in general by a system of large *en echelon* fault scarps. The metamorphic basement is exposed at the foot of the Saimo fault scarp and on the Kito Pass.

Kamasia eastern foot-hills: A series of basalts, trachytes and sediments (groups III and IV) occupy the area between the main Kamasia fault-scarps and the Quaternary lava-field of the Rift centre. They are typically strongly faulted and flexed and should be considered as part of the general tectonically positive area that includes the main range.

Quaternary Rift-centre volcanics: These comprise both flood-lavas, in the south, and trachyte–basalt central volcanoes. They are virtually un-eroded but have been strongly affected by closely spaced minor faulting. In the Baringo area the Quaternary fault-belt is midway between the Kamasia and Laikipia escarpments.

Laikipia escarpment: The name is properly applied only to a small part of the eastern Rift-margin in this area but has come to be applied generally to the whole structure. In contrast to the Elgeyo scarp the eastern wall of the Rift is formed by a series of *en echelon* step-faults, and is structurally more akin to the Kamasia fault system.

North of the Baringo basin the Kamasia main range continues into the Tiati area as a strongly assymetric faulted anticline exposing large areas of basement rocks on its shallow western limb. The Tiati range itself was a centre of phonolitic and trachytic activity for the group II volcanics, while its eastern flanks, equi-valent in position to the Kamasia eastern foot-hills, are composed of a series of huge low-angle trachyte volcanoes (group IV) which have been tilted to the east where they are overlapped by the Quaternary volcanics (Webb & Weaver 1976).

Southwards both the Elgeyo and Kamasia structures die out and volcanics of group IV (chiefly tuffs in this area) lap around the southern end and extend west-wards into the beginnings of the Kavirondo depression.

The Quaternary flood lava field rises gradually to the south, more or less in accord with the regional slope of the Rift floor.

Radiometric age-determinations

Elgeyo Escarpment

Age determinations on the phonolite lavas capping the Elgeyo escarpment form a coherent sequence and give support to the lithological evidence that these phonolites are the equivalents of the Tiim Phonolites Formation in the Kamasia sequence (Fig. 14:3). The only individual flow-members which can be correlated on a basis of lithological similarity are the basal trachyphonolite unit of the Elgeyo (= Atimet member of the Tiim phonolites) and, possibly, the aphyric flow of member 5 of the Elgeyo sequence which strongly resembles the top flow of the Tiim phonolites in the northern Kamasia.

There is thus only a younger age limit for the Tambach sediments which underlie the phonolite flows, but it seems possible that the argillaceous part of the sequence, may correspond to the sedimentary units within the Sidekh Phonolites Formation of the Kamasia.

The Chof phonolites are restricted to the Elgeyo escarpment sequence and, apart from an approximate time equivalence their relationship to events in the Kamasia is not known.

The two dates available (Table 1) for the Elgeyo basalts (Walsh 1969) suggest a rough age equivalence to the lowest part of the Kamasia sequence but it is conjectured that these autobrecciated basic volcanics were produced by the general basaltic eruptive episode which produced the widespread Miocene basalts of northern Kenya (Baker *et al.* 1971; Williams 1972).

TABLE 1: *Potassium argon ages–Elgeyo Escarpment*

Sample	%K	%atmos.^{40}Ar	rad^{40}Ar nl/g	Age & error (my)
Uasin Gishu				
phonolites				
MB/9 (member 7)	4.25	72.2	2.0400	12.0 ± 0.3*
9/623 (member 3)	4.17	61.3	2.2456	13.5 ± 0.3
MB/8 (member 2)	4.53	18.6	2.4750	13.6 ± 0.5*
Chof phonolites				
9/469	4.42	63.1	2.6432	15.0 ± 0.5
Elgeyo				
–	–	–	–	15.1 ± 3.2†
–	–	–	–	15.6 ± 3.2†

* Bishop, Miller & Fitch, 1969.
† Baker, Williams, Miller & Fitch, 1971.

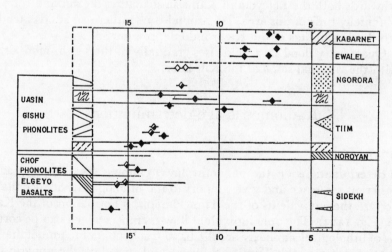

FIG. 14:3. Chronostratigraphic correlation of Cenozoic sequences on the Elgeyo Escarpment (left) and the Kamasia (right) ◆ K/Ar, whole rock, Kamasia; ◇ K/Ar, separated sanidine, Kamasia; ■ K/Ar whole rock, Elgeyo Escarpment. Horizontal scale million years.

Kamasıa main range (Table 2)

Age determinations on the plateau phonolites which dominate the succession in the main range have not given consistent results. However, Fig 14:4 indicates that the general span of ages can be set at between at least 16 m.y. (Sidekh phonolites) and about 8 m.y. (Ewalel phonolites). This time span is three times as long as that indicated for the Miocene plateau phonolites from areas outside the Rift (Baker *et al.* 1971) which are dated consistently at between 11 Ma and 13.5 m.y. This fact is of considerable tectonic and volcanological significance since it shows that phonolitic eruptions first occurred within the Rift, at centres such as Sigatgat (McClenaghan 1971), and that widespread phonolite sheets of the flanking plateaux are essentially an overspill phenomenon (Lippard 1973). The thick

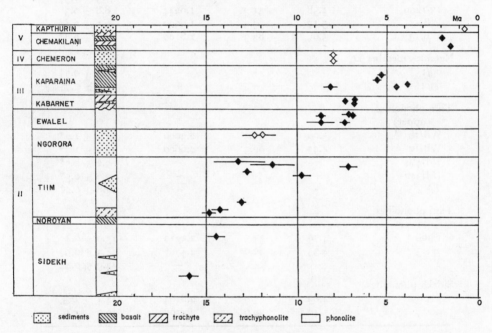

FIG. 14:4. K/Ar calibration of the lithostratigraphic sequence, central and northern Kamasia ◆ whole rock; ◇ separated sanidine. Horizontal scale million years.

Tiim Phonolites Formation of the Kamasia is clearly the equivalent of the flows outside the Rift. Later, as tectonism became dominant over volcanic accumulation, the last flows were again confined within the Rift.

Samples taken specifically to place firm dates on the fossiliferous Ngorora Formation (Bishop & Chapman 1970; Bishop & Pickford 1975) have given a disconcertingly wide range of results, particularly from the overlying Ewalel phonolites which are represented by a single very thick flow (230 m maximum). Results obtained scarcely allow one to say more than that the Ngorora Formation must be dated between 12 m.y. and 8 m.y. Separated sanidine crystals from primary tuffs of member D of the formation have given K–Ar ages of 12.3 m.y. and 11.9 m.y. which should, of course, be taken as no more than a maximum age for the sediments.

TABLE 2: *Potassium argon ages—Kamasia Main Range*

Sample	%K	%atmos.^{40}Ar	rad^{40}Ar nl/g	Age & error (my)
Kabarnet trachytes				
JM/734	–	–	–	7.3 ± 0.3*
		–	–	7.3 ± 0.3*
1/714	4.23	43.9	1.1369	6.7 ± 0.3
Ewalel phonolites				
9/545	4.70	68.6	1.3720	7.3 ± 0.3
13/1209	4.38	51.9	1.6244	8.8 ± 0.3
WB/11	5.17	73.5	1.3900	6.7 ± 0.1*
WB/13	5.09	83.3	1.8100	8.8 ± 0.8*
Ngorora, member D				
69/1	–	–	–	12.3 ± 0.7†
69/2	–	–	–	11.9 ± 0.7†
Tiim phonolites (upper)				
WB/10	5.00	75.3	2.3000	11.4 ± 1.2*
WB/12	4.95	82.5	2.6600	13.2 ± 1.4*
JM/474	–	–	–	7.2 ± 0.5*
JM/346	–	–	–	9.9 ± 0.5*
T/8	–	–	–	12.8 ± 0.2‡
Tiim phonolites (lower)				
1/368	4.79	34.5	2.5319	13.0 ± 0.4
2/315	4.35	42.0	2.4787	14.3 ± 0.5
T/3	–	–	–	14.9 ± 0.4‡
Sidekh phonolites				
3/514	4.89	43.0	2.8164	14.4 ± 0.5
2/217	5.40	39.0	3.4508	16.0 ±0.6

* Miller, whole rock analysis.
† Miller, separated sanidine anlysis.
‡ Mussett, whole rock analysis.

A possible explanation for the variation in age determinations on the phonolites, which in whole-rock chemistry and mineralogy are relatively close to the apparently reliable trachytes, is sought in the frequent presence in the former of secondary calcite and analcime, and the incipient alteration of nepheline to cancrinite.

The consistent radiometric ages given by the Kabarnet trachytes (Walsh 1969), coupled with their persistence as a recognisable unit over an area of 1750 km², has enabled investigators to accept this formation as a very useful stratigraphic marker over the entire Kamasia Range. In view of the angular unconformity and deep weathering horizon developed on the phonolites beneath the Kabarnet trachytes, the overlap in the radiometric dates of the Kabarnet and Ewalel units is somewhat surprising and has led to the acceptance of the older Ewalel dates as being the more likely for that unit.

Kamasia eastern foot-hills (Table 3)

Kabarnet trachytes. Conspicuous inliers of Kabarnet trachytes, caused by fault-defined arches and plunging horsts, provide the vital link between the main range sequence and the basaltic/sedimentary sequence of the eastern foot-hills; there is, at Yatya, one small outcrop of Ewalel phonolites beneath the trachytes. Sample 2/214 was collected from the lower flow of Kabarnet trachyte at Yatya. Here, also, is unequivocal evidence of the stratigraphic position of the Mpesida beds, lying between the two main trachyte flows, which are thus possibly the most firmly dated of all the fossiliferous sedimentary units in the Baringo basin.

Kaparaina Basalts Formation. South of the Kaparaina Range, near the Marigat–Kabarnet road, basalt lavas of this formation rest directly on the Kabarnet Trachytes Formation. North of 0°45′N, however, a basal fossiliferous sedimentary unit, the Lukeino Formation (Pickford 1975) lies unconformably on the Kabarnet trachytes and under the Kaparaina basalts.

The basaltic lavas have a maximum thickness of approximately 550 m, and consist of up to 50 flows of alkaline olivine–basalt and related lavas, many separated by 'bole' horizons. Distinctive flows of porphyritic anorthoclase trachyte occur above the middle of the formation in the north where they are almost the highest units of the formation to survive erosion.

Because of their low potassium content and the known tendency from other areas of Kenya (Baker *et al.* 1971) for basalts to give anomalous K–Ar ages, the predominant basalts of the formation were not sampled initially. Sample 2/125, a mugearite from the base of the lavas in the north, yielded an anomalously old date of 8.2 m.y. which might be explained by the presence of rare xenocrystic feldspars, visible in some thin-sections. Trachytes from the upper part of the formation in the north have given ages of 5.3 m.y. and 5.4 m.y. but recent determinations (Dagley *et al.* 1976) from the most complete continuous sequence of the basalts, in the Ndau River, 0°30′N, have given ages of 3.96 m.y. and 4.67 m.y. from basalts near the base of the formation in that locality.

Chemeron Formation (McCall *et al.* 1967). This thick (200 m) sedimentary formation has yielded a rich fauna of fossil vertebrates, including a hominid (Martyn 1967; Bishop *et al.* 1971). Two sanidine feldspar ages of 8.07 m.y. and 8.02 m.y. are the only determinations from this sedimentary formation. A more reasonable estimate of its age is given by considering the apparent ages of the bracketing lavas, that is, a probable youngest age of about 4 m.y. for the underlying Kaparaina basalts and determinations of 2.0 m.y. and 1.5 m.y. for the overlying mugearitic and trachytic lavas (Table 5)

Kaperyon Formation (Bishop *et al.* 1971). In the northern part of the eastern foot-hills this 130 m sedimentary formation occupies an analogous stratigraphic position to the Chemeron in the south, resting, with pronounced unconformity, on Kaparaina and Kabarnet Formations. It is overlain by an outlying trachyte

TABLE 3: *Potassium argon ages—Kamasia, eastern foothills*

(a) North of latitude 0°30′ N.

Sample	%K	%atmos.^{40}Ar	rad^{40}Ar nl/g	Age & error (my)
Chemeron Formation				
69/4	–	–	–	8.0 ± 0.2*
69/5	–	–	–	8.0 ± 0.2*
Ribon trachyte				
2/79	4.23	75.0	0.8203	4.9 ± 0.2
2/22	4.03	79.0	0.5873	3.7 ± 0.2
Kaperyon				
69/3	–	–	–	11.2 ± 0.2*
Kaparaina				
K5	–	–	–	3.9 ± 0.1†
K3	–	–	–	4.7 ± 0.1†
2/227	4.99	48.0	1.0480	5.3 ± 0.2
2/M5	4.63	46.0	0.9971	5.4 ± 0.2
2/125	2.45	74.0	0.8043	8.2 ± 0.4
Kabarnet trachytes				
2/214	4.61	31.0	1.2446	6.8 ± 0.4
Ewalel phonolites				
2/210	4.77	65.1	1.3433	7.1 ± 0.4

(b) South of latitude 0°30′ N.

Sample	%K	%atmos.^{40}Ar	rad^{40}Ar nl/g	Age & error (my)
Eldama Ravine tuff				
14/125	4.07	61.5	0.6960	4.3 ± 0.1
Kapkut volcano				
14/678	2.33	29.5	0.7148	7.6 ± 0.2
9/364	3.93	48.5	1.0360	6.6 ± 0.2
Kaparaina				
KP/5–6A	1.32	84.4	0.3620	6.9 ± 0.2
KP/5–6B	1.04	77.8	0.2780	6.7 ± 0.3

* Miller, separated sanidine analysis
† Mussett, whole rock analysis

flow from the Ribkwo volcano, the Ribon trachyte, which has given K–Ar ages of 3.7 m.y. and 4.9 m.y. The sediments can be traced northwards from 1°N as a tongue within the flank lavas of Ribkwo.*

Ribkwo volcano (in press). This is a very large, deeply eroded, low-angle trachyte volcano, one of a group of such volcanoes in southern Turkana. It appears to rest everywhere on a substrate of group III basalts and, in the west, abuts against the eroded flanks of the Sigatgat–Tiati massif while its eastern flanks are tilted sharply down towards the rift centre in conformity with the general structural style of the Kamasia eastern foot-hills to the south. The volcano has yielded a large number of K–Ar ages, not detailed in this contribution, between 4.5 m.y. and 5.5 m.y.

At the southern end of the Kamasia, group IV is again represented by a trachyte volcano, Kapkut, at the head of the Kerio Valley, and a unit of pyroclastic sediments, the Eldama Ravine Tuffs (Walsh 1969) which have given an age of 4.3 m.y. It is not possible by direct observation to deduce the stratigraphic relationship of the Eldama Ravine Tuffs to the Chemeron Formation.

Kapthurin Formation (McCall *et al.* 1967; Leakey 1969). A 200-m thick formation of red earths, silts and gravels, this was once thought to be younger than the Lake Baringo trachyte but more recent work has shown (Tallon 1976) that the lower of two units of pumiceous tuffs contained within the formation was probably erupted from the same source as the trachyte. The Kapthurin rests on an erosion surface cut across Chemeron and Kaparaina Formations and can be traced, via interfluve outliers, to piedmont gravel fans at the foot of the Kamasia main-range scarps.

A separated sanidine feldspar from a tuff in the formation gave potassium–argon ages of 0.66 m.y. and 0.68 m.y.

Quaternary Rift-centre volcanics (Table 4)

Among the young volcanics of the Rift centre those of most immediate relevance to the sedimentary history of the Baringo basin are the mixed lavas (basalts, mugearites and trachytes) overlying the Chemeron Formation, and those lavas on the opposite (Laikipia) side of the Rift known to be younger than the fossiliferous Chemoigut Formation sediments (Bishop *et al.* 1975). Most important of the former are the Chemakilani lavas (2.0 m.y.) and the Ndau mugearite 1.5 m.y.); these, together with the Marigat trachyte (undated) and the Kapsolop phonolite near Ratat (1.6 m.y. and 1.8 m.y.) crop out on the extreme western

* *Editorial Footnote*: Pickford (1975, 282) states that the Kaperyon Formation as originally mapped (Chapman 1971) comprised part of three sedimentary units of different ages. The oldest, which is equivalent in age to the Mpesida Beds, (about 7.0 m.y.) is in places separated from the overlying sediments by a flow of Kabarnet trachyte and elsewhere by an unconformity. The upper sediments belong to the Chemeron Formation (aged between 2 and 4 m.y.) as shown by tracing marker horizons (e.g. the characteristic Cheseton Lapilli Tuff) and by the fauna of large mammals found at a number of localities including the original Kaperyon type area. The remaining sediments can be mapped as continuous with the Lukeino Formation (aged about 6.5 m.y.). They contain characteristic Lukeino fossil mammals even in the original type area of the Kaperyon Formation. Pickford (personal communication) believes that use of the term Kaperyon Formation should be discontinued.

edge of the Quaternary lavas belt and suffered considerable faulting and tilting before the eruption of later units typified by the Loyamarok trachyphonolite (0.5 m.y.) and the Lake Baringo trachyte (0.25 m.y.).

TABLE 4: *Potassium argon ages—Quaternary Rift-centre Volcanics*

(a) North of latitude 0°30′ N.

Sample	%K	%atmos.^{40}Ar	rad^{40}Ar nl/g	Age & error (my)
Lake Baringo trachyte				
1/9	–	–	–	0.26*
Loyamarok phonolite				
JM/52	–	–	–	0.50*
Kapthurin Formation				
–	–	–	–	0.67*
Ndau mugearite				
–	–	–	–	1.5†
Songoiwa mugearite				
–	–	–	–	2.0†

(b) South of latitude 0°30′ N.

Sample	%K	%atmos.^{40}Ar	rad^{40}Ar nl/g	Age & error (my)
Hannington phonolites				
13/1640	3.95	93.7	0.0433	0.3 ± 0.1
13/1252A	5.22	71.0	0.2110	1.0 ± 0.1
13/851C	5.02	82.0	0.2096	1.0 ± 0.1
13/951B	5.29	83.3	0.3380	1.6 ± 0.2
Kapsalop phonolite				
13/KP–1	5.71	46.6	0.3650	1.9 ± 0.2
13/KP–2	5.73	48.6	0.2950	1.6 ± 0.1
13/129	4.44	62.0	0.3236	1.8 ± 0.1
Ainapno mugearites				
13/511	4.59	49.0	0.3479	1.9 ± 0.1

* Miller, whole rock analysis.
† Evernden & Curtis, whole rock analysis.

A large number of K–Ar ages, not detailed in this account, of between 0.3 m.y. and 1.5 m.y. have been obtained for the various lava units within the Lake Hannington trachytes and phonolites ('Dispei–Lake Hannington phonolites' of McCall 1967) which dominate the Rift floor to the south of Lake Baringo.

Laikipia Escarpment (Table 5, Fig. 14:5)

Although the eastern side of the Rift is in many ways structurally similar to the Kamasia main-range there is nothing on this side of the Rift resembling the eastern foot-hills of the Kamasia. Thus on the Laikipia, Quaternary volcanics overlie, or lap up against the Miocene sequence.

Samburu basalts. These are the oldest rocks exposed on the Laikipia escarpment no metamorphic basement being seen. Several of the basalt outcrops appear to expose the remnants of multi-centre basalt volcanoes, with swarms of parallel dykes. The three samples determined gave ages of between 11 and 21 m.y. The youngest is anomalous in view of the determinations on the overlying phonolites but the two older ages are comparable with other determinations from the widespread Miocene basalts of northern Kenya (Baker *et al.* 1971).

TABLE 5: *Potassium argon ages—Laikipia escarpment*

Sample	%K	%atmos.^{40}Ar	rad^{40}Ar nl/g	Age & error (my)
Emsos mugearite				
13/1475D	4.60	52.0	0.4769	2.5 ± 0.1
Kinodo phonolites				
13/1548C	4.70	64.0	0.7920	4.2 ± 0.1
13/1548D	4.57	80.0	2.1200	11.7 ± 0.3
Tasokwan trachyte				
10/943	6.03	64.7	1.3224	5.5 ± 0.1
13/1521	4.41	42.0	1.3406	7.6 ± 0.2
Rumuruti phonolites				
10/590 (Marmanet)	4.53	51.0	1.9081	10.6 ± 0.3
10/650 ,,	4.48	18.0	2.0820	11.6 ± 0.3
10/115 (Ngelesha)	4.87	45.6	2.2660	11.6 ± 0.3
10/511 ,,	4.64	48.0	2.0050	10.8 ± 0.3
12/1100 (Uaso Narok)	4.78	33.4	1.9690	10.3 ± 0.3
Samburu basalts				
13/1453	1.95	33.0	0.9230	11.8 ± 0.5
13/1367	1.98	54.0	1.1290	14.2 ± 0.4
13/1540	2.42	20.0	2.0010	20.7 ± 0.6

Rumuruti phonolites (Shackleton 1946). This group has now been divided into several separate formations with complex interdigitating relationships. Its maximum thickness is 750 m. It should be pointed out that the K–Ar ages obtained, ranging from 10.3 m.y. to 12.0 m.y., are not completely internally compatible with the group's mapped stratigraphy, which is not detailed here. These ages indicate that, like the Uasin Gishu phonolite sequence, these flows are the time-equivalents of the middle part, only, of the Kamasia sequence. They are slightly younger, on the K–Ar evidence, than the Uasin Gishu phonolites but it is obvious that at about this time the flood lavas overflowed both Rift shoulders.

Tasokwan trachytes. A moderate amount of faulting and block tilting occurred before the extrusion of these lavas which now occur chiefly on the back-slopes of tilted blocks in the escarpments east of Lake Hannington. The two ages determined,

6.6 m.y. and 7. 6 m.y. are close to the range of Kabarnet trachyte ages from the Kamasia and it seems that it is possible to extend the areal range of trachytes of this episode to the eastern side of the Rift.

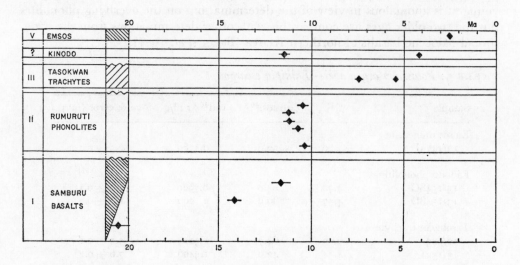

FIG. 14:5. K/Ar (whole rock) calibration of the lithostratigraphic sequence on the Laikipia escarpment. (Key as Fig. 14:4.) Horizontal scale million years.

Later lavas. A major episode of faulting occurred after the eruption of the Tasokwan trachytes but its exact dating is uncertain. A thin unit of phonolites, the Kinodo phonolites, occurs stratigraphically above the Tasokwan trachytes, but has given problematic ages, one of which is 4.2 m.y. but four others on one sample are between 11.5 m.y. and 12.0 m.y.

A minor basaltic phase is represented by the Ildamnoi basalts, undated, but possibly equivalent to the Kaparaina basalt in age. Above these basalts the Emsos mugearites have given a date of 2.5 m.y., suggesting correlation with the early Quaternary intermediate lavas west of Lake Baringo.

Rates of volcanism and sedimentation

It is obvious that individual lava flows are negligible as time-representative lithosomes; the eruption of a flow and its 'arrival' at a particular point in space are events to be measured in days rather than years. The outstanding question becomes the estimation of how much of the time range is represented by weathering and erosion and how much by sedimentation.

In Fig. 14:6 is shown the stratigraphic column, with formation thicknesses to scale, compared with the indicated time scale on which individual lava flows are shown as single lines. Ages of critical 'bracketing' lavas have been assigned by reasonable assumptions made on the basis of Fig. 14:4, and sedimentation, where it occurred, is assumed to have occupied all the available time between lavas. On this premise sediments would make up about half the chronostratigraphic sequence and sedimentation rate for the major units would be approximately 15 cm/10³ years.

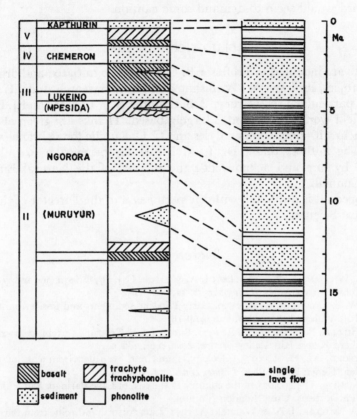

FIG. 14:6. Comparison of Kamasia lithostratigraphy with apparent K/Ar chronostratigraphy, with particular reference to sedimentary units. Vertical scale million years.

However, it is equally valid to argue, firstly, that exact attribution of ages to lava flows is not justifiable at all on the ambiguous evidence and secondly that sedimentation is induced in any particular area by tectonic circumstances rather than by available time, and might be much more rapid than indicated above. There are on average, at the northern end of the Kamasia, twelve flows of phonolite present in the group II sequence. Indicated time-span for the group is about 8 million years and average frequency of flow eruption is thus 0.73 million years. If the 360 m (maximum) of the Ngorora Formation were deposited in this interval, instead of the 2.5 million years postulated in Fig. 14:6 then sedimentation rate would be in the order of 50 cm/10³ years—arguably a not unreasonable figure.

While the K–Ar method has succeeded in providing the essential chronological framework for this volcanic province, its resolution is not at present sufficiently fine to provide answers to some of the more detailed questions. The outstanding problems remain the determination of the rates of weathering and erosion (pediplanation over large areas of the Baringo basin) and rates of sediment accumulation. Until these points are clarified the interpretation of the palaeontological record would seem to demand some caution.

Acknowledgements

The authors are indebted to Professor B. C. King who initiated and organised the EAGRU project, Professor W. W. Bishop and former research students at Bedford College, in particular J. N. Carney, P. S. Griffiths, S. J. Lippard and J. E. Martyn, on whose field work the account is largely based. Thanks are given also to H. S. Lloyd for help with potassium analyses and C. C. Rundle for assistance with argon analyses. The authors thank Dr J. A. Miller for permission to publish dates established by him and acknowledge the support of the Natural Environment Research Council.

This account is published with the permission of the Director of the Institute of Geological Sciences.

References

BAKER, B. H., WILLIAMS, L. A. J., MILLER, J. A. & FITCH, F. J. 1971. Sequence and geochronology of the Kenya rift volcanics. *Tectonophysics*, **11**, 191–215.
BISHOP, W. W. & CHAPMAN, G. R. 1970. Early Pliocene sediments and fossils from the northern Kenya Rift Valley. *Nature, Lond.* **226**, 914–18.
——, ——, HILL, A. P. & MILLER, J. A. 1971. Succession of Cenozoic vertebrate assemblages from the northern Kenya Rift Valley. *Nature, Lond.* **233**, 389–94.
—— & PICKFORD, M. H. L. 1975. Geology, fauna and palaeoenvironments of the Ngorora Formation, Kenya Rift Valley. *Nature, Lond.* **254**, 185–92.
CARNEY, J. N. 1972. The geology of the country to the east of Lake Baringo, Rift Valley Province, Kenya. Ph.D. thesis, Univ. London (unpubl.).
——, HILL, A., MILLER, J. A. & WALKER, A. 1971. Late Australopithecine from Baringo District, Kenya. *Nature, Lond.* **230**, 509–14.
CHAPMAN, G. R. 1971. The geological evolution of the northern Kamsaia Hills, Baringo District, Kenya. Ph.D. thesis, Univ. London (unpubl.).
DAGLEY, P., MUSSETT, A. E. & PALMER H. C. 1976. Preliminary observations on the palaeomagnetic stratigraphy of the area west of Lake Baringo, Kenya. (This volume Paper 15).
GREGORY, J. W. 1921. *The rift valleys and geology of East Africa.* Seeley Service, London.
GRIFFITHS, P. S. (in prep.) The geology of the area around Lake Hannington and the Perkerra River, Rift Valley Province, Kenya. Ph.D. thesis, Univ. London.
KING, B. C. & CHAPMAN, G. R. 1972. Volcanism of the Kenya Rift Valley. *Philos. Trans. R. Soc.* **A271**, 185–208.
LEAKEY, M. G. *et al.* 1969. An Acheulian industry with prepared core techniques and the discovery of a contemporary hominid mandible at Lake Baringo, Kenya. *Proc. Prehist. Soc.* **35**, 48–76.
LIPPARD, S. J. 1972. The stratigraphy and structure of the Elgeyo Escarpment, Southern Kamasia Hills and adjoining regions, Rift Valley Province, Kenya. Ph.D. thesis, Univ. London (unpubl.)
—— 1973. Plateau phonolite lava flows, Kenya. *Geol. Mag.* **110** (6), 543–9.
LOGATCHEV, N. A., BELOUSSOV, V. V. & MILANOVSKY, E. E. 1972. East African Rift development. *Tectonophysics*, **15** (1/2 'East African Rifts'), 71–81.
McCALL, G. J. H. 1967. Geology of the Nakuru–Thomson's Falls–Lake Hannington area. *Rep. geol. Surv. Kenya*, **78**.
——, BAKER, B. H. & WALSH, J. 1967. Late Tertiary and Quaternary sediments of the Kenya Rift

Valley. *In;* Bishop, W. W. & Clark, J. D. (eds) Background to evolution in Africa. University of Chicago Press.

McClenaghan, M. P. 1971. The geology of the Ribkwo area, Baringo District, Kenya. Ph.D. thesis, Univ. London (unpubl.).

Martyn, J. E. 1967. Pleistocene deposits and new fossil localities in Kenya. *Nature, Lond.* **215**, 476–80.

—— 1969. Geology of the country between Lake Baringo and the Kerio River, Baringo District, Kenya. Ph.D. thesis, Univ. London (unpubl.).

Pickford, M. 1975. Late Miocene sediments and fossils from the Northern Kenya Rift Valley. *Nature, Lond.* **256**, 279–84.

Sceal, J. S. C. 1974. The geology of Paka volcano and the country to the east, Baringo District, Kenya. Ph.D. thesis, Univ. London (unpubl.).

Shackleton, R. M. 1946. Geology of the country between Nanyuki and Maralal. *Rep. geol. Surv. Kenya,* **11**.

Tallon, P. 1976. The stratigraphy, palaeo-environments and geomorphology of the Pleistocene Kapthurin Formation, Kenya. Ph.D. thesis, Univ. London (unpubl.).

Walsh, J. 1969. Geology of the Eldama Ravine–Kabarnet area. *Rep. geol. Surv. Kenya,* **83**.

Webb, P. K. & Weaver, S. D. 1976. Trachyte shield volcanoes: a new volcanic form from the Kenya Rift. *Bull. Volc.* **39**.

Williams, L. A. J. 1970. The volcanics of the Gregory Rift Valley, East Africa. *Bull. Volcan.* **34**, 439–65.

—— 1972. The Kenya Rift volcanics: a note on volumes and chemical composition. *Tectonophysics* **15** (1/2 'East African Rifts'): 83–96.

15

PETER DAGLEY, ALAN E. MUSSETT
& H. C. PALMER

Preliminary observations on the palaeomagnetic stratigraphy of the area west of Lake Baringo, Kenya

Over 600 palaeomagnetic samples have been collected from igneous and sedimentary units of the succession exposed west of Lake Baringo in the Tugen Hills and surrounding area. The magnetic polarity has been measured for all but a few of the samples which span the period 16 m.y.–0 m.y. and some preliminary potassium-argon ages have been determined. The palaeomagnetic and petrological identifications have been used, with previous detailed geological mapping, to build up a preliminary palaeomagnetic stratigraphy. When the dates have been refined and some outstanding problems of magnetisation solved it may be possible to identify individual transitions in the period 0 to 5 m.y. For the older parts of the succession the transitions will aid stratigraphic and palae-ontological correlation.

Introduction

During 1973 we sampled most of the igneous units in the sequence exposed in the Tugen (Kamasia) Hills, west of Lake Baringo and in the Elgeyo escarpment along the road to the Fluorspar Company of Kenya's mine. In addition we have made measurements on samples from various members of several sedimentary formations and related lavas collected by Professor W. W. Bishop and his colleagues from localities east and west of Lake Baringo.

This report is concerned solely with the geomagnetic polarity information which is presented in the hope that it will prove useful as a stratigraphic tool: a full report will appear elsewhere.

Sampling

Locations of the collecting areas in Figs. 15:3a to 15:3c are referred to the coordinates of the 1:50 000 survey maps produced by Chapman (1971) and Martyn (1969) except for the few outlying areas which are well known by their site names. All the localities except Kilombe are shown in Fig. 15:1 and the generalised stratigraphic column is given in Fig. 15:2. Except for group M at least four independently oriented cores were obtained from each igneous unit using portable diamond drills. For these the orientation was determined relative to the horizontal and to true north by sun-sighting either directly or by means of an intermediate geographical sighting.

The samples of group M were obtained as blocks oriented relative to magnetic north and 2 or 3 cores were drilled from each in the laboratory.

The sediment samples were collected by carving a pedestal, using non-magnetic tools, at the chosen horizon. This was capped by a cubic plastic box with 2 cm sides which was then oriented with respect to the horizontal and to magnetic north, before being cut away with its contents.

Where possible the post-formational tilt of each unit was measured at the sampling location. In other cases the regional dip values recorded in Martyn (1969) and Chapman (1971) and elsewhere have been used. In total 72 sediment cubes, 13 blocks from 9 units and 570 cores from 136 units were collected.

Magnetic treatment

Standard palaeomagnetic methods have been used. All the samples have been magnetically cleaned in alternating fields and the directions of magnetisation

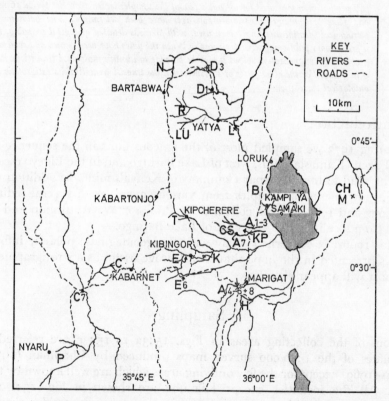

FIG. 15 :1. Map of the Baringo area showing the sampling localities. Only those features relevant to this work are shown. Letters A, B, etc. show the location of samples recorded in Figs. 15 :3a to c.

corrected for tectonic tilting of the lavas or sedimentary strata after magnetisation. Where the tilt corrections are based on regional values of dip a small uncertainty due to local variations must remain. However, most of the tilting in the area is due to faults with trends parallel to the north–south axis of the Rift Valley. As the magnetic inclinations at the latitude are shallow, even quite large tilts barely affect the direction of magnetisation. Thus provided a stable direction of magnetisation can be obtained the polarity is unambiguous. A possible exception to this arises in any case where the material has been magnetised by lightning strike. Magnestisation in these cases is often quite strong and stable, masking the primary thermal or depositional remanence. Studies are still in progress for several samples where magnetisation by lightning is suspected.

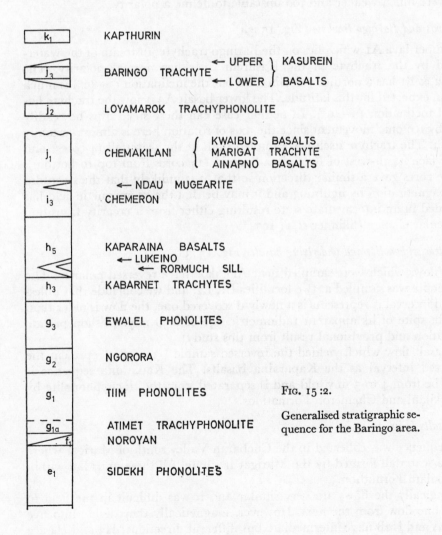

FIG. 15 :2.

Generalised stratigraphic sequence for the Baringo area.

Results

Throughout this section the results are described in sequence from youngest to oldest. The reader should refer to Figs. 15:1 and 15:2 for locational data and for more detailed stratigraphic relationships to other papers in this volume by Chapman & Brook, Pickford, Tallon, and Bishop *et al.*

Kapthurin Formation (Fig. 15:3a)

Five samples were obtained from the lower tuffs which overlie the Upper Kasurein basalt. Samples KP4 and KP6 from the green sandy tuff and KP7 from the red clay have a relatively strong and stable normal magnetisation. The magnestisations of KP3 and KP5 from a calcareous band of interbedded green-grey tuff were much weaker and too unstable to define a polarity.

Kasurein lavas and Baringo trachyte (Fig. 15:3a)

The upper lava A1 which lies on the Baringo trachyte upstream of the waterfall formed by the trachyte in the Kasurein river has normal polarity. The trachyte A2 & B1 has a northerly declination but the inclination ($+40°$) is much steeper than expected for this latitude. The lower basalt A3 below the trachyte has an upward inclination ($-49°$). In neither case can these steep dips be readily explained by tectonic movement since the axis of rotation here is almost certainly north–south. The trachyte itself was sampled both at the waterfall (4 cores) and at the cliff face north-west of Kampi-ya-Samaki (14 cores from top to bottom). All of these cores gave a similar direction so that it is unlikely that the steep dip is due to magnetisation by lightning and it may be that the geomagnetic field has been recorded in an intermediate state resulting either from a polarity transition or large secular change (Skinner *et al.* 1974).

Loyamarok trachyphonolite and underlying basalt (Fig. 15:3a)

These flows which were sampled near Alotakok have reversed polarity. The trachyphonolite was sampled at two localities L1, L2, and we conclude that unless the reversed interval it represents is a newly discovered one, the flow is older than 0.69 m.y. in spite of its apparent radiometric age of 0.58 m.y. (Bishop private communication and provisional result from this study).

The basalt flow which yielded the reversed sample L3 cannot represent the same reversed interval as the Kaparaina basalts. The Kaparaina sequence is thought to be from 4 to 5 m.y. old and is separated from the trachyphonolite by the Chemakilani and Chemeron Formations.

Kwaibus basalts (Fig. 15:3a)

This sequence was collected in the Chebaran Valley south of Marigat where it rests on a waterfall formed by the Marigat trachyte. All these units fall within the Chemakilani Formation.

Petrologically the flows are very similar and it was difficult in the field to distinguish one flow from the next. However, magnetically they divide into five groups. H2A and H2B have intermediate but different directions; H3 and H4 are normally magnetised and differ by 10° in direction; H5 and H6 have virtually identical reversed directions; but H7 although reversed has a significantly different direction; H8 and H9 are both intermediate.

Marigat Trachyte (Fig. 15 :3a)

This flow was sampled both in the Chabaran Valley, H1, where it lies below the Kwaibus basalts and beside the bridge, south of Marigat, A4. The polarity is reversed and the directional agreement between the two sites very good. A5, A6 and A8 were also collected near the bridge in the belief that they were outcrops of the Ainapno basalts which lie below the trachyte but petrological examination shows that they too are trachytic. A8 is probably the chilled base of the Marigat trachyte although the directional agreement with A4 is only to within 10°. A5 and A6 differ more markedly in direction and petrologically resemble trachytes found in the Kaparaina Formation.

Radiometric ages on the Marigat trachyte (2.0 m.y.) and Ndau mugearite (1.5 m.y.) which is interbedded with the underlying Chemeron Formation sediments, place these units in the Matuyama reversed interval. Therefore the normal interval represented by the Kwaibus basalts H3 and H4 is possibly the Jaramillo interval.

Chesowanja area (Fig. 15 :3a)

Samples from the sequence of lavas and sediments in this area, which lies to the east of Lake Baringo, were collected by Professor W. W. Bishop and his colleagues. The nearby Chepchuk trachyte, M6, is reversed and has been dated at 1.1 m.y. (Carney *et al.* 1971) but its relation to the sediments is not known. The available evidence (Bishop *et al.* 1975) suggests an age range between 0.7 and 2.4 m.y. for the Chesowanja Formation plus the Chemoigut Formation. It seems probable that this sequence and the Chemakilani one overlap in time.

(i) *Karau Formation.* The uppermost unit M26 is a welded tuff with reversed polarity; beneath this is the Karau trachyte M14, M25, followed by another welded tuff, M13, with clearly reversed polarity. The trachyte, however, was sampled at two places, one yielding a good reversed direction, the other a well defined intermediate direction. The reason for this discordance is not clear and further measurements are being made. A radiometric age of 0.54 m.y. has been reported for the trachyte (Bishop *et al.* 1975) and of 0.71 ± 0.07 m.y. for the Chesowanja basalt (Bishop *et al.* this volume) which underlies the Karau Formation.

Twelve out of 14 samples (CH1–8, CH36–40B) from the tuffaceous facies of the Karau sediments, having a thickness of 3 metres, have normal polarity, but samples 3 and 8 are reversed. Samples were taken from the same levels during successive visits so that samples 3 and 38, and 8 and 37 were expected to agree. All of the sediment 'cubes' gave good stable results on de-magnetisation so that further laboratory tests and possible additional samples are required including some more from the level of samples 3 and 8.

(ii) *Chesowanja Formation.* The Karau sediments lie unconformably on a palaeosol from the surface of which Acheulian artifacts have been recovered and below this there are two lava flows the upper of which yielded the date of 0.71 m.y. In the gorge at the type site where the basalts overlie the windows of fossiliferous sediment the upper dated flow gave reversed polarity (M30) and the lower normal polarity (M31). However, at other localities near Chesowanja samples M2, M11 and M3, from basalt flows thought to be equivalent to those at the type site (Fig. 15 :3a) yielded contrasting polarities. Tests are being made to try to resolve this problem. Reversed polarity has been established by Brock (Bishop *et al.* 1975) for baked sediments below the lower flow. A basalt flow (sample M12) is normal and lies above the Chesowanja basalts while a phonolite (sample M24) which is reversed underlies them.

(iii) *Chemoigut Formation.* Samples from the sediments of the Chemoigut Formation, which occurs below the Chesowanja basalts, were not as strongly magnetised and did not give such closely-grouped results on the whole as those of the Karau Formation. Samples from yellow silts and pumiceous tuffs (CH13), channel silts (CH20, 21), a red clay band (CH17, 18, 19) and the buff silts (CH30, 33, 34, 35)

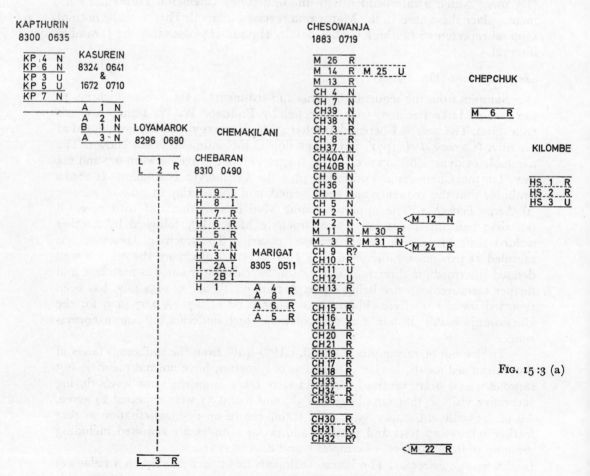

FIG. 15 :3(a). Palaeomagnetic stratigraphy of localities indicated in Fig. 15 :1.

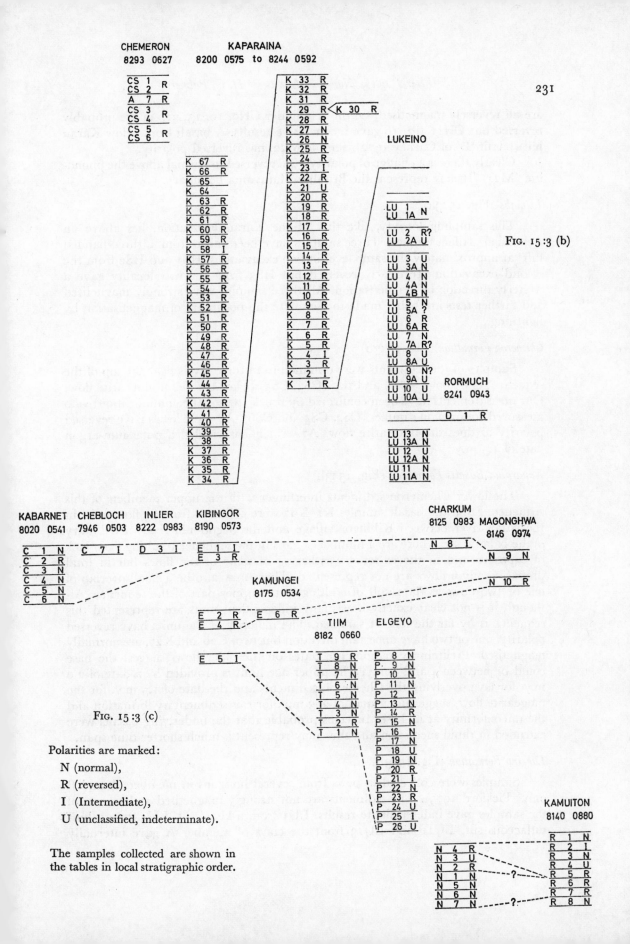

CHEMERON
8293 0627

KAPARAINA
8200 0575 to 8244 0592

231

CS 1 R
CS 2
A 7 R
CS 3 R
CS 4
CS 5 R
CS 6

K 33 R
K 32 R
K 31 R
K 29 R ← K 30 R
K 28 R
K 27 N
K 26 N
K 25 N

LUKEINO

K 67 R
K 66 R
K 65
K 64
K 63 R
K 62 R
K 61 R
K 60 R
K 59 R
K 58 I
K 57 R
K 56 R
K 55 R
K 54 R
K 53 R
K 52 R
K 51 R
K 50 R
K 49 R
K 48 R
K 47 R
K 46 R
K 45 R
K 44 R
K 43 R
K 42 R
K 41 R
K 40 R
K 39 R
K 38 R
K 37 R
K 36 R
K 35 R
K 34 R

K 24 R
K 23 I
K 22 R
K 21 U
K 20 R
K 19 R
K 18 R
K 17 R
K 16 R
K 15 R
K 14 R
K 13 R
K 12 R
K 11 R
K 10 R
K 9 R
K 8 R
K 7 R
K 6 R
K 5 R
K 4 I
K 3 R
K 2 I
K 1 R

FIG. 15:3 (b)

LU 1 N
LU 1A
LU 2 R?
LU 2A U
LU 3 U
LU 3A N
LU 4 N
LU 4A N
LU 4B N
LU 5 N
LU 5A ?
LU 6 R
LU 6A N
LU 7 N
LU 7A R?
LU 8 U
LU 8A U
LU 9 N?
LU 9A U
LU 10 U
LU 10A U

RORMUCH
8241 0943

D 1 R

LU 13 N
LU 13A N
LU 12 U
LU 12A N
LU 11 N
LU 11A N

KABARNET
8020 0541

CHEBLOCH
7946 0503

INLIER
8222 0983

KIBINGOR
8190 0573

CHARKUM
8125 0983

MAGONGHWA
8146 0974

C 1 N
C 2 R
C 3 N
C 4 N
C 5 N
C 6 N

C 7 I

D 3 I

E 1 1
E 3 R

N 8 I

N 9 N

KAMUNGEI
8175 0534

N 10 R

E 2 R
E 4 R

E 6 R

TIIM
8182 0660

ELGEYO

FIG. 15:3 (c)

E 5 I

T 9 R
T 8 N
T 7 N
T 6 N
T 5 N
T 4 N
T 3 N
T 2 R
T 1 N

P 8 N
P 9 N
P 10 N
P 11 N
P 12 N
P 13 N
P 14 R
P 15 N
P 16 N
P 17 N
P 18 U
P 19 N
P 20 R
P 21 I
P 22 N
P 23 R
P 24 U
P 25 I
P 26 U

Polarities are marked:

N (normal),

R (reversed),

I (Intermediate),

U (unclassified, indeterminate).

The samples collected are shown in
the tables in local stratigraphic order.

KAMUITON
8140 0880

N 4 R
N 3 U
N 2 R
N 1 N
N 5 N
N 6 N
N 7 N

?

?

R 1 N
R 2 I
R 3 N
R 4 U
R 5 R
R 6 R
R 7 R
R 8 N

are all reversely magnetised. Of the remainder CH9, 10, 15, 31, 32 are probably reversed but CH11, 12, 16 gave inconclusive results. A basalt from below Karau hill (basalt 'D' of Carney 1972), sample M22, has reversed polarity.

Clearly there is a change of polarity from reversed to normal above the phonolite (M24). It may represent the Brunhes-Matuyama boundary.

Kilombe (Fig. 15 :3a)

This sampling locality, like that of the Karau Formation, lies above an Acheulian artifact horizon. Three samples have been taken from a 'three-banded tuff' at approximately the same level in two excavations. HS1 and HS2 from the second excavation are clearly reversed but HS3 from the first locality gave a westerly direction of magnetisation. All three samples were strongly magnetised and further tests are being made to eliminate the possibility of magnetisation by lightning.

Chemeron Formation (Fig. 15 :3b)

Samples of the sediments were obtained from three levels near the top of the sequence; one above (CS1) and two (CS3, CS5) below the Ndau mugearite flow. Our measurements have been confirmed by Brock (personal communication) who measured companion samples (CS2, CS4 and CS6). All these levels have reversed polarity as does the mugearite flow, A7, which has yielded a potassium-argon date of 1.5 m.y.

Kaparaina Basalts Formation (Fig. 15 :3b)

The lower Chemeron sediments interfinger with the upper members of this sequence. From the basalt samples K1–K67 were collected from 65 flow units in the Ndau river between Kibingor village and the Segotionin river, a tributary of the Ndau river. There are a number of gaps in the sequence resulting from lack of exposure and to the extreme weathering of some of the flows but in total probably only 6 flows are not represented. There may also be a small overlap of one or two flows as the result of faulting in the upper part of the sequence. Although it is not clear exactly how many independent units are represented this sequence is by far the longest sampled. The majority of the units have reversed polarity, one or two have uncertain direction but two, K26 and K27, are normally magnetised. Preliminary radiometric dates on the older flows suggest the base could be between 4 and 5 m.y. The upper age limit is provided by a date of 2.0 m.y. for lava overlying the Chemeron sediments and the date of 1.5 m.y. for the mugearite flow, suggesting a span of 2–3 m.y. for this sedimentary formation and the unconformity at its base. It seems probable that the underlying basalts were extruded in rapid succession and thus may represent a much shorter time span.

Lukeino Formation (Fig. 15 :3b)

Samples were collected in pairs from several horizons in members A and B only (Pickford 1975). The sediments are not strongly magnetised and several of the samples gave indeterminate results. LU1 from a khaki silt, LU4 from a buff tuffaceous silt, LU11 and LU13 from the clays of member A gave internally consistent results and are normally magnetised. L6, a conchoidal red clay, is reversed. LU2 from a grey tuff is reversed. LU3 from a brown silt and LU12 from grits and silts in member A are normal on the basis of the result from one of a pair of cubes, the other of which was indeterminate.

Samples from the middle and bottom of a red earth, LU8 and LU10 respectively, are completely indeterminate but the results from a pair of samples, LU7, from the top of the same member disagree.

The Rormuch sill sampled at another locality, D1, lies within member B and has reversed polarity.

Kabarnet Trachyte Formation (Fig. 15 :3c)

Specimens C1 to C6 were collected in sequence working down into the valley to the north of the road from Kabarnet to Chebloch Bridge. Specimen C7 is from the fresh-looking massive flow forming the gorge at Chebloch Bridge on the Kerio river. Samples C5 and C6 are probably from the same flow as they look alike petrologically and have almost identical palaeomagnetic directions. C4 has a similar petrology but a different palaeomagnetic direction. Although a preliminary age of 4.6 m.y. has been obtained by us for C7, which has intermediate direction, the lower flows of the Kabarnet trachytes, which have normal polarity, have been consistently dated at about 7.0 m.y. which is in keeping with their stratigraphic position (Bishop 1971, Chapman & Brook, this volume).

No exact relationship can be given between these flows and the Kabarnet trachytes sampled elsewhere. However, the inlier D3, E1 from the Ndau river and N8 (Charkum lavas) all have intermediate or steeply dipping directions of magnetisation and they may be related. E3 from the Ndau river is reversed and N9 from near Makonghwa is normal.

Ewalel Phonolite Formation (Fig. 15 :3c)

This formation was sampled in the Ndau river (E2 and E4) and by the roadside near Kamungei (E6), where it can be seen to be capped by a member of the Kabarnet Trachyte Formation. N10 is from the Sumet phonolite forming the river gorge at Makonghwa which is considered equivalent to the Ewalel phonolites. All these units have reversed polarity but not identical directions. It has been generally thought that these phonolites are 8–10 m.y. old and according to the Heirtzler time scale (Heirtler 1968) reversed intervals occur between 8.2–8.3 m.y. and 8.5–8.8 m.y. However, we have obtained a provisional K–Ar age of 6.6 m.y. for E6.

Tiim Phonolite Formation (Fig. 15 :3c)

The samples T1–T9 represent the part of the sequence exposed in the river valley and escarpment west of Kipcherere; T2 and T9 are reversely magnetised and the remainder are normal. We have obtained provisional ages of 12.8 m.y. (T8), 0.8 m.y. (T6), 0.8 m.y. (T5) and 14.6 m.y. (T3). For dates established previously see Chapman & Brook (this volume) and Pickford (this volume).

Uasin Gishu phonolites and Elgeyo basalts (Fig. 15 :3c)

The opening of a road from Kimwarer at the base to Nyaru at the top of the Elgeyo Escarpment, to serve the Fluorspar Company of Kenya's Mine, provided some new exposures but the quality of the material was disappointing as weathering and alteration made sampling difficult.

The upper flows P8–P13 are normally magnetised with provisional ages for P8 of 11.0 m.y. and P13 of 15.6 m.y. These are flows within the Uasin Gishu phonolites two flows from which have been dated previously (Bishop *et al.* 1969) giving 12.0 m.y. for the top and 13.5 m.y. for the base of the sequence. The sequence is equivalent to the Tiim sequence but no palaeomagnetic correlations have been established.

The remainder of the sequence is a mixture of phonolites, basalts and agglomerates. The directions for the agglomerates below P20 are not reliable but all the other units except P14 are normally magnetised; P14 is reversed, but shows no similarity in direction to either T2 or T9 of the Tiim sequence.

Charkum sequence (Fig. 15 :3c)

This sequence along a river in the area near Charkum includes the Atimet trachyphonolite, N4, which has reversed polarity, the top and basal Noroyan flows N3 (indeterminate) and N2 (reversed) respectively and part of the Sidekh phonolite group N1, N5, N6, N7 (normal).

Kamuiton sequence (Fig. 15 :3c)

This collection, made to the east of Terenin, largely overlaps the Charkum sequence. R1 is a fubarite, R2, 3, 4, 5 and 8 are samples of the Atimet trachyphonolite, R6–7 are phonolites, presumably the Sidekh group, but just possibly part of the Tiim sequence downfaulted. The Atimet samples have reversed (R4, R5), normal (R3, R8) and intermediate polarities (R2) and there is no clear palaeomagnetic link between them and N4 from the Charkum sequence. R6 and R7 are reversed.

Palaeomagnetic stratigraphy

The magnetic stratigraphy is illustrated in Figs. 15 :3a to c where all the units collected at one locality are shown in the same vertical column. The position of a unit within a column represents its relative age as indicated by superposition but no attempt has been made to preserve a uniform scale of either time or thickness.

The linking of units and sequences has been made with reference to a condensed stratigraphic column prepared for us by Dr G. R. Chapman (private communication) together with the first-hand knowledge of other workers in the field. Correlations known certainly either from the mapping or petrological examination (Chapman private communication) are shown. In only the few cases already mentioned can a firm palaeomagnetic link be made based on identical directions.

Though this work is the most comprehensive palaeomagnetic survey of the region so far, our sampling of the igneous units was not complete and not all the sediments sampled provided useful material so that there are inevitably many gaps in the preliminary stratigraphic columns presented in Figs. 15 :3a to 3c.

Acknowledgements

Our thanks are due to Professor B. C. King and his colleagues for providing information acquired through the work of the EAGRU (East African Geological Research Unit). Dr G. R. Chapman carried out the petrological examination and gave generously of his time and knowledge in numerous discussions. Drs M. Pickford and P. Tallon gave us much geological and logistical information as well as invaluable guidance in the field. Mr S. Rhemtulla of the Fluorspar Company of Kenya provided facilities in his area.

We are particularly indebted to Professor W. W. Bishop for his interest, enthusiastic support and contributions to the project which was funded by a grant from the Natural Environment Research Council.

References

BISHOP, W. W., MILLER, J. A. & FITCH, F. J., 1969. New Potassium-Argon age determinations relevant to the Miocene Fossil sequence in East Africa. *Amer. Jour. Sci.* **267**, 669–99.

—— 1971. *The late Cenozoic history of East Africa in relation to Hominoid Evolution*, in, *The late Cenozoic Glacial ages*. Ed. K. K. Turekian, Yale Univ. Press.

—— PICKFORD, M. & HILL, A. 1975. New evidence regarding the Quaternary geology, archaeology and hominids of Chesowanja, Kenya. *Nature, Lond.* **258**, 204–8.

—— HILL, A. & PICKFORD, M. 1977. *Chesowanja revisited*. This volume page 8.

CARNEY, J., 1972. *The geology of the country to the east of Lake Baringo, Rift Valley Province, Kenya*. Ph. D. University of London (unpublished).

—— HILL, A., MILLER, J. A. & WALKER, A., 1971. Late Australopithecine from Baringo District, Kenya. *Nature, Lond.* **230**, 509–14.

CHAPMAN, G. R., 1971. *The Geological evolution of the North Kamasia Hills, Baringo District, Kenya*. Ph.D. theses, University of London (unpublished).

HEIRTZLER, J. R., DICKSON, G. O., HERRON, E. M., PITMAN III W. C. & LE PICHON, X. 1968. Marine magnetic anomalies, Geomagnetic field reversal and motions of the Ocean Floor and Continents. *Jour. Geophys. Res.* **73**, 2119–36.

MARTYN, J. 1969. *The geological history of the country between Lake Baringo and the Kerio River, Baringo District, Kenya*. Ph.D thesis, University of London (unpublished).

PICKFORD, M., 1975. Late Miocene sediments and fossils from the Northern Kenya Rift Valley. *Nature, Lond.* **256**, 279–84.

SKINNER, N. J., ILES, W. & BROCK, A. 1974. The recent secular variation of declination and inclination in Kenya. *Earth and Planet. Sci. Lett.* **25**, 338–46.

16

MARTIN H. L. PICKFORD

Geology, palaeoenvironments and vertebrate faunas of the mid-Miocene Ngorora Formation, Kenya

The Ngorora sedimentary sequence is the only major fossiliferous unit between 9 and 12 million years in sub-Saharan Africa that has been mapped and studied in detail. Consequently many of the taxa from it are new. Much of the fauna resembles material in the Siwaliks of the Indian subcontinent and the widespread 'Tethyan' faunas of southern Europe. It appears that there was a widespread homogeneous faunal province during the middle Miocene which may be viewed as an expanded 'proto-Ethiopian' zoogeographic region. During the mid-Miocene the ancestral Palaearctic zoogeographic region appears to have had a more northerly limit than that of the present day.

In addition, the sediments at Ngorora contain a variety of evidence contributing to an understanding of local and regional palaeoenvironments, tectonic and volcanic histories and sedimentology in a Rift Valley setting. During the period of deposition the Ngorora area seems to have been as varied as the Baringo/Hannington area at the present day. Habitats ranged from lacustrine, through fluviatile, gallery forest, open woodland and possibly grassland. The climate may have been seasonal, with wet and dry periods. It was probably never arid, but may have approached semi-aridity from time to time.

The occurrence of other sedimentary units above and below the Ngorora Formation enhances its usefulness in correlation. Comparisons can be made through time; changes can be observed in the faunas of the Baringo succession; and a picture is emerging of gradual faunal change as the result of local evolution plus migration from other areas. Earlier ideas concerning periods of rapid faunal turnover must be rejected. Boundaries between land mammal ages should be drawn with caution for Miocene and younger strata in East Africa.

Introduction

Definition

The Ngorora[1] Formation (Fig. 16:1) is defined as that body of sediment deposited between the Tiim Phonolite Formation and the Ewalel Phonolite Formation, both of which present lava contacts. The uppermost flow of the Tiim Formation is flow-banded and non-porphyritic. The lowermost Ewalel Phonolite is a feldsparphyric lava which when weathered appears light grey with white specks. In fresh samples it is dark grey.

[1] The name Ngorora is derived from the administrative location of Ngorora, Baringo District, Kenya, in which much of the Formation lies. Ngorora is the Tugen word for the 'wait-a-bit' thorn (*Acacia mellifera*).

The Ngorora Formation, which crops out in the scarp and dip slopes of the Tugen Hills, is presumed to extend eastwards from the Tugen Hills faults towards the axis of the Rift Valley where it is covered by various younger formations. To

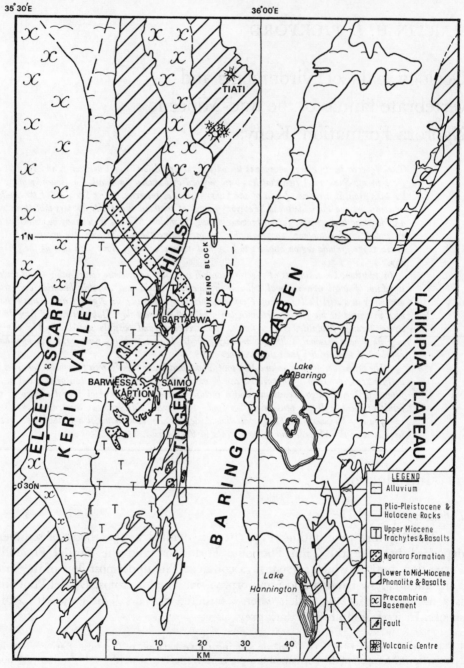

FIG. 16:1. Regional geology of the Baringo area.

the north, the sediments thin towards Tiati Volcano (Fig. 16:1) and the original edge of the sediment has been eroded (McClenaghan 1971). To the south, the sediments wedge out south of Kamungei (Lippard 1972) and they do not occur in the head of the Kerio Valley, where the Ewalel Phonolites lie unconformably on the Tiim Phonolites (Lippard 1972). The position of the western boundary of the formation may be estimated more accurately as it is probable that the Elgeyo Escarpment was the topographic feature against which the sediments were impounded.

The formation is exposed in a long narrow north-south belt near the crest of the Tugen Hills (Area Ib). Other major outcrops occur in the type area at Kabarsero, with its associated outcrops near Chepkesin (Area Ia), and in two large erosional embayments in the dip slope of the Tugen Hills (the Poi and Kapkiamu areas (Area II)). There are other smaller outcrops to the south in the Kibingor area (Area III) where the sediments are cut by numerous faults (Fig. 16/17).

Previous work

Bishop and Chapman (1970) published a preliminary account of the Ngorora Formation and a more complete stratigraphic context was given in Bishop *et al.* (1971) together with ages of the overlying and underlying lavas. Bishop and Pickford (1975) presented a more complete account of the geology of the formation. Other information relevant to the study of the sediments and associated volcanic rocks is in unpublished Ph.D. theses by Martyn (1969), Chapman (1971), McClenaghan (1971), Lippard (1972) and Pickford (1974).

The fauna has been studied by several workers and lists are given in Bishop and Chapman (1970), Bishop *et al.* (1971) and Bishop and Pickford (1975). Hyaenidae were examined by Crusafont and Aguirre (1971); Rhinocerotidae by Hooijer (1971); Pisces by Van Couvering (1972); Orycteropodidae by Pickford (1975); Carnivora, Hyracoidea and Suidae by Pickford (in preparation); Proboscidea by Maglio (1974); Potamidae by Morris (1976).

Palaeogeomorphology and Basin formation

The depositional surface of the Plateau Phonolites is often very flat (Lippard 1974, Walsh 1963) and land surfaces in the Ngorora area are thought to have been almost planar at the close of Tiim phonolite extrusion, apart from the volcanic centre of Tiati to the north (McClenaghan 1971) (Fig. 16:1). Surface expression of the Rift Valley at this time was neglible until faulting along the Elgeyo and Tugen Hills fractures about 12 m.y. ago initiated local depressions which received sediment as erosion commenced on uplifted areas. In the Baringo area these basins were located in the western portion of the area represented by the present-day Rift Valley. The rift at this stage was an assymmetric structure, with a monocline or fractured monocline in the region of Laikipia in the east and major, or growing, scarps in the Elgeyo and Tugen Hills area to the west (Fig. 16:2). Downwarping was centred in the Tugen Hills–Elgeyo area, with the greatest downwarp along the Elgeyo Fault west of Barwessa (Lippard 1972).

The Elgeyo Fault had a greater throw than the Tugen Hills Faults, and sediments appear to have ponded against the Elgeyo Scarp. In contrast sedimentation resulted in burial of most of the surface expression of the Tugen Hills fault scarps. The Saimo and Sidekh areas however, were probably emergent throughout deposition of the Ngorora Formation. At present the Rift Valley floor slopes from about 1800 m (6000 ft.) in the vicinity of Lake Nakuru 100 km south of Ngorora, to 365 m (1200 ft.) near Lake Turkana in the north. Tiati Volcano

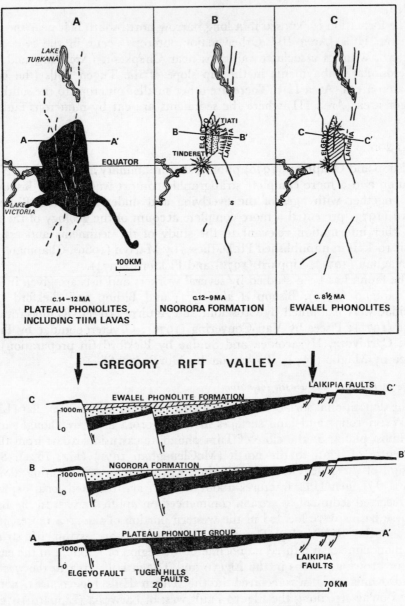

FIG. 16:2. Volcanic and sedimentary history of the Tugen Hills area.

formed a major barrier within t⸍ ⸍ʲeloping rift and acted as the northern boundary of the Ngorora Basin. Analogous barriers in the Rift Valley at present are the volcanoes of Silali, Emuruangogolak and 'The Barrier', all of which have sediments on their southern flanks. During the formation of Members A and B a volcanic centre at Kaption (Area II) supplied tuffs, agglomerates and lavas to the growing pile of Ngorora sediments, and eventually formed a substantial topographic feature within the watershed. It was not covered during Ngorora time but was eventually capped by Ewalel Phonolites.

The palaeogeomorphic setting of the Ngorora Formation was relatively simple at the beginning of sedimentation with emergent areas at the Saimo Horst, the Sidekh Tilt Block and the Kaption centre. This situation was to change with time as the Tugen Hills became a more positive feature in the topography and as the Kapkiamu Graben developed. In particular the area south of Saimo (Area III) appears to have been rather complexly cut by faults where the Saimo Horst bifurcates southwards and passes into a series of sub-parallel horsts and grabens (Fig. 16:5). Area III was exposed to subaerial processes until Member C time, after which sediment began to cover the weathered lavas of the Tiim Phonolite Formation.

Sedimentation

Sediments were derived from a variety of sources, in particular from the north and east (Tiati and Laikipia areas respectively). There was a major source in the volcano at Kaption, which supplied much of the sediment of Members A and B. The Saimo Horst supplied material to parts of the basin, in particular to the Kapkiamu Graben during Member C time and to area III during deposition of Members D and E.

Some of the Ngorora tuffaceous material may have been derived from Tiati where Webb (1971) records a thick sequence of Lelgrong Tuffs above the Tiim Phonolites. The Ewalel Phonolites (Fig. 16:2) flowed from the south-west following a pre-existing lineament to spread out and cover the Ngorora sediments (Lippard 1972). They ponded against the Elgeyo Scarp but nowhere crossed it, yet their source lies at a higher altitude (c 2700 m, 9000 ft.) than the crest of the Elgeyo Scarp (c 2100 m, 7000 ft.).

During Member A and early Member B times, the Kaption Centre was supplying much debris to the Ngorora basin, including possibly the lahars which form an important part of the sequence (see Fig. 16:4A to F for reference to this section). The predominant type of sedimentation throughout the later parts of Members A and B was fluviatile floodplain deposition. From time to time a lahar was emplaced giving rise to a series of alternating lahar/fluviatile units.

The eastern toes of the lahars show signs of erosion and have been reworked to provide some of the fluviatile material. The lahars may have been triggered by faulting along the Tugen Hills Faults (Kito Pass, Saimo and Cherial Faults). Sediment thickness differences between the upthrow and downthrow sides of the Kito Pass Fault prove that there was intermittent movement during sedimentation.

During the later part of Member B time there was rhythmic deposition of tuffs and clay/grit horizons, the latter often topped by palaeosols. Some of the tuffs were airfall deposits, but others appear to be fluviatile. Soil formation and numerous fossil plants in positions of growth indicate sub-aerial exposure throughout the deposition of Member B. This may explain why there are so many fossils in the unit.

Member C saw the beginning of fully lacustrine deposition with well-laminated shales, often suncracked. Clastic dykes indicate fissuring at the surface, possibly as a result of faulting along the Kito Pass Fault. The shales are white, and have been partially silicified as in several other lacustrine deposits of various ages in the Gregory Rift Valley. Occasional coarser bands occur, indicative of lake margin or fluviatile deposition. Clayey palaeosols within Member C contain mammalian faunas.

During Member D a return to floodplain conditions took place, and a suite

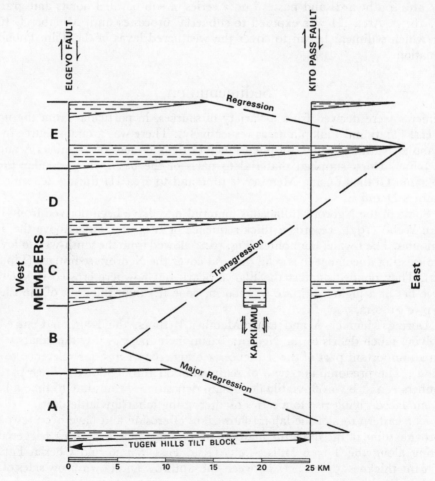

FIG. 16:3. Major patterns of sedimentation, Ngorora Formation.
Broken lines indicate lacustrine conditions.

of coarser clay/sand/grit horizons was deposited. Tuffs are more orange and golden when compared with those of Member B, but otherwise the two members are similar. Members B and D contain abundant mammalian fauna in gritty channel deposits.

Member E represents a return to fully lacustrine conditions. Shales similar to those of Member C were deposited until unit E4 which records a change to fluviatile conditions with deposition of conglomerates and gritty tuffs. Substantial overbank deposits in Unit E4 are silty and tuffaceous, with fossil fish and leaves. Contemporary clastic dykes indicate fissures open to the surface. Sedimentation was terminated by the eruption of the Ewalel Phonolite which entirely blanketed the Ngorora Basin.

At the start of Member A time there may have been a widespread lake which probably extended from near the Elgeyo Fault in the west to beyond the Kito Pass Fault in the east. It underwent a major regression shortly after, and possibly as a result of, the arrival of the first lahars (Fig. 16:4A). The Kaption eruptive centre filled a large part of the basin with lavas, pyroclastic rocks and lahars. It is possible that tectonic activity was partially responsible for the regression, but this is difficult to prove. During Member B time, the lake was small and situated to the west of the crest of the Tugen Hills Faults. The situation shown in Figure 16:4 is a reconstruction of the area immediately prior to the deposition of the accretionary lapilli tuff (unit B4).

Fig. 16:3 shows that a major eastward transgression began towards the end of Member B times which continued until Member E time. However, the sequence of events was complicated during Member C time by the formation of the Kapkiamu Graben with its saline/alkaline lake, and the impounding of a lake to the east of the Kito Pass Fault (see also Figs. 16:4C, 16:7).

The transgression continued through Member D time during which the lake extended eastwards almost to the crest of the Tugen Hills fault scarp. During Member D, the lake to the east of the fault ceased to exist and fluviatile sediments were deposited in the type area. The major transgression continued and the lake once more spread over the Bartabwa Gap into the type area after which the shales of Member E were deposited (Fig. 16:4E 1–4). At the end of deposition of unit E4, major faulting and uplift in the Tugen Hills area resulted in fragmentation of the lake into two basins, separated by the elevated crest of the Tugen Hills Tilt Block (Fig. 16:4E 5–6) which began to undergo erosion.

Stratigraphy, Age and Correlations of the Ngorora Formation

Type section

A geological section through the type succession at Kabarsero is reproduced in Fig. 16:6. 415 m (1400 ft.) of sediment overlies a thick massive non-porphyritic phonolite flow with apparent conformity and is capped by another phonolite of porphyritic variety again with an apparently conformable junction.

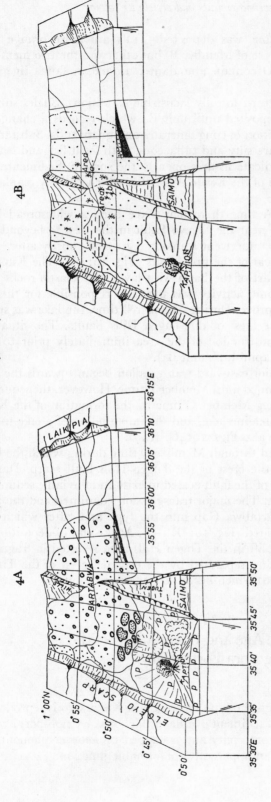

Figs 16:4 (A–F). Tectonic—sedimentary history of the Ngorora Basin.

Units A2 to B3 (6 lahar/clay beds)

Major movement along the Elgeyo fault (west) and minor faulting along the Tugen Hills and Laikipia trends results in initiation of sedimentation in western half of youthful Rift Valley.

Eruption of Kaption Centre (phonolite flows, agglomerates, tuffs) provides a major source of debris for the basin, including up to 6 lahars to the Bartabwa area; otherwise sediments mainly derived from the Laikipia area (fluviatile deposits).

Saimo and Sidekh structures begin to form and do not undergo sedimentation.

Elgeyo scarp large and provides western bounding feature of the basin.

Lacustrine sediments of Unit Al not shown but lahars may have caused a major regression of the unit Al Lake by filling the basin.

Units B4 to B6 (3 rhythmic units)

Lahar activity dies out but airfall tuffs still occur indicating possible late activity at Kaption Centre.

Tectonic setting similar to previous figure, although all structures more pronounced. Evidence of lacustrine conditions near Bartabwa (unit B3) with well vegetated surrounding country (palaeosols) consisting possibly of open woodland or grassland (based on faunal interpretation). Numerous small streams in type area and area Ib, containing abundant fossils.

4D

4C

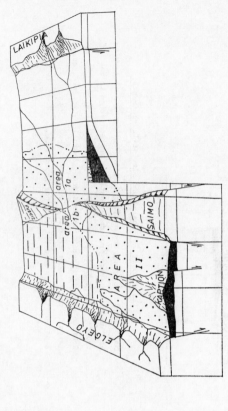

Member D (3 rhythmic units)

Lake Kabarsero dries up and Kapkiamu Basin filled. Lacustrine
conditions west of area Ib with floodplain and lake marginal
environments in areas Ia, Ib and II. Continued subsidence in
Kapkiamu with deposition keeping pace; thick sediments.
Fauna indicates open woodland or grassland over much of the area.
Saimo and Sidekh increasing in size. Chemnagoi wedge obscured,
and emergent crest of Kito Pass Fault covered by continued
sedimentation. Several streams in areas Ia and Ib.
Kaption still exposed and eroding, although lower slopes are being
covered by sediments.

Member C (Shales)

Major faulting along Elgeyo, and Tugen Hills trends results in
lacustrine basins being formed in downthrow areas. Kabarsero lake
fresh and confined to east of Kito Pass fault towards the end of
Member C time, although in early C time it extended westwards into
area Ib.

Formation of Kapkiamu Graben with very limited watershed. Lake
Kapkiamu saline/alkaline (some gypsum bands) with Tilapia, algae
and petroliferous limestones (oolitic and algal).

Waril Lake freshwater with rich insect and fish fauna. Abundant
leaves preserved in shales.

Chemnagoi wedge begins to form, Sidekh and Saimo increasing in
height, area Ib uplifted towards the end of Member C and undergoes
gentle erosion. Kaption Centre extinct and being eroded.

4E 1-4

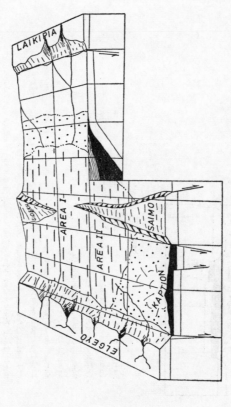

Units E1 to E4 (Shales)

Massive faulting along Elgeyo trend and inception of widespread lacustrine (freshwater) conditions, over much of areas I and II. Saimo, Sidekh and Kaption still emergent, the latter much reduced in topographic importance although the two former structures are increasing in size.

4E 5-6

Units E5 to E6 (Shales and Fluviatile sediments)

Major faulting along Tugen Hills Faults. Uplift of area Ib with exposure and severe erosion. Continued sedimentation to east of faults (area Ia) and near Elgeyo.

Continuous ridge from Saimo to Sidekh separating two lake basins (area Ia and west of area Ib).

Kaption still exposed but of minor importance.

The succession dips westwards at between 18° and 24° and five lithological subdivisions are recognised.

Member E   Diatomites, paper shales, conglomerates   160 m (530 ft.)
            and tuffs.
Member D   Rhythmic units with fossiliferous channel   66 m (220 ft.)
            conglomerates, clays and sands.
Member C   Porcellaneous cream and green silty tuffs   60 m (200 ft.)
            with pumiceous horizons.
Member B   Rhythmic units with fossiliferous channel   42 m (140 ft.)
            conglomerates, clays and sands.
Member A   Clays and coarse bouldery tuffs.   92 m (310 ft.)

(Tables describing in greater detail the stratigraphy and sedimentation of the Kabarsero; Poi–Kapkiamu; and Kibingor areas (Areas I, II and III), are lodged at the Library of the Geological Society, London, and the National Museums of Kenya, Nairobi.)

Correlation between Areas I, II and III

Correlation between Areas I and II is tentative because of the lack of marker horizons common to both areas and poor exposure in Area II. It has been based upon comparison of lithologies and relative positions of the strata. Unfortunately, phonolites are nowhere exposed in Area II, and close to Area I the overlying

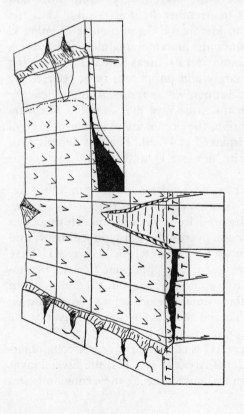

Ewalel Phonolite

Flooding of Ngorora Basin with extensive phonolite from the south covers all but the highest topographic features (Saimo and Sidekh), and abruptly ends sedimentation. (TTTT- Tiim phonolite or equivalent LLLL- Ewalel Phonolite. Black-Ngorora Formation).

FIG. 16:4F.

phonolite is also missing, having been eroded away, together with some of the sediment underlying it.

The Kaption Complex is composed of coarse primary agglomerates and tuffs with interbedded lavas (Figs. 16:4A and 16/17). Overlying this is a sequence of bedded tuffs outcropping in parts of Nuregoi Hill, and to the north at Kerelwanin. This exposure also contains accretionary lapilli tuffs, but it would be unwise to correlate these with unit B4 of the type section. At Barsawe and in the river cutting at Ndurum similar rocks are seen and a sequence of agglomerates and accretionary lapilli tuffs over 60 m (200 ft.) thick is exposed. One of these accretionary lapilli tuffs may possibly be equivalent to unit B4 at Kabersero but it is impossible to be certain. Below these tuffs is found a jumble of rocks, consisting of disrupted and broken beds of tuff and agglomerate, some blocks of which are up to 6 m (20 ft.) across and disposed in a highly haphazard way. Occasional blocks, up to 30 m (100 ft.) long appear to have been carried in the deposit which may represent the proximal portions of a lahar from the Kaption area. This unit could be equivalent to one of the lahars in the type area at Kabersero. The outcrops of lahar-like deposits near Ndurum are confined to a narrow deep valley which cuts through them.

Stratigraphically above the primary tuffs and agglomerates of Kaption, occurs a widespread deposit of shales and well-bedded fine-grained sediments. At Waril they lie unconformably on the Kaption lavas and tuffs. If the Kaption Complex is equivalent to Member A, and the accretionary lapilli tuffs and agglomerates of Ndurum are equivalent to Member B, it is possible that the shales which overlie them are equivalent to Member C. In particular Member C at Terenin (Area Ib, south) contains bituminous limestone nodules, which are a common feature of the shales in the Kapkiamu and Poi areas (Area II). Overlying the Kapkiamu shales is a series of silty, sandy and pumiceous tuffs, referred by Martyn (1969) to the Poi Tuffaceous Sandstones. These represent higher energy conditions than the underlying shales, in the same way that Member D differs from Member C in the type area. In addition, the predominant colour of the Poi sequence of tuffs is golden-orange, well displayed at Waril. The first deposit with this distinctive colour in the type area is in Member D, and the two successions are correlated tentatively on this basis.

Over much of Area II the Poi Tuffaceous Sandstones (Martyn 1969) are the uppermost sediments exposed, partly because erosion has removed the cover and partly because the sediments are obscured by scree and soil. However, at Moigutwa (north of Area II) shales and clays overlie the Poi type sediments and have been equated tentatively with Member E. The outcrops in Area III have both the overlying and underlying lavas exposed. The upper one (Ewalel phonolite) appears to represent a firm time-stratigraphic horizon, but the lower (Tiim phonolite) was exposed at the surface for a long period before sedimentation commenced. At Ngeringerowa and Chepbarserit there is a weathered layer at the top of the underlying phonolites about 45.5 m (150 ft.) thick.

The upper part of the sequence in Area III is composed of coarse conglomerates and tuffs, unlike the fine-grained shales of Areas II and I. If the Ewalel lavas were extruded onto the Ngorora Formation without a time lag these conglomerates and tuffs must be equivalent to Member E, at least in part. The underlying sediments are fine-grained marls, reworked soils and clays. These have been correlated with Member D. Underlying these rocks is a sequence of suncracked shales similar to those of Member C at Kabarsero. These are poorly exposed in Area III and are seen only in the lowest parts of valleys at Ngeringerowa and their total thickness is unknown. In other parts of Area III, only Members D and E, or E alone, are present. (Map. Fig. 16/17.)

Age and correlations

The Ngorora Formation lies between the Tiim Phonolite Formation and the Ewalel Phonolite Formation, both of which have been dated radiometrically, although the spread of dates is wide. Dates from 'derived' feldspar crystals have been obtained from tuffs of Members B and D of the Ngorora Formation.

Bishop and Chapman (1970) and Bishop *et al.* (1971) have discussed the radiometric dating of the Ngorora Formation, but since then more samples of the Plateau Phonolites (in which the Tiim and Ewalel Phonolites are included) have been dated and these are listed in Table 1. In East Africa the Ngorora fauna is more closely related to the Fort Ternan assemblage (c. 14 m.y.) than to any other. The ruminants are generically identical although they appear more advanced and slightly younger at Ngorora. This accords with the younger age (12–9 m.y.) suggested for the Ngorora sediments.

Tentative correlation based on radiometric and faunal evidence for the Baringo succession and other East African localities and for European sequences, based upon evidence from Van Couvering (1972), Gentry (1970) and Berggren and Van Couvering (1974), suggests that the Ngorora Formation correlates with the Upper portion of the Vindobonian, the Vallesian, and Lower parts of the Turolian Land Mammal Ages of Europe and with part of the Chinji/Nagri succession of the Siwaliks of the Indian subcontinent.

Hipparion appears to have arrived in East Africa during the deposition of the Ngorora Formation, and consequently, the upper Ngorora beds are probably younger than about 11.0 million years, the present best estimate for the *Hipparion* 'Datum' (Berggren and Van Couvering 1974). Hooijer (1975) has described and discussed the Ngorora *Hipparion*, which is identical with the original old world *H. primigenium* found from China to Spain.

The Ngorora hyracoid is a new species of *Parapliohyrax* Lavocat, otherwise known only from Beni Mellal, considered by some to be about 14 m.y. old on faunal evidence. *Parapliohyrax* is clearly more advanced that the East African lower Miocene hyracoids from Rusinga, Mfwangano etc. (Whitworth 1954) but more primitive than the extant *Procavia*.

Maglio (1974) has discussed the Ngorora gomphothere (now known to be a choerolophodont) stating that it is possibly the direct ancestor of the Elephantidae. The latter family is known from Mpesida (7 m.y. Bishop *et al.* 1971) and younger localities in Kenya. The specimens are only slightly more advanced than gomphotheres from Fort Ternan.

The Ngorora *Mellivora* (Pickford in prep.) is the earliest known member of the subfamily Mellivorinae and possesses some primitive features, although it is surprisingly advanced for a mustelid of its age.

Protragocerus and *?Pseudotragus* are found in Chinji and Nagri deposits, and the former genus is common in the Vindobonian of Europe. An early *Nyanzachoerus* occurs at Ngeringerowa and *Listriodon* in Member B at Kabarsero.

The Ngorora assemblage marks the establishment of many of the modern groups of East African mammals, and is similar in some respects to faunas of contemporary Eurasia. Southern Eurasia was apparently part of an expanded African faunal province during Ngorora time. *Hipparion* spread rapidly throughout the old world. In all three areas there followed the rapid diminution in abundance and eventual extinction of *Listriodon*. It is possible that *Hipparion* invaded the niche occupied by *Listriodon*. Analagous feeding adaptations include long snout,

TABLE I: *Potassium argon ages related to the Ngorora Formation (whole rock lava ages unless otherwise stated)*

FORMATION	SAMPLE	LABORATORY	AGE & ERROR m.y.
KABARNET	2/214	B	6.8 ± 0.4
TRACHYTE	1/714	B	6.7 ± 0.3
	JM/734	C	7.3 ± 0.3
	" "	C	7.3 ± 0.3
EWALEL	2/210	B	7.1 ± 0.4
PHONOLITES	9/545	B	7.3 ± 0.3
	13/1209	B	8.8 ± 0.3
	WB/11	C	6.7 ± 0.1
	WB/13	C	8.8 ±.0.8
NGORORA	69/1	C	12.3 ± 0.7
FORMATION	69/2	C	11.9 ± 0.7
(Derived feldspar crystals from sediment)			
TIIM PHONOLITES	WB/10	C	11.4 ± 1.2
(Upper)	WB/2	C	13.2 ± 1.4
	JM/474	C	7.2 ± 0.5
	JM/346	C	9.9 ± 0.5
	T/8	A	12.8 ± 0.2
TIIM PHONOLITES	1/368	B	13.0 ± 0.4
(Lower)	2/315	B	14.3 ± 0.5
	T/3	A	14.9 ± 0.4

LABORATORIES: A. Liverpool University, A. Mussett.

B. Institute of Geological Sciences, N.Snelling.

C. Cambridge University, J.A. Miller.

For references see: Chapman and Brook, Paper 14 this volume: Dagley, Mussett and Palmer, Paper 15 this volume.

strong incisor battery and flared symphysis. Limb bones in *Listriodon*, where known, indicate more cursorial adaptations than other Suidae.

The presence of hominoids at Ngorora is of interest, but the limited nature of the specimens prevents reliable interpretation of the material. Similar hominoids occur at Fort Ternan, Chinji, Nagri and numerous European localities of Vindobonian/Vallesian age. They appear to have been members of a widespread group, as with many other elements of the fauna.

Structure and Tectonics

Early fault movement in the Northern Kenya Rift Valley was insufficient for the formation of extensive sedimentary basins (Fig. 16:2), and eruptions of large volumes of phonolite were able to 'iron out' irregularities in the surface of the proto-rift in the Baringo area. At the end of Tiim Phonolite time, there was major movement along the Elgeyo fault which initiated the formation of the longer lived Ngorora Basin, which remained tectonically active through much of the period of Ngorora sedimentation. Figure 16:5 illustrates the major structural features of the Baringo area. In the west the Elgeyo Scarp forms an impressive topographic feature over a mile high in places. The fault dies out southwards and at the head of the Kerio Valley the edge of the Rift Valley is marked by the Equator Monocline (Lippard 1972) without major interrupting faults. However, to the north there has been major faulting giving rise to a steep escarpment facing east and a gentler dip slope facing west. Most of the movement which has resulted in the formation of the Tugen Hills can be ascribed to two faults, the Kito Pass Fault in the north and the Saimo Fault in the south. Other major faults are indicated in Figure 16:5.

The structural contours of the basement complex in the western half of the Baringo graben are shown in Figure 16:5 (slightly modified and extended after Lippard 1972). The contours show that the zone of greatest uplift along the Kito Pass Fault is at Kito Pass, where the Basement Complex is exposed in an anticlinal structure plunging northwest (McClenaghan 1971). The Bartabwa area lies in a synclinal flexure between the Kito Pass and Saimo 'half-domes' in the region where the two major faults meet. The area where the faults meet is rather complex structurally as several faults show movement at different times. For instance, early fault movements during Ngorora times occurred along the Kagilip and Panwa faults (Fig. 16:5): later movement was along the Bartabwa and Saimo faults. If faulting had continued to take place along the former faults, the type area of the Ngorora Formation would now be about 300 m (1000 ft.) below the surface.

To the east of the Tugen Hills lies the Baringo graben of the Gregory Rift (Fig. 16:5) containing lakes Baringo and Hannington. The former lies almost in a straight line with the Kerio depression and the Saimo 'half-dome', suggesting that the local zone of greatest depression within the Rift has migrated eastwards with time (Martyn 1967). East of the Baringo graben and forming the eastern wall of the Rift Valley, are the Laikipia faults, separating the Baringo graben from the Laikipia plateau. The Rift Valley in this region is about 70 km wide (including the Tugen Hills tilt block) and the Baringo graben about 50 km wide.

All the major features shown in Figure 16:5 were present in subdued form during Ngorora times. The Laikipia area was probably a faulted monocline, while the Elgeyo Scarp was already a major relief feature. The Tugen Hills were beginning to form, although in the Bartabwa area their surface expression was buried under sediments which covered the scarp as it formed. The sections in

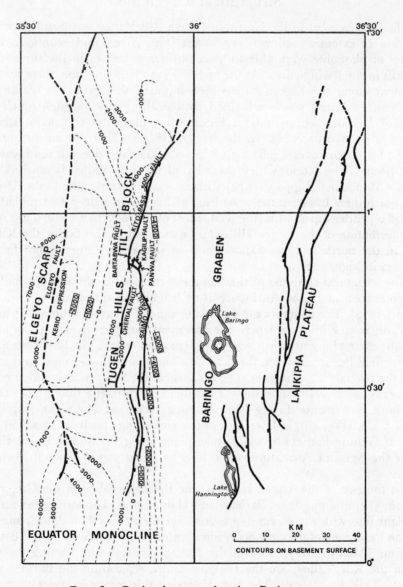

FIG. 16:5. Regional structural setting, Baringo area.

Figure 16:2 illustrate the assymmetry and evolution of the rift during Ngorora time.

Local Structure

In Area Ib the slope of the Tiim Phonolite is usually of the order of 90–120 m (300–400 ft.) per kilometre, but at Kamuiton and Charkum there are very much steeper slopes of the order of about 300 m (1000 ft.) per kilometre. These two areas represent the beginning of the structures which culminate in the half-domes of Saimo and Kito Pass respectively. In the Bartabwa area, there is no steep slope analagous to those leading to Sidekh and Saimo, and the surface is more or less uniform from the Kapitan fault to the Bartabwa fault (southern extension of the Kito Pass fault). Throws along the Kito Pass fault range from about 600 m (2000 ft.) in the south to over 910 m (3000 ft.) east of Sidekh. Other major faults are the Bartabwa fault, throw about 670 m (2200 ft.), Kagilip fault, about 515 m (1700 ft.), Saimo fault, about 730 m (2400 ft.) at the southern margin of the map dying out in the north, and the Panwa fault, throw about 910 m (3000 ft.).

Sediment Thicknesses

Sediment thicknesses vary according to the structures outlined above. They

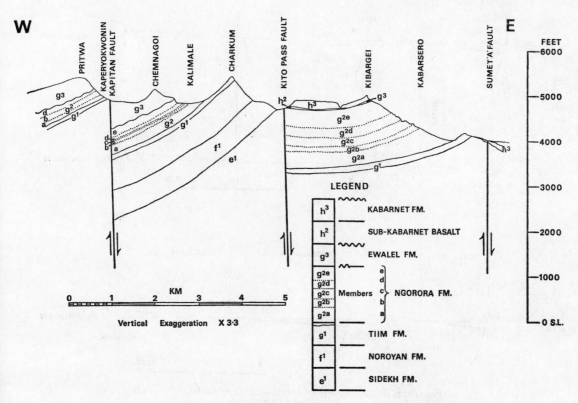

FIG. 16:6. Detailed geological section; Kabarsero, Kalimale, Pritwa.

thin rapidly towards the Saimo and Sidekh highs (Fig. 16:4e) and are thicker on the downthrow sides of major faults.

The triangular fault block at Chemnagoi steepens towards Charkum and was active during Ngorora time (see Fig. 16/17) and the sediments thin over it. The Kapitan fault which bounds the southeastern part of the Lolotwa block leads southwards to form the western edge of the Kapkiamu graben which, with the Kaption tilt block, are the major structural features of Area II. The former is bounded on the east by the west facing Cherial fault scarp (600 ±m (2000 ft.) throw), to the north by the south facing Rimo fault now obscured by Kabarnet trachyte cover. To the west lie the Kapitan and Kaption faults the latter of which is thought to swing eastwards and connect with the Cherial fault in the south. Here, the fault is also obscured by lava cover, but there is a suggestive lineation in the river valleys in the area, indicating the possible trace of the fault. In the middle of the western boundary of the Kapkiamu graben, where the Kapitan fault dies out and the Kaption fault begins the edge is a monoclinal flexure with several small faults. Figure 16:4 summarises the sedimentary and tectonic history of the Ngorora area.

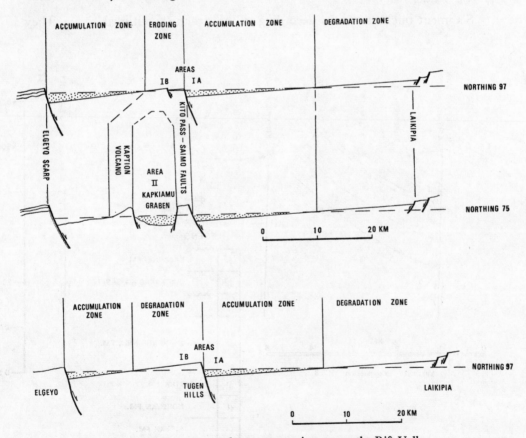

FIG. 16:7. Reconstructed east-west sections across the Rift Valley during Member C and Member D times.

Tectonics

In the Rift Valley tectonic and/or volcanic activity is essential to the formation of sedimentary traps of the size of the Ngorora basin. Subsidence initiates the basins which are then filled by sediments and lavas. The topographic effect of early subsidence in the Baringo area was often counteracted by the extrusion of extensive lava flows which filled the depressions as they formed. The early sedimentary units of the Baringo sequence, such as the Kapkiai shales and the Surkai shales (Martyn 1969) are thin intercalations within a predominantly lava sequence, but in Ngorora time volcanic activity was subdued or dormant long enough to allow the accumulation of a thick sedimentary sequence between the bracketing lavas.

Palaeoenvironments

Detailed palaeoenvironmental reconstructions are recorded in charts lodged at the Geological Society, London. Through the sediment pile, where evidence is available, there appear to have been periodic fluctuations of climate. There is often a strong overprint suggestive of semi-arid conditions, for example numerous oscillations in lake level combined with innumerable suncracked horizons. The alkaline Lake Kapkiamu probably had a limited input of water and no outlet, but its contemporary, Lake Kabarsero (Area Ia) was fresh, possibly reflecting its larger watershed (Fig. 16:4C). Similar juxtaposed fresh and saline lakes occur in the Gregory Rift at the present day. The shoulders of the Rift were not as elevated during the mid-Miocene as they are now, and it is unlikely that a rain shadow was present in the youthful rift. It is possible that the climate over the whole of Kenya was drier during the mid-Miocene than at present, but it is unlikely that arid conditions prevailed at Ngorora.

The soil in the Ngorora Formation is of two principal varieties, calcrete-bearing and clay palaeosols. The former can be formed under a positive evaporation and transpiration budget in areas of high rainfall such as the Ahero Swamp near Lake Victoria, and are not necessarily indicative of hot, dry climate but are usually formed where the water table is near the ground surface. The Ngorora clay palaeosols are quite different from the predominantly lateritic profiles formed on the Kabarnet Trachyte (7 m.y.) and younger formations. In the Baringo area a considerable change occurred in the soil forming processes about 7 million years ago. Whether this change was related to the establishment of rain shadow conditions and major escarpments within the rift, or was a more widespread phenomenon, is still to be established. Pedogenic laterites are widespread in Kenya, and are still forming, but are seldom demonstrably older than 7 million years.

Much of the vegetational reconstruction is interpreted from the faunal evidence. Fossil leaves and grass have been found, but they are limited in extent and indicate forest herb layers (H. Osmaston pers. comm.) possibly close to streams (?gallery forest). The herbivores suggest more open country, predominantly open woodland, and the majority of the remainder of the fauna confirms this. Ostrich and Marabou Stork in Member C, *Mellivora* in Member D, *Orycteropus* in Member B are forms which prefer open woodland or grassland to thick forest.

Against this evidence one must bear in mind that few of the fossils were found in their place of death, and may derive from environments some distance from their final resting place. The study of pollen samples collected in 1975 should provide useful evidence of regional vegetation conditions at Ngorora.

Comparison of the frequency of *?Pseudotragus* (a grassland form) with *Protragocerus* (a woodland form) (Gentry 1970, p. 310) suggests that a distinct change in regional environment occurred during deposition of Members B to E (Fig. 16:8). In Members B and C, *?Pseudotragus* was rare compared with *Protragocerus*. During Member D they were present in about equal quantity, but during Member E *?Pseudotragus* was dominant. This may indicate a gradual change from woodland towards grassland during Ngorora time, but this must be compared with the transgression and regression of the lakes. The change may be the result of facies migration in the vegetation, in phase with the major transgression which occurred from Member B until unit E4 (see Fig. 16:3) and may not have a regional significance. However, in the Lukeino Formation (c. 6½ m.y.) (Pickford, this volume) grass appears to be the dominant vegetation type and the change seen in the ratio of *Protragocerus* and *?Pseudotragus* may have real palaeoenvironmental significance.

The environments suggested by the sediments are similar to those which exist in the Rift Valley near Baringo today. The climate appears to have been seasonal, but predominantly hot. Semi-arid conditions may have existed during the deposition of Member C, but vigorous plant growth in Member B suggests a more equable climate. The present day Baringo–Hannington basin is probably the best analogue for the Ngorora conditions, with its juxtaposed fresh and alkaline lakes, low-lying swampy areas between the lakes and fault bounded sediment traps. In such a setting, a great variety of localised environments exists, in which a varied fauna and flora can flourish. Some of the habitats and niches although of limited extent may leave a strong impression in the geological record; for

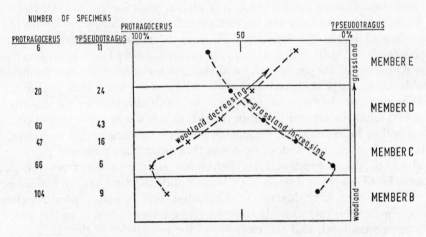

FIG. 16:8. Change in the ratio of *Protragocerus* to *?Pseudotragus* through the period of deposition of the Ngorora Formation from Member B to Member E.

example, the rich fauna of Member B appears to be associated with a flood plain environment in which gallery forests occurred.

Fauna and Flora

Mollusca

One species of bivalve and at least eight species of gastropod are found in the Formation. Most of them are aquatic but two pulmonate gastropod species occur. Some localities yielded abundant phosphatised molluscs intermingled with remains of arthropods and fish. Preservation is usually excellent, although some localities yield crushed specimens only.

Arthropoda

Ten species of arthropod are known and there is a possibility that more will be collected in future at Waril (Area II, locality 1/1017). They are preserved as impressions in tuffaceous shales similar to those in the Florisant deposits of America. They occur in profusion with numerous small (3 cm) fish and a variety of leaves. Crustacea (*Potamon* sp.) occur in phosphatic concretions at Kabarsero, and in clays associated with molluscs at Kalimale. Ostracoda are common in a few localities but preservation is poor.

Pisces

Five or more species of fish occur in the Ngorora Formation. Comminuted remains are extremely common throughout the formation and complete specimens occur at some localities, for example in the Kapkiamu basin. Catfish (*Clarias* sp.) are common but are seldom complete. *Tilapia* are ubiquitous and range in length from 3 cm to 16 cm, any single deposit being characterised by a restricted size range. At Waril all the specimens are less than 6 cm long while at Kapkiamu they are mostly longer than 10 cm. At Waril there are fragments of fish vertebrae up to 5 cm across and large teeth were collected occasionally.

Reptilia

Remains of Crocodile and Chelonia are fairly common in the Ngorora Formation but lacertids are rare. Remains are usually disarticulated but a com-

TABLE 2: *Distribution of invertebrates through Ngorora Formation*

	Members				
	A	B	C	D	E
Bivalvia		X******X			X
Gastropoda		X******X******X******X******X			
Decapoda			X******X		
Ostracoda				X	
Coleoptera		X******X			
Diptera			X		
Lepidoptera			X		
Isoptera			X		
Annelida		X******X******X			

plete tortoise preserved in unit D1 at Kabarsero, is 1.5 m (5 ft.) from head to tail and slightly smaller than a turtle eroding out of a hillside at Ngeringerowa (Area III). The latter specimen is broken but was probably 2 m (6.5 ft.) in diameter and has a carapace nearly 15 cm (0.5 ft.) thick. It is filled with comminuted fish remains. Smaller chelonians are common and widespread and belong to three genera, *Testudo*, *Pelusios* and *Trionyx*. Crocodile remains are usually broken. They are all assigned to the genus *Crocodylus*, although the remains are too broken for certain identification. Squamata are represented by Monitor Lizards (Varanidae), but all the specimens differ from the extant East African form.

Aves

A marabou stork, an Ostrich and a Diver (*Anhinga* sp.) have been recovered from Ngorora, but bird bones are rare.

Mammalia

A large number and variety of Mammalia have been collected from the formation. Four carnivore species are known, all of them new. (Pickford in prep.) A gomphothere, ancestral to the Elephantidae (Maglio 1974) occurs, and Deinotheriidae occur transitional in size between *Prodeinotherium hobleyi* and *Deinotherium bozasi*. A new species of giant hyracoid (Pickford in prep.) is common and a new species of Ant Bear (*Orycteropus chemeldoi* Pickford 1975) is recorded.

	Members				
	A	B	C	D	E
Clariidae	?	?	X	X	X
Cichlidae	?	?	X	X	X
Cyprinidae	?	?	X	?	?
Testudinidae				X	
Trionychidae	?	X		X	X
Pelomodusidae	X	X		X	X
Squamata		X		X	
Crocodilia	X	X	X	X	X
?Ophidia		X			
Struthionidae			X		
Ciconiidae			X		X
Anhinga					X
Hominoidea (Hominidae?)			X		
Pongidae		X			
Cercopithecoidea		X	X		?
Mustelidae				X	
Canidae	X	X		X	
Hyaenidae			X	X	
Deinotheriidae		X		X	
Gomphotheriidae	X	X		X	X
Procaviidae		X	X	X	X
Equidae				X	X
Rhinocerotidae	X	X	X	X	X
Suidae		X	X	X	X
Hippopotamidae				X	
Giraffidae		X	X	X	X
Climacocerus		X			
Bovidae	X	X	X	X	X
Tragulidae		X			
Orycteropodidae		X			
Rodentia		X		X	
?Pedetidae		X			
Insectivora		X			

TABLE 3: *Distribution of vertebrates through the Ngorora Formation.*

Perissodactyls are common, rhinocerotids are represented by three genera and *Hipparion primigenium* Von Meyer (Hooijer 1975), occurs in Members D and E (its earliest record in Sub Saharan Africa). Among the artiodactyls are found the earliest known true hippopotamids, so far only found in Member D. Five species of suid occur; *Lopholistriodon kidogosana* (Pickford and Wilkinson 1975); a true *Listriodon*; *Nyanzachoerus*; a small suine; and a large suoid reminiscent in several respects of hippopotamids. At least five bovids are known from the Ngorora Formation but two genera (*Protragocerus* and *?Pseudotragus*) make up 60% of the specimens. The rest are represented by a few horn cores and teeth of early *Cephalophini* (Duikers), Neotragini and the earliest known Tragelaphini. Three giraffoids are present; *Palaeotragus* is common and a larger species (*? Samotherium*) and a Palaeomerycid are present but rare. Eight or nine genera of Rodentia are known (Jaeger pers. comm.) most of which are new genera. In addition insectivores and bats are now known from several specimens and a hylobatid was collected in 1975.

Flora

Many specimens of wood and leaves were found and Brett (pers. comm.) has recognised palms, lianas and other dicotyledons among the woody remains. The leaves examined by Dr H. Osmaston (pers. comm.) are representative of about four dicotyledon species and one grass from Waril Locality (1/1017) and about seven dicotyledon species associated with about two monocotyledons from locality 2/114 at Kabarsero. Diatoms are numerous in Members C and E and Terry (pers. comm.) recognised *Melosira* sp.

Taphonomy

Seven types of fossil deposits occur at Ngorora. Channel deposits (43 localities) yielded the most varied fauna, many specimens of which are disarticulated and have been damaged by rolling, polishing or breaking. The fossils represent sampling and concentration of bones and teeth from a large area, possibly from a variety of habitats. Many non or poorly fossiliferous channels were located but there were no apparent lithological differences between those which were fossiliferous and those which were not. Some have yielded fine specimens of skulls and limb bones (e.g. localities 2/10, 2/11). There are often traces of pyrolusite in these deposits and the cement is invariably calcite.

Overbank deposits make up 12% of the fossiliferous deposits of the Ngorora Formation (15 out of 122 localities). Vertebrate fossils are usually well-preserved and sometimes articulated (e.g. at locality 2/2). Fossil leaves and fish in these deposits are frequently well-preserved and numerous Gastropoda occur in some of them. In this group have been included the earthy marls of Ngeringerowa which have yielded equids and hippopotamus-like suids.

Lacustrine deposits make up a large proportion of the sedimentary pile of the Ngorora Formation, and 42 sites have yielded fossils. Several localities in alkaline shales at Kapkiamu and silty shales occurring at Waril have yielded complete fish. Insects and leaves also occur at Waril.

Subaerial landsurfaces have yielded a number of fossils, but they are scattered and uncommon (18 examples). They are important in that they often yield articulated material, as at localities 2/72 and 2/74. Many animals die on subaerial land surfaces but for a variety of reasons, most of their bones and teeth are not preserved as fossils. Animals seldom die close together, so that there is little probability that this kind of deposit will yield rich faunas. Three categories of subaerial deposits are recognised; first, eroding/weathering horizons, which are poorly fossiliferous; second, clay palaeosols which are common at Ngorora and have yielded important faunas (e.g. locality 2/72); third, calcretes which are usually poorly fossiliferous (e.g. locality 2/49).

A deposit of calcareous tufa has yielded pulmonate molluscs at two localities. The deposits which are similar to recent tufas in the Lake Hannington area are rare in the Ngorora Formation. A scree deposit at Cheprimok contains numerous small wood fragments. Primary tuffs, although useful in sealing deposits rapidly, seldom contain fossils, unless the tuff settles round them, as at locality 2/114.

Note: Further charts and diagrams giving details of the stratigraphical palaeontology and palaeoenvironments have been lodged in the Library of the Geological Society, Burlington House, Piccadilly, London, and the National Museums of Kenya, Nairobi, where they may be consulted.

Conclusions

The Ngorora Formation (12–9 m.y.) records a long period in the history of the youthful Rift Valley. The detailed reconstructions which can be made from the sedimentary sequence yield evidence concerning the nature of volcanicity and tectonism within the developing rift. The sediments are a sensitive meter providing geological readings over a substantial time span. Lava flows because of their instantaneous nature are extremely useful for correlation and radiometric calibration. From the sediments it is possible to determine that localised fault movement took place in numerous small increments, and that there were periods of rapid displacement interspersed with periods of minimal activity.

Abundant fossil flora and fauna allow checks to be made on palaeoenvironmental reconstructions based on sedimentary evidence. At Ngorora they generally confirm them. The environments envisaged for Ngorora are as varied as those of the present day Baringo area, and indeed the Baringo/Hannington basin provides a good analogy for Ngorora time. There is evidence to suggest that the climate was seasonal and hot. Much of the fauna is new and is providing abundant palaeontological data for taxonomists and biostratigraphers. As an example, the discovery of *Hipparion* in the Ngorora Formation has led to a reappraisal of the palaeodistribution of the group in Africa. The hypothetical ecological barrier (?proto-Sahara) thought to separate North and East Africa may be little more than an imaginary prerequisite based on an inadequacy in the fossil record. In this context it is as well to remember that 'Absence of evidence is not evidence of absence'. In addition, the many similarities between the Ngorora fauna and those of similar time-spans in Southern Europe and Asia indicate very strong ecological similarities and a lack of long lasting regional barriers between the three areas.

The Ngorora mammals belong to a widespread, fairly homogeneous mid-Miocene Eurasian-African faunal complex.

Much of the modern East African fauna can trace its history back to the Ngorora Formation (e.g. *Hippopotamus*, *Mellivora*, *Orycteropus*, Elephantidae, *Thryonomys*, Cephalophini, Neotragini, Tragelaphini, *Ciconia* and *Anhinga*) but much of the Ngorora fauna is extinct (e.g. Listriodontinae, *Climacoceras*, Amphicyoninae, *?Pseudotragus* and three genera of Rhinocerotidae).

As one of a sequence of six sedimentary units in the Tugen Hills Area, the Ngorora Formation provides an important part of the Neogene history of East Africa.

Acknowledgements

I thank all those who have helped in the production of this paper. Government authorities of Kenya provided permits to study in the Baringo area and the National Museums of Kenya gave valuable assistance. Financial support was supplied through grants to Professor W. W. Bishop from the Natural Environment Research Council (NERC Grant No. GR3/835), the Boise Fund of Oxford University and the Wenner-Gren Foundation for Anthropological Research, New York. I am grateful for grants from the Government of Nova Scotia and the Canadian Student Loan Committee. Research facilities were provided by Bedford College and Queen Mary College, University of London, where the writer held a NERC Research Assistantship to Professor W. W. Bishop and now holds a Post Doctoral Research Fellowship.

References

BERGGREN, W. A. & VAN COUVERING, J. A. 1974. The Late Neogene. *Palaeogeography, Palaeoclimatology, Palaeoecology.* **16**: 1–216.

BISHOP, W. W. & CHAPMAN, G. R. 1970. Early Pliocene sediments and fossils from the Northern Kenya Rift Valley. *Nature, Lond.* **226**: 914–18.

——, CHAPMAN, G. R., HILL, A. P. & MILLER, J. A. 1971. Succession of Cainozoic vertebrate assemblages from the northern Kenya rift valley. *Nature, Lond.* **233**: 389–94.

——, MILLER, J. A. & FITCH, F. J. 1969. New potassium-argon age determinations relevant to the Miocene fossil mammal sequence in East Africa. *Am. Journ. Sci.* 267: 669–99.

—— & PICKFORD, M. 1975. Geology, fauna and palaeoenvironments of the Ngorora Formation, Kenya Rift Valley. *Nature, Lond.* **254**: 185–92.

CHAPMAN, G. R. 1971. Ph.D. thesis, unpublished, Univ. of London.

CRUSAFONT-PAIRO, M. & AGUIRRE, E. 1971. A new species of *Percrocuta* from the middle Miocene of Kenya. *Abh. hess. L-amt. Bodenforsch.* **60**: 51–8.

GENTRY, A. W. 1970. The Bovidae (Mammalia) of the Fort Ternan fossil fauna. *Fossil Vertebrates of Africa.* **2**.

HOOIJER, D. A. 1971. A new Rhinoceros from the late Miocene of Loperot, Turkana District, Kenya. *Bull. Mus. Comp. Zool.* Harvard, **142**.

——, 1975. The *Hipparions* of the Baringo Basin sequence. *Nature, Lond.* **254**: 193.

KING, B. C. & CHAPMAN, G. R. 1972. Volcanism of the Kenya Rift Valley, *Phil. Trans. Roy. Soc. London.* **A. 271**.

LIPPARD, S. J. 1972. Ph.D. thesis, unpublished, Univ. of London.

—— 1973. Plateau Phonolite lava flows, Kenya. *Geol. Mag.* **110**.

MAGLIO, V. J. 1974. A new Proboscidean from the Late Miocene of Kenya. *Palaeontology.* **17**, (3): 699–705.

MARTYN, J. M. 1967. Pleistocene deposits and new fossil localities in Kenya. *Nature, Lond.* **215**.

—— 1969. Ph.D. thesis, unpublished, Univ. of London.

McCLENAGHAN, M. P. 1971. Ph.D. thesis, unpublished, Univ. of London.

MORRIS, S. F. 1976. A new fossil freshwater crab from the Ngorora Formation of Kenya. *Bull. Br. Mus. nat. Hist.* (Geol.) **27**, 4: 295–300.

PICKFORD, M. 1974. Ph.D. thesis, unpublished, Univ. of London.

—— 1975. New fossil Orycteropodidae (Mammalia, Tubulidentata) from East Africa. *Neth. J. Zool.* **25** (1).

—— & WILKINSON, A. 1975. Stratigraphic and phylogenetic implications of new Listriodontinae from Kenya. *Neth. J. Zool.* **25**: 128–37.

VAN COUVERING, J. 1972. Ph.D. thesis, unpublished, Cambridge Univ.

VAN COUVERING, J. A. 1972. Calibration of the European Neogene., *in* BISHOP, W. W. & MILLER, J. A. (Eds.), *Calibration of Hominoid Evolution.* Scott. Acad. Press, Edinburgh.

WALSH, J. 1963. Geology of the Ikutha Area. *Geol. Surv. Kenya Report.* **56**.

WEBB, P. 1971. Ph.D. thesis, unpublished, Univ. of London.

WHITWORTH, T. 1954. The Miocene Hyracoids of East Africa. *Fossil Mammals of Africa.* Brit. Mus. (Nat. Hist.), 1–58.

17

MARTIN H. L. PICKFORD

Stratigraphy and mammalian palaeontology of the late-Miocene Lukeino Formation, Kenya

The Lukeino Formation is one of a series of volcanic and sedimentary units in Baringo District, Kenya, which together span much of the Neogene and Quaternary periods. Outcrops of the Formation occur over an area of about 450 km² to the west and north-west of Lake Baringo, and east of the Tugen Hills range.

The Formation contains an abundant and varied fauna, including Hominidae and Cercopithecidae, dated as upper Miocene on both radiometric and faunal evidence. It is equivalent to the upper parts of the Turolian Land Mammal Age, and is the only known fossiliferous unit of its age in sub-Saharan Africa.

Over 140 metres of sediment, exposed in the type section, comprise principally diatomaceous silty tuffs and shales which are considered to have accumulated slowly under lacustrine conditions. However, there are substantial piedmont and fluviatile facies as well as some primary volcanic material in the sequence.

It has been possible to reconstruct details of the palaeogeomorphology of the basin as well as the palaeoenvironmental, sedimentary and structural history of the area. The climate is considered to have been dominated by seasonal variation in rainfall in an area with both large trees and bushland/grassland. The lake was probably fresh throughout its existence, although the presence of algal mats in lake margin sediments might indicate slightly saline conditions for part of the time.

The general conditions of basin formation, sedimentation, volcanicity, burial and subsequent exposure of the Lukeino succession have parallels in several other units in the Baringo area and elsewhere in the Rift Valley.

In the Baringo succession lavas provide excellent datable marker horizons between sedimentary bodies, and a combined study of both is essential to the understanding of the processes leading to the accumulation of, and the stratigraphic relationships between, the individual units of the Neogene sequence.

Introduction

The Lukeino Formation has recently been discussed by Pickford (1975b). This paper complements previous work and deals particularly with structure and palaeoenvironments. Previous work on the unit includes preliminary geological study by Martyn (1969), Chapman (1971) and McClenaghan (1971) and initial faunal studies by Bishop *et al.* (1971, general fauna), Maglio (1970, Proboscidea), Hooijer and Maglio (1973, Equidae), Cooke and Ewer (1972, Suidae), Pickford (1975c Tubulidentata) and (Coryndon this volume—Hippopotamidae).

A definition and description of the type section appears in Pickford 1975a.

The formation has been divided into four members, and has been intruded by three sills. It is underlain almost entirely by the Kabarnet Trachyte Formation (Walsh 1967) except in the north where the Barpelo Basalt (McClenaghan 1971) forms the floor of the basin. It is overlain predominantly by Kaparaina Basalt (Martyn 1969) but in the north is capped by Ribkwo Trachyte (MacClenaghan 1971) and/or the Chemeron Formation (Martyn 1967). Upper and lower boundaries are unconformable. To the west, the rising Tugen Hills tilt block formed a barrier against which the sediments were impounded. More gentle slopes to the north, east and south were overlain by sediments which exhibit facies changes possibly controlled by interplay of climatic and tectonic influences. Tectonism played an important part in the formation and subsequent history of the Lukeino Basin. Volcanoes were the primary source of much of the sediment, although some of the early sediments and piedmont facies deposits were derived from weathered products of pre-existing formations.

Fauna is common throughout the unit, but mammals are generally restricted to lake margin and fluviatile facies of Members A and B. Fish, grass and leaves are very abundant in the lacustrine shales, but rare in the fluviatile facies. A variety of sedimentary environments is indicated for the Lukeino area. These include an extensive freshwater lake, bounded to the west by the youthful Tugen Hills. The escarpment was about 450 m (1500 ft.) high with substantial piedmont deposits at its base, merging into fluvio-lacustrine flats near the lake margin. These former lake flats provide the bulk of the mammal fauna, and were probably covered in grass and perhaps gallery forest. The postulated variety of ecological conditions could explain why there is such a rich and varied fauna in the formation (Pickford 1975b).

Sedimentation, stratigraphy, age and correlations

Sedimentation

In the Lukeino Basin sediment accumulation followed subsidence in the area to the east of the Tugen Hills (Fig. 17:1). The Kabarnet Trachyte which had been subaerially weathered was locally downwarped and sediments began to accumulate. The basal deposits at Lukeino comprise sandy to gritty trachyte weathering products (Martyn 1969) which are thicker in the south and give way northwards to red and purplish shales and clays. Laminated shales indicate early lacustrine conditions in the north.

There was a rapid southward transgression of the lake at the beginning of Member B time. Deposition was predominantly lacustrine, and the sediments are pale, cream, grey, yellow and white silts and shales. However, in the south, close to the base of the Tugen Hills scarp, deposition of sandy to gritty redbeds continued. These possibly represent piedmont deposits at the base of the Saimo escarpment similar to those accumulating at present. The southward lacustrine transgression at the beginning of Member B time is recorded in this area by the presence of two and sometimes three layers of diatomaceous shales and some algal limestones intercalated in the redbeds.

Although lacustrine conditions prevailed for the remainder of Lukeino time, areas near the Tugen Hills escarpment and close to the southern margin of the basin received fluviatile sediments and were subaerially exposed from time to time. This is well documentated for Member C time by deposition of grey tuffs varying from a massive, coarse-grained facies in the south near Kapcheberek, to a well-bedded, finer-grained, lacustrine, sometimes fossiliferous, facies in the north. A similar but less marked facies change occurs in the uppermost member (D), which in the south is composed of weathered silty and sandy tuff; in the north it is well laminated, contains numerous fish and grass impressions, and is thicker.

Eruptions from the Kobuluk Centre during Member D time, provided large quantities of tuffaceous and agglomeratic material to the basin. The trachytic plug at Cheparchelom to the north-east of Kobuluk represents the source of the Kobuluk agglomerates. There are pronounced depositional dips in many outcrops of the Kobuluk tuffs, and there was probably a significant topographic feature at the vent near Cheparchelom. The Lukeino lake later covered much of the volcanic cone and laminated shales with limestone bands (containing fish, ostracods and gastropods) were deposited above the agglomerates and tuffs. The top of the cone was probably never covered by water.

The Lukeino Formation has been intruded by three sills. The two lower sills were emplaced before eruptions of the Kobuluk Centre, as shown by the presence within the agglomerates, of fragments of the Rormuch and Kobuluk Sills. Near Cheparchelom, however, there are dykes of material identical to that of the Rormuch Sill which indicate that the sills were emplaced almost contemporaneously with the eruption of the agglomerates. The doleritic sills, the trachytic plug of Cheparchelom and the Kobuluk Agglomerates are thus closely related in time although not in lithology. The plug of Cheparchelom is situated on the eastern boundary of the Lukeino Block, and the fault plane could have acted as a suitable passageway for magma to rise to the surface. Such fault controlled vents are common in the Rift Valley.

The third, or Cheseton Sill cuts sediments overlying the vent facies and must therefore be later than the eruptions. Whether the third sill was emplaced during or after the close of Lukeino sedimentation has not yet been determined, although the former seems more likely.

Stratigraphy:

Member A. The base of Member A is not exposed at Kobuluk and the only part of it seen comprises 12 m (40 ft.) of red silts and grey silty tuffs. Elsewhere Member A is up to 30 m (100 ft.) thick and contains channel grits, weathered horizons and algal layers. It is frequently fossiliferous.

Member B. Member B, 50 m (168 ft.) thick in the type section, is composed of the following units:

Top	1 m (3 ft.)	red silts
	2.5 m (8 ft.)	diatomaceous shales
	3 m (10 ft.)	bedded buff tuffs with silicified top
	4.5 m (16 ft.)	cream and grey tuffs, poorly exposed
	11.5 m (37 ft.)	Rormuch Sill

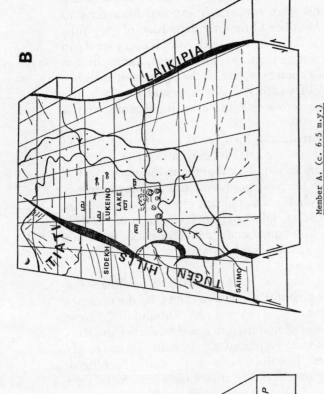

Member A. (c. 6.5 m.y.)

Continued faulting along Tugen Hills Faults resulted in the formation of a major basin on the downthrow side of the faults, bounded to the west by the east-facing scarp. Sedimentation initiated.

Southern margin of lake east of Sumet, with well-developed algal mats, numerous closed bivalves and abundant hippopotamid remains, in silty red sediment. Lake present to north-east of Sumet, while diatomaceous shales in south giving way northwards to pink shales. Fairly abundant chelonia in lake sediments.

Major drainage probably from Laikipia area and from the south, although a major component came from the Saimo area.

Fossil wood indicates the presence of trees; some fragments appear 'chewed' by insects (?termites). Rich and varied mammalian assemblage (including Hominidae and Cercopithecidae) in fluviatile grits, sands and silts south of the lake.

Extrusion of Kabarnet Trachyte (c. 6.7 m.y.)

Major topographic features at Saimo and Tiati truncated to east by Tugen Hills Faults (Saimo, Sumet and Kito Pass Faults). Scarp at eastern edge of Rift Valley, (Laikipia).

Eruptions of Barpelo Basalts in north and Kabarnet Trachytes from south cover much of the area, but topographic highs remain exposed. In the Lukeino area an initially planar surface existed, only disturbed by the youthful Tugen Hills to the west and northwest. Weathering and soil formation (lateritic profiles). Some erosion exposing underlying sediments and lavas in places.

Fig. 17:1 A–D. Volcanic, tectonic and sedimentary history of the Lukeino Basin.

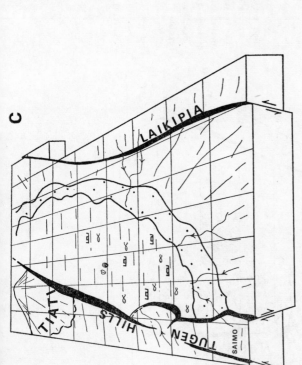

Member B to early Member D

Major transgression of lake to south (and ? east) for about 10km, probably due to continued movement along Tugen Hills faults.

Facies shift in sediment type to the south. White shales with diatoms, small cyprinids and rarer gastropods, give way southwards to red lake marginal sediments with closed bivalves, algae, oolitic limestone lenses and numerous hippopotamid remains. Further south development of fluviatile (floodplain sediments) with rich and varied mammal fauna, especially in grit-filled channels. Evidence of contemporary trampling of sediments by large mammals (? hippo or proboscidea).

Member C represented by grey (basaltic? tuffs) massive and coarse in south, giving way to laminated and fine facies in north. Lacustrine gastropods in north indicate deposition under freshwater conditions. Few mammals in southern coarse facies. Derivation of tuff from south?

Member D continues with lacustrine conditions similar to Members B and C. Very little evidence of dessication or regression of lake throughout that time.

Eruption of Kobuluk Centre

Intrusion of Lukeino sediments by two extensive sills (Rormuch and Kobuluk Sills), and penecontemporaneous eruption of brown/yellow agglomerates and tuff from a large volcanic edifice at Kobuluk (and Plug at Cheparchelom) which probably formed an island or headland in the Lukeino Lake. Tuffs pinch out very rapidly to the south, but in the north are extensive and thick. Lacustrine conditons persist. Kobuluk centre covered by diatomaceous shales containing very abundant grass and fewer ferns. (Cheparchelom Plug remained exposed). Cyprinids extremely abundant (2cm and 15cm length), in many horizons.

In extreme south continued deposition of lake marginal red silts with grit-filled channels. Fossils rare. Disruption of part of sediment pile of Member E, giving rise to extensive (12km north to south) pseudo-intrusive tuff (?evidence for tectonic activity).

Intrusion of diatomites overlying Kobuluk tuff and agglomerate by third Sill. (Cheseton).

1 m (3 ft.)	grey tuff
10 m (32 ft.)	diatomaceous tuffs and grey tuffs, poorly exposed
0.25 m (1 ft.)	yellow calcified tuff
7 m (22 ft.)	cream shales
1 m (3 ft.)	well-bedded grey tuffs
4 m (13 ft.)	diatomaceous shales

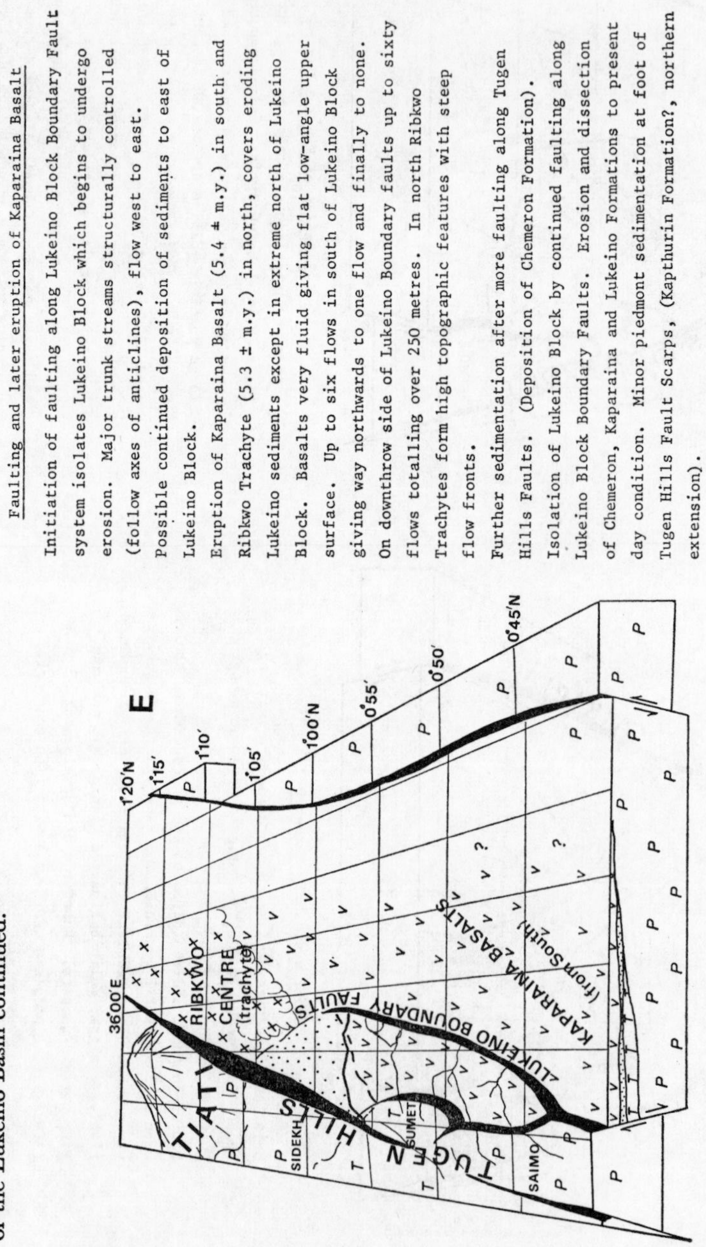
Faulting and later eruption of Kaparaina Basalt
Initiation of faulting along Lukeino Block Boundary Fault system isolates Lukeino Block which begins to undergo erosion. Major trunk streams structurally controlled (follow axes of anticlines), flow west to east.
Possible continued deposition of sediments to east of Lukeino Block.
Eruption of Kaparaina Basalt (5.4 ± m.y.) in south and Ribkwo Trachyte (5.3 ± m.y.) in north, covers eroding Lukeino sediments except in extreme north of Lukeino Block. Basalts very fluid giving flat low-angle upper surface. Up to six flows in south of Lukeino Block giving way northwards to one flow and finally to none.
On downthrow side of Lukeino Boundary faults up to sixty flows totalling over 250 metres. In north Ribkwo Trachytes form high topographic features with steep flow fronts.
Further sedimentation after more faulting along Tugen Hills Faults. (Deposition of Chemeron Formation).
Isolation of Lukeino Block by continued faulting along Lukeino Block Boundary Faults. Erosion and dissection of Chemeron, Kaparaina and Lukeino Formations to present day condition. Minor piedmont sedimentation at foot of Tugen Hills Fault Scarps, (Kapthurin Formation?, northern extension).
E
1°20'N
1°15'
1°10'
1°05'
1°00'N
0°55'
0°50'
0°45'N
36°00'E
RIBKWO CENTRE
(trachyte)
TIATY
TUGEN HILLS
SIDEKH
SUMET
SAIMO
LUKEINO BOUNDARY FAULTS
KAPARAINA BASALTS
(from South)

FIG. 17:1 E. Volcanic, tectonic and sedimentary history of the Lukeino Basin continued.

0.25 m (1 ft.)	bedded grey tuffs
1.25 m (4 ft.)	silty tuff with kunkar at top
2 m (6 ft.)	diatomaceous shales
4.5 m (16 ft.)	creamy silty tuff with calcrete at the top
0.25 m (1 ft.)	suncracked diatomaceous shales
6 m (20 ft.)	cream silty tuff
Bottom 3 m (11 ft.)	diatomaceous shaly tuff

Member C. This is a thin member comprised of grey tuffs. In the type section there are 11.2 m (36 ft.) tuffs of which the lowermost 7 m (22 ft.) is bedded pumice tuff which forms a prominent outcrop, upon which the Bartabwa–Yatya road has been built. Overlying this there is a 2 m (6 ft.) layer of red silt and 2.2 m (8 ft.) of grey tuff.

Member D is composed of up to 12 m (46 ft.) of diatomaceous shales with narrow bands of grey and yellow tuffs, succeeded by a thin bed of pale brown silty tuff. It is overlain unconformably by a mugearite of the Kaparaina Basalt Formation. The diatomaceous shales have been 'intruded' by a silty tuff 19 m (61 ft.) thick.

Age

Potassium-argon dates determined on bracketing lavas of the Lukeino Formation suggest an age of about 6½ million years for the basal layers (Members A and B). Dates from the underlying Kabarnet Trachytes and the overlying

TABLE 1: *Potassium-argon ages related to the Lukeino Formation (all whole rock lava ages).*

FORMATION	SAMPLE	LABORATORY	AGE & ERROR m.y.
KAPARAINA	K5	A	3.9 ± 0.1
BASALTS	K3	A	4.7 ± 0.1
	2/227	B	5.3 ± 0.2
	2/M5	B	5.4 ± 0.2
	2/125	B	8.2 ± 0.4
LUKEINO FORMATION			
KABARNET	2/214	B	6.8 ± 0.4
TRACHYTE	1/714	B	6.7 ± 0.3
	JM/734	C	7.3 ± 0.3
	" "	C	7.3 ± 0.3

LABORATORIES: A. Liverpool University, A. Mussett.

                       B. Institute of Geological Sciences, N. Snelling.

                       C. Cambridge University, J.A. Miller.

For references see: Chapman and Brook, Paper 14 this volume;

                      Dagley, Mussett and Palmer, Paper 15 this volume.

Kaparaina Basalts are tabulated from a variety of sources (Table 1), (Martyn 1969; Chapman 1971; McClenaghan 1971; Bishop *et al.* 1971) and more samples have been obtained in the hope of refining the age determinations for the Lukeino sediments lying between them.

Correlations

The fauna is most closely related to that of Lothagam I (Patterson *et al.* 1970). It is of interest that the sediments at Lothagam have also been intruded by a sill. Among the Suidae, most of the *Nyanzachoerus tulotos* Cooke and Ewer from Lukeino is inseparable from that of Lothagam (Cooke pers. comm.) although some of the specimens from the Kapgoyu locality (2/215) at Lukeino appear to be slightly more primitive (Cooke and Ewer 1972, p. 229). Two of the Proboscidea at Lukeino are the same as those from Lothagam (Bishop *et al.* 1971), the Equidae are the same (Hooijer pers. comm.) and the Hippopotamidae are similar (Coryndon this volume).

Bishop *et al.* (1971) point out the similarity between the Reduncine from Lukeino and one from the Dhok Pathan Formation in the Siwaliks. The fauna from Lukeino is distinctly more advanced than that of Ngorora (c. 12–9 m.y.) but is more primitive than that of Chemeron, East Lake Turkana and other Plio-Pleistocene East African localities. There is little reason to doubt the age of the sediments deduced from radiometric data.

Structure

Structural aspects of special interest in the study of the Lukeino Formation are illustrated in figures 17:2 and 3. Sections through the Yatya and Kisitei anticlines and the northern syncline are shown (Fig. 16–17, page 262).

It is not possible to determine whether the anticlines and synclines within the Lukeino Block started to form during Lukeino time. It seems probable that the structures were of only minor importance during the period of deposition, because sediments are continuous on either side of them, and there is no marked thinning of deposits across the axes of the anticlines. However, the structures had certainly been initiated prior to the deposition of the Kaparaina Basalts, and probably formed in conjunction with the faulting episode which isolated the Lukeino Block from the downthrow area to the east.

The Kabarnet Trachytes, which underlie the Lukeino sediments, behave in an unusual manner when compared with other formations in the Baringo area. At Lukeino they are more prone to folding than to fracturing, so that several folded structures occur in the Lukeino Block, some of them suggesting compressive forces. The Lukeino Block lies east of the Bartabwa Gap (Sumet) area in which the Kito Pass and Saimo Faults meet (Fig. 17:3). The Bartabwa gap lies midway between the Kito Pass and Saimo half-domes, and the ends of the Lukeino Block coincide approximately with the culminations of these structures.

The eastern edge of the block is approximately an arc of a circle, and developed along the margin are five anticlines with axes arranged normal to the bounding faults. From south to north these are known as the Aragai, Yatya,

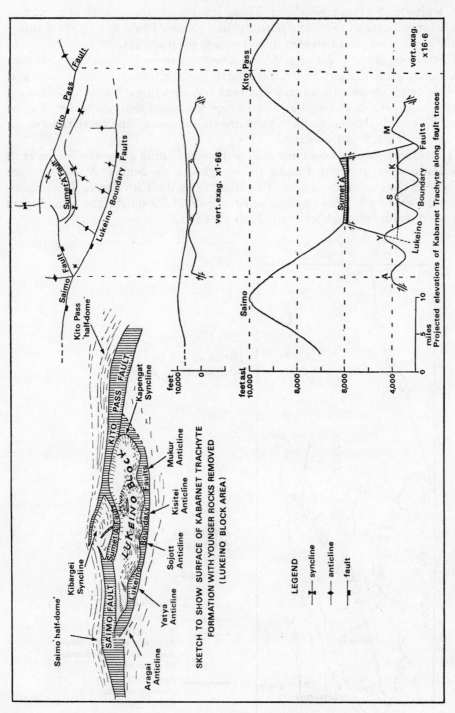

FIG. 17:2 (a).

Perspective view and structural map of the
Lukeino Block.

FIG. 17:2 (b).

Projected elevation of Kabarnet Trachyte
along fault traces.

Sojott, Kisitei and Mukur anticlines. There is a remarkable degree of symmetry in the arrangements of the folds, both in the Lukeino Block and in the Saimo–Kito Pass structures, and between these two sets of structures.

To the west of the Sojott and Kisitei structures there are faults, so that they are in reality horsts. Sojott and Kisitei horsts are symmetrical fault arches and have little or no plunge. In contrast the next two structures (Yatya and Mukur) not only have greater throw, but also plunge westward and are wider. To the west at Kipkwaitit there is another halfdome dipping west, and there is a graben between it and the Sojott anticline.

Overprinted onto this complex is a component of drag along the Tugen Hills Faults. This is best developed along the margins of the Sumet 'A' Fault where the trachytes dip eastwards at up to 70°. Drag along the Kito Pass fault is responsible for the gently plunging syncline at Koitugum in the north. The axis plunges gently southwards sub-parallel to the Kito Pass Fault.

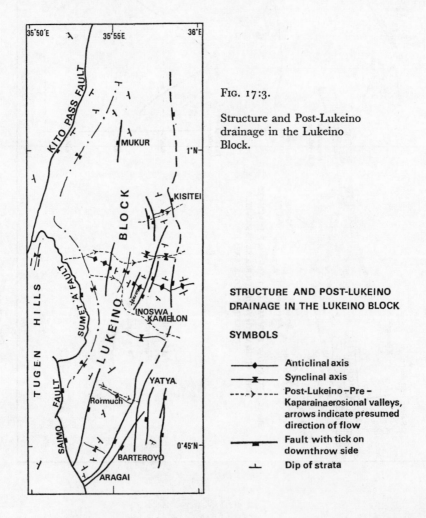

Fig. 17:3.

Structure and Post-Lukeino drainage in the Lukeino Block.

Genesis of structures in Lukeino Block

Any hypothesis explaining the structures exhibited in the Lukeino Block must account for the components of shortening in the north-south as well as in the east-west sense.

Since the Kabarnet Formation was deposited, the Lukeino Block has been down-thrown approximately 750 m (2500 ft.) relative to the Tugen Hills Block. The Lukeino Block has shortened by about 600 m (2000 ft.) in a north-south sense and about 500 m (1600 ft.) in an east-west sense. Thus it would appear that strike-slip and dip-slip were prevented, probably by the mass of rock on the down-throw side of the Lukeino Block. The result has been the folding of the block, which now occupies less area than it occupied when the Kabarnet Trachytes were deposited.

Structural controls of drainage

The streams in the area at the present day display strong structural controls (see maps Fig. 16/17, 17:3). Major trunk streams occur along synclinal axes at Ketpotunoi, Kobuluk and Kamsorar; along anticlinal axes at Kisitei, Mukur, Yatya and east of Chepirimor, while at Chepanda the stream is oblique to a syncline.

Valleys cut into the Lukeino sediments prior to the deposition of the Kaparaina Basalt (Fig. 17:3), indicate that there was important structural control on drainage at that time. The anticlinal and synclinal folds of the Lukeino Block probably began forming at the end of Lukeino time, perhaps as a result of the faulting along the Lukeino Block boundary faults which effectively stopped sedimentation on the now relatively uplifted block. Some of the streams cut deep valleys, and one at least, removed all the Lukeino sediment in its path and cut into the underlying Kabarnet Trachyte Formation. Others cut down to the Kobuluk and Rormuch sills proving that those two intrusions were emplaced before the end Lukeino time, and indicating that they were intruded at shallow depths. The maximum amount of sediment seen over the sills is 33 metres in the case of the Kobuluk sill and 70 metres for the Rormuch sill.

Palaeogeomorphology, palaeoenvironments, fauna and flora

Palaeogeomorphology

Figure 17:1 illustrates a reconstruction of the palaeogeomorphology of the Lukeino Basin. The Saimo Horst was at least 450 m (1500 ft.) high and its base was covered by piedmont deposits. The Sumet area was also elevated above the basin, although not to the same extent as Saimo, and further north, the Kito Pass area was also an area of high relief. It was against this subdued ancestor of the present Tugen Hills that the sediments of the Lukeino Formation were ponded. The floor of the basin dipped gently westwards towards the Tugen Hills. In the north, the floor of the basin rose towards the Tiati area, and the sediments pinch out under the Ribkwo volcanic complex. To the south, it is more difficult to reconstruct the setting of the basin, but where the underlying rocks can be seen, there are either thin sediments (Riwo Beds of Martyn 1969), or weathered trachytes of the Kabarnet Formation. This indicates a rise in the floor of the basin southwards, and in fact the sediments thin rapidly towards Aragai, at the junction of the boundary faults of the Lukeino Block with the Saimo Fault. The eastern margin of the Lukeino Basin is more difficult to reconstruct, because of the cover of younger rocks, but the floor presumably rose towards Laikipia in the east.

Palaeoenvironments

The palaeoenvironments of Members A to D are also reconstructed in Figure 17:1. The rapid transgression of the lake in Member B time is clear, and the piedmont deposits at Kapcheberek indicate the position of a small portion of the lake shore. Most of the fauna collected from the Lukeino Formation came from Members A and B, and in all the localities a mixture of terrestrial and water-loving animals has been collected. Most of the sediments in Members B, C and D are lacustrine, but there is evidence of suncracking, development of palaesols and calcrete nodule horizons which suggests periodic dessication.

Many of the weathering horizons and soils in the Lukeino Formation are ferruginous in contrast to those of the older Ngorora Formation, which are predominantly manganiferous. The Lukeino sediments are generally less well calcitised than those of the Ngorora Formation. Near major faults they are frequently silicified. It seems that a change in weathering and soil forming conditions occurred during the period between Ngorora and Lukeino time and it may be that more pronounced rain-shadow conditions developed in the Rift Valley about 7 million years ago. Prior to this the walls of the Rift were probably too low to have been of regional meteorological significance. Fossil ferns indicate moist, shaded areas, and the paucity of dicotyledons may reflect their virtual absence from the area. Leaves similar to those of *Acacia mellifera* occur but identification is not certain. However, wood in Members A and B indicates the presence of trees within the watershed, possibly on the hills of Saimo and Sidekh to the west of the lake. Some of the wood appears to have been chewed and 'shaped' by termites, which would indicate that the climate was possibly seasonal and not too arid. The dominance of Tragelaphini within the bovid fauna would indicate bushland rather than grassland and possibly a mixed vegetation existed during Lukeino times. Abundance of fossil grass is indicative of grass near the lake at Lukeino.

It is possible that the grass represents reworked *Hippopotamus* faecal matter, carried to deep and quiet parts of the lake by gentle currents. This occurs at the present day at Mzima Springs in Kenya.

Fauna and Flora

An up-to-date faunal list is presented in Table 2. New fauna collected in 1975 includes Lagomorpha, a second species of bird and new rodents. Close affinity with the faunas of Lothagam and the Mpesida Beds is indicated by the assemblage rather than with the Ngorora and older formations. A large faunal

TABLE 2: *Detailed fauna lists, Lukeino Formation, Kenya.*

Member D

Ostracoda		
Pisces	Cyprinidae	

Member C

Mollusca	Gastropoda	*Melanoides tuberculata*
	Bivalvia	
	Cyprinidae	
Pisces	Cyprinidae	
Mammalia	Hippopotamidae	*Hippopotamus* sp.

Member B

Mollusca	Bivalvia	
	Cyprinidae	
Pisces		
Reptilia	Trionychidae	
	Pelomedusidae	
	Testudinidae	
	Crocodilidae	
	Varanidae	
Aves		
Mammalia	Rodentia	*Hystrix*
	Carnivora	*Enhydriodon*
		Felidae
	Proboscidea	*Anancus* sp.
		Stegotetrabelodon
		Primelephas
		Deinotherium
	Perissodactyla	*Hipparion* cf. *stifense*
		Chalicotheriidae
		cf. *Ceratotherium*
	Artiodactyla	*Nyanzachoerus tulotos*
		Hippopotamus sp.
		Giraffa
		Tragelaphini
		Reduncini
		Hippotragini
		Antilopini
		Neotragini

Member A

Mollusca	Bivalvia	
Pisces	Cyprinidae	
	Cichlidae	
	Clariidae	
Reptilia	Trionychidae	
	Pelomedusidae	
	Crocodilidae	
Mammalia	Cercopithecidae	
	Hominidae	
	Lagomorpha	
	Rodentia	*Hystrix*
		Others
	Canidae	
	Viverridae	cf. *Ichneumia* sp.
	Hyaenidae	cf. *Crocuta*
	Felidae	2 spp.
	Tubulidentata	*Orycteropus* 2 spp.
	Probiscidea	*Anancus*
		Stegotetrabelodon
		Primelephas
	Perissodactyla	*Hipparion* cf *stifense*
		cf.*Ceratotherium*
	Artiodactyla	*Nyanzachoerus tulotos*
		Hippopotamus
		Giraffa
		Tragelaphini
		Reduncini
		Neotragini
		Antilopini cf *Aepyceros*
		Gazella
		Alcelaphini
		Cephalophini
		Hippotragini

turnover occurred between upper Ngorora time and Lukeino time, and that from the younger formation is much more modern in aspect than that from Ngorora. The hominid from Lukeino is the earliest known occurrence of undoubted Hominidae and predates the next oldest from Lothagam, by about a million years. Figure 17:4 shows a geological sketch map of locality 2/219 from which the hominid came.

Taphonomical aspects

In the total of 45 fossiliferous localities at Lukeino there are 10 lacustrine localities, 31 fluviatile or lake marginal ones and 2 palaeosols. Two localities yielded surface specimens only. Three lake-margin algal limestones yielded fossils, predominantly hippopotamids, but also a complete rhinoceros and a few isolated bovid and hyaenid fragments. In lake margin silts there were several localities which yielded bivalves in closed position, and these localities also often yielded mammals at slightly higher levels. Channel grits were by far the most varied and richest localities for mammals, and five occurrences yielded about three-quarters of the mammalian fauna from Lukeino. Fossils from these deposits are usually rolled, polished and disarticulated.

Lacustrine shales yielded large quantities of fish and plant material. The number of localities could be significantly increased by further search, but the fish fauna appears from present evidence to be restricted to an abundance of a single species of small (2 cm.) cyprinid, with possibly another less common species of larger size (15 cm.). Clariids are not found in the shales, but fragmentary specimens occur in the fluviatile sediments. Some of the lake marginal silts (e.g.

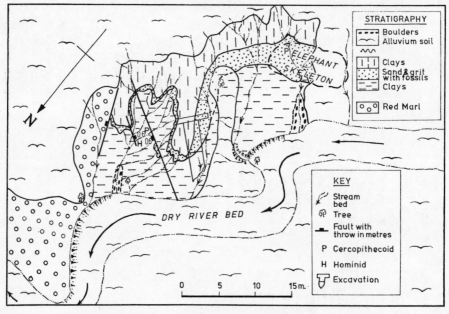

FIG. 17:4. Geological sketch map of hominid locality (2/219), Lukeino Formation.

locality 2/225) appear to have been trampled by animals after deposition. From the size of the 'footprints' it would appear that the animals responsible were either Hippopotamidae or something of similar size. Analogous disturbance of sediments is evident along the shores of present day Lake Baringo.

Close to the Lukeino lake shore, algal masses formed. Most of these are spheroidal masses joined laterally into mats, but some of them appear to be 'spherulitic' (spastolithic?) similar to extensive limestones in upper Miocene sediments of the Suguta Valley in north Kenya. They contain ostracods and bones in places.

Concentration of mammal remains at Lukeino appears to have been mostly by fluvial activity. Skeletal and dental parts were probably derived from several habitats, with deposition near lake margins. Some fossil deposits were probably formed on emergent lake flats in stream channels, but others were probably formed at the lake margin or in shallow water near the shores of the lake. Mammals are rare in still-water (laminated) deposits, although fish are abundant.

Note: Further charts and diagrams giving details of the stratigraphical palaeontology and palaeoenvironments have been lodged in the Library of the Geological Society, Burlington House, Piccadilly, London where they may be consulted.

Conclusions

The Lukeino Formation, which was deposited about six and a half million years ago, following a period of weathering and slight erosion, is represented by up to 130 metres (400 ft.) of sediment. It is unconformably overlain by lavas of the Kaparaina Basalt Formation. During deposition of the sediments a volcanic vent became established near Kobuluk, and the plug of Cheparchelom represents the centre from which the Kobuluk tuffs and agglomerates were erupted. Continued sedimentation after extinction of the centre resulted in the deposition of shales and limestones on the flanks of the cone, but the upper slopes were probably never covered by water. Tectonism played an important role in the initial formation of the Lukeino Basin, in its subsequent sedimentary history and in the eventual cessation of deposition at the end of Lukeino time. More recently it has led to the re-exposure of the sediments. The present-day fluvial system is strongly controlled by the structure of the Lukeino Block. Rivers both in the past and at present show major control at the axes of anticlines and synclines.

During deposition of the formation various animals were preserved as fossils. Reconstructions of the palaeogeomorphology and palaeoenvironments show that there were hills to the west of the sedimentary basin, and that there were probably both woodlands and grasslands in the neighbourhood. The climatic indicators that have been noted, suggest a seasonal climate such as occurs in the Tugen Hills–Baringo area today. The area was not arid but it is possible that rain shadow conditions within the Rift Valley developed immediately prior to the onset of Lukeino sedimentation.

Most of the mammalian fossils from Lukeino have been recovered from fluviatile sediments deposited near former lake margins and are predominantly isolated fragments of teeth and skeletal parts. Fossil fish on the other hand are usually preserved as complete impressions in paper shales which were deposited under quiet water conditions.

Detailed study of the Lukeino sediments has helped in the interpretation of other sedimentary bodies in the Rift Valley, because many of the controls on sedimentation appear to have been common to other basins. The tectonic and volcanic controls are of major importance, and many Rift Valley basins appear to have gone through similar histories.

Acknowledgements

I thank all those who have helped in the production of this paper. Government authorities of Kenya provided permits to study in the Baringo area and the National Museums of Kenya gave valuable assistance. Financial support was supplied through grants to Professor W. W. Bishop from the Natural Environment Research Council (NERC Grant No. GR3/835), the Boise Fund of Oxford University, and the Wenner-Gren Foundation for Anthropological Research, New York. I am grateful for grants from the Government of Nova Scotia and the Canadian Student Loan Committee. Research facilities were provided by Bedford College and Queen Mary College, University of London, where the writer held a NERC Research assistantship to Professor W. W. Bishop and now holds a Post Doctoral Research Fellowship.

References

BISHOP, W. W., CHAPMAN, G. R., HILL A. & MILLER, J. A. 1971. Succession of Cenozoic Vertebrate Assemblages from the Northern Kenya Rift Valley. *Nature, Lond.* **233**: 389–94.

CHAPMAN, G. R. 1971. Ph.D. thesis unpublished. University of London.

COOKE, H. B. S. & EWER, R. F. 1972. Fossil Suidae from Kanapoi and Lothagam, northwestern Kenya. *Bull. Mus. Comp. Zool.* Harvard, **143** (3).

HOOIJER, D. A. & MAGLIO, V. J. 1973. The earliest *Hipparion* south of the Sahara, in the late Miocene of Kenya. *Koninkl. Nederl. Akad. van Wetensch*, **76**: 311–15.

MAGLIO, V. J. 1970. Early Elephantidae of Africa and a tentative correlation of African Plio-Pleistocene deposits. *Nature, Lond.* **225**, 328–32.

MARTYN, J. E. 1967. Pleistocene deposits and new fossil localities in Kenya. *Nature, Lond.* **215**: 476–9.

—— 1969. Ph.D. thesis unpublished. University of London.

McCLENAGHAN, M. P. 1971. Ph.D. thesis unpublished. University of London.

PATTERSON, B., BEHRENSMEYER, A. K. & SILL, W. D. 1970. Geology and fauna of a new Pliocene locality in NW Kenya. *Nature, Lond.* **226**.

PICKFORD, M. 1975a. Another African Chalicothere, *Nature, Lond.* **253**.

—— 1975b. Late Miocene sediments and fossils from the Northern Kenya Rift Valley. *Nature, Lond.* **256**: 279–84.

—— 1975c. New fossil Orycteropodidae (Mammalia, Tubulidentata) from East Africa. *Neth. J. Zool.* **25** (1), 57–88.

WALSH, J. 1969. Geology of the Eldama Ravine–Kabarnet Area. *Geol. Surv. Kenya Rept.* **84**: 1–56.

18

SHIRLEY CAMERON CORYNDON

Fossil Hippopotamidae from the Baringo Basin and relationships within the Gregory Rift, Kenya

More than one thousand specimens representing fossil hippopotamus have been recovered from deposits in the Baringo Basin. This collection, the greater part of which consists of isolated teeth, spans a time range covering the last ten million years and contains the earliest recognizable hippo remains in the world. Comparison with other hippopotamus faunas reveals a striking conservatism until about four million years ago. Specimens from later deposits are much more numerous, with speciation and diversification in the East African forms becoming apparent. In the Baringo area this diversification is first seen in hippos from the Chemeron Formation which can be favourably compared with specimens from Kanapoi and the early levels of the Omo and East Lake Turkana. In the later Pleistocene deposits of Baringo the hippopotamus is close to the extant Hippopotamus amphibius *and associated with permanent water. It is possible that the earlier levels of the Baringo Basin may hold the key to the origin of the Family Hippopotamidae, the ancestry of which is at present obscure.*

Introduction

Systematic geological mapping of the Baringo Basin, Kenya, by personnel of Bedford College, London over a number of years has resulted in a considerable collection of fossil faunas from deposits covering many square miles and some fourteen million years (Fig. 18:1). The significance of these collections, particularly the mammal faunas, lies not only in the taxonomic and palaeoecological interest of the species *per se*, but also in the fact that the earlier fauna-bearing Baringo deposits span a time sequence unknown in East Africa. Thus for the first time a link between the Miocene faunas of Fort Ternan (about 14 m.y.) and those of the later Lake Turkana faunas (earliest about 5 m.y.) can be tentatively forged. These Baringo animals can help to fill a major gap in previous knowledge of faunal presence and evolution in Africa during this time (Bishop *et al.* 1971) (Fig. 18:2). Except for a few specimens of exceptional quality, most of the material, representing many taxa, is unfortunately very fragmentary. The fragmentary state of specimens representing the Hippopotamidae makes interpretation of the speciation and evolution of the family very difficult. The majority of specimens comprise isolated teeth and unassociated post cranial elements; a few specimens are a little more complete but virtually no cranial material is preserved. From a total of over 1000 specimens representing various hippopotamus species, only 18 specimens comprise more than one tooth or post cranial element (Table 1). All the material is housed in the National Museum, Nairobi, Kenya, where it is registered with the prefix KNM– followed by two letters representing the locality and the individual number.

Stratigraphy

The history of sedimentation in the Baringo area indicates a succession of small depositional basins within the Gregory Rift, with the older sediments on the boundary of the present rift valley and younger sediments closer to the centre of the basin deriving material within the rift as well as from the boundary scarps. As

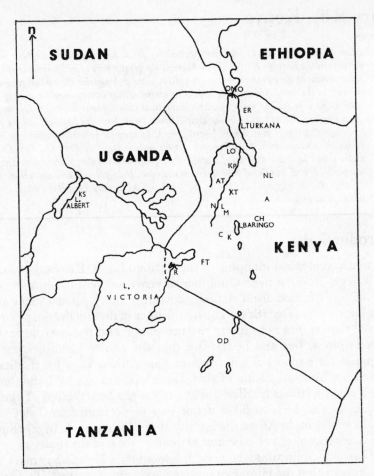

FIG. 18:1. Sketch Map of East Africa showing some fossil hippopotamus localities in the Gregory Rift.

Key: AT = Aterir, KT = Karmosit, NL = Nakali, A = Alengerr, N = Ngorora, L = Lukeino, M = Mpesida, CH = Chesowanja, C = Chemeron, K = Kapthurin, ER = East Lake Turkana, LO = Lothagam, KP = Kanapoi, KS = Kaiso, OD = Olduvai, FT = Fort Ternan, R = Rusinga.

TABLE 1: *Distribution of Specimens among the Hippopotamid-bearing sediments of the Baringo Basin*

LOCALITY	NUMBER OF SPECIMENS	SPECIMENS WITH ASSOCIATED TEETH
Loboi	40	–
Kokwob	3	1
Kapthurin	18	1
Chesowanja	15	1
Karmosit	16 (all broken)	–
Chemeron	c500	10
Aterir	40	2
Toluk	5 (all broken)	–
Lukeino	278	3
Mpesida	61	–
Nakali	1	–
Ngorora	14 (all broken)	–
Muruyur	–	–

c = Approximately

in other East African localities, tectonic and sedimentary activity occurred over a long period of time causing a succession of changes of drainage and accumulation within the basin. The pyroclastic elements provide minerals from which some radiometric ages of the sediments have been obtained. The majority of the dates have been established for bracketing lavas. The sediments contain fluviatile, lacustrine and sub-aerial facies, indicating rivers flowing into small lakes which fluctuated in size and mineral content, the latter causing a certain degree of

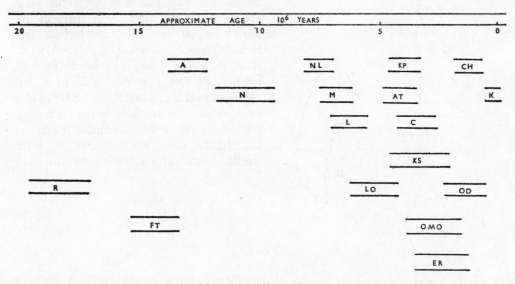

FIG. 18:2. Temporal Relationships of Fossiliferous Deposits in the Baringo Basin and other localities in East Africa. Key: As in Fig. 18:1.

alkalinity from time to time (Pickford 1975). The overall picture is of small basins whose size and sedimentation rates fluctuated with tectonic activity, subsequent basins not necessarily following form or position of previous ones.

The oldest mammal bearing deposits in the Baringo region are to the north-east of Lake Baringo at Alengerr and to the west at Muruyur. Several radiometric dates of more than 12 m.y. and about 14 m.y. are closer in age to the important Fort Ternan deposits in the Kavirondo area than any sites so far recognised. Successively younger deposits in the Baringo region are seen not only to fill an unknown age group in East African fossil-bearing levels, but also to interdigitate in age with other localities outside the Baringo Basin in the upper levels; this enables comparisons to be made between various sedimentary formations and contained faunal assemblages and ecologies of similar ages outside the Baringo Basin.

The hippopotamus fauna

The oldest fossiliferous sediments in the Baringo Basin at Alengerr and Muruyur may overlap in time with the Fort Ternan deposits, but none has yet yielded any hippopotamid material.

The Ngorora Formation (Bishop & Chapman 1970; Bishop & Pickford 1975), with deposits older than nine million years and younger than twelve contains the oldest evidence of hippopotamus so far obtained in the world. Of fourteen specimens, from six sites within the Formation, none are complete and all but one represent isolated teeth. An upper incisor, KNM–BN 126 from 2/12s, represents the least broken tooth in the collection; it has a small band of enamel on the anterior face, is slightly curved, cylindrical and has a single wear facet (broken) on the tip; these features are found in all early hexaprotodont (six incisors) hippopotamus (Fig. 18:3A). The most diagnostic specimen from Ngorora consists of the posterior half of a left upper second premolar, KNM–BN 129 also from 2/12s. The tooth is large, with very rugose, pustulate enamel and is preserved sufficiently well to compare it with similar but complete teeth from the later

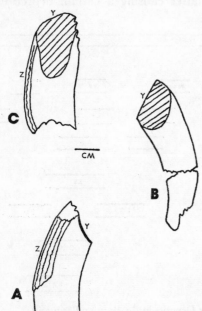

FIG. 18:3.

Hippopotamid Upper incisor teeth. A. Ngorora KNM–BN 127; B. Nakali NAK 69.70; C. Mpesida KNM–MP 107. Y = Occlusal wear facet. Z = Enamel.

deposits of Lothagam, west of Lake Turkana and dated at around five million years (Fig. 18:4D). The earliest hippopotamus cranial material of any significance has been obtained from Lothagam, together with the slightly later deposits of Kanapoi, and indicate that these primitive hippos were nearly as large as, but more slender than the extant *Hippopotamus amphibius*; the anterior dentition was hexaprotodont, and the premolars were large and robust when compared with the molars, so much so that in the Lothagam hippopotamus the length of the premolar row equals the length of the molar row, whereas in later forms the premolars tend to diminish in size in comparison with the molars. Other morphological characters of the skull and dentition indicate a possible ancestral form from which the Asiatic *Hexaprotodon* was derived. The hippopotamus from Kanapoi is only a little more advanced than that from Lothagam. The considerably older material from Ngorora strengthens the hypothesis that the hippopotamus arose in Africa, later migrating to Asia before becoming extinct on that continent in the late Pleistocene.

The locality of Nakali, to the north-east of Lake Baringo has been very tentatively dated at around eight million years, and has produced a fragmentary mammalian fauna; the only nearly complete specimen representing a hippopotamus consists of an upper incisor tooth, NAK 69/70, a cast of which was sent to the author by Dr Aguirre. Although slightly smaller than the incisor from Ngorora, it is morphologically similar, with a wear facet on the tip in one place only, indicating direct tip-to-tip occlusion with the lower incisor(s), as in all early hippopotamus (Fig. 18:3B). One or two other specimens from Nakali might represent hippopotamus, but are too fragmentary for certain identification.

A firm radiometric age of seven million years has been obtained from the Mpesida beds, from which sixty-one remains of hippopotamus have been recovered—all regrettably representing isolated teeth and unassociated post cranial elements, with no cranial material. These remains do, however, show an even stronger similarity to specimens from Lothagam than do those from the earlier beds of Baringo. Although only isolated teeth are preserved in the Mpesida collections, a picture can be drawn of a presumably hexaprotodont hippopotamus; the incisors are similar to those from Ngorora (Fig. 18:3C). and the canine teeth (not represented from the earlier deposits) are robust with the uppers displaying a deep posterior groove; the lowers are equally large and have smooth enamel. A well-preserved upper second premolar is almost identical with the preserved parts of the Ngorora tooth, and confirms the strong similarity with teeth from Lothagam. Other premolars, both upper and lower, are large and pustulate with marked cingula. The molar teeth are low crowned, with tapered lobes wearing to a triangular enamel pattern (Fig. 18:4B, C, E).

The post cranial elements from Mpesida include carpal, tarsal and phalangeal bones, an astragalus with deep articular grooves and a greater overlap between navicular and cuboid, indicating feet which must have been more slender than those of the living hippopotamus, making it more efficient for faster movement.

In adjacent deposits the slightly younger beds of the Lukeino Formation (six and a half million years) are exposed. The collection of hippopotamus remains is more abundant from this area, and comprises two hundred and seventy-eight specimens, of which three represent two or more teeth in association. These more complete specimens include a broken mandible, KNM–LU 170–5, with three molar teeth and sockets for six incisors, a fragment of left maxilla containing d^4 –M^2 preserved (KNM–LU 385) and a left mandible with d_4 –M_2 (KNM–LU 751). Thus for the first time in specimens from the Baringo deposits we have clear evidence of hexaprotodonty in the hippopotamus fauna. One element missing from the Ngorora, Nakali and Mpesida collections but present from Lukeino is the upper fourth premolar. This tooth demonstrates evolutionary trends in the

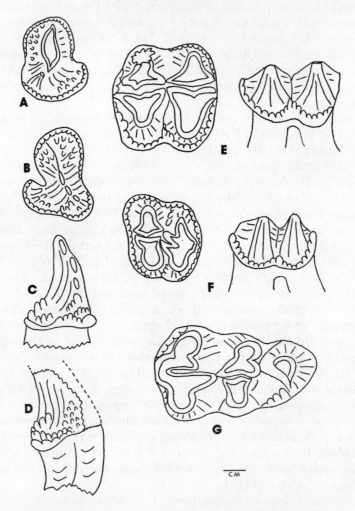

FIG. 18:4. Premolars and Molar teeth of Hippopotamidae from the Gregory Rift.
Upper Second Premolars: A. Lothagam KNM–LT 402, Occlusal; B. Mpesida KNM–MP114, Occlusal; C. Mpesida KNM–MP 114, Posterior; D. Ngorora KNM–BN 129 (broken), Posterior.
Upper Second Molars, Occlusal (left) and Buccal (right): E. Mpesida KNM–MP 199B; F. Chemeron KNM–BC 386.
Lower Third Molar, Occlusal: G. Chesowanja KNM–CH 180C.

morphology which help to pin-point the species involved (Coryndon & Coppens 1973). In the case of the Lukeino specimens these are very similar to upper fourth premolars from Lothogam. The main features of the P⁴ in early forms of hippopotamus show two main subequal cones, buccal and lingual, surrounded by a strong cingulum. The trend in later forms is towards a reduction in the lingual cone and an overall reduction in size by comparison with the molar teeth (Fig. 18:5A).

Other isolated teeth from Lukeino are very like those from Mpesida in size and morphology, with the exception of four specimens. The molar teeth are essentially similar to, but are rather higher crowned and more lophodont than those from Mpesida. The upper and lower first premolars, not known from the earlier sediments, have two roots, unlike the living species in which the first premolar has a single root and is often not retained in the permanent dentition. This bi-rooted condition of the first premolar is also seen in the Lothagam species, but was previously only known among hippopotami from the Asian *Hexaprotodon* and is a diagnostic character of that genus (Fig. 18:5B).

Among the specimens from Lukeino are four teeth and some tarsal bones of a slightly smaller animal which may represent a second species. The third lower molar is smaller and lower crowned than others from this site and the tarsal bones indicate greater mobility. At Lothagam, although the large hexaprotodont hippo

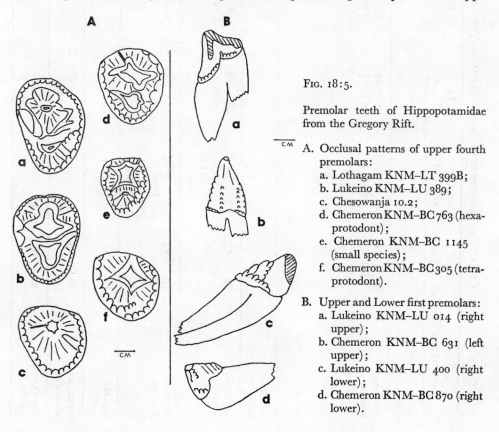

Fig. 18:5.

Premolar teeth of Hippopotamidae from the Gregory Rift.

A. Occlusal patterns of upper fourth premolars:
   a. Lothagam KNM–LT 399B;
   b. Lukeino KNM–LU 389;
   c. Chesowanja 10.2;
   d. Chemeron KNM–BC 763 (hexa-protodont);
   e. Chemeron KNM–BC 1145 (small species);
   f. Chemeron KNM–BC 305 (tetra-protodont).

B. Upper and Lower first premolars:
   a. Lukeino KNM–LU 014 (right upper);
   b. Chemeron KNM–BC 631 (left upper);
   c. Lukeino KNM–LU 400 (right lower);
   d. Chemeron KNM–BC 870 (right lower).

TABLE 2: *Dimensions of various isolated hippopotamid teeth from the Baringo deposits*

A UPPER TEETH	I1		C		P2		P3		P4		M1		M3	
	AP	LT	AP	LT	L	B	L	B	L	B	L	B	L	B
CHESOWANJA KNM–CH	–		–		–		–		34 32b 10.2		–		–	
CHEMERON KNM–BC	20	14 248	40	44 ♂ 397 34 45 ♀ 522 27 38a 864	33	23 1199	35	29 1426	28 31 1385 22 20a 1144 31 21b 236		36	36 565	46 44 1162 38 33a 717 50.5 49b 394	
ATERIR KNM–AT	–		–		–		43	29.5 122	–		37	c33 123	–	
LUKEINO KNM–LU	19	17.5 422	53	68 ♂ 135 37 47 ♀ 769	43	36 044	40	34 146	38	44 074	44.5	41.5 385	54	59 471
MPESIDA KNM–MP	19	23 105	63	68 ♂ 199A 52 52.5 ♀ 097	–		45	34 114	–		–		54	49 109
NAKALI NAK	17	15 69,70	–		–		–		–		–		–	
NGORORA KNM–BN	20.5	19.5 127	–		–	35+ 129	–		–		–		–	
LOTHAGAM 57–67K	31	21	44	51	42.5	27	36	26	30	40	29	39	46	47

LOWER TEETH	I₁ AP	I₁ LT	C AP	C LT	P₂ L	P₂ B	P₃ L	P₃ B	P₄ L	P₄ B	M₁ L	M₁ B	M₃ L	M₃ B
CHESOWANJA KNM–CH	31.5 (180A)	42b	78.5 (180A)	51b	–	–	–	–	–	–	50 (180C)	36b	84 (180oc)	c47b
			42 (180E)	28d										
CHEMERON KNM–BC	25 (1182)	26	49 (822)	31	35 (534)	20	34 (475)	22	34 (1201)	26.5	41 (393)	30	66 (1203)	40.5
			33 (1186)	21a									58 (242)	37a
ATERIR KNM–AT	e18 (141)		50 (143)	35	–	–	–	–	–	–	42.5 (145)	c25	–	–
LUKEINO KNM–LU	25 (029)	29	54.5 (445)	34	38.5 (387)	24	40 (093D)	23	44 (028)	28.5	42 (174)	29	66 (756)	40
			38.5 (387)	24a					35 (587)	25a			58 (398)	33a
MPESIDA KNM–MP	25 (117)	26	56.5 (096)	33	–	–	49 (088)	25	42 (112)	47	47 (106)	34	–	–
NAKALI	–		–		–		–		–		–		–	
NGORORA KNM–BN	25 (126)	28												
LOTHAGAM 246–67K	c18	c18	46	33	33.5	24	37	26	41	30	36	32	65	40
EAST LAKE TURKANA KNM–ER 798	36	40	40	25d	37	25	40	25	41	29	48.5	38.5	78	52

KEY: AP = Antero/Posterior, LT = Lateral, L = Length, B = Breadth, a = pygmy species, b = tetraprotodont form, d = East Lake Turkana form, e = estimate, c = approximately. The registered number of each tooth is given below the measurements.

is by far the more common, there is also an undoubted pygmy species which with the Lukeino specimens could represent the beginnings of a pygmy element in the hippopotamus fauna of East Africa.

Dimensions of teeth from various levels are given in Table 2. Five hippopotamid specimens from the Toluk beds, a little younger than Lukeino, are extremely fragmentary but appear to be similar to those from Lukeino, without the pygmy element.

The Aterir beds in the north of the area are dated at a little younger than four million years. The specimens are rather weathered and fragmentary, but appear to be similar to those from Lukeino. A small species may be present, but this is not yet certain.

Hippopotamus specimens from the slightly younger beds at Karmosit, consist of sixteen isolated teeth, all broken.

The richest vertebrate bearing deposits in the Baringo Basin are those within the Chemeron Formation whose age is derived from a capping lava dated at about two million years, with deposits probably in the region of three million. These deposits have yielded a large faunal assemblage (Bishop *et al.* 1971) including many (c500) specimens of hippopotamus; this hippopotamid material contains mostly isolated teeth and post cranials, but also contains a few more complete specimens. It appears that in this formation the hippo fauna experiences an expansion with at least three separate species recognisable. One of these species is more advanced than but similar to those from earlier levels and at Lothagam, indicating a large hexaprotodont rather similar to specimens from Kanapoi in the Lake Turkana Basin with typical large, posteriorly grooved upper canines, large premolars and low-crowned molar teeth. The specimens of this taxon, which is the basic type of hippopotamid from which all others must have derived, in the Chemeron collections indicate evolutionary trends which result in slightly smaller premolars and molar teeth that are slightly higher crowned than those from Lothagam or Kanapoi.

A second species at Chemeron can be separated from the 'primitive' form. Although of similar size, differences compared with the primitive hexaprotodont can be seen in (a) a more 'advanced' upper fourth premolar, in which the lingual cusp is very reduced (Fig. 18:5A), (b) a shorter premolar row with each tooth of a more simple pattern and lacking the pustulate enamel of the former species, (c) upper canines with a shallow posterior groove, (d) molar teeth generally larger and higher crowned. These characters are indicative of the tetraprodont (four incisors) genera of hippopotamus, including the extant *Hippopotamus amphibius*, and may represent a species in the Chemeron deposits close to *H. kaisensis* from the Kaiso Formation in Uganda.

A third species from Chemeron deposits is best illustrated by a mandible, KNM–BC 19 from site JM507, representing a small hexaprotodont hippopotamus with cheek teeth similar in size and morphology to those of *Hippopotamus imagunculus*, also from the Kaiso Formation (Cooke & Coryndon 1970). It is not yet clear if the small species from the earlier deposits of the Baringo series is the same as that from Chemeron. The Chemeron species is clearly very different from another pygmy hippo, *Hippopotamus aethiopicus* found in later deposits at Omo and East Lake

Turkana to the north (Coryndon & Coppens 1975). Neither *H. imagunculus* nor *H. aethiopicus* can be shown to have any connection at all with the living pygmy hippopotamus of West Africa, *Choeropsis liberiensis*.

Post cranial bones, none in association, tend to support the probability of at least three distinct hippo species from the Chemeron Formation.

It can be postulated that fairly stable conditions prevailed during the deposition of beds from the time of the Ngorora Formation until Aterir times, with a hippopotamus fauna consisting of one large hexaprotodont and a less common pygmy form. There need not have been large lakes or great stretches of water to support the type of hippopotamus living at that time, as the greater mobility of these early forms would have allowed for the exploitation of a more varied habitat

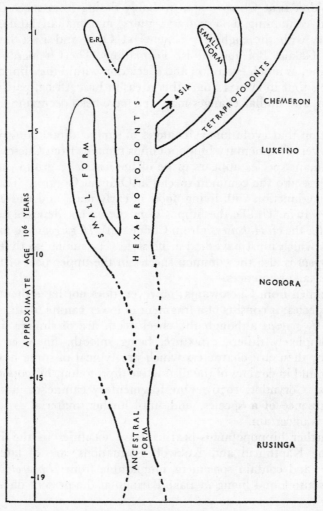

Fig. 18:6. Suggested Phylogeny of the Hippopotamidae. ER = East Lake Turkana form.

including bush savannah. During the deposition of the Chemeron Formation however, the hippopotamus remains indicate changing conditions favouring larger and more stable areas of water with new species emerging; these conditions would favour the establishment of the amphibious hippopotami such as *H. kaisensis*, *H. gorgops* and the living *H. amphibius*, all three of which are closely related tetraprotodonts (Fig. 18:6).

The history of mammalian faunas in the Baringo Basin is not known during the period for about 1.5 million years after the close of the Chemeron episode. This is the period during which deposits containing very rich mammalian remains were preserved in the Omo and East Lake Turkana areas. Chemeron deposits are roughly contemporary with the lower levels at these northerly sites. There are some areas in the Baringo Basin where deposits of post-Chemeron age are preserved. The oldest is at Chesowanja, probably dating from more than a millon years ago and containing a very fragmentary mammalian fauna. These beds equate in time with the highest sediments at Omo and East Lake Turkana, excluding the Galana Boi and Kibish Formations. The Chemeron–Chesowanja gap is tantalising, as much evolution and speciation within the Hippopotamidae is seen during this time in faunas from other areas in East Africa, particularly in the Turkana Basin, and similar changes may well have been occurring in the Baringo Basin.

A suggestion that evolutionary changes of similar dimensions did occur can be seen in the very fragmentary hippopotamus remains from Chesowanja. At this level the dominant species appears to be of the *amphibius* group, most probably *H. gorgops*; this is also the common species at Olduvai Gorge in Tanzania where it is found in conjunction with living floors of early man, and was clearly an important dietary item. Unlike the hippos from the earlier deposits in the Baringo Basin (except for the cf. *H. kaisensis* from Chemeron), *H. gorgops* was an amphibious hippopotamus which must have had a life style very similar to that of the living species. *H. gorgops* is also the common species in the upper part of the Omo and East Lake Turkana sequences.

One specimen from Chesowanja, however, does not have the characteristics of the above species; it consists of a fragment of lower canine tooth, much smaller than that of *H. gorgops* although the cheek teeth are of similar size, and with enamel of completely different texture, being smooth, finely crenellated and rather pig-like; these are characters which are typical of some hexaprotodonts, and are also found in canines of the distinctive diprotodont hippopotamus of East Lake Turkana (Coryndon 1976). One fragmentary canine cannot confirm the presence or absence of a species, and until further material is recovered the identification is uncertain.

Three further hippopotamus-bearing fossil localities in the Baringo area, Loboi and the Kapthurin and Kokwob Formations, are of late Pleistocene/Holocene age, and contain specimens inseparable from *Hippopotamus amphibius*. This species is also found living in Lake Baringo at the present day.

Problems of hippopotamus origins

One of the constant problems relating to hippopotamid phylogeny lies in the fact that no clear ancestor of the hippopotamus has yet been discovered. Previous workers have attempted to derive the Hippopotamidae from the Suidae or the Anthracotheriidae, and whilst I would support the latter hypothesis on present evidence, a review of the Tertiary Suidae and Anthracotheriidae must be undertaken before any clear picture can emerge.

The Ngorora Formation with deposits older than nine million years contains not only the oldest evidence so far of hippopotamus but also some isolated bunodont teeth which look superficially like a very small hippopotamid. These are broadly similar to, though smaller than, the teeth of a maxillary specimen from the earlier deposits of Rusinga Island, Kavirondo, Kenya, and may indicate the presence of a creature very close to the basal stock from which true hippopotamus was derived. The Ngorora teeth, together with the single specimen from Nakali found by Dr Aguirre are simple, low crowned and rather pig-like; they most probably represent a genus of the Suidae similar to *Bunolistriodon* (Fig. 18:7).

Although many specimens from African and Eurasian Neogene and Palaeogene deposits have been examined for possible affinities and ancestry to the Hippopotamidae, each has had to be rejected for one reason or another. To date, the true ancestry of the Hippopotamidae remains obscure.

Conclusions

The Baringo Basin is clearly a crucial area in the study of hippopotamus evolution, with the earliest known members of the family in the lower levels, evidence of wide speciation in the middle levels, and emergence of the living species in the latest deposits. These changes reflect not only the evolutionary pattern of the Hippopotamidae but also the changing conditions within the Baringo Basin during the past ten million years.

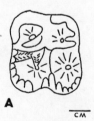

A

CM

FIG. 18:7.

A. Upper second molar from Rusinga, possibly representing an ancestral hippopotamus.

B

B. Upper second molar from Ngorora, possibly representing a *Bunolistriodon*.

Acknowledgements

The author is grateful to Professor W. W. Bishop of Queen Mary College, London for making the Baringo specimens available for study and to the Leakey Foundation of Los Angeles for a travel grant to visit Nairobi. The staff of the National Museums of Kenya gave freely of their time and hospitality and are thanked with gratitude.

References

BISHOP, W. W. & CHAPMAN, G. R. 1970. Early Pliocene Sediments and Fossils from the Northern Kenya Rift Valley. *Nature, Lond.* **226**, 914–18.

—— —— HILL, A. & MILLER, J.A. 1971. Succession of Cainozoic Vertebrate Assemblages from the Northern Kenya Rift Valley. *Nature, Lond.* **233**, 389–94.

—— & PICKFORD, M. H. L. 1975. Geology, fauna and palaeoenvironments of the Ngorora Formation, Kenya Rift Valley. *Nature, Lond.* **254**, 185–92.

COOKE, H. B. S. & CORYNDON, S. C. 1970. Pleistocene mammals from the Kaiso Formation and other related deposits in Uganda. In: *Fossil Vertebrates of Africa Vol. II.* L. S. B. Leakey and R. J. G. Savage, Eds, Academic Press, London, 107–224.

CORYNDON, S. C. 1976. Fossil Hippopotamidae from Pliocene/Pleistocene Successions of the Rudolf Basin. In: *Earliest Man and Environments in the Lake Rudolf Basin.* Y. Coppens *et al.*, Eds, University of Chicago Press, 238–50.

—— & COPPENS, Y. 1973. Preliminary Report on Hippopotamidae (Mammalia, Artiodactyla) from the Plio/Pleistocene of the Lower Omo Basin, Ethiopia. In: *Fossil Vertebrates of Africa Vol. III.* L. S. B. Leakey, R. J. G. Savage and S. C. Coryndon, Eds, Academic Press, London, 139–57.

—— —— 1975. Une espèce nouvelle d'Hippopotame nain du Plio-Pleistocène du bassin du lac Rodolphe (Ethiopie, Kenya). *C. R. Acad. Sc.* Paris **280**, Ser. D. 1777–80.

PICKFORD, M. 1975. Late Miocene sediments and fossils from the Northern Kenya Rift Valley. *Nature, Lond.* **256**, 185–92.

19

ALAN W. GENTRY

The fossil Bovidae of the Baringo Area, Kenya

A preliminary report is given of bovid fossils from the Ngorora Formation and other sedimentary units in the Baringo area, Kenya. The Ngorora Formation contains boselaphines and a caprine similar to those of Fort Ternan, but showing some interesting changes. The Mpesida Beds contain early examples of extant tribes of antelopes. The Lukeino Formation has a reduncine similar to a Siwalik form, and the Karmosit beds contain two species similar to ones from the Mursi Formation, Ethiopia.

Introduction

Several Miocene to Recent vertebrate assemblages have been found in the Baringo area of the northern Kenya Rift Valley since 1965. An outline account of the geology and faunas was given by Bishop *et al.* (1971), and further information is given in Bishop & Pickford (1975), Pickford (1975), and this volume.

The deposits from which bovids have come and their most likely ages are as follows:

Ngorora Formation	an appreciable time span between 9 and 12 m.y.
Mpesida Beds	probably about 7 million years
Lukeino Formation	About 6.0–6.7 million years
Chemeron Formation	several localities older than 2 million years

The above constitute part of the Tugen Hills succession. To the north are:

Aterir beds	a little younger than 4 million years
Karmosit beds	probably a little over 3.4 million years

East of the Tugen Hills is:

Chemoigut Formation	probably between 1 and 1.5 million years.

The first collections of Bovidae from the sedimentary units of the Baringo area are here reported in greater detail than hitherto. I studied them while they were on loan to Dr W. W. Bishop in London, but they are the property of the Kenya National Museums and have now been returned to Nairobi. Each fossil is referred to by its registered number, which always begins with KNM. For Ngorora material a more detailed provenance is given, as, for example, in the phrase 'from 2/1 level B5' which means from locality 2/1 in bed 5 of member B. In reference to comparative material the abbreviation BM(NH) stands for British Museum (Natural History).

Measurements are given in millimetres.

In the course of the paper, comparisons are made with fossils from other localities, as follows:

Fort Ternan, Kenya, dated to about 14 million years (Gentry 1970; Bishop, Miller & Fitch 1969).

Beni Mellal, Morocco, of Miocene age (Lavocat 1961; Jaeger *et al.* 1973).

Marceau, Algeria, of Miocene age (Arambourg 1959).

Langebaanweg, Cape Province, South Africa, of Pliocene age (Hendey 1973, 1974).

Laetolil, Tanzania, of Pliocene age (M. D. Leakey *et al.* 1976).

Mursi Formation, Omo, Ethiopia, where a basalt overlying the fossiliferous levels has been dated to 4.05 million years (Butzer & Thurber 1969).

Shungura Formation, Omo, Ethiopia, with a time span from about 3 until a little less than 1 million years (Coppens *et al.* 1976).

Olduvai Gorge, Tanzania, where Beds I to IV have a time span from 2.1 until 0.6 million years (L. S. B. Leakey 1965; M. D. Leakey 1971).

Siwalik Hills of India and Pakistan. It is likely that the Chinji Formation predates and most of the Nagri Formation postdates the appearance of *Hipparion* (Hussain 1971). The Dhok Pathan Formation in the area of its type locality may be late Miocene to Pliocene. Time-transgressive facies changes in the Siwalik deposits introduce difficulties into faunal correlations (pers. comm., M. H. L. Pickford).

Pikermi & Samos, Greece, the second of which is about 8 million years old (Van Couvering & Miller 1971; Gentry 1971).

The classification of bovids used in this paper is modified from Simpson (1945) and Ansell (1971):

Family BOVIDAE

Subfamily Bovinae		
Tribe Tragelaphini	bushbuck, kudu, eland and allies	
„ Boselaphini	Indian nilgai, four-horned antelope and many extinct forms	
„ Bovini	cattle and buffaloes	
Subfamily Cephalophinae		
Tribe Cephalophini	duikers	
Subfamily Hippotraginae		
Tribe Reduncini	reedbuck, kob, lechwe, waterbuck	
„ Hippotragini	roan, sable, oryxes, addax	
Subfamily Alcelaphinae		
Tribe Alcelaphini	wildebeest and hartebeest group	
Subfamily Antilopinae		
Tribe Neotragini	dik dik group	
„ Antilopini	gazelles, springbok group [also including Saigini]	

bfamiSuly Caprinae
Tribe 'Rupicaprini'   goral, serow group
   ,,   Ovibovini   muskox, takin and many
                                extinct forms
   ,,   Caprini   sheep and goats.

Ngorora Formation

Boselaphini, *Protragocerus* sp.

Numbers of tooth remains from Ngorora represent a species of boselaphine. A right maxilla, KNM–BN 52 from 2/1, level B5, with P^3–M^3 is very like the Fort Ternan *Protragocerus labidotus* Gentry (1970: 247), and can be identified as *Protragocerus* sp. Its teeth show the same brachyodonty and somewhat rugose enamel. The molars have very small basal pillars, and the postero-medial lobe is still separate anteriorly from the other lobes at quite a late stage of wear, just as at Fort Ternan. Occlusal lengths are M^1–M^3 43.6, M^2 16.6 and P^4 9.4, which are within the range of *P. labidotus* but larger than the means, as shown by the numerical data for the Fort Ternan species below:

	Number measured	Mean	Range	Standard deviation	Standard error
Length M^1–M^3	8	41.5	37.8–45.5	2.4	0.84
Length M^2	20	15.3	12.6–17.5	1.2	0.27

To the same species one may reliably assign:

KNM–BN 53 left upper molar, occlusal length 12.7, from 2/11, level D3
   ,,   54 right upper molar, occlusal length 15.7, from 2/1, level B5
   ,,   55 left upper molar, occlusal length 16.6, from 2/1, level B5
   ,,   56 fragment of right mandible with a damaged M_3, occlusal length 21.4, from 2/10, level D1
   ,,   57 right lower molar from 2/11, level D3
   ,,   58 fragment of left mandible with a rather worn molar from 2/1, level B5
   ,,   61 left P_2, occlusal length 8.7, from 2/1, level B5 ⎫ perhaps from
   ,,   62 left P_3, occlusal length 12.3, from 2/1, level B5 ⎬ one individual
   ,,   63 left P_4, occlusal length 12.5, from 2/1 ,level B5 ⎭ (Fig. 19:1)
   ,,   64 right upper molar from 2/11, level D3
   ,,   65 fragment of left mandible with two damaged molars from 2/1, level B5
   ,,   66 right M_3 in jaw fragment, occlusal length 19.8, from 2/14, level B5
   ,,   67 right M_3, occlusal length 20.3, from 2/1, level B5 (Fig. 19:1)
   ..   68 left lower molar, occlusal length 13.4, from 2/1, level B5
   ,,   69 most of a right lower molar from 2/1, level B5 (Fig. 19:1).

The P_4 KNM–BN 63 is in middle wear and shows a separate entoconid and entostylid, just as in Fort Ternan P_4s. This is a boselaphine character and can be contrasted with the fusion of these cusps normally seen in contemporaneous Caprinae at Fort Ternan and Ngorora. The fusion occurs in early wear in these caprines and is connected with reduction in size of their premolars. KNM–BN 63 differs from P_4s of *P. labidotus* by a less projecting hypoconid and better separation of the paraconid and parastylid.

Two right upper molars, KNM–BN 59 and 60 from 2/1 and 2/14, both level

B5, are slightly larger but perhaps conspecific. KNM–BN 59 has an occlusal length of 18 and KNM–BN 60 of 17.7.

Boselaphini/Tragelaphini, sp. or spp. indet.

There are also dental remains of a larger boselaphine or tragelaphine at Ngorora. The best piece is a left mandible KNM–BN 99 from 2/10, level D1, with P_3–M_2 (Fig. 19:1). The teeth are well worn and their occlusal lengths are: P_3 12.8, P_4 14.3 and M_2 18.6. The basal pillars are small to moderate sized rather than tiny to small, the lateral lobes of M_2 are elongated transversely, and there are no constrictions across the middle of the central cavities of M_2. The P_4 is like that of *Mesembriportax acrae* from the Pliocene of Langebaanweg, South Africa (Gentry 1974) in the dimpled medial edge of the metaconid not projecting medially beyond the level of paraconid and entoconid, the parallel-sided valley between metaconid and entoconid, and the laterally projecting hypoconid. Other pieces belonging to this species are a left mandibular fragment in matrix with most of M_3 and the back of M_2 (KNM–BN 97 from 2/14a, level B5), a left mandibular piece surrounded by much matrix with a little worn M_3 at 23.4 and the back lobe of M_2 (KNM–BN 95 from 2/11, level D3), and most of a left lower molar (KNM–BN 98 from 2/1, level B5).

Later-collected dental material in this size group, especially the mandibles

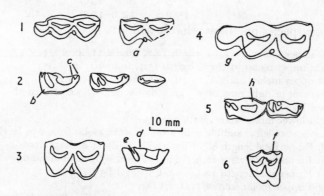

Fig. 19:1. Occlusal views of some teeth from the Ngorora Formation. The lateral side is towards the foot of the page and the anterior side towards the right.

1. *Protragocerus* sp. M_3, KNM–BN 67, on the left and lower molar, KNM–BN 69, on the right. a = basal pillar.
2. *Protragocerus* sp. P_4, P_3 and P_2, KNM–BN 63, 62 and 61 (left teeth drawn as if from the right side). b = hypoconid, c = paraconid (on the left) separate from parastylid.
3. Boselaphini/Tragelaphini sp. M_2 and P_4 from the mandible KNM–BN 99 (left teeth drawn as if from the right side). d = medial surface of metaconid, e = valley between metaconid and entoconid.
4. ?*Pseudotragus* sp. nov. M_3, KNM–BN 89. g = constriction in central cavity.
5. ?*Pseudotragus* sp. nov. P_4 and P_3, KNM–BN 90. h = fused paraconid and metaconid.
6. ?Cephalophini sp. M^2 from maxilla, KNM–BN 96. f = rib between parastyle and mesostyle.

KNM–BN 1071 and 1235 from 2/10 in level D1, have teeth differing from those of the *Protragocerus* by their flatter medial walls, straighter sides to the lateral lobes and straighter course of the central cavities on the molars, and paraconid–metaconid fusion on P_4. These characters are all like Tragelaphini, and are the basis for the designation used here and in Bishop & Pickford (1975, table 1). It seems that tragelaphine-like teeth occur earlier in the fossil record than tragelaphine horn cores, of which the earliest known to me are in the Mursi Formation, Ethiopia, and at Langebaanweg, South Africa, both sites being less than 5 million years old. No tragelaphine-like horn cores are known from the Ngorora Formation and the implication is that one or more lineages of Boselaphini evolved advanced characters earlier in their teeth than in their horn cores. Classification at tribal level may well be difficult when the Ngorora material comes to be fully described.

?Cephalophini, sp. indet.

A small antelope is represented by a right maxilla, KNM–BN 96 from 2/1, level B5, with the M^1 broken, M^2 and M^3. The M^2 (Fig. 19:1) has an occlusal length 10.7 and the M^3 10.9. The teeth are too brachyodont and rather small to fit a gazelle. Primitive features are the quite rugose enamel and a pronounced rib on the lateral wall of the molars between parastyle and mesostyle. The maxilla may belong to a duiker, but is distinguished from living duikers by its lower crowned molars with flatter lateral walls between mesostyles and metastyles. Arambourg (1959: 127, pl.17 figs. 8, 8a, 8b) provisionally referred a small M_3 from the Miocene of Marceau, North Africa, to *Cephalophus*, but its identity is not clearcut. Marceau is likely to have an age around 7 million years (J. J. Jaeger, pers. comm. to M. H. L. Pickford), and hence to be younger than Oued Hammam (= Bou Hanifia) from where most of the fossils described by Arambourg (1959) derive.

What appear to be similar teeth to those of KNM–BN 96 were described from Beni Mellal, Morocco, by Lavocat (1961: 93, figs 20–2). This site is sometimes accepted as around 10–11 million years old (Van Couvering & Miller 1971, fig. 2) but is now considered to be between $13\frac{1}{2}$ and 14 m.y. (Jaeger, Michaux & David 1973), which would make the teeth of equivalent age to the Ngorora maxilla or older. However, links between the Miocene bovids of North and East Africa have not hitherto been striking, and it could be misleading to invoke a connection on the basis of such primitive teeth. Present-day duikers show considerable speciation and do not occur north of the Sahara Desert.

Caprinae, ?*Pseudotragus* sp. nov.

DESCRIPTION. The most complete bovid fossil in the early collection from the Ngorora Formation is a skull KNM–BN 100 from locality 2/11, level D3, found by Andrew Hill. It lacks only the distal parts of the horn cores and the snout region; while among the teeth there are preserved the left P^3–M^3 and the right M^2 and M^3. The skull resembles in almost every aspect the Fort Ternan and Siwalik (supposedly Nagri Formation) antelope ?*Pseudotragus potwaricus* (Pilgrim) discussed by Gentry (1970: 284, pls. 12–14). Of this species there is known the holotype frontlet (Pilgrim 1939, pl. 2 figs 1, 2), and from Fort Ternan a cranium with horn cores, a left horn core, and assigned palate, mandibles and postcranial bones. There was no direct association at Fort Ternan between the horn cores on the one hand and the dental and postcranial remains on the other.

Similarities of the Ngorora to the Fort Ternan species are the continuing and even accentuated widening of the braincase, the uprightness of the horn core insertions, the absence of raised frontals between the horn core bases, the strong nuchal crest, and the degree of cheek tooth hypsodonty which appears to have changed but little. The supraorbital pits cannot be discerned, so presumably they are still extremely small; whether they were set against the pedicel as on the Fort Ternan specimen cannot be known.

The Ngorora skull is different from ?*P. potwaricus* principally by its larger size. Other differences may be allometrically linked with the size difference. The horn cores appear to be slightly less medio-laterally compressed (Fig. 19:2). The central cavities of the molars have a slightly more complicated outline which is almost certainly linked with the increased overall size. The anterior tuberosities of the basioccipital are larger (which may be connected with the greater onto-genetic age of the Ngorora specimen), and the posterior ones narrower and less prominent (Fig. 19:2). The preorbital fossa is very extensive. I had stated (Gentry 1970: 284) that on the Fort Ternan cranium the preorbital fossa was probably small, but it is more likely that I was wrong than that the fossa had increased in size by the time of the Ngorora Formation. Fig. 19:2 also indicates that the occipital surface was relatively lower, which would be expected from allometry in a

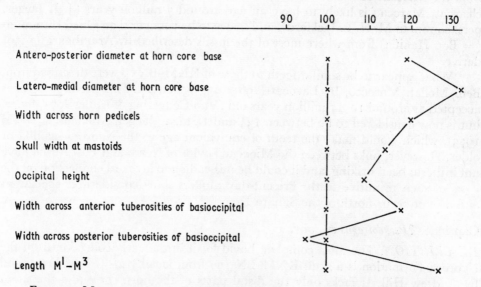

Fig. 19:2. Measurements on the skull KNM–BN 100 of ?*Pseudotragus* sp. nov. expressed as percentages of the same measurements on ?*P. potwaricus*. The standard line at 100 per cent is constructed from measurements on the Fort Ternan cranium KNM—FT 2748 of ?*P. potwaricus*, except for the last measurement which is based on the mean of the molar row lengths of the Fort Ternan dentitions KNM–FT 1032 and 1047.

larger antelope; however, the poor preservation of this region of the Fort Ternan cranium diminishes the reliability of this difference.

A much damaged frontlet with horn core bases, KNM–BN 88 from locality 2/11, level D3, also belongs to this species. A damaged base of a right horn core KNM–BN 101, again from locality 2/11 and level D3, is strongly medio-laterally compressed, more obliquely inserted than the horn cores of the complete skull, and with a lower backward curve. It is probably the same species, and shows something of the extent of sexual or ontogenetic variation.

There are a few dental remains which may be assigned to this species. A rather worn right mandibular piece KNM–BN 89 (Fig. 19:1) has M_2 and M_3 measuring 19.7+ and 29.3 respectively, and there is a left M_3 at 30.1 in a mandibular fragment with a depth of 43.5 ,KNM–BN 94. Both these pieces are from 2/11 and level D3. A fragment of mandible with the damaged parts of two left molars KNM–BN 93 comes from 2/14, level B5. The teeth of these fossils show small to tiny basal pillars, and the M_3 of KNM–BN 89 has medio-lateral constrictions in the middle of its central cavities. The central constrictions suggest an approach to Alcelaphini, although the lower teeth of some smaller-sized alcelaphines from Langebaanweg are without such constrictions. A right mandible with P_3 and P_4 and the alveolus for P_2, KNM–BN 90 (Fig. 19:1) found at 2/3 in member E, is doubtfully assigned to this species. It shows paraconid–metaconid fusion on P_4 despite being only in early wear, and this must be accounted an advanced character. Neither of the premolars is large, nor is P_3 large relative to P_4 as would be expected in a boselaphine or tragelaphine. The occlusal length of P_3 is 11.0 and of P_4 13.0, and the depth of the ramus below P_4 is 26.2. Such a great depth would also not be expected in a boselaphine or tragelaphine.

MEASUREMENTS. Measurements on the skull are listed below. The second column of readings are from the Fort Ternan cranium KNM–FT 2748 (field number 62.2517) of ?*P. potwaricus*, except for the last two readings which are the means of M^1–M^3 and M^2 on two upper dentitions KNM–FT 1032 (63.3424) and KNM–FT 1047 (64.627).

Skull width across posterior side of orbits	130.0	c.108.0
Antero-posterior diameter at base of horn core	50.5	42.6
Medio-lateral diameter at base of horn core	38.7	29.3
Minimum width across lateral surfaces of horn core pedicels	c.99.4	83.0
Maximum braincase width	c.72.7	61.6
Skull width across mastoids immediately behind external auditory meati	c.88.5	77.8
Distance from rearmost point of occlusal surface of M^3 to back of occipital condyles	125.0	–

Occipital height from top of foramen magnum to top of occipital crest	c.40.4	c.37.2
Width across anterior tuberosities of basioccipital	20.9	17.7
Width across posterior tuberosities of basioccipital	26.8	28.2
Occlusal length M¹–M³	63.8	50.1
Occlusal length M³	25.2	19.2

CLASSIFICATION. Gentry (1970: 287) stated that the best suprageneric placing for the Fort Ternan and Nagri species was within the Caprinae but of indeterminate tribe. This was a decision in favour of 'horizontal' classifying at the Fort Ternan time level because of the great resemblances of the antelopes at that site to their Eurasian contemporaries. There has been so little advance in the Ngorora skull that it too may be referred to an unknown tribe of the Caprinae. It is of interest that the Nagri frontlet agrees in size with the Fort Ternan rather than with the Ngorora species.

The Ngorora skull differs from the well-known Pikermi and Samos caprine genera *Protoryx* Major and *Pachytragus* Schlosser by the frontals not being raised between the horn core bases, the braincase widening posteriorly, the occipital surface being entirely in one plane facing backwards, and the stronger nuchal crests.

The reasons for the queried assignation of Pilgrim's trivial name *potwaricus* to the Samos genus *Pseudotragus* Schlosser as set out by Gentry (1970: 288) have become less satisfactory when applied to the Ngorora skull. Only the upright horn core insertions in side view and the pedicels being taller at the front than the back remain as resemblances to *Pseudotragus*, and they are not very convincing. Even with the Fort Ternan species there are no grounds for taking it as on the same lineage as the Samos *P. capricornis* Schlosser which probably lived 5 or 6 million years later. There was simply a possibility that they could have shared a common ancestry. However, I am reluctant to launch a new genus for the African lineage. There is still a lot of uncertainty about antelope evolution between the late Miocene of Fort Ternan and the late Pliocene and Pleistocene of Omo, Olduvai Gorge and numerous other sites.

A suggestion made for the Fort Ternan *Oioceros tanyceras* and ?*P. potwaricus* that they are possibly linked with later Alcelaphini (Gentry 1970: 315) also applies to the Ngorora fossil. The tooth characters just considered reinforce this suggestion. The increase of size which took place on this lineage, and the flattening of the backward curvature of its horn cores, already apparent at Fort Ternan and especially marked in KNM–BN 101, lead me to guess that it may prove to be ancestral to the large Pleistocene alcelaphine *Megalotragus* Van Hoepen.

COMPARISONS. With such a paucity of African Miocene and Pliocene antelopes, there are few other fossils with which the new skull can be compared. The horn core of *Praedamalis deturi* Dietrich (1950: 30, pl. 2 fig. 23) from Laetolil in Tanzania is slightly smaller but quite similar. It was probably fairly long, had the same amount of medio-lateral compression (its basal diameters being 45.5 and 34.1) and a similar shape of cross-section, had no transverse ridges, was slightly curved backwards, inserted above the back of the orbits, was probably inserted uprightly and was not very divergent. A really clear difference from the Ngorora specimen is the acquisition of internal hollowing of the frontal bone. Moreover Laetolil is a site considerably younger than Ngorora. *Praedamalis deturi* remains rather enigmatic for the present, and may not be an alcelaphine at all.

A small alcelaphine cranium from Laetolil, 1959.277 in the National Museum, Nairobi, is probably linked with the *Parmularius/Damaliscus* group of alcelaphines. It differs from the Ngorora skull in having a low parietal boss on the braincase roof, a tendency to a postero-lateral swelling at the base of the horn cores, longer pedicels, and a strong lateral-facing component of the occipital surface.

At Langebaanweg there are one or perhaps two species of primitive Alcelaphini, neither of them quite as large as the Ngorora skull. The frontals between the horn core bases are higher, the horn core pedicels internally hollowed, the nuchal crests are weaker, and the teeth are more alcelaphine-like, all of which befit fossils of a later date than Ngorora. The commoner species has a braincase more strongly angled on the facial axis, a lateral-facing component of the occipital surface, and the long cross-sectional axis of the horn core bases set at a considerable angle to the longitudinal axis of the skull. There is little likelihood of them being related to the Ngorora skull.

?Neotragini

A fragment of a right maxilla KNM–BN 92 from 2/1, level B5, is small enough to belong to this tribe. The deciduous fourth premolar and first molar are present; they are more hypsodont than in the supposed duiker maxilla KNM–BN 96, and the M^1 has a poorer rib between parastyle and mesostyle, and a stronger mesostyle. The damaged base of a left horn core and part of a frontal KNM–BN 91 from locality 2/14, level B5, could belong with this tooth row. The horn core pedicel has an internal hollow, but I am not completely sure that it was not caused by postmortem damage. The horn core is very little compressed.

Antilopini, *Gazella* sp.

A small antilopine, presumably *Gazella*, is represented by horn cores KNM–BN 257, 694 and others. Their orientation is such that the lateral surfaces face partly anteriorly, and this is one of the differences of these horn cores from those of the Fort Ternan gazelle (Gentry 1970: 292). There are also some antilopine tooth remains, of which the best is a left mandible with P_3–M_2, KNM–BN 1362. Differences from the Fort Ternan gazelle are that the medial walls of the lower molars are rather flattened and that the metaconid on the P_4s is not turned anteriorly.

Mpesida Beds

Tragelaphini

Part of a left lower molar, KNM–MP 071, has an occlusal length of 16 or 17 and a basal pillar of small to moderate size. It is almost as large as M_1 on the Karmosit tragelaphine mandible KNM–KM 008 (see below), and is likely to have come from an antelope larger than a bushbuck.

Part of a right mandible with shallow horizontal ramus and broken teeth, KNM–MP 072, may also be tragelaphine.

?Alcelaphini

A worn left upper molar, KNM–MP 077, has an occlusal length of 17.9.

A much damaged basal part of a left horn core, KNM–MP 068, shows a deep postcornual fossa and slight internal hollowing in the frontal. It could be *Aepyceros* but the basal antero-posterior and medio-lateral diameters at c.30 × 26 are rather too large. It can be taken as questionably alcelaphine.

Antilopini

A right mandibular piece with partial M_2 and M_3, KNM–MP 129, is probably *Gazella*. The medial walls of the molars are quite flattened, there are no basal pillars, the rear (third) lobe of M_3 is large and the mandibular ramus shallow. The occlusal lengths of M_2 and M_3 are c.11 and c.15 respectively, and the depth of the ramus below M_3 is 15.9.

Tribe indet

Three horn core pieces from Mpesida are perhaps conspecific. KNM–MP 069 is a damaged base of a left horn core with basal antero-posterior and medio-lateral diameters of 36.3 and 31.3. There is no internal hollowing of the frontal, the lateral surface of the horn core is less convex than the medial, and there is an approach to a postero-lateral keel. KNM–MP 070 is a segment of another large horn core. KNM–MP 075 with basal diameters of 37.6 and 30.3 curves backwards and has no transverse ridges.

Lukeino Formation

Tragelaphini/Boselaphini

A left P^3 or P^4, KNM–KY 049, has an occlusal length of 8.9 mm. It is brachyodont and probably belongs to one of the above tribes.

Reduncini

A frontlet with the base of the horn cores, KNM–LU 011, has the appearance of being reduncine. The antero-posterior and medio-lateral diameters at the base of the better-preserved right horn core are 38.0 and 28.5, and the distance across the lateral walls of the supraorbital pits is 39.9. The horn cores are thus somewhat compressed. They are inserted above the orbits at a fairly low inclination in side view, are quite strongly divergent and have a slight and smooth backward curvature. The frontlet agrees closely with some slightly smaller fossils from the Siwalik Deposits. These are a left horn core of *Dorcadoxa porrecticornis* (Lydekker) BM (NH) M.15473 with basal diameters at 36.1 × 28.8 from near Hasnot and supposedly in the Dhok Pathan Formation; a right horn core of ?*Gazella superba* Pilgrim, M.15474 with basal diameters of 31.2 × 24.3 also from Hasnot; and a cast of the holotype frontlet of (cf. *Indoredunca*) *theobaldi* Pilgrim possibly from either the Tatrot or Dhok Pathan Formations. The cast has the registered number M.15827 and its basal diameters are 33.2 × 34.3. All three Asian specimens could well be conspecific with one another and with the Lukeino frontlet.

A proximal right metacarpal, KNM–LU 010, is very probably reduncine by the fairly large size of the unciform facet and its angled anterior edge. It has a transverse width across the top articular surface of about 27.5 mm.

Tribe indet

A piece of a horn core, KNM–KY 046, has basal diameters of 38.1 and 31.7, but cannot be orientated. It appears to have incipient keels, unlike the three undetermined horn cores from Mpesida.

KNM–KY 043 is a fragment of another horn core.

Chemeron Formation

The fossils from this formation come from a number of different sites.

JM 85

A number of alcelaphine teeth come from this locality:
KNM–BC 83 a right upper molar with an occlusal length of 22.3
KNM–BC 84 a left lower molar
KNM–BC 85 part of a left upper molar
KNM–BC 86 fragment of a tooth
KNM–BC 111 part of a left M_3, somewhat smaller than nos. 83–85.

JM 489

Part of a bovine right lower molar, KNM–BC 99, is of a size appropriate for the *Syncerus* lineage.

JM 493

A tragelaphine right M_3 with some adherent bone, KNM–BC 105, has an occlusal length of 26.6 and is slightly smaller than the kudu in the upper part of the Shungura Formation.

A large atlas vertebra, KNM–BC 113, is either tragelaphine or bovine.

Two left upper molars, KNM–BC 104 and 107 with occlusal lengths of 18.4 and 15.2 mm respectively are probably reduncine.

Part of a left maxilla, KNM–BC 102, with unerupted P⁴, M¹ and M² is alcelaphine. The occlusal length of M² is 24.7, which matches the size of the middle group of alcelaphine teeth from the Shungura Formation at Omo.

JM 494

A right maxilla, KNM–BC 112, with P⁴–M³ and part of P³ is an alcelaphine about the size of the middle group from the Shungura Formation at Omo. The occlusal length of M^1–M^3 is 67.5 and of M^2 25.5.

The lower part of a damaged right horn core, KNM–BC 88, comes from *Aepyceros*. There is little medio-lateral compression, no clear sign of transverse ridges, it curves gently and evenly backwards and outwards, is inserted above the orbits, there is a moderate-sized and deep postcornual fossa, and the frontals show internal hollowing around the horn pedicel base.

JM 511

There are some tragelaphine teeth from this site: a right mandibular fragment with M_2 at 20.8 and M_3 at 27.4 KNM–BC 96, a right lower molar and two other tooth scraps KNM–BC 97, and part of a left M_3 in a piece of mandible KNM–BC 106. These are all of a size equal to the kudu in the upper part of the Shungura Formation.

Site indet

Part of a tragelaphine right horn core from Chemeron, KNM–BC 90, shows a postero-lateral keel but not an anterior one, and some degree of antero-posterior flattening. It may be designated *Tragelaphus* cf. *spekei*, like similar sized horn cores at other Plio-Pleistocene sites.

Aterir Beds

Part of a large right upper molar, KNM–AT 079, is probably bovine or boselaphine. The lateral wall is missing, and the anterior lobe has been distorted antero-medially from its true alignment with the posterior lobe. The occlusal length of the complete tooth would have been about 29 to 30. The identification of the tooth is based on its large size, a constriction across the walls of the medial part of the anterior lobe, and the rugosity of the enamel.

Karmosit Beds

Tragelaphini, *Tragelaphus* sp.

A small part of a tragelaphine right horn core, KNM–KM 13, has postero-lateral and anterior keels, the former being the stronger of the two, and shows some compression antero-posteriorly. It is about the same size as fragments of tragelaphine horn cores YS 4–10 and YS 1968.2078, collected by the American and French parties at the Yellow Sands locality of the Mursi Formation, Omo. YS 4–10 has a weak anterior keel and a stronger postero-lateral one just as in the Karmosit horn core, but the first is weaker and the second stronger than in KNM–KM 13 and the horn core is more flattened antero-posteriorly. YS 1968.2078 is a horn core base with antero-posterior and medio-lateral diameters of 38.7 and 44.3. Its divergence increases from the base upwards as in kudus (*Tragelaphus imberbis*, *T. strepsiceros* and extinct relatives), so that the postero-lateral keel lies along the concave edge. In this it agrees with the other Mursi and Karmosit fragments and contrasts with most other *Tragelaphus*, including the abundant Shungura species *T. nakuae*, in which this keel lies along a convex edge. A third horn core from the Mursi Formation, YS 4-6, has a stronger anterior keel and is less antero-posteriorly compressed than YS 4-10, so is presumably from the more distal part of a horn core. Yet another piece, YS 4-4, is also less antero-posteriorly compressed, but its keels are not very clear. The Karmosit horn core can provisionally be referred to the same species as the Mursi Formation pieces, that is to an unnamed kudu probably ancestral to *T. gaudryi* (P. Thomas) of North Africa and members E to G of the Shungura Formation as well as to the living kudus.

Part of a right mandible, KNM–KM 008, with P_4–M_2 belongs to the Tragelaphini and is slightly smaller than *T. gaudryi* as it is known from members E to G of the Shungura Formation. The paraconid has fused to the metaconid on P_4 which is in fairly late wear, and the molars show gentle outbowings on their medial walls, fairly small basal pillars and no goat folds. Occlusal lengths are: P_4 14.4, M_1 15.3, M_2 18.0.

Reduncini

A right M_3, KNM–KM 011, with an occlusal length of c.24.4, and a left upper molar, KNM–KM 012, with an occlusal length of 16.9 are both reduncine. Neither is much worn. KNM–KM 006 is the tip of a horn core which appears to be reduncine.

Alcelaphini

A damaged distal right humerus, KNM–KM 002, is probably alcelaphine.

A right and a left horn core, KNM–KM 009 and 010, belong to *Aepyceros*, a genus usually included in the Antilopini but which I believe to belong to the Alcelaphini. They show little medio-lateral compression, a less rounded section on the lateral than on the medial side, and an approach to a posterior and even to an anterior keel. They are inserted over the orbits, curve gently and steadily backwards, have a small to moderately sized and deep postcornual fossa, and show internal hollowing in the frontals. The left one has two transverse ridges and the right one has only traces of them. The antero-posterior and medio-lateral basal diameters of the left and right ones are 29.6 × 25.4 and 32.4 × 25.9 respectively. The longer right horn core shows less lyration than in impala horn cores from the Shungura Formation and compares well with two Mursi Formation horn cores, YS 4-1 and YS 4-3. The left horn core is inserted less uprightly than in the living impala.

Chemoigut Formation

Nearly all the bovid fossils from the Chemoigut Formation (Bishop *et al.* 1975) are isolated teeth, but there are some maxilla and mandible fragments. As well as specimens on loan to Bedford College, some specimens were seen in Nairobi in January 1971. The following groups were represented:

Tragelaphini

KNM–CE 004, 005   Parts of two teeth, the size of *Tragelaphus strepsiceros*. Also four other unnumbered pieces.

?Tragelaphini

KNM–CE 028–032   Four tooth fragments and one complete tooth.

Bovini

KNM–CE 007   Left M_3 with occlusal length 38.0, and a tooth fragment.

Reduncini, *Kobus* sp.

KNM–CE 001–003, 033   One complete tooth and parts of three others. Other, unnumbered, pieces may reach a size equal to the largest Shungura and Olduvai *K. sigmoidalis*, but a P_4 on the most complete one shows that they are not hippotragine.

Alcelaphini, *Megalotragus kattwinkeli?*

KNM–CE 006 a + b   Right mandible with P_4–M_3 in early wear. The ascending ramus is more upright than in the South African *M. priscus* which occurs in later deposits than *M. kattwinkeli*. The latter is found in Olduvai middle Bed II to Bed IV, and a similar or identical species in later members of the Shungura Formation. Occlusal lengths: M_1–M_3 86.7, M_2 29.4, M_3 31.1, P_4 17.2.

Alcelaphini, size of *Megalotragus* or *Connochaetes*

KNM–CE 013–020   Three teeth and parts of five others.

Alcelaphini, smaller than above

KNM–CE 010, 021   Part of right mandible and maxilla.

KNM–CE 011, 022, 008   Two teeth and parts of four
009, 023, 024   others. Also eighteen unnumbered teeth or parts thereof.

Antilopini, ?*Antidorcas* sp.

KNM–CE 025	Maxilla with occlusal lengths M^1–M^3 44.6, M^2 16.0.
KNM–CE 026–027	Complete and fragmentary teeth.

Conclusions

When I first studied the bovids of Fort Ternan, I was impressed by their obvious relationship to contemporaneous boselaphines and caprines in Eurasia. Later it seemed that this might not preclude them, or some of them, from being also ancestral to the more recent, non-boselaphine and non-caprine antelopes of Africa (Gentry 1970: 315). Possibly the Fort Ternan deposits dated from a period before the bovid faunas of Eurasia and African had become so strongly different from one another. The bovids described in this paper tend to support my later view. So far as they are known, the Ngorora Formation bovids are similar at tribal and generic level to those of Fort Ternan. The teeth of smaller boselaphine are broadly similar to those of *Protragocerus labidotus*, a larger boselaphine (or tragelaphine?) has appeared, and the ?*Pseudotragus* has evolved just far enough to be conveniently referred to a different species. The new ?*Pseudotragus* and the new boselaphine/tragelaphine maintain and even strengthen the idea that alcelaphines and tragelaphines differentiated from the same stocks as caprines and boselaphines.

A second feature of interest is that although the Ngorora bovids are like those of Fort Ternan as far as bald systematics are concerned, they appear to show a different ecological balance. At least two species of the boselaphine/tragelaphine group are represented, and conversely there is only one caprine, and that is the less abundant of the two caprine lineages represented at Fort Ternan. Furthermore the rather strong differences of the Ngorora antilopine from what I originally supposed was a gazelle at Fort Ternan suggest that it is on a different lineage and would have had different ecological requirements. More detailed work has to be done on faunal changes over the several million years represented within the Ngorora Formation, but a preliminary conclusion could be that the bovids indicate less open environments at certain levels than at Fort Ternan (see also Pickford, this volume, on the palaeoenvironments of the Ngorora Formation).

In the Mpesida Beds is a tooth which is doubtfully accepted as alcelaphine. These beds are thought to be around 7 million years old, which is slightly later than the upper limit of the Ngorora Formation, and it could be from this time onwards that the problem of drawing tribal boundaries diminishes.

The Lukeino reduncine is extremely interesting by reason of its resemblance to the Dhok Pathan forms, and draws attention to the Tertiary links between the antelopes of Africa and the Siwaliks.

It is interesting that the Karmosit *Tragelaphus* and *Aepyceros* both resemble fossils in the Mursi Formation, although one would hesitate on the amount of information available here to postulate any very precise coincidence of age.

Acknowledgements

I thank Professor W. W. Bishop for inviting me to study the fossils on which a preliminary report has been given in this paper. Dr Andrew Hill, Dr Martin Pickford and several former research students of Bedford College gave me help and various facilities.

References

ANSELL, W. F. H. 1971. Artiodactyla. In Meester, J. & Setzer, H. W. (Eds) *The mammals of Africa: an identification manual.* Part 15: 1–93. Washington D.C.: Smithsonian Institution Press.

ARAMBOURG, C. 1959. Vertébrés continentaux du Miocène supérieur de l'Afrique du Nord. *Mém. Carte géol. Algérie (n.s. Paléont.)* Algiers **4**, 1–159.

BISHOP, W. W., CHAPMAN, G. R., HILL, A. & MILLER, J. A. 1971. Succession of Cainozoic vertebrate assemblages from the northern Kenya Rift Valley. *Nature, Lond.* **233**, 389–94.

——, HILL, A. P. & PICKFORD, M. H. L. 1975. New Evidence regarding the Quaternary geology. archaeology and hominids of Chesowanja, Kenya. *Nature, Lond.* **258**, 204–8.

——, MILLER, J. A. & FITCH, F. J. 1969. New potassium-argon age determinations relevant to the Miocene fossil mammal sequence in East Africa. *Am. J. Sci.* New Haven, **267**, 669–99.

—— & PICKFORD, M. H. L. 1975. Geology fauna and palaeoenvironments of the Ngorora Formation, Kenya Rift Valley. *Nature, Lond.* **254**, 185–92.

BUTZER, K. W. & THURBER, D. L. 1969. Some late Cenozoic sedimentary formations of the Lower Omo Basin. *Nature, Lond.* **222**, 1138–43.

COPPENS, Y., HOWELL, F. C., ISAAC, G. L. & LEAKEY, R. E. F. (Eds). 1976. *Earliest man and environments in the Lake Rudolph basin: stratigraphy, paleoecology and evolution.* 615 pp. Chicago.

DIETRICH, W. O. 1950. Fossile Antilopen und Rinder Aquatorialafrikas. *Palaeontographica*, Stuttgart **99A**, 1–62, 7 pls.

GENTRY, A. W. 1970. The Bovidae (Mammalia) of the Fort Ternan fossil fauna. In: Leakey, L. S. B. & Savage, R. J. G. (Eds) *Fossil Vertebrates of Africa*, **2**, 243–324. London: Academic Press.

—— 1971. The earliest goats and other antelopes from the Samos *Hipparion* fauna. *Bull. Br. Mus. nat. Hist.* (Geol.), **20**, 229–96, 6 pls.

—— 1974. A new genus and species of Pliocene boselaphine (Bovidae, Mammalia) from South Africa. *Ann. S. Afr. Mus.* Cape Town, **65**, 145–88.

HENDEY, Q. B. 1973. Fossil occurrences at Langebaanweg, Cape Province. *Nature, Lond.* **244**, 13–14.

—— 1974. The late Cenozoic Carnivora of the south-western Cape Province. *Ann. S. Afr. Mus.* Cape Town, **63**, 1–369, 78 figs.

HUSSAIN, S. T. 1971. Revision of *Hipparion* (Equidae, Mammalia) from the Siwalik Hills of Pakistan and India. *Abh. bayer. Akad. Wiss.* Munich, **147**, 1–68.

JAEGER, J. J., MICHAUX, J. & DAVID, B. 1973. Biochronologie du Miocène moyen et supérieur continental du Maghreb. *C.r. hebd. Séanc. Acad. Sci.* Paris **277**D, 2477–80.

LAVOCAT, R. 1961. In: Choubert G. and Faure-Muret, A. Le gisement de vertébrés Miocenes de Beni Mellal. *Notes Mém. Serv. Mines Carte géol. Maroc.* 155, 1–122.

LEAKEY, L. S. B. 1965. *Olduvai Gorge 1951–61. I. Fauna and Background.* 118 pp., 97 pls. Cambridge.

LEAKEY, M. D. 1971. *Olduvai Gorge. 3. Excavations in Beds I and II, 1960–63.* 306 pp., 41 pls. Cambridge.

——, HAY, R. L. CURTIS, C. H., DRAKE, R. E., JACKES, M. K. & WHITE, T. D. 1976. Fossil hominids from the Laetolil Beds. *Nature Lond.* **262**, 460–66.

PICKFORD, M. 1975. Late Miocene sediments and fossils from the northern Kenya Rift Valley. *Nature, Lond.* **256**, 279–84.

PILGRIM, G. E. 1939. The fossil Bovidae of India. *Palaeont. indica.* Calcutta (n.s.), **26**, 1–356.

SIMPSON, G. G. 1945. The principles of classification and a classification of mammals. *Bull. Am. Mus. nat. Hist.* New York, **85**, 1–350.

VAN COUVERING, J. A. & MILLER, J. A. 1971. Late Miocene marine and non-marine time scale in Europe. *Nature, Lond.* **230**, 559–63.

20

WILLIAM BISHOP, ANDREW HILL
and MARTIN PICKFORD

CHESOWANJA: A revised Geological Interpretation

Chesowanja is located in the north Kenya Rift Valley, to the East of Lake Baringo, between Tangulbei and Mukutan (36°12′E, 0°39′N). An important succession of Quaternary sediments occurs in this vicinity. Some of the sedimentary units contain abundant fossils and intermittent evidence of hominid activity over a period of perhaps 2 my. Remains of Australopithecus sp. have also been found. Artifacts appear to belong to the Oldowan/Developed Oldowan Cultural Complex, the Acheulian tradition, and to later stone industries. This paper outlines the geological setting of the faunal and artifact assemblages.

Introduction

The Chesowanja area was first investigated geologically by John Carney, as part of a programme of regional mapping carried out by the East African Geological Research Unit, based on Bedford College, University of London, and directed by Professor B. C. King. Carney recovered fossil vertebrates from the Chemoigut Formation and during a follow-up faunal survey with A. H., a partial cranium of *Australopithecus* was found (Carney *et al.* 1971). A revisit to Chesowania by the three of us in 1973 resulted in a reassessment of the geological situation, the discovery of an extensive sequence of artifact horizons, and more hominid material (Bishop *et al.* 1975). This paper reports the results of more detailed work by a further expedition in 1974.

Geology

The geological setting of the Chesowanja area originally proposed (Carney *et al.* 1971; Carney 1972) was substantially revised in the light of more detailed work (Bishop *et al.* 1975). Further details are provided here, along with information concerning the regional geology.

Succession

This account should be read in conjunction with the map (Fig. 20:1) and the composite section (Fig. 20:2).

Miocene rocks. The oldest surface rocks are un-named Miocene basalts and phonolites, which are exposed in the Engelesha and Aruru escarpments 8 km to the east of Chesowanja. They also crop out extensively to the south as a series of

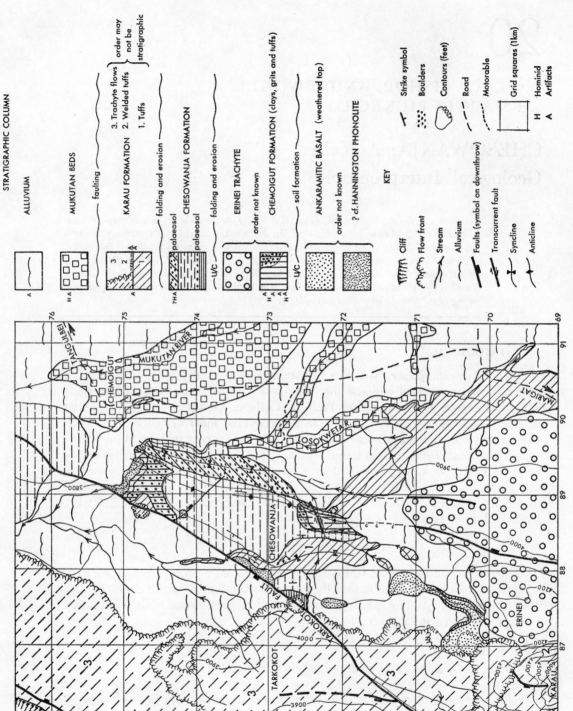

FIG. 20:1. Geological map and stratigraphical column of the Chesowanja area.

en echelon platforms rising southwards above Pliocene to Recent rocks. Sediments within the phonolites contain silicified wood and diatoms, but are otherwise poorly fossiliferous. The phonolites can be broadly correlated with the Plateau phonolites (Group II of King and Chapman 1972) of the western shoulder of the Rift Valley, which are dated to between 10.7 my and 13.5 my (Baker *et al.* 1971). These rocks do not appear in the area of the map (Fig. 20:1) but are thought to underlie it. Phonolite cobbles comprise an important fraction of all conglomerates within the Chesowanja area, and were favoured as a raw material for artifact manufacture.

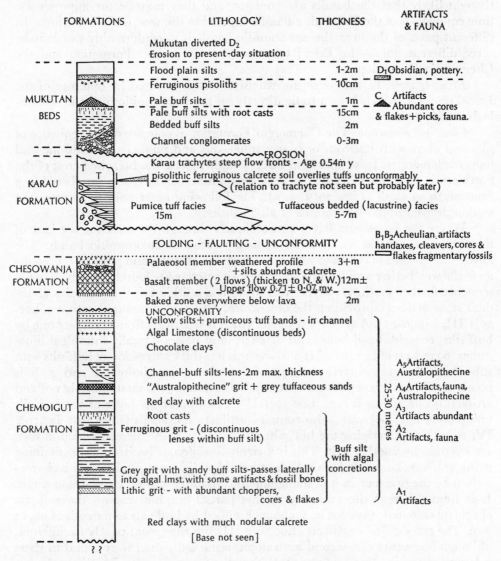

FIG. 20:2. Detailed stratigraphical column of the Chesowanja area showing the location of artifacts, fossil hominids and fossil mammal assemblages

Phonolite and ankaramitic basalt. The oldest locally exposed rocks are either phonolites or a deeply weathered ankaramitic basalt. The phonolite is not named here, but was presumed equivalent to the Lake Hannington Phonolites (McCall 1967) by Carney *et al.* (1971). The Lake Hannington Phonolites have recently been shown to contain rock units of various ages (Griffiths 1977), and the correlation of the phonolites in this area with any of these should be regarded as only provisional. These phonolites crop out in grid square 8873, and the ankaramatic basalt is found exposed extensively to the north (GR 8770, GR 8975). The relationship between the phonolite and the basalt was nowhere observed, but it is thought likely that the basalts are younger, and they may be the approximate time equivalents of the Kaparaina Basalts exposed to the west of Lake Baringo. In different parts of the area the ankaramitic basalt is unconformably overlain by three different units—the Erinei trachyte, the Chemoigut Formation and the Chesowanja Formation.

Erinei trachyte. In the south an extensive area is covered by outcrops of the Erinei trachyte, which is seen to lie directly on the ankaramitic basalt in streambeds at GR 8770.

Chemoigut Formation. The Chemoigut Formation is composed of a sequence of silts and clays with horizons of coarse tuffaceous and pumiceous sandstone and fine conglomerates. It is exposed in three windows along the breached crest of the north-south trending Chesowanja anticline. These windows are numbered 1–3 from north to south (Figs. 20:3, 4, 5). The following notes refer to a composite section derived from observations at all localities.

Where the Losokweta River cuts obliquely through the anticline the base of the formation can be seen resting upon a weathered ankaramitic basalt. The lowest unit consists of red clays with much nodular calcrete, and possibly represents an *in situ* weathering profile. Similar clays with calcrete are found at the base of the sequence further south. Overlying this stratum is a lithic grit that contains abundant stone artifacts, providing the oldest local evidence of hominid activity (Loc. 10/1 III, window 1). A second grey grit, which is separated from the lower one by buff silts, contains fossil bones and artifacts, and passes laterally into algal limestones (60 m west of Loc 10/1 III). Above this is a thick succession of buff silts with sub-spherical algal concretions, exposed extensively in windows 1 and 2. It is poorly fossiliferous except where lenses of ferruginous grits containing bones and artifacts occur within it (eg.. Loc 10/1 II). The uppermost layer of the buff silt unit, containing root casts, also contains artifacts and fossils (Locs 10/1 Ia, 10/2 IV, 10/2 IVa). Overlying the buff silts are red clays with calcrete, which in turn are overlain by another grit. This is a richly fossiliferous horizon also containing many artifacts (Locs 10/2 I, 10/2 II, 'artifact ridge' (Fig. 20:4)). The grit is characterised by the presence in it of manganiferous pisoliths and grey tuffaceous sands. It is from this that the australopithecine partial cranium came. Above it are chocolate coloured clays cut by a channel infilled by buff silts to a depth of about 2 m. The clays contain artifacts (Loc 10/2 III) but they seem poorly fossiliferous, although fragments of a second australopithecine individual were found in them ('Australopithecine Gulley' fig. 20:4). Near the top of the clays is a discontinuous horizon of algal limestone overlain by yellow silty and pumiceous tuffs in channels.

These represent the uppermost sediments of the Chemoigut Formation. The lower basalt flow of the Chesowanja Formation, which rests unconformably on the Chemoigut Formation, has baked the underlying sediments to a depth of about 2 m.

In the north it is possible to estimate a thickness of about 50 m for the Chemoigut Formation, although the exposure is not good. The estimate may be slightly misleading, as it is measured close to an anticlinal/synclinal flexure, and there may be some distortion of strata. Only 25 m to 30 m are exposed further south, but here the upper surface is eroded.

Unfortunately the relationship between the Chemoigut Formation and the Erinei trachyte is unknown. One possibility is that the sediments predate the

TABLE I: *Lithological sequence in the Chesowanja area*

Alluvium A variety of sediments from numerous sources, some local, others more distant. Maximum thickness observed 2 metres.

Mukutan Beds Type area GR 8972 and 9072. Outcrops in the bed of the Losokweta River and a 'dry' tributary which formerly acted as the main channel of the Mukutan River. Observed thickness up to 6 metres. Lower division with sheets and lenses of conglomerate overlain by bedded red clays and silts with rootlet casts; upper division with pale buff silts containing numerous artifacts; capped by a 1–2 m silt horizon, with pisolithic horizon yielding obsidian artifacts, preserved in remnant mesas along the former course of the Mukutan River.

Unnamed beds Thin pisolithic, ferruginous calcretes developed on bedded tuffs of the Karau Formation and yielding artifacts with prepared striking platforms.

Karau Formation Type area Karau volcano, south of the map area of Fig. 20:1, described by Carney (1972). Three sub-units: pumice tuffs, welded tuffs and trachyte lavas. The order of formation is most probably that given above, although there may be interdigitating relationships in places.

Chesowanja Formation Type area immediately south of Chesowanja Hill and type section in a small gorge at GR 885723. Two basalt flows both have palaeosols developed on their upper surfaces. The upper flow (6 metres thick) is an olivine basalt with dictytaxitic texture. It has vesicular pipes and lenses, and its fine structure consists of randomly orientated plagioclase laths; microphenocrysts of olivine pseudomorphed by red/brown iddingsite; ore mineral skeletal, may be ilmenite; secondary or deuteric calcite plates. The lower flow is also an olivine basalt (5 metres thick) with a trachytic fissility in outcrop and frequently weathered. It is a plagioclase-rich taxitic basalt with iddingsitised olivine phenocrysts; magnetite/ilmenite ore minerals in clinopyroxene, glassy mesostasis; much acicular ilmenite in groundmass which may be a quench texture.

Chemoigut Formation The type area is the breached anticline (GR 8872) where 25–30 metres of sediments are exposed. Elsewhere up to 50 metres are known. Type sections measured in the northern (lower portion) and central of the three windows (upper portion). Section given in Bishop *et al.* 1975 and in Fig. 20:2.

Erinei trachyte Type area at Erinei Hill described by Carney 1972. Relationship to Chesowanja basalt and Chemoigut Formation not known.

Ankaramitic basalt (unnamed) Outcrops in Losokweta river bed (GR 8975) where it is seen to underlie the Chemoigut Formation. Larger outcrops occur south-west of Chesowanja Hill.

? Hannington phonolite This group of rocks has recently (Griffiths in prep.) been shown to contain rocks of different ages. The outcrop in the map area is only provisionally correlated with the Hannington phonolite, and it may not be possible to correlate this isolated outcrop with any unit.

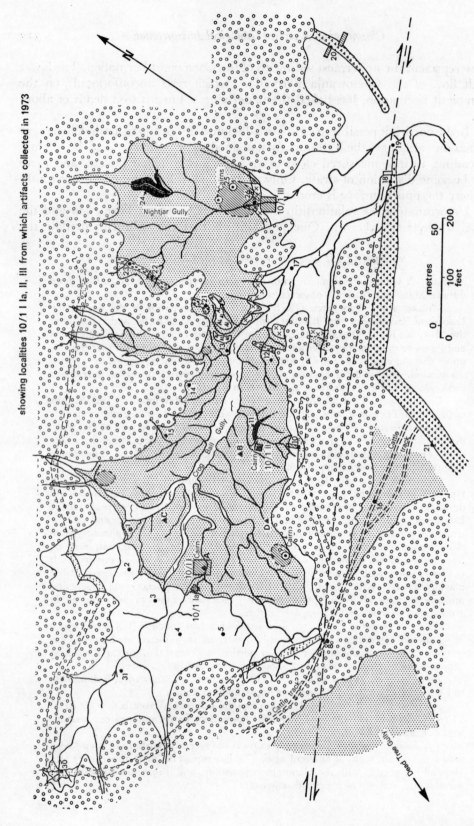

showing localities 10/1 I Ia, II, III from which artifacts collected in 1973

Fig. 20:3. Geological map of Chesowanja, Window I

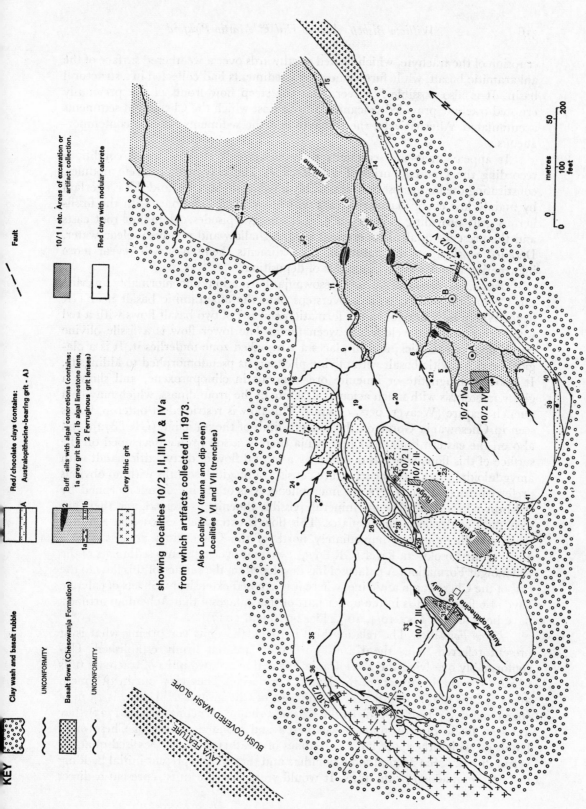

KEY

▨ (wavy)	Clay wash and basalt rubble
	UNCONFORMITY
▨ (dotted)	Basalt flows (Chesowanja Formation)
	UNCONFORMITY

showing localities 10/2 I, II, III, IV & IVa
from which artifacts collected in 1973.

Also Locality V (fauna and dip seen)
Localities VI and VII (trenches)

▨	Red/chocolate clays (contains: Australopithecine-bearing grit - A)
A	
▨ 1a — — 1b	Buff silts with algal concretions (contains: 1a grey grit band, 1b algal limestone lens, 2 ferruginous grit lenses)
▨ 2	
×××× ××××	Grey lithic grit

– – –	Fault
10/1 1 etc.	Areas of excavation or artifact collection.
▨	Red clays with nodular calcrete
□ ▪	

Fig. 20:4. Geological map of Chesowanja, Window 2

eruption of the trachyte, which flowed northwards over a weathered surface of the ankaramitic basalt, while further north the sediments had collected in a structural basin. It is also possible however, that the steep flow fronts of the previously erupted trachyte provided volcanic dams against which the Chemoigut sediments accumulated. Alternatively, the trachyte and the sediments may be contemporaneous.

It appears that the sediments represent largely shallow water conditions recording the gradual burial of a landscape. The clay with calcrete nodules, constituting the lowermost unit, is possibly a palaeosol horizon, and this is overlain by mainly low energy lacustrine silts with some algal limestone, with the fossiliferous grits providing evidence of higher energy episodes. Abundant root casts and the presence of algal concretions suggest shallow and relatively clear water. Detailed work upon the nature of the sediments themselves will reveal more information about the environments of deposition.

Chesowanja Formation. The Chesowanja Formation unconformably overlies the Chemoigut sediments, and oversteps onto the ankaramitic basalt near GR 8770. Where fully developed the formation consists of two basalt flows with a red weathering profile developed between them. The lower flow is a fissile olivine basalt up to 10 m thick, and in places a 2 m baked zone underlies it. It is a plagioclase rich taxitic basalt with olivine phenocrysts pseudomorphed to iddingsite. It also has magnetite or ilmenite ore minerals in clinopyroxene, and shows a glassy mesostasis with much acicular ilmenite in the groundmass, which may be a quench texture (Weaver, pers. comm.). This flow is restricted in outcrop, but is seen in Chesowanja Gorge on the western limb of the anticline (GR 8872), and also on the eastern limb. At Chesowanja there is a soil profile developed on the surface of this flow, which is overlain by a second flow of dictytaxitic basalt with amygdaloidal pipes. Its fine structure consists of plagioclase laths with no obvious preferred orientation. There are microphenocrysts of olivine pseudomorphed by red-brown iddingsite. The ore mineral, possibly ilmenite, is skeletal, and there are secondary or deuteric calcite plates. This flow is extensively exposed in a north-south trending outcrop immediately north of the Chesowanja area, and also north of the Losokweta River (GR 8975), where it oversteps onto sediments of the Chemoigut Formation. A palaeosol has developed on the surface of this flow to the east of the Chesowanja anticline, where it is exposed extensively. Sheets of calcrete have also formed. It is in the upper part of this palaeosol that Acheulian artifacts have been found (Locs 10/4, 10/5, Figs 20:6 and 20:7).

Karau Formation. The relationships between the units comprising what is at present referred to as the Karau Formation are not firmly established. Unconformably overlying the Chesowanja Formation are two units of trachytic tuffs which form the basal part of the Karau Formation. They crop out in two main areas as tuffs of differing facies. To the west of Chesowanja Hill there is 15 m of coarse massively bedded tuff. In the axis of a shallow syncline (GR 881727) they are seen to lie directly upon a weathered and deflated surface of Chesowanja basalt on which there are large corestones of weathered lava. Localised quaquaversal dips are seen around some boulders and they may represent initial bedding or later compaction of the tuffs. It would seem that the tuffs represent a direct

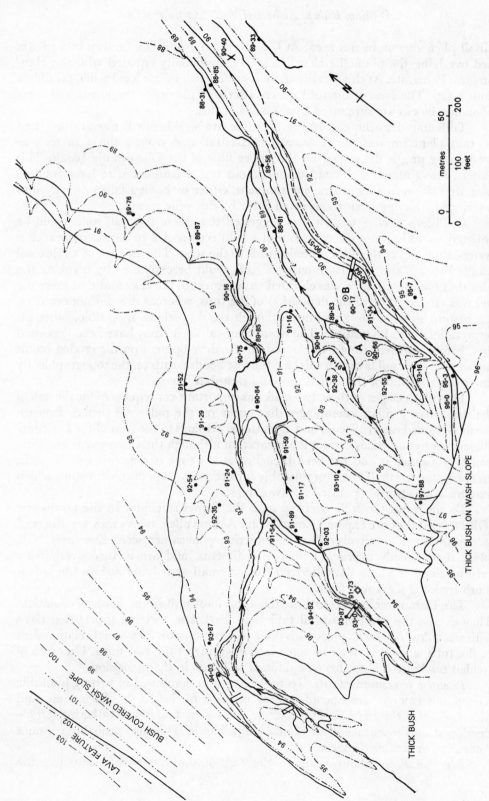

FIG. 20:5. Chesowanja, Window 2, showing height data and gully pattern

airfall phenomenon in this area. At GR 881733 the tuffs can be seen in a stream bed overlying the phonolite that has been provisionally equated with the Hannington Phonolite. At this locality there is evidence of minor local relief, possibly a fault scarp. The base of the tuffs covers a scree of phonolite fragments and some phonolite blocks are supported by a matrix of tuff.

Tuffs found to the east of the anticline are well-bedded, fine-grained, and contain plant impressions. They are horizontal, and overlie the 3 m to 4 m weathering profile developed on the upper flow of the Chesowanja basalt. This eastern fine-grained, well-bedded, micaceous tuff is considered to have resulted from deposition in a lacustrine environment, either by having fallen directly into a lake or by having been transported there by fluviatile means.

In Bishop *et al.* (1975) it was suggested that these two tuff units could be equated, in which case the eastern unit would represent a penecontemporaneous water-lain facies of the airfall pumice tuffs to the west. The absence of a thick soil profile beneath the western outcrop of tuffs could be explained by invoking the idea that the western area was elevated with respect to the east, and that there the soil was eroded prior to the deposition of the tuffs, whereas this did not occur on the eastern limb of the anticline. Although this hypothesis is feasible, other interpretations are also possible. The western air-fall facies may have been deposited first, before sufficient time had elapsed for the development on the eroded basalt surface of a deep soil layer. In this case the fine bedded-tuffs on the topographically lower eastern limb of the anticline would post-date them.

The relative ages of these two tuffs has important consequences for the age of the Acheulian artifact assemblages that occur on the palaeosol surface beneath the fine-grained eastern facies, and as scattered artifacts to the west of the anticline. Whichever hypothesis is correct it appears that although the Chesowanja anticline had been formed it had not been breached at the time of deposition of the tuffs, except in the north. Thus it presumably formed a low topographic feature which may have effectively separated the two areas.

To the south the tuffs overstep onto the Erinei trachyte. In the north-west (GR 881738) they are exposed along the dry Agulu valley, where they are observed to overlie a tuffaceous clay, but no underlying lavas are seen. Elsewhere (GR 8874) they directly overlie Chemoigut sediments, and are in turn overlain by very localised trachyte flows derived from a small vent lying astride the north-south trending Tarkokot fault.

The Karau trachytes constitute the uppermost unit in the Karau Formation. They overlie the western air-fall tuffs near Tarkokot, forming superposed flows with steep flow fronts. In places it is difficult to determine their exact relationships to the tuffs, and there may be some interfingering of the two units. Outcrops of welded trachytic tuff are likewise difficult to locate in the succession.

Pisolithic ferruginous calcrete. To the east a unit composed of rubbly pisolithic ferruginous calcrete lies unconformably on tuffs of the Karau Formation, and oversteps onto the very similar calcrete that is part of the weathering profile developed on the surface of the Chesowanja basalt. This also probably contains Acheulian or similar artifacts.

Mukutan Beds. About 1 km to the east of the Chesowanja anticline the

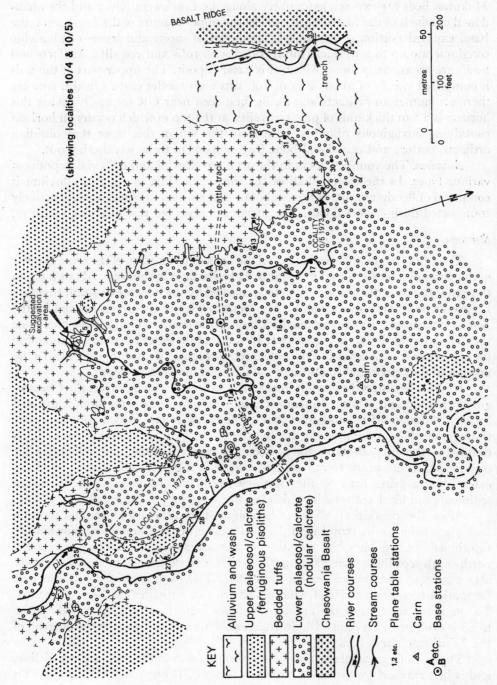

FIG. 20:6. Geological map of the Acheulian artifact locality, Chesowanja

KEY

	Alluvium and wash
	Upper palaeosol/calcrete (ferruginous pisoliths)
	Bedded tuffs
	Lower palaeosol/calcrete (nodular calcrete)
	Chesowanja Basalt
	River courses
	Stream courses
1,2 etc.	Plane table stations
△	Cairn
⊙ A etc. B	Base stations

(showing localities 10/4 & 10/5)

BASALT RIDGE

metres 0   100   200
feet   0   50

LOCALITY 10/5 1973

LOCALITY 10/4 1973

Suggested excavation area

cattle-track

Cairn

Mukutan Beds are exposed extensively along the Losokweta River and the abandoned channels of the Mukutan River. In the upper reaches of the Losokweta, the basal exposed portion of this unit is composed of sheets and lenses of phonolite conglomerate up to 3 m thick, interbedded with tuffs and red silts. Artifacts and fossils are occasionally found within this basal deposit. The upper part of the unit is comprised chiefly of about 2 m of buff silts with rootlet casts. On this stratum there are numerous artifacts and fossils, best seen near GR 9072. Overlying this horizon is a 1 m thick unit of pale buffs silts, at the top of which occurs an horizon containing ferruginous pisoliths. It is from the top of this layer that obsidian artifacts, pottery and occasional bone fragments seem to be weathering out.

Alluvium. The youngest rocks in the area are alluvial and colluvial deposits of various types. In the eastern half of the map area (Fig. 20:1) this alluvium is composed of floodplain silts of the Mukutan River which in places are still actively transported during sheet floods. The maximum observed thickness is about 2 m.

Structure

The Chesowanja area is situated on the back-tilted surface of the most easterly of the series of fault steps that descend to Lake Baringo in the west. About 8 km to the east is the shoulder of the Rift Valley, marked by the long north-south trending Engelesha and Aruru escarpments. The western edge of the step is formed by the Karau escarpment. Its present height is largely the result of fault movements since the deposition of the Karau trachytes. In the south a series of *en echelon* phonolite tilt blocks rise from a cover of Pliocene to recent rocks. Their plunge is towards the north and they presumably underlie the Chesowanja area at depth.

It is this back-tilted step that formed the sediment trap in which the Chemoigut Formation and other sediments accumulated. At present the area is drained by the Mukutan River and its tributary the Losokweta, which join and pass through the Karau escarpment via the deeply incised Losokweta Gorge. Being drained in this way, the area today is only a rather shallow, although extensive trap for sediments, at the termination of a westerly sloping wash plain. There is an extensive low lying area to the north-east of Chesowanja covered by recent sediments and black cotton soils which become waterlogged during periods of rain.

After the deposition of the Chemoigut Formation, folding and erosion took place prior to the eruption of the basalt flows of the Chesowanja Formation. Continued folding after the eruption of the Chesowanja basalts, produced a north-south trending assymetrical anticline. The western limb of the fold dips at about 11° and the eastern one at about 60°. The exposures of the Chemoigut Formation occur as a series of windows where this anticline has been breached. To the south the axis of the anticline plunges beneath alluvium, but the same trend is exhibited by a fault with downthrow to the east, which cuts the outcrop of the Erinei trachyte.

The axis of a parallel syncline passes about 300 m to the east of the anticline, and is best exposed west of the Losokweta River (GR 8973 and GR 8974). This syncline is lost southwards but the structure may continue into a west facing fault scarp developed in the Erinei trachyte. In effect there seems to a graben in the south which passes northwards into a monoclinal structure. Possibly the presence of

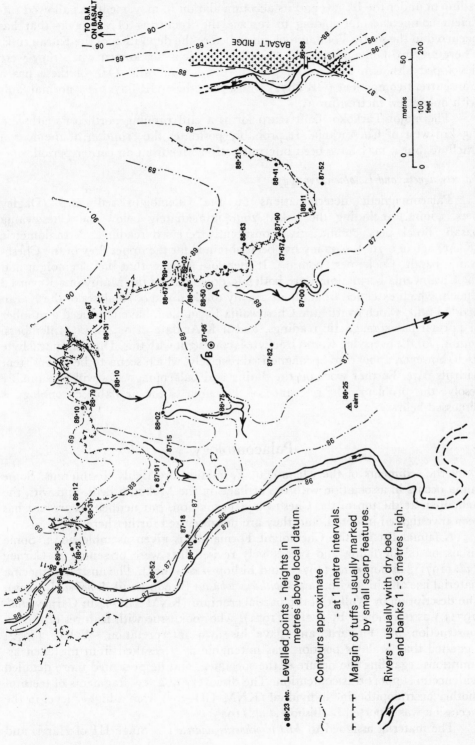

FIG. 20:7. Chesowanja Acheulian locality showing height data, river valleys and gully pattern

sediment under the basalts and its accommodation to movement has allowed sufficient competence for folding to replace the fracturing of the lavas that has occurred in the south. Folding took place before the deposition of the Karau tuffs. There are two faults cutting the anticline, but it is not known when they were developed although both could predate the Karau tuffs. One of these has a transcurrent component (GR 8872) whilst the other (GR 8874) is a normal fault with only a few metres throw.

The major Tarkokot fault scarp forms a cliff running northeast-southwest, 1.5 km west of Chesowanja. It probably postdates the eruption of the Karau trachyte, but it may have been intermittently active from an earlier period.

Palaeomagnetic and Isotopic chronology

Palaeomagnetic determinations on the Chemoigut sediments (Dagley pers. comm.) including the baked zone immediately below the Chesowanja basalt (Brock pers. comm.) give consistently reversed readings. A radiometric (K/Ar) age of 0.71 ± 0.07 my has been obtained for the upper flow of the Chesowanja basalt (Cooke pers. comm.). It therefore appears that the Chemoigut and the Chesowanja Formations were both laid down during the Matuyama Reversed Epoch, which extended from about 2.4 my to 0.7 my ago. Analyses of the Karau bedded tuffs, which overlie the Chesowanja Formation, have produced a number of normal palaeomagnetic readings, and a K/Ar date of 0.54 my (Miller pers. comm.) for the overlying Karau trachyte is consistent with this. The Karau trachyte itself, however, appears to be magnetised reversely, which seems to be inconsistent with its date. Further work on the dating and palaeomagnetics will presumably resolve this problem. The relevance of the fauna to the probable chronology is discussed below.

Palaeontology

Only the sediments of the Chemoigut Formation are richly fossiliferous. Some fauna occurs in association with the artifacts in the Mukutan Beds, and with the Acheulian at the top of the Chesowanja Formation, but neither occurrence has been investigated in detail, and they are not discussed further here.

A faunal list for the Chemoigut Formation is given as table 20:2. Some discussion of the fauna and successively revised lists were presented in Carney *et al.* (1971), Bishop *et al.* (1971), and Bishop *et al.* (1975). The australopithecine material has affinities with *Australopithecus robustus* (Broom) and *A. boisei* (Leakey). The description by Walker of the partial cranium (KNM–CH–1) in Carney *et al.* (1971) was questioned by Szalay (1971), who took issue with features of the reconstruction put forward, and gave his own interpretation. Walker (1972) suggested that Szalay's position was untenable as it resulted from mistaken assumptions regarding the nature of the specimen, and he presented more detailed evidence to support this contention. The discovery of a few fragments of teeth of another australopithecine individual (KNM–CH–302) from a higher level in the succession was reported by Bishop *et al.* (1975).

The material assigned to *Metridiochoerus andrewsi* is Stage III of Harris and

White (in *Ms*) and *M. hopwoodi* is a large form. This situation occurs in the Shungura Formation of the Omo area between members G and J, in the middle portion of Middle Bed II at Olduvai, and just below the Koobi Fora Tuff in area 103, East Lake Turkana. The latest known occurrences of *Deinotherium bozasi* is in the Shungura Formation below tuff J (Coppens and Howell 1974), and at Olduvai in the middle portion of Middle Bed II (Harris 1976).

The *Kobus sp.* is as large as the largest *K. sigmoidalis* from the Omo and Olduvai deposits, and the extant *K. ellipsiprymnus*. The middle size group of the Alcelaphini is about the size of *Damaliscus niro* or *Parmularius angusticornis* (Gentry pers. comm.). *Megalotragus kattwinkeli* has not been found in sediments below member G of the Shungura Formation, and nor has *Equus sp.*

This faunal evidence suggests that the Chemoigut Formation is the equivalent of that part of the Shungura Formation between tuffs G and J, and taking the Omo radiometric dates to be correct, was therefore laid down between 1.93 my and 1.34 my ago. This suggestion is in agreement with the Olduvai chronology.

Environmentally, modern representatives of some of the genera present in the assemblage are associated with a bushed grassland habitat, with riverine and lacustrine elements. It seems possible that such a local environment existed at the time the Chemoigut Formation was being deposited. However, the same can be said of most Plio-Pleistocene fossil vertebrate sites in eastern Africa, and more palaeoecological work will be necessary to refine this very general view.

Archaeology

Archaeological material comes from at least eight horizons within the succession. Harris and Bishop (1976) refer these to three main groups, consisting of:

  (i) from the Chemoigut Formation; an industry having affinities with the Developed Oldowan.
  (ii) from the top of the Chesowanja Formation; an Acheulian industry.
  (iii) from the Mukutan Beds; a later stone industry, and obsidian tools and pottery.

Further archaeological work is to be undertaken by J. W. K. Harris and J. Gowlett. and hence only a few tentative comments will be made here. Investigations are also in progress on the nature of the varied materials from which the tools are made. This petrological work may supply interesting information concerning the minimum distances from which various raw materials must have come.

  (*i*) *Industry from the Chemoigut Formation.* Artifacts occur abundantly within the Chemoigut Formation, weathering out or observed *in situ* at five levels, including the two from which have come remains of *Australopithecus* (A_1–A_5 Figs. 20:2–5). J. W. K. Harris has provided an analysis of a surface collection (*in* Bishop *et al.* 1975; Harris and Bishop 1976). It comprised 220 specimens from four main localities, and they seem to belong to a single cultural entity, the Oldowan/ Developed Oldowan Cultural Complex. Following the nomenclature of M. D. Leakey (1971) the collection can be allocated to the categories shown in Table 20:3. Amongst the choppers are side, end, and two-edged forms, some of which show signs of chipped edges that may indicate utilisation. A feature of the assemblage that is atypical of Oldowan/Developed Oldowan occurrences known elsewhere, is the presence of two large symmetrical discoid/core specimens. They have bifacially worked edges which extend entirely around their circumference, and are biconvex in cross section. There are also polyhedrons, discoids, protobifaces, and light-duty scrapers made on flakes. No bifaces have been found. The maximum dimension of debitage flakes has a mean of 48 mm (22 mm to 86 mm). They have simple platforms, 44 per cent have cortex on the dorsal surface, and they are usually irregular. Seventy-six per cent are end struck rather than side-struck.

The age of this industry is probably somewhere between 1.93 my and 1.34 my.

(*ii*) *Industry from the Chesowanja Formation.* An industry attributable to the Acheulian is found in the upper parts of the weathering profile developed on the upper basalt member of the Chesowanja Formation (B Figs. 20:6, 7, Fig. 20:2). Artifacts occur at the foot of a pronounced basalt feature caused by a sharp flexure forming the eastern limb of the anticline. This surface is overlain by well-bedded tuffaceous strata placed in the Karau Formation. These display characteristics typical of tuffs deposited in standing water, and it is possible that such a lake may have existed earlier, to the east of the Acheulian sites.

J. Gowlett (pers. comm.) has analysed a surface collection of hand axes which display a great range of size and shape. Some show a high degree of control in manufacture, and are finished with finely resolved flaking. Some are symmetrical and flaked to a sharp edge around the whole outline, having the place of maximum thickness nearer to the point than the place of maximum breadth. Two well finished cleavers have the edge length slightly shorter than the maximum breadth, and at right angles to the long axis. There are also bifacially flaked choppers retaining areas of pebble cortex, and scrapers. In general character, the assemblage is one that invites comparison with the Late Acheulian collection from the Kapthurin sites to the west of Lake Baringo (M. G. Leakey 1969). A date of 0.23 my has been suggested for these sites. To this extent the suggested date of earlier than 0.54 my for the Chesowanja material based on a date from the Karau trachyte is surprising, though not inconceivable. The relationship of the Karau trachyte to the Karau tuffs in the west is not certain and they may interdigitate. Similarly, doubt exists as to whether the tuffs on the west of the anticline are contemporaneous with those on the east, which overlie the Acheulian sites. In fact the Acheulian at Chesowanja could be later than has been supposed, and it would be wise to reserve judgement on this matter until more stratigraphic information about the Karau Formation is available.

Another surface collection of Acheulian artifacts may be weathering out from the later 'Pisolithic ferruginous calcrete'. This assemblage also includes 'Levallois' cores.

(*iii*) *Industry from the Mukutan Beds.* A later stone industry (C Fig. 20:2) is eroding from the pale buff silts of the Mukutan Beds, which probably represent a former flood plain of the Mukutan. Some artifacts occur in very large mounds, which probably constitute factory sites. They consist of heavy duty core tools, including double ended picks, and associated flakes, made from water rounded

TABLE 2: *Fauna from the Chemoigut Formation*

Mollusca	
Gastropoda	*Bellamya sp*
Bivalvia	
Pisces	
Siluriformes	
Reptilia	
Chelonia	
Trionychidae	
Pelomedusidae	
Crocodilia	*Crocodylus sp*
Mammalia	
Primates	
Cercopithecidae	*Theropithecus sp* sp. indet.
Hominidae	*Australopithecus sp*
Proboscidea	
Elephantidae	*Elephas cf recki* Dietrich
Deinotheriidae	*Deinotherium bozasi* Arambourg
Perissodactyla	
Equidae	*Equus sp*
Rhinocerotidae	*Ceratotherium sp*
Artiodactyla	
Suidae	*Mesochoerus limnetes* (Hopwood)
	Metridiochoerus andrewsi (Hopwood)
	M. hopwoodi (Leakey)
Hippopotamidae	*Hippopotamus cf amphibius* Linn
	Hippopotamus sp.
Giraffidae	
Bovidae	
Tragelaphini	*Tragelaphus sp*
?Tragelaphini	
Bovini	
Reduncini	*Kobus sp*
Alcelaphini	*?Megalotragus kattwinkeli* Schwarz
	?Connochaetes sp
	small species
	smaller species
Antilopini	*?Antidorcas sp*

cobbles of phonolite, and are in a very fresh condition. Gowlett notes (pers. comm.) that there is a prominent flake-blade element in the surface sample collected.

Also in the Mukutan Beds is an assemblage of obsidian flakes and pottery (D Fig. 20:2). They appear to be eroding out from an horizon slightly higher in the sequence then the artifact mounds. There are no good estimates of the age of the Mukutan Beds and the tools found there need not be very old.

TABLE 3: *Surface Artifacts fram the Chemoigut Formation*

Tools	
Choppers	17
Polyhedrons	8
Discoids	4
Proto-bifaces	3
Scrapers	6
Sundry tools	5
TOTAL TOOLS	43
Debitage	
Whole flakes	117
Snapped flakes	20
Split flakes	2
Broken flakes	2
Angular fragments	21
Core and cobble fragments	13
Battered cobbles	1
Broken choppers	1
TOTAL WASTE	177
OVERALL TOTAL	220

(after Harris and Bishop 1976)

Acknowledgements

We thank the Government of Kenya for research permission, and Mr R. E. Leakey and the Trustees of the National Museums of Kenya for making research facilities available. The work was part of a programme of research financed by the Natural Environment Research Council, the Wenner-Gren Foundation for Anthropological Research, the L.S.B. Leakey Foundation, and the Boise Fund of the University of Oxford. Assistance with field work and valuable discussion was provided by Dr S. Weaver and R. Foley. Palaeomagnetic analyses have been carried out by Professor A. Brock and Dr. P. Dagley, and radiometric determinations by Dr P. Reynolds and Dr J. A. Miller. Dr J. M. Harris and T. White gave advice concerning the pigs, and we are also grateful to Dr. Harris for his comments on the manuscript. We thank J. W. K. Harris, J. Gowlett and Dr Mary Leakey for their comments on the archaeological material.

References

BAKER, B. H., WILLIAMS, L. A. J., MILLER, J. A. & FITCH F. J. (1971) 'Sequence and geochronology of the Kenya Rift volcanics' *Tectonophysics.* **11** 191–215.

BISHOP, W. W., CHAPMAN, G. R., HILL, A. & MILLER, J. A. (1971) 'Succession of Cainozoic vertebrate assemblages from the northern Kenya Rift Valley.' *Nature* **233** 389–394.

—— PICKFORD, M. & HILL, A. (1975) 'New evidence regarding the Quaternary geology, archaeology, and hominids of Chesowanja, Kenya.' *Nature* **258** 204–208

CARNEY J. (1972) The Geology of the area to the East of Lake Baringo, Rift Valley Province, Kenya Ph.D. thesis University of London.

—— HILL, A., MILLER, J. A., & WALKER, A. (1971) 'Late Australopithecine from Baringo District, Kenya.' *Nature* **230** 509–514.

COPPENS, Y., & HOWELL, F. C. (1974) 'Les faunes de mammifères fossiles des formations Plio-Pleistocenes de l'Omo en Ethiopie (Proboscidea, Perrissodactyla, Artiodactyla)' *C. r. hebd. Séanc. Acad. Sci. Paris* **278**D 2275–2280

GRIFFITHS, P. (1977) The Geology of the area around Lake Hannington and the Perkerra River, Rift Valley Province, Kenya Ph.D. thesis University of London.

HARRIS, J. M. (1976) 'Cranial and dental remains of *Deinotherium bozasi* (Mammalia: Proboscidea) from East Rudolf, Kenya.' *J. Zool.*, Lond. **178** 57–75.

—— & WHITE, T. D. (in MS) 'Evolution of the Plio-Pleistocene African Suidae'.

HARRIS, J. W. K. & BISHOP, W. W. (1976) 'Sites and assemblages from the early Pleistocene beds of Karari and Chesowanja.' *in* Clark, J. D. and Isaac, G. 'The earlier industries of Africa' abstracts of *Colloque V, IX Congrès de UISPP*, Nice.

KING, B. C. & CHAPMAN, G. R. (1972) 'Volcanism of the Kenya Rift Valley' *Phil. Trans. R. Soc. Lond. A.* **271** 185–208.

LEAKEY, M. G. (1969) 'An Acheulean Industry with prepared core technique and the discovery of a contemporary hominid at Lake Baringo, Kenya.' *Proc. Prehist. Soc.* **35**, 48–76.

LEAKEY, M. G. (1971) '*Olduvai Gorge Vol. III*' Cambridge University Press 306 pp.

McCALL, G. J. H. (1967) Geology of the Nakuru—Thomson's Falls—Lake Hannington Area *Rep. Geol. Surv. Kenya* **78**, 1–22

SZALAY, F. S. (1971) 'Biological level of organisation of the Chesowanja robust australopithecine.' *Nature* **234**, 229–230

WALKER, A. (1972) ''Chesowanja australopithecine.' *Nature* **238**, 108–9.

21 (A)

WALTER W. BISHOP

Geological framework of the Kilombe Acheulian archaeological site, Kenya

The geological setting of the Kilombe Acheulian site is described. A single horizon yields a prolific artifact assemblage from an erosion gully floor 200 m N–S and E–W. The sequence is underlain by deeply weathered phonolite lava upon which a red-brown clay wash has accumulated. The artifact horizon lies on the almost planar clay surface and is sealed beneath pale weathered pumice tuff exhibiting weak palaeosols. The upper part of the sequence commences with a three banded tuff, that appears to have been deposited in still water, succeeded by reworked tuffs cut by several deep channels of fluviatile origin and overlain by a coarse ash flow tuff. The underlying phonolite is dated at 1.7 m.y. by K–Ar and the palaeomagnetic polarity of the three banded tuff is reversed suggesting an age in excess of 0.7 m.y. The Acheulian assemblage lies between these two limiting dates.

Introduction

An erosion gully complex yielding a rich assemblage of Acheulian handaxes and cleavers was discovered in 1972 by Dr W. B. Jones in the course of geological mapping for the East African Geological Research Unit, based at Bedford College, London University. Following two preliminary visits, one in company with Dr Mary Leakey, I made a preliminary geological map of the locality in February 1973 with the assistance of Peter and Diana Tallon. Detailed investigation of the geology and archaeology commenced in August 1973 and continued during 1974. The archaeological excavations were directed by John A. J. Gowlett and are described elsewhere in this volume (Paper 21(B)).

The main Kilombe artifact site is at 2000 metres above sea level on the western flank of the rift valley slightly south of the Equator (35°53'E, 0°06'S) (see Gowlett 1976, Fig. 21b:1). The site lies 4 kilometres south-east of the caldera of the extinct Kilombe volcano. Much of the surrounding farmland is well vegetated with secondary acacia and euphorbia bush. Two periods of erosion have exposed steep cliffs. An early cycle of pedimentation produced extensive flats which now survive on the spurs between lines of easterly flowing drainage. At the back of the pediment notch cliffs 20 metres high are preserved beneath a resistant cap rock.

Below the pediments erosion systems which are still active have produced a series of badland amphitheatres with steep gully heads up to 10 metres high. The gully system which contains the Acheulian artifact horizon has exhumed a weathered lava substrate from beneath an extensive cover of tuffs. The main gully has been adapted for watering stock by building an earth dam to produce a pond. This has resulted in impedence of the wash on the floor of several active gullies and sedimentation is beginning to veneer the most easterly spreads of derived artifacts. Several resistant strata have helped to protect the softer beds which contain the *in situ* artifacts. These survive in two lines of outliers to the east of the main gully head (Fig. 21a:1).

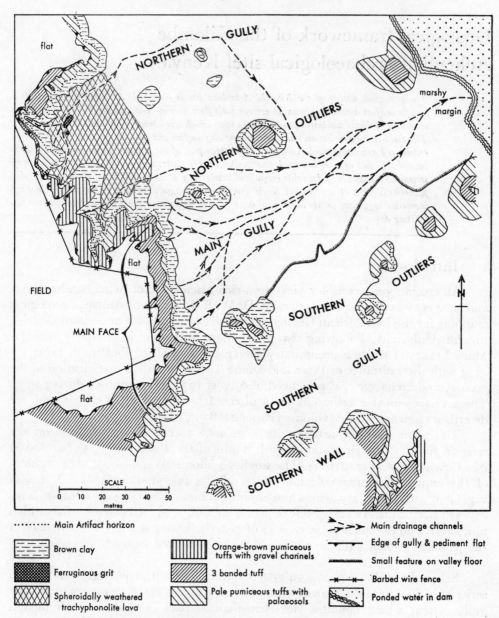

Legend:

.... Main Artifact horizon

Brown clay

Ferruginous grit

Spheroidally weathered trachyphonolite lava

Orange-brown pumiceous tuffs with gravel channels

3 banded tuff

Pale pumiceous tuffs with palaeosols

Main drainage channels

Edge of gully & pediment flat

Small feature on valley floor

Barbed wire fence

Ponded water in dam

FIG. 21a:1. Geological map of the Kilombe artifact site.

Local lithological sequence

The general geology of the country surrounding Kilombe Mountain has been
mapped by Jones (1975). Field work in 1973–4 enabled local stratigraphic details
to be established (Fig. 21a:1) as follows:

9.	Ferruginous pisolithic gravel.	30 cm to 3 m
8.	Massive ash flow tuff.	7 m
7.	Orange-brown pumiceous tuffs with gravel channels.	20 m
6.	'Three-banded tuff'—gritty, brown lapilli tuff.	2 m
5.	Pale pumiceous tuffs with palaeosols.	1.5 to 2 m
4.	Main Acheulian artifact horizon.	< 20 cm
3.	Brown clays—largely derived by wash from weathering of trachyphonolite lava (includes a yellow tuff lens with grass imprints).	Seen to 2.5 m
2.	Local patches of ferruginous grit.	up to 40 cm
1.	Weathered trachyphonolite. (Trachyte flows from Kilombe volcano are broadly contemporary with the trachyphonolite and outcrop to the north west of the artifact gully).	Seen to 4 m locally

The succession is best seen on the N–S trending face at the head of the main
gully (Fig. 21a:2). Trachyphonolite crops out on a low knoll at the north end of the
face but is very rotten and spheroidally weathered. Mapping round the artifact
gully together with excavations of pits within the site show that the surface of the
weathered lava is highly irregular.

The overlying brown clays, which become highly mobile when wet, outcrop
along the foot of the main face and at the base of the outliers. Excavations show
that the clays are derived largely from breakdown of the trachyphonolites. Occasional fragments of the lava occur through the clay and several broken pieces of
corroded bone and teeth of fossil mammals have been found. They include: a
shattered molar of a loxodont elephant and an elephant rib; molars of a large
bovid c.f. *Pelorovis* sp. Random yellow flecks in the clay represent deeply rotted
pumice granules. The brown clay is not an *in situ* weathering product, except in
deep hollows and clefts in the lava. This is shown by the thin ferruginous grit
which survives in patches on the lava surface (Fig. 21a:2) and the gently dipping
lens of bedded fine tuff with plant impressions that occurs within the clay.

In contrast to the irregular junction between the lava and the clays, their
upper surface is very flat when viewed on the N–S face although it descends slightly
from N to S at a gradient of 1 in 150 (Fig. 21a:2). In outcrops on the flanks of the
outliers the clay surface dips consistently to the ENE at 1 in 40 until it descends
below the gully floor 50 metres west of the margin of the dam.

The main artifact horizon occurs on the flat upper surface of the clays and
excavations showed that about 80 per cent of the artifacts occur in a thickness of
only 15 to 20 cms (Gowlett 1977). The outliers enable the extent of the surface
originally covered by the artifact spread to be computed as a minimum of 15 000
sq. metres and very much greater if allowance is made for its continuation beneath
overburden to the east, south and west.

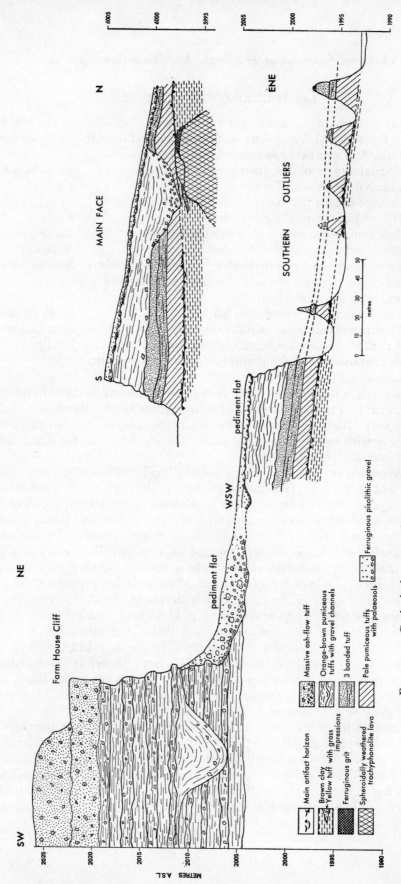

Fig. 21a:2. Geological cross sections north-south and east-west across the Kilombe artifact site.

Legend:

- Main artifact horizon
- Brown clay / Yellow tuff with grass impressions
- Ferruginous grit
- Spheroidally weathered trachyphonolite lava
- Massive ash-flow tuff
- Orange-brown pumiceous tuffs with gravel channels
- 3 banded tuff
- Pale pumiceous tuffs with palaeosols
- Ferruginous pisolithic gravel

SW · NE · N · S · WSW · ENE

Farm House Cliff · pediment flat · MAIN FACE · pediment flat · SOUTHERN OUTLIERS

METRES A.S.L.

0 10 20 30 40 50 metres

The main artifact horizon is overlain by a pale weathering pumice tuff which contrasts sharply in colour and composition with the brown clays below the horizon. The pumices are deeply weathered and two palaeosols, recognised by development of interstitial clay, record periods of weathering separating two phases of deposition. No brown clay elements or natural trachyphonolite fragments occur in the pale tuffs although a scatter of small artifacts occurs (Gowlett 1977). The sharp lithological contrast marked by the influx of pumice clasts suggests that the catchment area received a sudden input of pumice tuff. This not only sealed in the artifacts, leaving them virtually undisturbed, but also buried all other trachyphonolite outcrops and their derived clay products throughout the drainage basin.

Studies of the orientation of long axes of the artifacts reveal a random distribution. Indeed the flat clay surface, which becomes very soft when wet, shows no evidence of any consistent direction of water movement that might impart a current alignment to the tools. A local departure from this situation is that several thin (20 to 40 cm) clean sandy lenses from 1 to 2 metres in diameter occur at the clay tuff interface. They are located both above and below the artifact level but with one exception are within 10 cms of the horizon or upon it. The lenses are seen only in sections which give restricted local observation of any possible trend of the sandy 'shoe strings'. Although they cannot be reconstructed in plan they appear to represent deposition by running water in ephemeral runnels and shallow channels on the clay surface. Some re-alignment of artifacts may have occurred along such drainage lines. In the erosion embayment at the north end of the main face (Figs. 21a: 1 and 2), a nest of lava boulders occurs on the clay surface at the base of a larger channel (25 cm deep) some 40 cm below the artifact horizon. This is the exception referred to above. The area was the subject of an archaeological excavation. The boulders range up to 30 cm in diameter and are too heavy (average 0.5 kilos) to have been moved by the current flow indicated by the grade of the sands which infill the channel.

The pumice tuffs and palaeosols are overlain by characteristic three-banded brown lapilli tuffs. They take their name and typical appearance from three resistant coarse bands which make a triple ledge at the base of the deposit. The three beds exhibit graded bedding and are well laminated. The unit appears to have been deposited in still water. This is further supported by the consistency of thickness and lithology of the unit for more than 1 kilometre from the artifact site. The upper metre of this unit is finer grained and less well bedded.

The three-banded tuff is overlain by a sequence 20 metres in thickness of orange to brown pumice tuffs interbedded with finer grained, clayey weathered tuffs. The base of the unit can be seen at the northern end of the main face to descend into a channel 4 metres deep and 50 metres across which cuts through the three-banded tuff and the pale pumice tuff to intersect the main artifact horizon (Fig. 21a:2). A further channel, well exposed on the Farm House Cliff above and to the south-west of the artifact gully, is developed in the middle of the unit and is 6 metres deep and 40 metres in diameter. The upper 5 metres of the unit consist of more earthy tuffs with fewer and thinner pumiceous bands. The deposits are subaerial tuffs with structures that indicate much re-working in fluviatile channels.

The Farm House Cliff is capped by a massive resistant ash flow tuff which contains large pale pumices and blocks of black scoraceous pumice. It forms prominent cliff features wherever it occurs.

Other localities

Elements of a sequence similar to that in the artifact gully can be mapped intermittently to two other major exposures in gullies respectively 500 metres to the NW and 1200 metres to the ENE of the main site. The latter locality exposes weathered trachyphonolite overlain by:

4.	Unstratified buff pumice tuffs and earthy beds	+ 2.5 m
3.	Typical three-banded lapilli tuff	1.6 m
2.	Two beds of coarse pumice tuff with numerous root-casts suggesting subaerial deposition	90 cm
1.	Thin red-brown earth.	75 cm

The lithological similarity of this sequence and the consistency of the thickness of the units allow similar conditions of deposition to be inferred over an area at least 2 kilometres in diameter from E to W.

Depositional history

One of the trachyte flows from the Kilombe volcano has been potassium-argon dated at 1.9 ± 0.15 m.y. while a sample of the trachyphonolite yielded an age of 1.7 ± 0.05 m.y. (Jones 1975). The relative ages of the two flows cannot be established on the ground by mapping owing to poor exposure. A phase of deep weathering and subaerial erosion followed the extrusion of the lavas. Weathering products from the trachyphonolite were redeposited as clays in low lying hollows on the irregular surface of the lavas. At least some of the erosion was accomplished by running water as indicated by remnant patches of ferruginous grit. Continued volcanic activity in the vicinity is suggested by occasional pumice granules in the clays and by the lens of bedded fine tuff.

The artifact horizon is interpreted as resting upon a clay wash surface of gentle gradient on which sand and grit lenses indicate the position of ephemeral runnels and shallow channels.

Subaerial conditions are again indicated by the pale pumice tuffs with root-casts and palaeosols which represent intervals between deposition of increments of pumice tuff. Some lateral redistribution of the pumices by water action may have taken place and the degree of weathering varies considerably from place to place. However, primary fall out is indicated at several localities away from the main artifact site where the deposit is less weathered.

The well-bedded and graded three-banded tuff unit appears to indicate shallow ponding over an area some 2 kilometres in diameter. However, similar consistency and sedimentary structures can be produced in primary tephra. The base of this unit has yielded reversed palaeomagnetic polarity readings at three localities which probably indicate an age in excess of 0.7 m.y., that is, older than the Brunhes–Matuyama Boundary (Dagley *et al.* 1977).

The overlying pumice tuff with channels indicates continuing volcanic deposition but with strong fluviatile action represented by cut and fill channels and reworking of the tuffaceous material. Finally a major volcanic eruption produced the massive ash-flow tuff which seals the local sequence. All of the units from the pale pumice tuff to the ash flow are included in the Lower Menengai Tuffs of Jones (1975).

The palaeoenvironment of the artifact horizon

The base of the three-banded tuff, like the artifact horizon, dips towards the ENE at an average gradient of 1 in 50. This may represent an original primary dip but the presence of faults in the area shows that tectonic activity cannot be ruled out. If the base of the three-banded tuff is made horizontal by applying an element of tilt towards the WSW, the artifact horizon would also become virtually a horizontal plane. Accurate levels have been recorded which permit the form of the base of the three-banded tuff to be accurately reconstructed from 22 points and of the artifact horizon from 21 points. A small fault of 35 cms downthrow to ESE is visible in the embayment at the north end of the main face but may result from local compaction of the underlying sediments as it is closely aligned with the steep south-east face of the lava knoll.

On the geological evidence of palaeoenvironmental setting it is difficult to suggest a reason why so many artifacts should have accumulated on a flat clayey wash plain as the result of hominid action. However, concentration by natural physical processes seems to be ruled out by the palaeogeomorphological evidence. Access to some local water supply may provide a reason for the concentration of hominid activity. If the three-banded tuff marks an extension of an earlier sheet of ponded water which lay outside the area in which sediments survive for study, it is still strange to find so much evidence of stone technology concentrated on this flat clay surface. To judge from present-day observations the surface would become waterlogged and treacherous for several days after rain. Most important of all the nearest source of fresh rock for use in making the artifacts must have been several kilometres distant from the site.

Further work is being undertaken on various aspects of the geochronology, micro-stratigraphy and petrology, including the geochemistry and petrology of the artifacts and the distribution of possible source rocks for the tools.

Geomorphology

A major phase of erosion is represented by the pediment which is developed at the foot of the Farm House Cliff. This planation surface was associated with development of a thin but persistent ferruginous pisolithic rubble which everywhere underlies it. In the pediment notch and locally in channels incised below the main surface, coarser sediments occur which contain blocks of earlier resistant strata as well as the typical concretionary pisoliths. The presence of obsidian flake tools in the deposit suggests a late Pleistocene date for its formation.

The recent phase of erosion is still proceeding in the main gully but in the southern gully the slopes and floor are becoming stabilised by vegetation. The survival of the Northern and Southern Outliers with their remnant caps of three-banded tuff has greatly facilitated a three-dimensional investigation instead of a sectional view of the site and has provided ideal situations for excavation with only minimal overburden. The artifacts themselves are in some measure protecting the site from erosion by armouring patches of the gully floor with derived tools. This is particularly seen along the minor feature connecting the bases of the Southern Outliers on their northern flank.

Acknowledgements

I am indebted to Peter and Diana Tallon for assistance with plane table surveys in 1973 and 1974. Mr Michael Butler helped in many ways while Dr Mary Leakey advised on methods of study. Miss Gillian Fulger and David Ilsley undertook reconnaissance and geological mapping in 1973. Dr W. B. Jones not only discovered the site but also assisted in setting local mapping into a wider context. The Kenyan Government provided permits to carry out field work and Mr Richard Leakey arranged for Kilombe to be a scheduled prehistoric site. The research was financed by grants from the Natural Environment Research Council, London and the Wenner-Gren Foundation for Anthropological Research, New York.

References

DAGLEY, P., MUSSETT, A. E. & PALMER, H. C. 1977. Preliminary palaeomagnetic stratigraphy in the Baringo area. In: W. W. Bishop (Ed.). Geological background to fossil man. *Geol. Soc. Lond.* Scottish Academic Press, Edinburgh.

GOWLETT, J. A. J. 1977. Kilombe—an Acheulian site complex in Kenya. In: W. W. Bishop, (Ed.). Geological background to fossil man. *Geol. Soc. Lond.* Scottish Academic Press, Edinburgh.

JONES, W. B. 1975. The geology of the Londiani area of the Kenya Rift Valley (Unpublished Ph.D. Thesis, London University).

21 (B)

JOHN A. J. GOWLETT

Kilombe—an Acheulian site complex in Kenya

Interim Report of the 1973 and 1974 Excavations

This paper reports on excavations which have been carried out at an Acheulian site complex situated in Kenya on the western flank of the Rift Valley north-west of Nakuru. There is a profusion of stone implements on the ground, and excavations supplemented by surface studies indicate that large numbers of hand-axes, cleavers, other artefacts and manuports are concentrated mainly on a single extensive horizon, interpreted as an occupation surface. Artefacts occur across an area of at least 14 000 sq. m., though density of remains is variable. Two excavations have produced artefact samples of similar composition, and the biface groups are remarkably similar, even though the trenches were sited 60 m apart. This homogeneity in material from the main horizon is not maintained in other occurrences, which include a flake horizon, an area with stone blocks in and around a hollow, and two low density sites with bifaces some distance from the main site and at a higher level.

The site is valuable not only for its size and horizontal continuity, but also for its superb context with artefacts lying undisturbed on a clayey surface, and for its strategic location between other Acheulian sites. In the conclusion, comparisons are drawn with some other sites. Palaeomagnetic determinations suggest that the site complex is likely to be older than 0.7 m.y.

Introduction

This article constitutes an interim report of excavations carried out at the Acheulian site of Kilombe in 1973 and 1974. It also includes some wider discussion of the questions raised by the study of Acheulian occupation surfaces, and of artefact variability between sites. The recent application of improved dating methods has shown that the Acheulian industrial tradition, characterised in the main by hand-axe industries, has a temporal span of one million years or more (Isaac & Curtis 1974; Leakey 1975). Although intensive study has been made of the Acheulian, especially at Olduvai, Olorgesailie, Isimila and Kalambo Falls in Eastern Africa, overall cover is inevitably rather thin in relation to the enormous spans of time and space involved. Consequently, we still have a limited idea of the variability present in archaeological material from early sites, and this in itself is sufficient justification for much further work.

The newly-discovered Acheulian site at Kilombe is extremely extensive and rich in artefact material. Potentially it is very useful for amplifying our knowledge, as it preserves ancient surfaces with artefacts along considerable lengths of exposures. With this continuity and lack of disturbance, it provides the opportunity to study local variation in occupation density and in artefact composition and style within the stone industry, together with some hope of reconstruction of the palaeo-environment.

Kilombe is in the Rift Valley Province of Kenya, about 30 km north-west of Nakuru. The main site (designated GqJh1 under the African site naming system) lies about 4 km south-east of the hill of the same name, about 10 km south of the equator (co-ordinates 0°06'S, 35°53'E.), and at a height of just under 2000 metres. It was discovered by W. B. Jones in the course of geological mapping in 1972, and subsequently visited by Dr W. W. Bishop, who noted the extensive spread of archaeological material on the surface, and began to organise further investigations. Preliminary geological survey and mapping of the localities was carried out by Dr Bishop assisted by Mr P. Tallon in February 1973. At the invitation of Dr Bishop, the writer began archaeological work at the site in August 1973, and the operations took place in conjunction with the detailed geological work reported in the preceding paper (Bishop 1978). Further study of surface artefacts took place in December 1973, and in 1974 excavations were carried out in June and September.

The site and the excavations

General description

The site lies on the western flank of the Rift Valley. The edge of the Rift is not defined by any one major scarp in this region, and topographically there is a fairly gentle slope from the bases of Mount Londiani and Kilombe Hill leading down into the central depression of the Rift Valley. At the present time much of the area is covered with acacia bush.

Other Acheulian sites are known from the general area (Fig. 21b:1). Kariandusi lies some 80 km further south-east, and the sites of the Kapthurin Formation near Lake Baringo (Margaret Leakey 1969) are about the same distance to the north.

The main physical feature of the Kilombe site (GqJh1) is a low erosion scarp some 7–8 metres high, running approximately from north to south for some hundreds of metres, and facing to the east. Part way up this scarp is exposed a pale tuff with associated palaeosols, and it is from this complex that the large concentrations of surface artefacts have been shown to be derived. A number of outliers with similar stratigraphy occur to the east of the main scarp, and some of these too are surrounded by artefacts. Hundreds of Acheulian hand-axes and cleavers are visible on the surface.

Although a small quantity of fossil bone has been found on the surface of the site, none was found in the excavations apart from a bovid scapula and tooth which came from a sounding into the brown clays which underlie the main archaeological horizon.

Account of excavations

Over the two seasons of 1973 and 1974 excavations were carried out in three separate parts of the main site. In 1973 work was concentrated upon an area designated EH, where small flakes had been observed eroding out as well as bifaces. The artefact horizon from which the majority of the surface material is derived was located, together with a scatter of artefactual material from other levels, and one well defined horizon of flakes at a higher level. These excavations gave a useful indication of the nature of the site, but the extensive exposures clearly

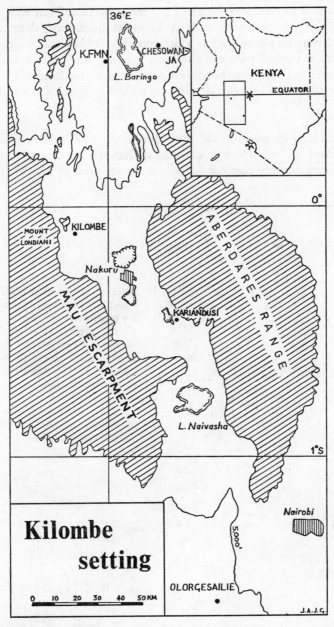

FIG. 21b:1. Kilombe, setting in relation to other Acheulian sites. K. FMN. = Kapthurin Formation. Land over 7000 ft (2134 m) is shaded.

made necessary a wider pattern of sampling, if the data preserved in the site were to be exploited more fully. Accordingly in 1974 further excavations were carried out in a sector some 60 metres north of EH excavation of 1973, at two localities separated by some 20 metres. In the first of these, Area AH, the main artefact horizon was uncovered over some 14 square metres, stratigraphically in a near-identical situation to the assemblage of the main horizon from EH.

The remaining excavations in Area AF attest much more dramatically to the variation that may be found within a site complex. There was revealed a channel situation, with a virtual pavement of stone blocks, but bifaces and other large tools were absent. There was possible evidence of surface modification by man. The finds are discussed in more detail under the headings of the individual sites.

EH Excavation

An initial trial trench into the sloping side of one of the outlying mounds, a few metres from the main scarp, confirmed that the area was likely to be productive. The excavation was carried out over an area of about 25 sq. m, and

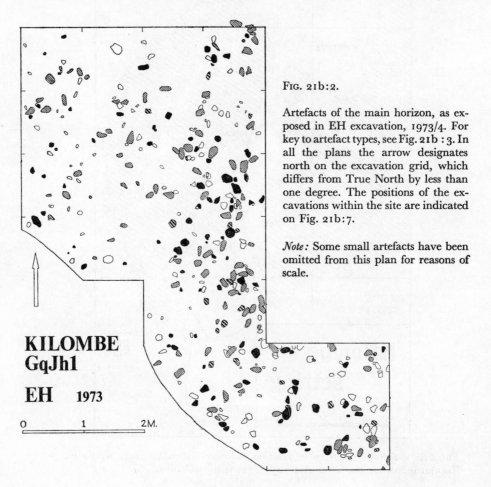

FIG. 21b:2.

Artefacts of the main horizon, as exposed in EH excavation, 1973/4. For key to artefact types, see Fig. 21b : 3. In all the plans the arrow designates north on the excavation grid, which differs from True North by less than one degree. The positions of the excavations within the site are indicated on Fig. 21b:7.

Note: Some small artefacts have been omitted from this plan for reasons of scale.

KILOMBE
GqJh1

EH   1973

0   1   2M.

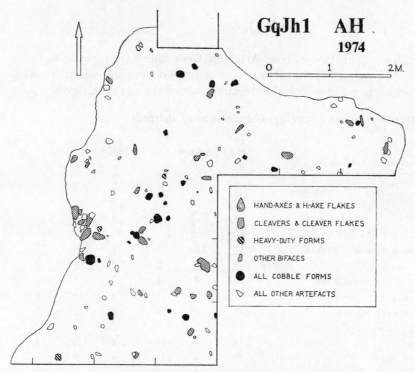

GqJh1 AH
1974

0   1   2 M.

	HAND-AXES & H.-AXE FLAKES
	CLEAVERS & CLEAVER FLAKES
	HEAVY-DUTY FORMS
	OTHER BIFACES
	ALL COBBLE FORMS
	ALL OTHER ARTEFACTS

FIG. 21b:3. AH Excavation, 1974: plot of artefacts from the main horizon. The edges of the excavation are defined by erosion gullies on the west and north sides. The top of a 1973 step trench abuts the excavation on the north side.

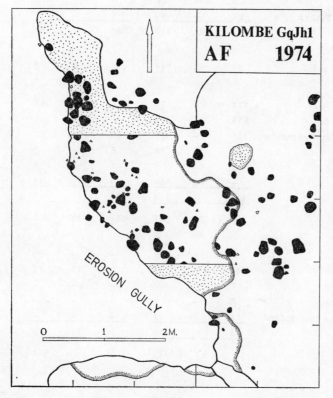

KILOMBE GqJh1
AF   1974

EROSION GULLY

0   1   2 M.

FIG. 21b:4.

AF Excavation: a channel site, with stone blocks, it underlies the main horizon. The channel bank is on the right, a recent erosion gully on the left. Areas of channel sand which were left unexcavated are shown lightly stippled, while close stippling marks the maximum extent of the sand channel filling before excavation. The isolated patch of sand on the right was only a few cm deep.

through a depth of up to two metres. Artefacts were found throughout the pale tuff and associated palaeosols (total thickness a little over 1 m) but there were found to be two horizons at which the density of artefacts was much greater.

TABLE I: KILOMBE GqJh1 (1973/4): *Classification of Artefacts*

Classification of Artefacts	EH Main Horizon			AH Main Horizon		
	Fresh or slightly abraded	Abraded	Total	Fresh or slightly abraded	Abraded	Total
TOOLS						
Large Cutting Tools						
Hand-axes & hand-axe flakes whole	65	18	83	19	–	19
broken	5	–	5	3	–	3
Cleavers & cleaver-flakes	19	4	23	2	–	2
Picks (small)	2	–	2	–	–	–
Other Bifacial Pieces	13	1	14	8	1	9
Heavy Duty						
Choppers	14	–	14	4	–	4
Cores/spheroids	11	–	11	–	–	–
Cores/choppers	7	1	8	–	–	–
Battered Modified Blocks	3	–	3	–	–	–
Light Duty						
Scrapers	26	–	26	8	1	9
Miscellaneous Trimmed Flakes	29	–	29	5	–	5
TOTALS ALL TOOLS	194	24	218	49	2	51
UTILISED						
Flakes, utilised	21	–	21	2	–	2
DEBITAGE						
Flakes	171	–	171	66	–	66
Flakes, broken or snapped	211	–	211	6	–	6
Angular debitage & Angular fragments	175	–	175	60	–	60
Cobble flakes	22	–	22	3	–	3
Core fragments	5	–	5	1	–	1
All forms, abraded	–	164	164	–	13	13
TOTALS, ALL DEBITAGE	584	164	748	136	13	149
TOTALS	799	188	987	187	15	202
COBBLE COMPONENT						
Cobbles (Manuports)	–	–	29	–	–	9
Cobbles, modified	43	55	98	2	11	13
Cobbles, modified, pitted	2	–	2	–	–	–
Cobbles, modified, ?pitted	5	–	5	–	–	–
Pitted cobbles, otherwise unmodified	4	–	4	–	–	–
TOTAL COBBLE COMPONENT	54	55	138	2	11	22
OVERALL TOTALS			1125			224

The lower or main horizon formed a well-defined floor, with numerous hand-axes, cleavers and modified cobbles. Most of the artefacts from this horizon lay on a clayey surface at the foot of the pale tuff, which had subsequently sealed off the finds (Fig. 21b:2). The association with a clay surface is of some importance, as most other East African Acheulian assemblages have been found in sandy deposits, one other exception being EF–HR at Olduvai (Leakey 1971). That site, however, represents an Early Acheulian and although the Kilombe site has not yet been dated, the material is closely comparable with that from some sites that have been ascribed to the Upper Acheulian, for example Olorgesailie.

Altogether 1096 artefacts, including modified cobbles, were recovered from the spits which represent this horizon, and a classification of the material is given in Table I. Including a few broken and some abraded specimens there are 127 hand-axes and cleavers from this horizon, the great majority of them made from fine grained lavas which may be phonolites. Chemical weathering has produced a pale 'patina' on these artefacts, sometimes to a depth of several millimetres. The remaining pieces are of a coarser grained lava which may be trachyte. The only other raw material known to be represented on the site is obsidian, used for one hand-axe that was found on the surface.

In the excavation there was also a component of 138 cobbles, many of them modified, but shaped heavy-duty tools were less common, amounting to 36 choppers, cores/spheroids, and related forms. Excluding the cobbles and abraded artefacts, flakes, utilised flakes and other debitage form about 76 per cent of the total from the main horizon. Although a substantial proportion, this is not enough to account for the manufacture of all the large tools within the area. Only a few

	Fresh or slightly abraded	Abraded	Total
TOOLS			
Flake Scrapers	2	–	2
Miscellaneous Trimmed Flakes	23	–	23
TOTAL TOOLS	25	–	25
UTILISED			
Flakes, utilised	10	–	10
DEBITAGE			
Flakes	27	–	27
Flakes, broken or snapped	88	–	88
Angular debitage and fragments	32	–	32
Cobble Flakes	–	–	–
Core Fragments	–	–	–
All forms, abraded	–	21	21
TOTAL DEBITAGE	147	21	168
Cobbles, modified	–	2	2

TABLE Ia: KILOMBE EH (1973) *Artefacts from the Upper (Flake) Horizon.*

flakes were found of the coarser-grained lava (which is distinctive as it is more
resistant to chemical weathering), suggesting that the bifaces of that material, at
least, were made elsewhere on the site, if not further afield. At a conservative
estimate some 4000 flakes would be produced from making the bifaces found in the
excavation, and many of these must lie elsewhere. Further sampling of the site may
reveal areas of the main horizon where debitage predominates, but all the work so
far has indicated that the horizon is in fact relatively homogeneous in general
composition.

Above the main horizon, in the middle zone of the pale tuff, there is a much
lower density of artefacts, and a much higher proportion of abraded pieces. It is
not certain that this sparse distribution represents archaeological material in
primary context, and it has been suggested that it may have washed in from higher
areas during deposition of the tuff (Bishop, this volume); the archaeological
evidence does not contradict this, and no analysis of the material is presented here.

However, at the top of the pale tuff is a palaeosol, again associated with a
denser concentration of artefacts, with a higher proportion of fresh and only
slightly abraded artefacts than at any other level of the excavation. Some of the
artefacts occur a little higher up in a thin brown tuffaceous horizon overlying the
grey-brown palaeosol, and it is clearly not possible to state with certainty that a
single occupation is involved. Nevertheless, all the material belongs to a similar
facies, characterised by an almost total absence of all heavy-duty forms, bifaces
and cores (Table 1a). The assemblage consists mainly of flakes, some of them with
broken edges. There is no way of knowing whether this last is a deliberate feature.

It is an interesting aspect of variability that such an assemblage should be
found stratified less than a metre above the horizon with bifaces. The very different
composition of the two horizons could be interpreted as a parallel to the occurrence
of separate facies of material at other sites, for example the Acheulian and 'Hope
Fountain' industries at Olorgesailie. However, the Hope Fountain industry itself
does not consist just of flakes (Jones 1949), nor does that at Olorgesailie (Posnansky
1959) or that at Broken Hill (Clark 1959), whereas at Kilombe the flakes make up
almost the whole assemblage and there were no choppers. The debitage from the
upper horizon is not notably different from that of the main horizon, and some
similarities are discussed later.

AH Excavation

In Area AH the main artefact horizon was uncovered over 14 sq. metres
(Fig. 21b:3). This assemblage, with hand-axes and other artefacts, was found
stratigraphically in a near-identical situation to the EH assemblage (from the
main horizon), that is, lying on a grey-green clay, and overlain by a pale coloured
tuff. At this locality, the pale tuff was almost completely sterile of artefacts,
emphasising the discrete nature of the artefact horizon.

There can be little doubt that the same horizon is represented in the two
localities. The two assemblages are therefore of the same age, and it could even
be argued that technically they are two samples of the same assemblage. A lower
density of material is evident from the AH excavations, where 33 bifaces were recovered from 14 sq m, in comparison with 127 specimens from the 25 sq m of

EH. Additionally, there is a lower proportion of cleavers within the AH assemblage (only 2), but a higher proportion of cruder 'other bifaces'. The bifaces themselves are fairly similar in general form to those from EH, but a better comparative sample can be assembled when the bifaces excavated from AH are lumped with those from surface collection nearby (AC and AD). As in EH, manuports were also found, and a considerable flake component. An analysis of the spatial structure of finds from the excavated floors is now under way. A deep sounding which penetrated three metres into the brown clays at the base of the AH trench showed that they contain occasional horizons with flakes, some fresh. A well-preserved bovid tooth and part of a scapula were found from the same sounding. Most of the fauna from the site has come from the brown clays, and a larger trench might allow recovery of a useful sample.

AF Excavation

The second excavation carried out in 1974, that in Area AF (Fig. 21b:4), has provided more notable evidence of the variation within a site complex. Sited only 20 metres from Area AH, the excavations over an area of 14 sq m have revealed a channel location with a dense distribution of stone blocks which in places amounts almost to a pavement, but bifaces and other large tools were quite absent. There were some small flakes in the base of the channel. The fine, but clean, sand may indicate a minor ephemeral channel situation rather than a closed hollow (W. W. Bishop, personal communication). However, the occurrence of archaeological material, including small flakes, exclusively at the base of the sand may suggest that the stream did not run with force sufficient to move stones. Many of the larger stones are actually imbedded in the clay of the channel base or channel bank, and would apparently be unlikely to be moved by water. But it is clearly important not to make dogmatic statements about such mechanics, while we lack sound experimental evidence.

The sharp interface between the channel sand and the underlying surface made it possible to follow the contours of the clay with considerable accuracy. The moulding of the surface, and the peculiar distribution of stones suggest at least the possibility that we are dealing with a hollow modified, if not made, by man. One of the larger depressions at the side of the main hollow contained three or four stone blocks, in such a way as to suggest that they could have anchored a structural support of some kind.

Although these finds are very close to the excavations in AH, a minor fault separates the two areas. Through the whole depth of the excavations in AH only one archaeological horizon was found, but in AF, apart from the channel and stone blocks, indications were found of a horizon with bifaces and manuports at a higher level. This is probably to be correlated with the main horizon elsewhere on the site, and indicates that the horizon with the stone blocks is slightly older. However, occupation of the main horizon may have been recurrent over a considerable period of time, and as a small channel may be cut and filled in from one season to another, it is still possible that these features are contemporary with parts of the main horizon.

Evidence of structures or of surface modification has been reported from the

Acheulian elsewhere, for example at Terra Amata (de Lumley 1969), and at site JK in Bed III of Olduvai Gorge (Leakey 1975) where a group of small pits was found. The evidence from Kilombe is not entirely conclusive, though the concentration of stones is certainly human in origin, and the artefacts demonstrate human presence. An extended excavation in this area might be rewarding.

Surface material

The surface of the site is littered with artefacts which have eroded out from the receding scarp. There is a dense distribution along most of the lower slopes of the scarp, extending out onto the near-flat erosion bench below. Many of the artefacts are near-fresh, others appear to have been on the surface for much longer, but the excavations seem to demonstrate beyond question that they all derive from the same beds. Since the material can have moved very little in relation to the dimensions of the scatter, it is suitable for the study, with due caution, of horizontal variation within the industry. However, this study must largely be restricted to the bifaces, which are the most conspicuous form, suffering less from weathering than the debitage.

During the 1973 work one small area of the surface scatter was plotted intensively, orientation data were taken, and a near-complete collection was made. Selective collection was made from one other ten-metre square. This is probably as large a sample as should be removed from the site for study, and it has since been possible to study all the bifaces from two other squares (AC and AD), without removing the finds from the site. This group of artefacts was contained within a single embayment of the main scarp, and therefore its original position can be estimated with some accuracy. Similar work has been started in another area of the site, and it is hoped that further studies of this nature can be carried out in the future.

Small surface collections were also made from two low density sites, GqJh2 (1200 metres ENE of the main site) and GqJh3 (500 metres to the NW of the main site). In each case, artefacts including bifaces were found *above* the 3 banded lapilli tuff which overlies the archaeological material on the main site.

The Artefacts

The artefacts from the excavations are here considered in more detail. The main aim in presenting further analysis here, is to assess how much the artefacts may be seen to vary from area to area of the site, particularly within the main horizon of GqJh1. It has been pointed out recently that studies of occupation surfaces may grow to have more importance in relation to studies of artefacts (Isaac 1972b); on the other hand, many occupation surfaces are themselves made up almost entirely of artefacts, as bone remains and surface features are often lacking. Hence, a study of the horizontal variation of artefacts may be one of the only ways of making detailed study of an extensive occupation surface.

Most of the discussion relates to the main horizon, but the material from the upper horizon of EH is also dealt with in the section on the flake component. The bifaces are considered first. These come from the main horizon assemblages of

EH and AH, and also from the surface sample recorded in squares AC and AD.

The typological classification employed is based on that worked out by Kleindienst (Kleindienst 1962), with some modifications. There are some unresolved difficulties in the application of typological schemes to East African Acheulian material, both within the biface groups, and particularly within the broader category of 'Heavy-Duty' forms, where the relationships between 'cores' and actual tool forms cannot be defined on functional grounds, and where the full variety of morphological forms is not yet established. For the moment, then, as explained further below, elements of the classification employed here must be taken as provisional. The category 'Battered Modified Block' has been adopted from M. D. Leakey's classification of the Oldowan (Leakey 1971), and some pieces that could possibly be classified simply as cores have been qualified further as 'cores/spheroids' and 'cores/choppers'.

The bifaces

Through metrical analysis the Kilombe bifaces can be compared in more detail, both internally, and with the bifaces from other Acheulian sites. The bifaces have been measured according to an attribute scheme similar to those devised and used by Roe and Isaac (Roe 1964, 1968; Isaac 1968; 1977) and some compilation of statistics has taken place by computer. The results obtained so far suggest some similarity among the bifaces from the main horizon of GqJh1, and a general description is given first.

The bifaces seem to be made almost entirely on large flakes. Technologically they are relatively sophisticated, with shallow invasive scars, and often with extensive use of step-flaking around the margins (Fig. 21b:5). Some pieces are very slender in transverse section, and hardly any are thick enough in relation to breadth to be termed picks. In plan, broad lanceolate forms predominate, with a few ovates, some of which are very finely finished; few examples are sharply 'pointed'. Some bifacially worked pieces appear to take the form of angled cleavers, with an angled 'cutting edge' formed by flaking, though technically they fall within the definition of hand-axes, and are treated as such here.

In all the areas that have been sampled so far, by excavation or by surface collection, cleavers and cleaver flakes are in the minority. The variation in numbers of cleavers in relation to hand-axes may be one criterion for distinguishing one region of the site from another in terms of bifaces. In area EH, cleavers have 27 per cent of the frequency of hand-axes; in area AH, this ratio is 9.5 per cent, but as the sample size is rather small, it is better to operate by including the surface finds from AC and AD; this then raises the frequency of cleavers to 13.4 per cent (20: 149 specimens), which is almost exactly half the frequency found in EH. It seems likely that cleavers do not reach much higher percentages than these figures at any place on this main floor, in spite of its large extent, which might have allowed much variation. Nevertheless, the specimens which have been found are sometimes very finely made, and in EH excavation they have a rather steady distribution across the floor (see Fig. 21b:2).

The dominance of hand-axes over cleavers, and the almost complete absence of picks, begs comparison with Olorgesailie where there is a similar situation (Isaac 1968). The ratio of cleavers to hand-axes is also roughly comparable with those from sites PDK and HEB/1, 2 and 3 from Bed IV, Olduvai (Leakey 1975).

Although the areas sampled are around 60 metres apart, the bifaces seem to conform to a generally similar pattern. This can be demonstrated from the means

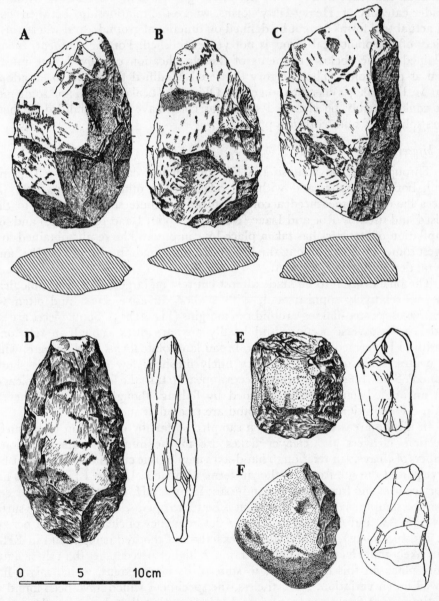

FIG. 21b:5. A, B, hand-axes from EH Excavation. C, cleaver from EH. D, hand-axe from AH Excavation. E, sub-rectangular chopper. F, side chopper, both from EH.

and standard deviations of dimensions and of key ratios (Table II). It is plain that the overall dimensions vary little, that the frequency curves for individual parameters are similar, embracing similar ranges, and that morphology is generally similar. The correspondence for the frequency distribution of length is particularly striking (Fig. 21b:6). There is no reason to expect an exact correspondence, but the bifaces from two widely separated parts of this main floor show very similar characteristics, and these findings contrast with those from Olorgesailie and Isimila, for example, where bifaces from sites that are *paenecontemporaneous* show marked variation from site to site. Surface inspection at Kilombe suggests that other areas of the site may not be markedly different, though separate areas are perhaps marked by small idiosyncracies.

Other bifacial forms

Though classification is necessary, the writer has been cautious about impelling the material from Kilombe into exactly the same categories that have been devised for other sites; it is, necessarily, slightly different material, and one should not hasten to define idiosyncratic pieces until having a good idea of the range of the material, especially when dealing with possible 'types' which have low frequencies of occurrence. It is plain, however, that there are some bifacially worked pieces which do not readily fall into the main established categories. Most of those from AH excavation are longer than broad, and could be regarded as very rude hand-axes (they have been included as such in the metrical analysis). Some other pieces are shaped rather like a trefoil pattern that sometimes appears in Gothic architecture—triangular with bowed sides. For the moment all these pieces are lumped into the same category, but a study of larger samples might establish type categories *relevant to this site*.

Heavy-duty material

Much the same may be said for the heavy-duty material: the present small size of the sample urges caution in interpretation. Choppers occur, but not with high frequency, and they do not appear standardised. The exception is a possible class of sub-rectangular bifacial choppers which are rare but well-finished, and which have been observed on different parts of the site. There are a number of pieces which can be termed spheroids, but no apparently standardised core-forms within the excavations. A few large cores have been observed on the surface elsewhere, and ten were collected for study. Although there are some large flake scrapers, very few of these can be termed heavy-duty.

Most interesting is the presence of some cobbles with pitting, from EH excavation, tentatively assigned to the 'pitted-anvil' class designated by Mary Leakey. These are most common from the Upper Bed IV sites at Olduvai.

The flake component

Overall size data are presented for flakes (excluding bifaces on flakes) from the main horizon excavations and from the upper horizon of EH (Fig. 21b:6). Utilised flakes and flake scrapers have been included here, because they cannot be separated from the rest by function with any certainty, and in any case slight chemical weathering of the Kilombe artefacts sometimes makes it difficult to assess the degree of retouch. All three assemblages show a general comparability with other Lower and Middle Pleistocene sites, with a modal value of flakes 2–3 cm long and with few flakes over 8 cm long. The AH excavation yielded a higher proportion of flakes 1–2 cm long, but it is difficult to tell whether this is significant in any way.

Since nearly all bifaces from the site are over 8 cm long, with a mode of around 15 cm, bimodality of this parameter for total finds from the main horizon is clearly demonstrated. In this respect, too, there is a very strong parallel with the situation at Olorgesailie. However, not all Acheulian sites match in this respect: for example, a very large assemblage of lava artefacts from Kariandusi, all in derived context and slightly abraded, has a considerable proportion of flakes which overlap in size with the bifaces.

The samples of the main horizon from EH and AH show a very similar ratio of total flake component to all bifaces, even though the density of finds was much greater in EH. This does not in itself establish a direct relationship between the bifaces and the flakes, particularly as the number of flakes is much too small to account for all the bifaces, but at least some of the flakes are highly likely to be hand-axe trimmers, as they exhibit such features as step-flaking near the margin.

No evidence was found of any bifaces from the flake horizon in EH, but the overall length parameters for flakes nevertheless show great similarity with those from the main horizon. Further analysis will be necessary before the separate horizons can be compared in more detail. Most of the measured whole flakes from the EH assemblages have simple platforms. The flakes are not much elongated and there are no actual flake-blades such as are characteristic of some of the Kapthurin assemblages.

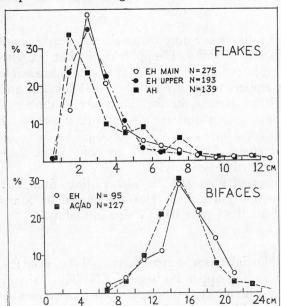

FIG. 21b:6.

Frequency diagrams for the lengths of flakes and bifaces (see text for details).

The artefacts: conclusions

As a summary of this fairly detailed discussion of the artefacts from the various excavations, a number of points may be made. Although density of distribution varies, the industry from separate parts of the main horizon has broadly the same composition. The same major components were present, in slightly varying proportions, in both of the excavations where a substantial sample was recovered from the main horizon. The bifaces from the separate areas were similar, and the same may be said for the flake component.

The other assemblages found in the excavations, namely the flake horizon from EH and the stone block horizon from AF, are clearly very different from anything so far found on the main horizon. Overlying and underlying the main horizon, as they do respectively, they belong to a similar time range, and without evidence to the contrary they must be interpreted as facies of Acheulian industries, regardless of the absence of bifaces.

The few bifaces found at other occurrences (GqJh2 and GqJh3) seem generally similar to those from the main horizon of GqJh1, though they occur at a higher level stratigraphically, overlying the banded tuffs of lacustrine deposition which themselves overlie the occurrences of GqJh1. The samples are very small, but with careful choice of appropriate statistical methods, it is hoped to test them in more detail against the ample samples from GqJh1.

Discussion

Some knowledge of the modulations of variation in artefacts through space and time is a prerequisite for interpretation of such variation, whether in cultural, functional or other terms. The stratigraphic approach to such problems is well enough known, and for the Acheulian the unparalleled opportunities at Olduvai Gorge have been used to assemble a series of assemblages from different points on a very long time scale. At Olorgesailie, on the other hand the assemblages are relatively close in time as well as in space. The same may probably be said for the Acheulian assemblages at Kalambo Falls. In concentrating his attention on the Olorgesailie grouping Isaac spoke of a conscious sampling decision (Isaac 1968). Geological circumstances made it difficult to determine the exact relationships between many of the sites, various of which in the Middle Set alone may be separated by anything from a few years to several thousand years. The detailed comparisons which were made were therefore partly based upon a certain model for regarding those time relationships, in which Isaac used the term *paenecontemporaneous* for their definition. A considerable amount of variation was noted between the sites, and possible explanations were offered for this.

At Kilombe a single artefact horizon extends over a wide area, and this is easily verifiable by both archaeological and geological means. The occupation may have occurred over many seasons, but if this suggests a *paenecontemporaneous* state from area to area, we are nevertheless on the limit of the use of this term: as far as can be determined from any *archaeological* situation, these artefacts may be regarded as contemporaneous. In contrast, then, with the time sequence from Olduvai, or even with the close cluster at Olorgesailie, we have here an extraordinarily 'tight' distribution in time as well as in space. The corollary of this is that it is impossible to assess time trends from the evidence at Kilombe (apart from the very limited opportunity provided by the outlying occurrences for comparisons with the main horizon), but that experimentally this may actually be seen as advantageous. With one dimension (time) fixed, we have ample opportunity for surveying spatial variations in the artefacts, and then of comparing the variation found here with that in other site groups where the time dimension is not so controlled.

The main result of admittedly limited sampling within this very 'tight' distribution at Kilombe is that we can see a considerable degree of homogeneity throughout the occupation area, which contrasts with the heterogeneity found from site to site at Olorgesailie, or through a long time sequence as at Olduvai.

What is especially interesting is that the key dimensions and ratios for the bifaces from Kilombe show a remarkable similarity to these figures for the total series from Olorgesailie. These similiarities are clearly demonstrated when the

TABLE II

BIFACE MEASUREMENTS

	Kilombe EH	Kilombe AH + AC	Kilombe AC	Kilombe AH
Number	95–106	144–156	119–127	25–29
L	153 ± 29.9	149 ± 32.4	150 ± 29.5	143 ± 43.5
T	43 ± 10.1	41 ± 9.8	42 ± 9.2	37 ± 8.9
B	94 ± 17.8	89 ± 15.3	89 ± 13.5	89 ± 21.6
T/B	0.47 ± 0.12	0.46 ± 0.10	0.48 ± 0.09	0.43 ± 0.10
B/L	0.62 ± 0.08	0.61 ± 0.07	0.60 ± 0.07	0.62 ± 0.08
BA/BB	0.86 ± 0.23	0.81 ± 0.19	0.82 ± 0.20	0.76 ± 0.17
TA/L	0.15 ± 0.03	(0.15 ± 0.04)		0.15 ± 0.04
TA/TB	(0.73 ± 0.17)*	(0.72 ± 0.22)		0.72 ± 0.22
PMB/L	0.44 ± 0.10	0.44 ± 0.10	0.44 ± 0.11	0.41 ± 0.06

	Kariandusi Lower Site (Lava)	Olorgesailie** Overall	
Number	84–110	674–1273	
L	163 ± 24.3	170 ± 46.6	
T	49 ± 9.1		
B	94 ± 11.1		
T/B	0.53 ± 0.11	0.48 ± 0.12	
B/L	0.58 ± 0.07	0.58 ± 0.09	
BA/BB	0.83 ± 0.20	0.80 ± 0.24	
TA/L	0.17 ± 0.05		
TA/TB	0.73 ± 0.25	0.72 ± 0.21	
PMB/L	0.43 ± 0.10	0.41 ± 0.10	

* Figures in brackets were calculated from only part of the series.

** The summary of Olorgesailie data is taken from Isaac (1968).

figures are tabulated (Table II), and they are notable because the two site complexes are about 200 km apart in space, while the Kilombe site could easily be separated by 100 000 years or more in time from any of the Olorgesailie occurrences. A sample of lava bifaces from one of the Kariandusi sites, which are about 60 km away, also shows very similar results, even though the sample was drawn from material in secondary context, and might thus have been vulnerable to distortion (data from this author's report, in preparation).

Such a situation calls for some attempt at explanation. The view has been expressed (Isaac 1968, 1972b) that some of the variability in the Olorgesailie assemblages cannot easily be accounted for by activity differences, raw material differences, or by long term trends in cultural evolution. Isaac advances the explanation that random or stochastic variation in the norms of local craft traditions may account for at least some of the variation found in size, style and morphology of artefacts.

If we accept this, a possible explanation of the Kilombe situation is that within the site this element of stochastic variation is suppressed because a relatively short time duration is represented. Plausible as this is, it does not in itself substantiate the theory of stochastic variation, which Isaac himself labels as in a sense a 'non-explanation', accounting only for a residuum of variability that cannot be explained by other means. Consequently, other factors need to be taken into consideration.

Although both the means and range of variation (expressed in standard deviations) for the dimensions of the Kilombe bifaces tally very closely in the separate samples, there is a considerable range of variation *within* each sample. We may observe that apart from the dimension of length (where the large s.d. reflects the summing of separate samples) the summed Olorgesailie samples have a very similar range of variation also, at least as shown by the ratios which are taken to express morphology. The individual Olorgesailie samples are often more idiosyncratic, morphologically, than those from Kilombe, but combined, they form a similar repertoire. Why should this be so? It could well be a matter of chance. On the other hand, one feels obliged to postulate that perhaps, in some circumstances, a more representative selection of an East African 'Acheulian repertoire' is expressed than in other circumstances. Some variation in biface assemblages might be accounted for in this way, but it is difficult to envisage the factors (minor differences in activity?) which might lead to such a controlled variety of expression.

If the theory of stochastic variation holds good, then it could be asked why the Kilombe bifaces conform so closely to the limits of the variation that have been established for Olorgesailie. One reasonable explanation is that if the two complexes belong to a similar time range, they might both belong to the same cultural microcosm, as the distance of about 200 km which separates them is not necessarily a long distance to a hunter-gatherer group. If this should seem too much of a coincidence, one might consider other factors that could restrict free play in the variation. There could, for example, be particular local functional requirements in this general area, but it is difficult to believe that these could exercise such an influence towards conformity, when there is such enormous variability in bifaces in other parts of the Old World.

It is apparent from all this that the patterns of variation amongst East African bifaces can be very, very subtle indeed. If there is random-walk or stochastic variation amongst these assemblages, then in this general area it seems to be confined within certain limits, which seem to operate almost as stringently within the general area of space and time, as within the individual site complex. Only very detailed testing of the results from these site complexes, and the addition of results from other sites and areas, will enable firmer conclusions to be reached. In summary, in metrical terms the Kilombe bifaces are astonishingly similar from area to area of the site, and they are surprisingly like the bifaces from Olorgesailie, and some of the bifaces from Kariandusi. Various reasonable explanations can be advanced for this situation, but they must all imply very strong controlling factors, of one sort or another, containing variation within limits, or else a highly unlikely degree of coincidence.

The occupation surface

The importance of the study of occupation surfaces is now widely recognised. As Kilombe offers a particularly fine opportunity for a study of this kind, the various aspects of it will be considered here. If ever present on the main horizon, bone was not preserved, and this has naturally concentrated attention upon the distributions of artefacts.

In the preceding discussion of the artefacts, reference has been made repeatedly to artefact horizons. Although firmly embodied in the literature as the 'smallest cultural stratigraphic unit that can be defined at any one place' (Clark & Kleindienst, in Clark 1974) the term horizon still carries implications of a plane surface or interface, something that is two-dimensional. Yet open sites are three-dimensional, just as are cave sites, and this applies as much to the geometry of artefact distribution as to that of the containing sediments. If we are to take horizons as integral units, then there should be some attempt to show that reasonable assumptions are being made. One of the most common problems of so-called settlement archaeology, of any period, lies in assessing to what extent a cultural palimpsest is represented on any continuous surface. This is certainly a problem for palaeolithic studies, and it is prefaced by the need to demonstrate whether or not artefacts are likely to be in their original context.

Context

Although it is not possible to establish with complete certainty whether or not artefact horizons have been disturbed by water, there are some tests that can be applied. The general geological circumstance at Kilombe, with a very wide spread of artefacts lying on a clayey surface in what was presumably a wide and almost plane gully floor, suggests that water flow would not have disturbed the artefacts substantially (W. W. Bishop, personal communication). The size of the distribution alone makes it unlikely that it has all been much moved by water, and the few channels that have been observed were probably small and ephemeral. Other indications may support this view, for example the constant association of very small flakes with the larger artefacts, and the results of an orientation experiment on artefacts from EH, which did not imply any significant preferred orientation, though one should add that such tests are far from conclusive. Additionally, for comparison the bifaces on the surface in squares AC and AD were plotted, and then relocated six months later after heavy rains, and it was found that their positions were almost identical. This may suggest that larger artefacts move very little except when they are actually in channels. Some of the floors from Kalambo Falls are similarly undisturbed (Clark 1969), but there are few other Acheulian localities in East Africa where this is so sure, particularly as most of the Acheulian occurrences from Olorgesailie and Olduvai have been found in minor channels. Various similarities between the Kilombe and Olorgesailie artefacts have been pointed out, but it is important to stress this difference in actual context.

Vertical diffusion

In each area where the main horizon was uncovered it appeared to occur at the interface of a grey-green clay (the top of the brown clays, altered in colour) and the overlying pale tuff. The plots of vertical distribution of the bifaces were made to check this association and in each case the result is something near to a normal distribution. In AH, 48 per cent of the bifaces come from a single 5 cm interval, and 81 per cent occur within the space of a 15 cm interval. In EH the distribution does not appear so close, as the interface itself slopes by around 20 cm from side to side of the excavation, but nevertheless 77.7 per cent of the bifaces come from a 20 cm span. Such figures cannot prove in any way that there was a single occupation of the site area, but they do indicate that the great majority of the finds, at least of the bifaces, is specifically associated with a single surface.

Lateral extent

A feature of chance which is of assistance in assessing the extensive main horizon of GqJh1, is the physical length of the present day exposures (Fig. 21b:7). These wind their way across the archaeological area, and additional exposures of artefacts are created by the island form of several small outliers. Although this does mean that considerable areas of the original extent have been destroyed by erosion, the outliers of Pleistocene strata partly compensate for this as they show whether or not artefacts were present at certain points. Apart from the problem of deposits that have been eroded out, there is also the difficulty involved in dealing with areas which have not yet been exposed.

The main framework that has been selected for making an estimate is in fact based upon three furthest points of surface exposures of artefacts, which define a triangle. This triangle is approximately equilateral, and of side 180 metres. Within this area falls the whole length of the main site scarp demonstrated to include artefact occurrences; the three excavations in their entirety; and numerous areas with surface distributions of artefacts. If the whole of this triangle is therefore regarded as being part of the site occupation area, then we are left with an estimate for area of some 14 200 square metres. It is true that the distributions cannot be checked in detail everywhere, but this is surely a valid estimate for the minimum *extent* of the occupation, even if it did not occur solidly throughout this area.

Density patterning

It is clear that there is much variation in the density of distribution of the artefacts and manuports strewn across this area. This is apparent within the relatively limited area of the excavations, even though they indicate (so far) a relative homogeneity in the composition of the industry. In EH excavation there

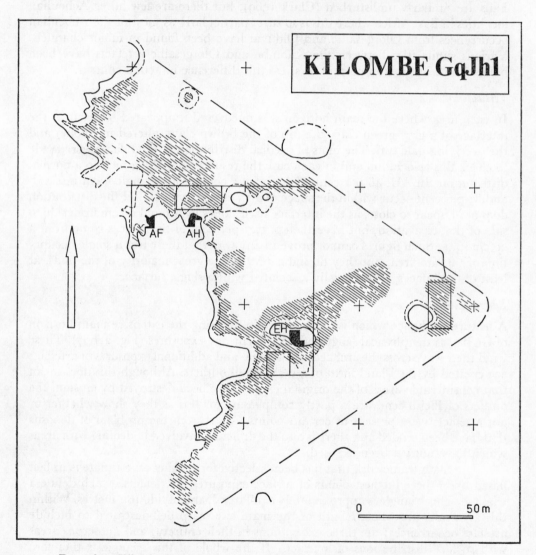

FIG. 21b:7. Kilombe GqJh1 : distribution of surface artefacts (indicated by hatching). Exposures of the artefact horizon are marked by a continuous line, and excavated areas are shown in black. The area in which the archaeological grid was laid is also shown, as are individual squares in which studies of surface artefacts were carried out. W. W. Bishop kindly allowed the use of the geological base map (q.v., Fig. 21a:1) for the preparation of this plan.

was an average density of five bifaces per sq metre, and about five cobbles or manuports. In AH excavation the average biface distribution was about half as dense, but within each excavation there was sufficient variation in the distribution that there is an overlap in the density for some metre squares. In general terms, two likely characteristics can be pointed out for the distributions:

(a) Density of finds is greatest in the 'centre' of the occupation area, and thins out towards the margins.

(b) In spite of this, there is local variation across the surface, with marked subconcentrations of artefacts separated by areas of thinner distribution.

Dense distributions as large as this have rarely been documented from Middle Pleistocene sites, and there are not many opportunities for comparison. Equally and even more extensive distributions have been reported, both from Isimila and Olorgesailie, but these are low density scatters, which do not necessarily reflect a similar concentration of human activity. Subconcentrations within sites, however, have been recorded at other sites, for example Arkin 8 in Nubia (Chmielewski 1968). At Olorgesailie, in contrast with Isimila, the concentrations often have abrupt edges, perhaps because of their occurrence in channels (Isaac 1966). At Kilombe there seem to be gradations between denser and sparser areas of distribution. In general, in the excavations of the main horizon, artefact density seems to vary with manuport density, and the concentrations may really represent areas of concentrated activity.

It is difficult to interpret this evidence in terms of human behaviour, while we are still not sure what the artefacts were used for. There could well be up to 30 000 bifaces on the Kilombe surface, many of them probably made elsewhere, and the total expenditure of man-hours and energy is perplexing. It was generally believed that bifaces were used in animal butchering, but more recently J. D. Clark has suggested that a simple tool kit is most commonly found at kill sites, while bifaces may have been utilised for other purposes (Clark 1972; 1975). On a site as large as Kilombe, it is perhaps unlikely that the bifaces were used exclusively for butchering carcases. Some very large concentration of resources is implied, but whether this was plant or animal remains uncertain.

It is legitimate to speak of the Kilombe distributions as an occupation surface, though it is not established that the main area was a 'camp site'. The important point is that much time was spent there, and at times the area probably constituted a 'home base'. The stone blocks uncovered in AF excavation in and around a depression form a different 'facies' of occupation, and tend to reinforce this idea. An estimate made a few years ago (Cook & Heizer 1965) suggested 10 sq m as the average area occupied per person by some modern hunter-gatherer populations in camp. This figure has often been used, largely because it fits neatly with our common conceptions about band size and settlement area. As an occupation by 1400 people at Kilombe would seem excessive to most people, it seems that the larger the site, the less happy we are with this sort of estimate. The possibility of reoccupation is an important variable, which has earned discussion already (e.g. Clark 1972), and it is not one which can be quantified in relation to population with any certainty. For example, a very large site may imply reoccupation, but equally it may imply a concentration of resources that will lead to the congregation of a larger than average group. At Kilombe, we could postulate a group of about a hundred people coming back seasonally fourteen times, but we could also choose other factors multiplying to 1400. As in practice we have very little idea of how the making and use of bifaces correlates with spending hours in camp, in this case the *extent* of the occupation may not be sufficient basis for detailed estimates, though such work as that outlined recently by Wiessner (1974) may eventually provide a better idea of the variables involved in dealing with large sites. To get much further, it will be necessary to obtain new sources of information. Much human activity has been concentrated on the land surface preserving the Kilombe main horizon, and it could be that sampling to the edges of the main artefact distribution would be fruitful.

Conclusion

This is merely an interim report, and the conclusions that are drawn are set forward tentatively. It is hoped that more work can be carried out at Kilombe, not primarily with the intention of recovering vast quantities of artefacts, which is easy enough, but with the aim of elucidating further the structure of the sites. The large spread of the main horizon seems denser in the centre, and the artefacts seem to thin out gradually towards the edges. It may well be profitable to obtain better 'transects' from centre to edge across this area, for if the main horizon itself is relatively homogeneous, the most useful strategy may be to investigate mainly the least 'normal' areas, which may even be off-centre in relation to the biface and manuport distribution.

At the beginning of this article attention was drawn to the long time-span of the Acheulian. The reversed palaeomagnetic readings obtained from Kilombe (Dagley *et al.* 1976) suggest that the site complex is likely to antedate the Brunhes-Matuyama transition, and is therefore over 700 000 years old. In the past, archaeologists might have referred the material to the 'Upper Acheulian' on typological grounds, but it now seems difficult to use this term with confidence in East Africa. For the moment, it seems that Kilombe is amongst the earliest Acheulian sites known where a variety of industrial expressions occurs very closely linked both stratigraphically and spatially. These facies will probably merit an addition to formal nomenclature (possibly a local 'phase' of the Acheulian), but such a step should not be taken until the variability of the material has been fully evaluated.

Kilombe is certainly one of the three richest Acheulian sites known in Kenya, but more important than this, it is a particularly fine experimental situation for studying the structure of a large occupation site or site group, and for producing further evidence of specific human behavioural patterns. At this interim stage the main emphasis has been on the physical details, of the artefacts, and of the actual sites. Ultimately, however, the value of this evidence will depend on the extent to which the human behaviour patterns involved can be explained.

Acknowledgements

My thanks are due to the following: to Professor W. W. Bishop for much assistance and co-operation in and out of the field during both the 1973 and 1974 seasons;

to R. A. Foley, Peter and Diana Tallon and Conran A. Hay for invaluable help in supervising and recording the work at various times; also to Kimolo Kimeu; and to O. Odak; to the Ministry of Natural Resources of Kenya, and to the National Museums of Kenya, for all assistance rendered and to the owners of the land, for every help. I thank also Mr M. A. Mirza of the Computer Department, University of Khartoum, who wrote a programme for processing the data from the bifaces. Dr Mary Leakey gave encouragement and valuable advice. I am very grateful to Glynn Isaac for reading through a draft of this paper and offering helpful comments and suggestions, and also for allowing me to refer to unpublished Olorgesailie data for comparison. At the time of the research I held a scholarship from the Department of Education and Science. The fieldwork was kindly supported by grants from the Cambridge University Worts and Smuts funds in 1973, and from the Smuts Fund, the Lambarde Fund of the Society of Antiquaries and the Boise Fund in 1974.

References

CHMIELEWSKI, WALDEMAR. 1968. Early and Middle Palaeolithic Sites near Arkin, Sudan. In: Wendorf, F. (ed.) *The Prehistory of Nubia, Vol. I.* Dallas, Texas: Southern Methodist University Press.

CLARK, J. D. 1959. Further Excavations at Broken Hill, Northern Rhodesia. *J.R.A.I.* Vol. 89, Part II. 201–32.

—— 1969. *Kalambo Falls Prehistoric Site, Vol. I*: The Geology, Palaeoecology and Detailed Stratigraphy of the Excavations. Cambridge: Cambridge University Press.

—— 1972. Mobility and settlement patterns in sub-Saharan Africa: a comparison of late prehistoric hunter-gatherers and early agricultural occupation units. In: Peter J. Ucko, Ruth Tringham and G. W. Dimbleby (Eds), *Man, Settlement and Urbanism.* Duckworth.

—— 1975. Africa in prehistory: peripheral or paramount? *Man* (N.S.), Vol. 10, No. 2. 175–98.

—— & HAYNES, C. V. 1970. An elephant butchery site at Mwanganda's Village, Karonga, Malawi, and its relevance for Palaeolithic Archaeology. *World Archaeology,* **1** (3), 390–411.

—— & KLEINDIENST, M. R. 1974. The Stone Age cultural sequence: terminology, typology and raw material. Chapter 4. In: Clark, J. D. 1974. *Kalambo Falls Prehistoric Site, Vol. II*: The Later Prehistoric Cultures. Cambridge: Cambridge University Press.

COOK, S. F. & HEIZER, R. F. 1965. The quantitative approach to the relation between population and settlement size. *Report of the Univ. of California Archaeological Survey,* 64.

DAGLEY, P., MUSSETT, A. E. & PALMER, H. C. 1978. Palaeomagnetic stratigraphy in the Baringo area. In: W. W. Bishop (Ed.), *Geological Background to Fossil Man,* Geol. Soc. Lond. Scottish Acad. Press., Edinburgh.

HOWELL, F. C., COLE, G. H. & KLEINDIENST, M. R. 1962. Isimila, an Acheulean occupation site in the Iringa Highlands, Southern Highlands Province, Tanganyika. In: G. Mortelmans, and J. Nenquin (Eds), *Actes du IVe Congres Panafricain de Prehistoire* (Tervuren, Belgium), 43–80.

ISAAC, G. LL. 1966. New Evidence from Olorgesailie, relating to the character of Acheulian occupation sites. *Proc. 5th. Pan-African Congress on Prehistory,* Teneriffe, 1963. 2:125–44.

—— 1968. The Acheulian Site Complex at Olorgesailie: A Contribution to the Interpretation of Middle Pleistocene Culture in East Africa. Doctoral Thesis, University of Cambridge.

—— 1969. Studies of early culture in East Africa. *World Archaeology,* **1,** 1–28.

—— 1972a. Chronology and the tempo of cultural change during the Pleistocene. In: W. W. Bishop and J. A. Miller (Eds), *The Calibration of Hominoid Evolution.* Edinburgh.

—— 1972b. Early phases of human behaviour: models in Lower Palaeolithic archaeology. In: Clarke, David L. (Ed.), *Models in Archaeology.* Methuen. 1972.

—— 1977 *Olorgesailie.* Chicago: Chicago Univ. Press.

—— & CURTIS, G. H. 1974. Age of early Acheulian industries from the Peninj Group, Tanzania. *Nature, Lond.* **249**, 624–7.

KLEINDIENST, M. R. 1962. Components of the east African Acheulian assemblage: An analytic approach, in G. Mortelmans and J. Nenquin (Eds), *Actes du IVe Congres Panafricain de Prehistoire* (Tervuren, Belgium), 81–112.

JONES, N. 1949. *The Prehistory of Southern Rhodesia.* Cambridge.

LEAKEY, M. G. *et al.* 1969. An Acheulian industry and hominid mandible, Lake Baringo, Kenya. *Proceedings of the Prehistoric Society* **35**, 48–76.

LEAKEY, M. D. 1971. *Olduvai Gorge, Vol. III*: Excavations in Beds I and II, 1960–3. Cambridge: Cambridge University Press.

—— 1975. Cultural Patterns in the Olduvai Sequence. In: *After the Australopithecines: Stratigraphy, Ecology, and Culture Change in the Middle Pleistocene.* Eds. Karl W. Butzer, Glynn Ll. Isaac. Mouton: The Hague, 477–94.

LUMLEY, H. DE. 1969. A Paleolithic camp at Nice. *Scientific American*, **220** (5), 42–50.

POSNANSKY, M. 1959. A Hope Fountain Site at Olorgesailie, Kenya Colony. *South African Archaeol. Bulletin*, Vol. 14: 83–9.

ROE, D. A. 1964. The British Lower and Middle Palaeolithic: some problems, methods of study and preliminary results. *Proceedings of the Prehistoric Society*, **30**, 245–67.

—— 1968. British Lower and Middle Palaeolithic hand-axe groups. *Proceedings of the Prehistoric Society*, **34**, 1–82.

WIESSNER, P. 1974. A functional estimator of population from floor area. *American Antiquity* **39**, 343–50.

22

PETER W. J. TALLON

Geological setting of the hominid fossils and Acheulian artifacts from the Kapthurin Formation, Baringo District, Kenya

The hominid from the Kapthurin Formation was found in the northeastern part of the basin of the Kapthurin River. The Formation is dominantly a sedimentary sequence with intercalated lavas and pyroclastic members. Evidence from lithology, fauna and algae has permitted reconstruction of the contemporary environment of the fossil hominid which has morphological affinities with some Algerian specimens of Homo erectus *from Ternefine. The artifact assemblage is Late Acheulian and exhibits a prepared core technique. The age of the hominid and the artifacts is between 0.7 and 0.23 m.y.*

During reconnaissance mapping west of Lake Baringo, J. Martyn discovered an incomplete hominid temporal bone, apparently derived from Member d of the Chemeron Formation (Martyn, 1969) and probably of Pliocene age. As a result of Martyn's discovery, an excavation was carried out by Dr Mary Leakey to ascertain the source of the bone and to see if any more hominid remains could be found. While the excavation was in progress, assistants prospected in the surrounding area and found an almost complete hominid mandible in the Kapthurin Formation on the south bank of the Bartekero river valley (Fig. 22:3). Associated with the mandible was postcranial material while Acheulian artifacts were found nearby. Excavations were carried out under the supervision of Richard and Margaret Leakey (Margaret Leakey *et al.* 1969). All material recovered from this excavation came from the Middle Silts and Gravels Member of the Kapthurin Formation.

Summary of the geology

The sediments of the Kapthurin Formation lie to the west and southwest of Lake Baringo (Figs. 22:1 and 2). They are bounded in the north and south by Lower to Middle Pleistocene basalts, mugearites and trachytes. In the north, Martyn named these lavas the Chemakilani Group (Martyn 1969), and in the south Griffiths called them the Ainapno Formation (Griffiths Ph.D. thesis in preparation). The western boundary of the Formation is formed by the Tugen Hills.

The Formation infills a fractured and downwarped basin floored mainly by tilted and faulted strata of the Chemeron Formation and the older Kaparaina Basalt Formation. The Kapthurin Formation was subdivided into five members by Martyn:

Member	Maximum Thickness
Upper Silts and Gravels	20 m
Bedded Tuff	12 m
Middle Silts and Gravels	40 m
Pumice Tuff	20 m
Lower Silts and Gravels	35 m

The Lower Silts and Gravels Member varies greatly in thickness because it infills many pre-existing fault troughs. The base is seldom exposed, but 35 m may be taken as an average figure for its thickness. The Lower Silts and Gravels are mainly coarse, fluviatile pebble beds. They exhibit a relatively high degree of sorting, which locally attains Stage III (mature) of Folk's stages of textural maturity (Folk 1951). The poor sorting of the modern river deposits in the Baringo area is the result of the highly seasonal contemporary climate. It seems possible that during the deposition of the Lower Silts and Gravels, the rainfall was more evenly spread throughout the year.

The Middle Silts and Gravels are separated from the lowest Kapthurin Member by the Pumice Tuff (Figs. 22:3 and 4). They consist mainly of light brown, poorly stratified silt similar in colour and texture to the fluviolacustrine sediments bordering Lake Baringo today. A few thin gravel horizons are present

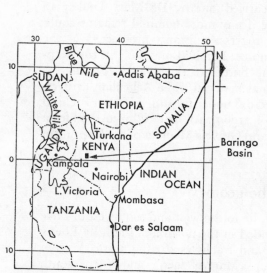

FIG. 22:1. East Africa – locality map. Grid of 10° squares lat. and long.

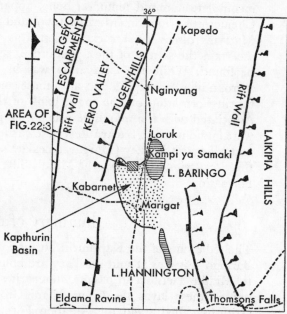

FIG. 22:2. Baringo Basin showing location of Fig. 22:3. Scale 1 cm = 10 km.

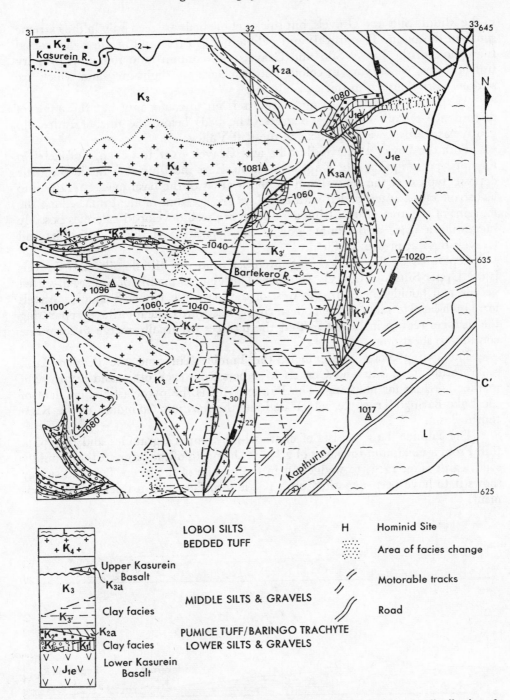

FIG. 22:3. Geology of the north eastern Kapthurin Basin showing the distribution of the lacustrine facies of the Middle Silts and Gravels Member. Grid of 1 km squares. Vertical scale in key 1 cm = 20 m.

in the Middle Silts and Gravels, but thick pebble beds are seen only in the extreme south of the basin. The gravels and pebble beds in this unit are less well sorted than those of the Lower Silts and Gravels and contain numerous thin primary tuff horizons, which vary in colour between dark and light grey and which are seldom more than 0.3 m thick.

Within the Middle Silts and Gravels is an unconformity but the angle of discordance seldom exceeds 5 degrees. The beds below the unconformity dip gently eastwards towards the axis of the Rift Valley.

Both the Lower and Middle Silts and Gravels exhibit a change of facies in the northeastern part of the Kapthurin Basin. Here the alluvial or fluviolacustrine deposits typical of the central and southern areas of the basin pass laterally into lacustrine clays. Interbedded with the clays are horizons of algal tufa. Along the margins of the area of lacustrine deposition are well developed calcretes possibly suggesting a more arid climate during the deposition of the Middle Silts and Gravels than at present in the area.

Separated by the Bedded Tuff Member from the Middle Silts and Gravels is the Upper Silts and Gravels Member. This consists mainly of coarse, ill-sorted pebble and boulder beds. In the west of the basin the Upper Silts and Gravels form high-level terraces of the modern rivers. Eastwards, the interfluves separating the terraces reduce in height so that the terraces eventually merge to form one continuous sheet of gravel.

The two pyroclastic Members, the Pumice Tuff and the Bedded Tuff, separate the three Silts and Gravels Members from each other and form useful marker horizons for mapping. The Pumice Tuff is the pyroclastic equivalent of the Lake Baringo Trachyte, which delimits the northern boundary of the Kapthurin Basin.

The Pumice Tuff consists of charcoal-grey pumice granules and fine pink tuff. From a maximum thickness of 20 m in the north of the basin, it thins rapidly southwards (Fig. 22:3) to wedge out in the middle of the basin. Detailed survey shows it to have been deposited on a flat landsurface with a thick soil development.

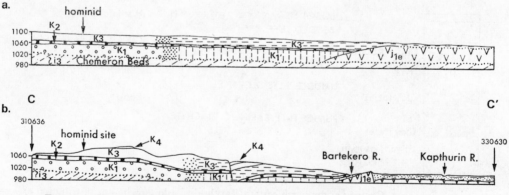

FIG. 22:4. Cross section (b) along line C-C' on Figure 22:3, and reconstruction of conditions at the end of K3 times (a), prior to faulting. Vertical scale metres above sea level. Length of section 2 km.

The upper, Bedded Tuff, extends further north than any other Kapthurin unit and has been identified about 35 km north of the centre of the Kapthurin Basin. Its maximum exposed thickness is 12 m and its lithology is markedly different from the Pumice Tuff, consisting of fine, grey to buff tuff often containing a large admixture of silt. The Bedded Tuff marks a period of sporadic volcanicity emanating from the north and resulting in a slow accumulation of tuffaceous material. It was deposited in a wide variety of facies interpreted as indicative of subaerial, alluvial, fluviolacustrine and lacustrine conditions. In the area to the north of the main Kapthurin Basin this unit was named the Kampi-ya -Samaki Beds by Martyn (1969), after the settlement on the western shore of Lake Baringo. The lower part of the unit is well bedded and fine grained, while the upper part is usually coarse and pumiceous.

The hominid locality

The hominid and associated artifacts were found on the south side of the Bartekero valley (Fig. 22:3). The terrain in this area is largely overgrown and there are few well exposed sections but a composite section (Fig. 22:5) has been drawn to illustrate the stratigraphic succession.

The lower part of the sequence comprises 12 m of red clay which marks a lateral change of facies of the Lower and Middle Silts and Gravels members. The clay overlies unconformably a vesicular basalt, which forms part of Martyn's Chemakilani Group. When dry, the surface of the clay hardens into discrete nodules, usually not more than 2 or 3 cm in diameter. Angular and semi-rounded lava fragments are scattered throughout the clay, but are not concentrated into bedded layers. The lava fragments do not exceed 3 cm in diameter and are mostly basaltic. The red clay is the most fossiliferous horizon in the Kapthurin Formation.

Within the red clay is the Pumice Tuff, which is here 6 m thick. Although very weathered, the tuff can be seen to be coarse grained and parallel bedded. The regularity of the bedding possibly suggests deposition in standing water.

Overlying the red clay is 8 m of black clay with the 'cotton soil' texture typical of soils developed from lavas. In part the black clay is a more southerly facies variant of the upper part of the red clay which it also overlies. The black clay is more homogeneous in lithology than the red and contains fewer lava fragments. Both clays include layers of algal mats and domes. The significance of these is discussed below.

About 5 m of rust-brown silty clay conformably overlies the black clay. The silt content increases upwards in the sequence, fossils become fewer and more fragmentary, algal deposits become more scarce and in their stead layers of calcrete become an important constituent which continues to the base of the Bedded Tuff Member.

The silty clay passes upwards into 18 m of massive brown silt, which in turn is overlain by the Bedded Tuff. The silt is permeated by calcified root casts, which are sometimes densely packed to form calcrete layers. The *in situ* hominid remains were found in the brown silt 2 m above a 1 m thick calcrete (Fig. 22:6).

This was at 313635, where the hominid horizon is approximately 3 m below the unconformity in the Middle Silts and Gravels. The plane of unconformity cannot be identified because of the poorly stratified nature of the brown silt.

Because the Upper Silts and Gravels do not extend north of the Kapthurin

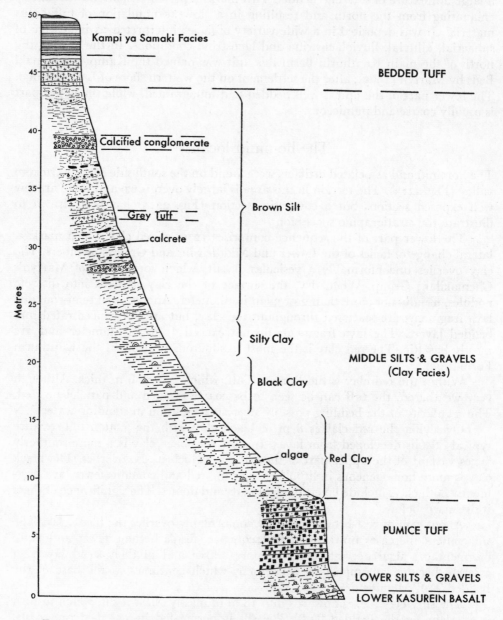

FIG. 22:5. Generalized sequence of the Kapthurin Formation in the area between the rivers Bartekero and Kapthurin: Between map refs: 315636 to 323636 (northern boundary) and 315625 to 323626 (southern boundary).

river, the uppermost unit in the hominid area is the Bedded Tuff which here is grey and parallel bedded, each bed averaging 3 cm in thickness and with a thin coating of calcium carbonate on each bedding plane. On closer inspection, each bed can be seen to exhibit parallel microbedding. The thickness of the whole Member in this area is 8 m. The Bedded Tuff here represents the most southerly extension of the lacustrine facies of the Kampi-ya-Samaki-Beds.

Palaeoenvironmental interpretation

The hominid was found in the transition zone between the fluviolacustrine silt beds typical of most of the Kapthurin Formation, and the clay area in the north-eastern part of the basin. Lithology, flora and fauna provide evidence of an environment in the northeast which was different from that of the rest of the Kapthurin Basin.

The red and black clays in the hominid area are similar in texture to the

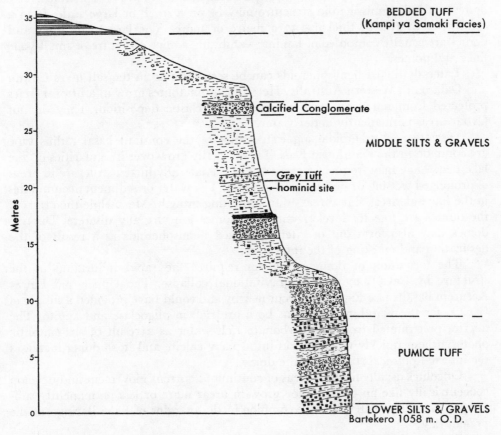

FIG. 22:6. Detailed sequence through the Kapthurin Formation hominid site. Bartekero river section, Ref. 313635.

blue-black clays being deposited now at the bottom of Lake Baringo. The differences in colour can be attributed to variations in local rock types and degree of oxidization. Fauna obtained from the red and black clays includes fragments of *Tilapia* sp., *Crocodylus* sp., *Hippopotamus amphibius* (Linn.) and shells of the gastropod *Melanoides* sp. Although this fauna does not conclusively prove the existence of lacustrine conditions, a similar association can be seen at the present day along the margins of Lake Baringo but with a smooth shelled gastropod having replaced *Melanoides* sp.

More information concerning the lacustrine deposits can be obtained by a study of the algal bioherms. The distribution of these matches closely that of the clays with which they are interbedded. A simple classification of algal deposits, based on their compound structure was devised by Logan *et al.* (1964), for the identification of hand specimens in the field. The structures are not mutually exclusive, but grade from one to the other. The two basic forms are spheroids and hemispheroids (oncolites).

Only hemispheroids are found in the Kapthurin Beds. These structures are subdivided into laterally linked hemispheroids and discrete hemispheroids. The dome and link hemispheroid structure may be on a small or large scale, from a centimetre or two in thickness to a metre or more. The large-scale links and domes are usually composed of laminae exhibiting a microstructure of small-scale links and domes.

Laterally linked hemispheroids can be seen forming in the salt lakes Cowan and Dalaroo in Western Australia. Here the stromatolites grow in a littoral facies protected from much current movement and clastic deposition. They do not favour areas permanently under water.

Discrete hemispheroidal algal structures of the constant basal radius type are common in the Kapthurin Beds. These usually grow over irregularities on the lake bed. They may frequently grow out of the laterally linked structure in areas of prolonged wetting or heavy sedimentation. The water or sediment accumulates in the low link areas (interareas) thus inhibiting growth. Meanwhile the crests of the domes are free to develop until the domes are virtually discrete. Discrete domes may also form out of laterally linked hemispheroids as a result of the dessication and cracking of the interareas.

The formation of dome structures requires the rapid induration of the structure by $CaCO_3$ to prevent gravitational collapse. The Upper and Lower Kasurein Basalts (see Fig. 22:3) occur nearby and could have provided a supply of $CaCO_3$ for the Kapthurin algae as both are rich in oligoclase and augite. The micrite precipitated from the carbonate rich water as a result of stromatolitic photosynthesis quickly recrystallizes into sparry calcite and in so doing increases the internal physical stability of the dome.

Oncolites usually form in areas of continuous current movement and develop concentrically like an oolite. They grow in areas more or less permanently submerged in shallow water, and are common in the margins of Lake Baringo at the present day.

Stromatolites thrive equally well in saline or fresh water conditions. In the case of the Kapthurin algae, the associated fauna precludes the possibility of a saline lake. Cloud, 1942, estimated the maximum depth for stromatolitic growth to be 10 m in fresh water.

It is now possible to summarize the environment that existed immediately to the north and east of the hominid locality. The lateral extent of the clay and algal horizons in the Kapthurin Beds defines the northeastern Kapthurin facies as forming an east–west trending lobe of shallow standing water which was almost certainly a westerly projection of the contemporary central Rift Valley Lake. (Fig. 22:7). The lobe was bounded on three sides by land and formed an area of heavy clastic deposition, and restricted current circulation.

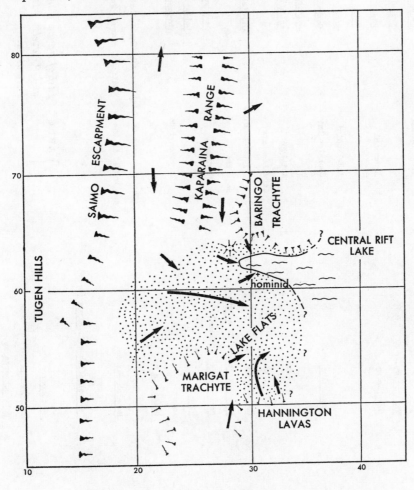

 MIDDLE SILTS & GRAVELS

 DIRECTION OF FLOW

Fig. 22:7.

Reconstruction of probable palaeo-environments during deposition of the Middle Silts and Gravels Member of the Kapthurin Formation. Scale 1 cm approx. 2 km.

MOLLUSCA
  Gastropoda   Melanoides sp.

PISCES   Tilapia sp.
                    Clarias sp.

REPTILIA
  Crocodilia   Crocodylus sp.

MAMMALIA
  Primates
    Cercopithecidae   Colobus sp.
                        Cercopithecus c.f. aethiops (Linn.)
                        Papio anubis (Lesson).
    Hominidae   Homo sp.
  Carnivora   indet.
  Proboscidea   Loxodonta sp.
  Hyracoidea
    Procaviidae   Heterohyrax sp.
  Perissodactyla
    Rhinocerotidae   Ceratotherium c.f. simum (Gray)
  Artiodactyla
    Suidae   Mesochoerus sp.
                Notochoerus sp.
                Potamochoerus sp.
                Tapinochoerus sp.
                Phacochoerus sp.

    Hippopotamidae   Hippopotamus amphibius (Linn.).
    Bovidae
      Tragelaphini   Tragelaphus c.f. scriptus (Pallas).
                      Tragelaphus sp.
      Reduncini   Redunca sp.
                   Kobus ellipsiprymnus (Buckley).
      Alcelaphini   Alcelaphus sp.
      Cephalophini   Cephalophus sp.
                      Sylvicapra sp.
      Antilopini   Antidorcas sp.
  Tubulidentata
    Orycteropodidae   Orycteropus sp.
  Rodentia
    Muridae   Oenomys sp.
                Rattus c.f. rattus (Linn.).
    Thryonomyidae   Thryonomys sp.

(The specimens on which Dr. John Harris based the above identifications are lodged in the Kenya National Museum, Nairobi)

TABLE I: *List of Fauna from the Middle Silts and Gravels Member. Kapthurin Formation.*

A substantial faunal assemblage was recovered from the clays in the north-eastern part of the Kapthurin Basin (Table 1). The fossils are mainly fragmentary, which suggests river transportation. The fossils recovered represent extant forms with the exception of the hominid and some of the Suidae. Apart from *Antidorcas* sp., which is now restricted to southern Africa, the full Kapthurin fauna can still be seen in East Africa.

The Kapthurin Formation contains fauna similar to that living in open bushland today. There is a small contingent of marsh-dwellers present such as *Kobus ellipsiprymnus* (Buckley) and *Antidorcas* sp. *Potamochoerus* sp. also favours marshy or woodland conditions. The areas of marsh were probably limited to the margins of the central Rift Lake, as they are today.

The physical palaeoenvironmental setting of the Kapthurin Basin

This summary of the environment in the Kapthurin Basin is made as through the eyes of the hominid if he had been standing at the location where his remains were found (313635).

One km to the north, would be the conspicuous flow front of the Lake Baringo Trachyte, and in the foreground the westerly lobe of the central Rift Lake. This lobe may well have been a watering place for the local fauna, and hence a potential source of food for the hominid. The Baringo Trachyte provided the raw material for making stone tools.

Southwards towards the main area of Kapthurin sedimentation, lay broad expanses of lake flats with only meagre soil development, probably similar to the modern lake flats surrounding Lake Baringo. In the distance 12 km away, low, dark hills formed by the Marigat Trachyte and Hannington Phonolite Formation, delimited the southern boundary of the basin (Fig. 22:7).

In the west, the escarpment of the Tugen Hills dominated the landscape as it does at present. The lake flats passed laterally into a wash plain of alluvial deposition, which extended as far west as the hills.

The climate was not greatly different from that of today. The hominid could have been standing on a calcrete soil profile (his remains were found just above one), formed by leaching during the dry season. Eastwards on the lake margin broad areas of algal growth probably flanked the shore line.

The hominid and associated artifacts

Detailed work on the morphology of the hominid and the nature of the artifact assemblage has been carried out (Margaret Leakey *et al.* 1969). Here only a very brief summary of that work will be given.

Artifacts are common scattered over the surface of the interfluve separating the rivers Bartekero and Kapthurin, but little artifact material was found associated with the *in situ* hominid remains. Only one bifacial core tool and 14 flakes were recovered from the site.

Artifacts have been found at virtually all stratigraphical levels in the Middle Silts and Gravels, but the vast majority of tools come from a level 9–10 m above that of the hominid, and 3–4 m below the base of the Bedded Tuff. This level yielded 42 flake tools and 13 core tools. The former include ovates and cleavers, while the latter are choppers. Also present were 10 retouched flake tools. Prepared cores are common in this assemblage. Of the 83 found, 26.7 % are blade cores. Eight hundred and seventy flakes were found of which 13.4 % are blades.

Another important site lies just to the south of the Kapthurin river at 312623. This is apparently a factory site and is at the same stratigraphical level as the hominid horizon. Only 3 core tools and 2 flake tools were found at the factory site. The rest of the material includes 18 cores and 514 flakes, of which 8.9 % are blades.

Nearly all the artifacts are made of Baringo Trachyte. The abundance of local raw material probably explains the large number of big flakes in the Middle Silts and Gravels. Other large prepared flake tool sites have been found in the Vaal River Basin (van Riet Lowe 1945), and in the Horn of Africa (Clark 1954).

The hominid surface material consisted of part of an ulna, some limb bone shaft fragments and the mandible. The *in situ* remains comprised two phalanges and one metatarsal. The mandible is almost complete and resembles three Middle Pleistocene Algerian mandibles from Ternifine (Tobias, *in* Margaret Leakey *et al.* 1969). The Algerian specimens were described as *Homo erectus mauritanicus* (Arambourg 1963). Although it fitted well within the range of known *Homo erectus* mandibles, Tobias was unwilling to ascribe the Kapthurin specimen to *Homo erectus* on the basis of mandible morphology alone. More cranial or postcranial material is required to permit a categorical identification.

Chronology

The Middle Silts and Gravels are bracketed between the two pyroclastic members both of which have been dated by ^{40}K–^{40}Ar as follows:

The Pumice Tuff;	
Series 1	0.66 ± 0.10 m.y.
	0.68 ± 0.10 m.y.
Series 2	0.85 ± 0.28 m.y.
	0.82 ± 0.28 m.y.
	0.87 ± 0.26 m.y.
The Bedded Tuff;	
	0.24 ± 0.12 m.y.
	0.23 ± 0.08 m.y.

(All dates were determined by Dr J. A. Miller in the Cambridge University Geochronology Laboratory.)

The Pumice Tuff exhibits normal polarity (Dagley, this volume), which suggests that it is younger than 0.7 m.y. B.P. and hence favours the younger dates. The ^{40}K–^{40}Ar ages indicate a latitude of half a million years for the deposition of the Middle Silts and Gravels.

The remains of the hominid were found half way up the Middle Silts and Gravels Member, 2 m above a calcrete and about 3 m below the unconformity. All that can be said about the antiquity of the specimen and the artefacts is that they are probably younger than 0.7 m.y. and older than 0.23 m.y.

Acknowledgement

I thank Dr J. A. Miller for permission to publish the potassium-argon dates and Dr J. Harris for identifying the mammalian fossils. I am indebted to my wife, Diana, for assistance with the field work. Dr A. C. Walker, Dr & Mrs A. Brock and Mr J. Morgan assisted in many ways. Professor W. W. Bishop supervised the research which was funded by grants from the Wenner-Gren Foundation for Anthropological Research, New York and the Natural Environment Research Council (NERC). The work was carried out while I held a Natural Environment Research Council Overseas Research Studentship.

References

ARAMBOURG, C. 1963. Le Gisement de Ternifine, *Arch. Inst. de Paléontologie Humaine, mem.* **32**: 37–190, Paris, Masson et Cie.

CLARK, J. D. 1954. *The Prehistoric Cultures of the Horn of Africa*, Cambridge University Press.

CLOUD, P. E. 1942. Notes on stromatolites. *Am. Jour. Sci.* **240**: 263–9.

FOLK, R. L. 1951. Stages of textural maturity in sedimentary rocks. *Jour. Sediment. Petrol.*, **21**: 127–30.

LEAKEY, MARGARET, TOBIAS, P. V., MARTYN, J. E. & LEAKEY, R. E. F. 1969. An Acheulian industry with prepared core technique and the discovery of a contemporary hominid at Lake Baringo, Kenya. *Proc. Prehist. Soc.* **35**: 48–76.

LOGAN, B. W., REZAK, R., & GINSBURG, R. N. 1964. Classification and environmental significance of algal stromatolites. *Jour. geol.*, **72**: 68–83.

LOWE, C. VAN RIET. 1945. The evolution of the Levallois technique in South Africa. *Man*, London, **45**: 37.

MARTYN, J. E. 1969. *The geological history of the country between Lake Baringo and the Kerio river, Baringo District, Kenya*. Unpublished Ph.D. thesis, Bedford College, London.

23

ROBERT J. G. SAVAGE &
PETER G. WILLIAMSON

The early history of the Turkana depression

The Turkana depression lies between the Kenyan and Ethiopian domes and is traversed by Lake Turkana. Within and around the depression are sequences of clastic and volcaniclastic sediments and lavas. In six areas around Lake Turkana the sediments have yielded fossils which indicate that some of the deposits are Miocene in age and others are considerably earlier, probably late Mesozoic. The post-Miocene history is not treated. Detailed study of the sediments in one area, Kajong, reveals two distinct cycles of deposition. A model is proposed which relates these to cycles of downwarping and deposition, uplift and erosion. The sediments of the downwarped depression are seen as alluvial fan deposits. The periodic uplift of the domes caused faulting along the margins that produced downfaulted blocks on which the sediments have been preserved. The sedimentary features, vertebrate faunas and radiometric dating are used to establish a tentative correlation of the early events in the depression.

The Turkana depression in Northern Kenya is a triangular lowland lying between the Kenyan and Ethiopian domal uplifts in the middle sector of the East African rift system (Fig. 23:1). Much palaeontological and geological interest in recent years has centred on the Plio-Pleistocene deposits of the area, but comparatively little is known in detail of the earlier history of this portion of the rift.

Several areas within the depression are known where pre-Pliocene sediments and volcanics occur, and a general picture of the early evolution of the area has emerged and become accepted.

Major uplift apparently occurred as early as the late Mesozoic on both the Kenyan and Ethiopian domes (Baker *et al.* 1972) but the Turkana depression between them is generally considered to have originated as a distinct feature during the early Miocene. This feature is thought to have been the result of the downwarping of the sub-Miocene surface on the north-west flank of the Kenyan dome. The broad, partly faulted depression thus formed extended between the two domes from north-east Kenya via Lake Turkana (formerly Lake Rudolf) to the southern Sudan (Baker *et al.* 1972). Early sedimentation within this depression consisted o coarse clastics derived from its edges; these rest unconformably on the Basement System on which the sub-Miocene surface had been cut (Dodson 1971, Joubert 1966).

Certain of these early sediments in the Turkana depression have yielded Miocene faunas, as have some volcaniclastic and sedimentary intervals in the overlying volcanic sequence. Subsequent to the Miocene period of volcanism and deposition, major submeridional faulting was considered to have cut the floor of the Turkana depression, dividing it into west-tilting blocks and defining the

Gregory graben which now bounds the area to the east. The major faulting is considered to extend from late Miocene to late Pliocene (Baker *et al.* 1972).

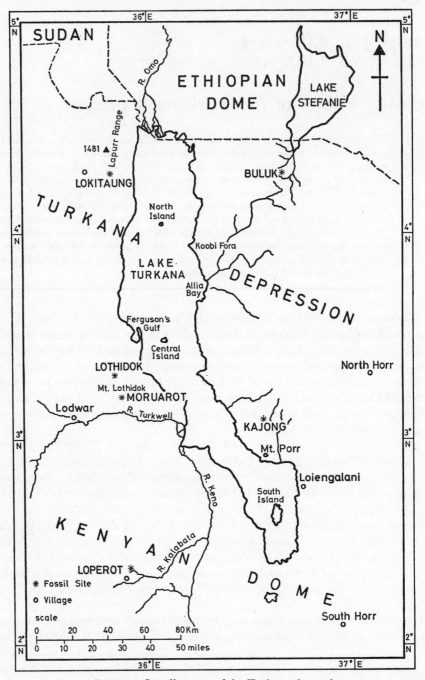

FIG. 23:1. Locality map of the Turkana depression.

Recent work however in the southeastern portion of the Turkana depression (Kajong), allied to the recovery of a Mesozoic faunal element in the deposits of the Lapurr Range[1] north-west of Lake Turkana (Arambourg & Wolff 1969), challenges a blanket allocation to the Miocene of all coarse clastic sediments with a basement contact. It suggests a more complex history and greater antiquity for the Turkana depression and many of its major structures than has hitherto been supposed.

Six areas within and around the Turkana depression have yielded vertebrate faunas. They are Lokitaung, Loperot, Lothidok, Moruarot, Buluk and Kajong. The authors have worked at Buluk and Kajong, but have neither worked at nor visited the other four areas.

Regional survey of areas around Lake Turkana

Lokitaung

The Lapurr range west of Lake Turkana rises to around 1500 m on Mount Lapurr, some 18 km NNW of Lokitaung. The geology of the sedimentary succession in the area is described in Arambourg (1933b, 1935, 1943); Arambourg & Wolff (1969); Fuchs (1934, 1935, 1939); Murray-Hughes (1933) and Walsh & Dodson (1969). Arambourg and Fuchs are responsible for the only extensive field work in the area, and that achieved under considerable difficulties over forty years ago.

Murray-Hughes (1933) described a sedimentary series in north-west Kenya as 'Turkana Grits'. The term has been applied by almost all later authors to a wide range of sediments, including volcaniclastic sediments, not only in north-west Kenya but further south, and at its widest is used for any sediment between the Basement System and the Tertiary lavas. Walsh & Dodson (1969) use the term in a more restricted way, applying it only to sediments which lack any volcanic content. Arambourg (1933b) introduced the term 'Lubur Series' for the non-volcanic sediments of the Lapurr range, but the term has not received wide acceptance. The Kenyan Geological Survey have always used the term Turkana Grits. Until we have studied the area we do not propose to recommend a preferred nomenclature.

The clearest account of the sediments of the Lapurr range is still the original account of Arambourg (1933b) where he described the sequence a few kilometres north of Lokitaung. The sediments there are around 200 m thick, comprising a detrital series of sandstones and conglomerates with dips of 15° to the SSW. They rest on the Basement System and are capped by Miocene lavas. The sediments contain no material of volcanic origin. There is a predominance of quartz with varying proportions of arkoses and occasional schist pebbles. The conglomerates are mostly made up of rounded quartz pebbles and the cement is siliceous, calcareous or ferruginous. Yellow and red colours predominate. The sandstones are crossbedded and alternate with conglomerates; in modern terminology they would

[1] Throughout the paper the spelling adopted for place names is that used on the 1:250 000 maps of Kenya.

be referred to as cycles of fining-up sequences. They are reported to wedge out rapidly north of Mount Lapurr and the maximum thickness of around 300 m is recorded a few miles south of the peak.

The fauna from the sediments of the Lapurr range is extremely sparse, comprising only fragments of wood and bone, and this has resulted in wide divergences of view concerning the age of the deposits. In the absence of any identifiable fauna, Arambourg (1933b) compared the succession with the Adigrat Sandstone of Ethiopia and so suggested a Trias-Jurassic age. Fuchs (1939) found large silicified tree trunks in the grits of Lokitaung gorge which were identified as *Dryoxylon*, a genus recorded from the Tertiary of India and Africa. Fuchs inferred from this that the sediments were of 'Oligo-Miocene' age. Arambourg (1943) argued that *Dryoxylon* gives a poor indication of age, and that Eocene or even Cretaceous was a more likely guess. More recently Arambourg & Wolff (1969) recorded the discovery of a dinosaur humerus in the grits at the east end of Lokitaung gorge. The bone is heavily silicified and only the distal end is preserved; it can be referred to the order Sauropoda, but not identified further. Sauropods are amongst the last of the dinosaurs to have become extinct in late Cretaceous times, from which it might be inferred that at Lokitaung some of the sediments are of Cretaceous age. But Walsh & Dodson (1969) report that sediments in this vicinity are reworked and it is not impossible that the fossil is derived.

Amongst the conflicting statements regarding these sediments, it is patently clear that Arambourg alone recognised two distinct periods of sedimentation in the area west of Lake Turkana, one Miocene and the other considerably pre-Miocene. Other authors have failed to appreciate this fundamental point.

Loperot

Murray-Hughes (1933) mapped Turkana Grits as occurring east of Loperot, and Champion (1937) also noted their presence. Dixey (1945) recorded Turkana Grits outcropping both north-east and south of Loperot. He noted that north-east of Loperot there are hills with 100 m of tabular white grits without a lava cap, while between these and Loperot ridge are low hills in which grits occur interbedded with lavas. South of Loperot Dixey recorded extensive areas of interbedded sandstones and lavas, and from these he obtained *Prodeinotherium hobleyi*. The area was visited by Leakey in 1948 when he collected a tragulid and a rhinocerotid (Clark & Leakey 1951). Prof. Bryan Patterson led Harvard University expeditions to the area between 1963 and 1965. The Kenya Geological Survey report by Joubert (1966) gives the only account of the geology of the area and the papers on the fauna are to be found in Hooijer (1966, 1971, 1973), Maglio (1969) and Mead (1975). Descriptions of some of the mammalian genera collected by Patterson's expeditions have yet to be published.

Joubert's (1966) report on the geology of the Loperot area gives only a generalised section in the Turkana Grits and none has yet appeared for the fossil sites. From Dixey (1945) it would appear that two cycles of sediments occur, clastic sediments without volcanic material and sediments with interbedded tuffs. The former could be the southern equivalent of Arambourg's 'Lubur Series'. A similar distinction was made by Mason & Gibson (1957) for sediments further

TABLE 1: *Mammal faunas*

	Loperot	Kajong	Buluk	Lothidok & Moruarot
PRIMATES				
Dryopithecus nyanzae				+
„ sp.	+			
RODENTIA				
Paraphiomys pigotti				+
„ sp.	+	+		
CARNIVORA				
Anasinopa leakeyi				+
Metasinopa sp.	+			
Hyaenodon andrewsi				+
Pterodon sp.			+	
Kichechia zamanae				+
CETACEA				
Ziphiid	+			
HYRACOIDEA				
Megalohyrax championi			+	+
Prohyrax sp.	+			
Hyracoidea indet.		+		
PROBOSCIDEA				
Prodeinotherium hobleyi	+	+	+	+
Gomphotherium angustidens				+
„ *kisumuensis*		+	+	
Platybelodon sp.	+			
Zygolophodon				+
RHINOCEROTIDAE				
Aceratherium acutirostratum				+
Dicerorhinus sp.			+	
Chilotheridium pattersoni	+			
Rhinocerotid indet.		+		
SUIDAE				
Bunolistriodon jeanneli				+
Lopholistriodon moruoroti				+
Listriodon sp.			+	
Suidae indet.	+	+		+
ANTHRACOTHERIIDAE				
Brachyodus sp.				+
Gelasmodon sp.	+			
Masritherium sp.	+			
Anthracotheriid indet.		+		
TRAGULIDAE				
Dorcatherium chappuisi				+
„ *parvum*				+
„ cf. *pigotti*	+	+		
GIRAFFOIDEA				
Canthumeryx sirtensis				+
Propalaeoryx sp.				+
BOVIDAE				
Bovidae indet.	+			

south; they use the term Turkana Grits for sediments with recognisable basement clasts and abundant quartz, and introduce the term Tiati Grit Series for sediments containing volcaniclastic material. Joubert gave a generalised section near Lakhapelinyang, 18 km north-east of Loperot; here he recorded 215 m of conglomerates, grits, sandstones, silts, limestones and tuffs. The sediments appear to show cyclic fining-up sequences; the matrix is often calcareous, especially near the base and ferruginous beds are recorded. Mammals occur in both the sandstones and limestones. The series rests on basement and is capped by lava.

Radiometric dating of the lavas close to the sediments is reported by Maglio (1969), Hooijer (1971) and Mead (1975). At the fossil sites the lava is said to rest on the sediments without any marked unconformity. Mead recorded that a lava 6 m above a fossiliferous level was dated at 16.7 ± 1.0 m.y. Maglio quoted a figure of 17.5 ± 0.9 m.y. for a similar level. Hooijer quoted slightly different figures, 17.5 ± 0.9 m.y. for a lava 1.5 m above the grit, 16.7 ± 0.8 m.y. for a lava 61 m above the grit and 15.8 ± 1.2 m.y. for a boulder in the grit at 12 m below the lava contact. It would seem that the fossiliferous beds are around 16 to 17 m.y.

The fauna of the beds comprises bivalves, gastropods, coprolites, fish (including *Protopterus* and *Polypterus*), crocodiles (including *Tomistoma*) and mammals. The mammals are tabulated in Table 1. Some elements are worthy of comment; the ziphiid whale (Mead 1975) is unique in Africa—indeed no other non-marine whale is known on the African continent. The *Platybelodon* (Maglio 1969) is likewise unique to Africa and appears to represent a very early member of a stock that diversified into Asia and North America. Unhappily only two taxa are specifically identifiable; the deinotheres are not good for close stratigraphical correlation and the rhinoceros is a *species novum*, though Hooijer (1973) stated that this species can be recognised at Ombo, Ngorora and possibly Rusinga.

Lothidok and Moruarot

These two localities are 11 km apart; each has two sites from which mammals have been recorded. The fossiliferous area was discovered by Arambourg in 1933. It was visited again by Phillips, Deraniyagala and Cooke in 1948 and their collections are in Berkeley and Colombo. Over 300 specimens are known from Moruarot and about 50 from Lothidok. It is mostly rather scrappy material and identification is difficult. The geology is described in Arambourg (1933a, 1933c, 1935, 1943) and Walsh & Dodson (1969). Papers mentioning the fauna are numerous; those with original contributions are Arambourg (1933c, 1947); Clark & Leakey (1951); Deraniyagala (1951); Churcher (1970); Hooijer (1966, 1968); Madden (1972); Savage (1965); Whitworth (1954, 1958) and Wilkinson (1976).

The geology appears to be similar in both localities and on the basis of the information available, separate descriptions are not warranted. Between 195 m and 312 m of interbedded lavas and volcaniclastic sediments are recorded; the latter comprise tuffs, tuffaceous sandstones, grits and conglomerates. Crossbedding is seen in the sandstones and fining up cycles appear to be present. The beds dip at 15° to 30° to the west. The lowest lavas rest on 'Lubur Series' which dips at 20°–25° to the east or north-east. Sediments are capped by analcime and augite lavas. Mammals mostly occur in fine purplish-red sandstones and are coated in a calcareous cement.

Radiometric dating on the overlying basalts gives ages that range from 32 to 14 m.y. Reilly *et al.* (1966) gave dates of 20.7 and 32.2 m.y. for lavas regarded as equivalent flows in the Maralal area, some 275 km to the south. Baker *et al.* (1971) gave a date of 14.0 m.y. for a lava at Lothidok, but did not state whereabouts within the section it occurs.

The faunas from the sites include bivalves (including perhaps *Etheria*), gastropods (Van Damme & Gautier 1972, Verdcourt 1963), reptiles and mammals. The mammal inventory given in Table 1 is based on examination of all the specimens in Nairobi and Berkeley. Some refinements may be achieved and perhaps a few new taxa added, but we cannot uphold a number of the taxa introduced by Madden (1972). There are a number of elements at both generic and specific level which occur in early Miocene sites near the Gulf of Kisumu. Of interest is the occurrence of *Zygolophodon*, which is also recorded from Moroto, Uganda (Bishop 1967), and *Listriodon*, which is known from Ngorora and Buluk but not in the Kisumu sites.

Kajong

During 1973 Mr Richard E. F. Leakey, Director of the Kenya National Museums, visited The Kajong area, north-east of Loiengalani and close to the south-east shores of Lake Turkana; here he noted a series of clastic and volcaniclastic sediments containing Miocene fossils. The sediments were discovered during the explorations of Count Teleki in 1888 and reported upon by von Höhnel *et al.* (1891), but as no fossils were found, their age was unknown. Over the past two seasons the authors have undertaken geological mapping and a limited amount of fossil collecting in the area. A large faulted basement horst is overlain by two major sedimentary cycles, each resting on a distinct erosion surface. These exposures cover a rectangular area of about 330 km² over a region centred on 3°03′ N 36°32′ E (see Fig. 23:1). The horst consists of a series of fault bounded blocks which are defined to the south of the area by an intersecting grid of E–W and NNW–SSE trending normal faults, and to the north by the NNW–SSE set alone. Some lateral movement, mainly dextral, has taken place on these faults, particularly on the E–W set. The NNW–SSE faulting parallels the lineation in the basement system. To the north of the area the blocks dip to the NW or NNW, but to the south, where grid faulting is well developed, dips are more variable. The sedimentary sequences originally deposited on the basement have been preserved by the faulting, either in the angles between adjacent fault-bounded blocks or when a change in direction of downthrow of two adjacent faults has produced graben-type structures within the horst.

The lowest sedimentary sequence, the 250 m thick Sera Iltomia Formation, rests nonconformably on basement and may be divided into four members. (Fig. 23:2). Member 1 is interpreted as lateritised, partly *in situ* debris lying on a low relief basement erosion surface. In places basement may be seen breaking down in situ to form this regolith; elsewhere lensoid conglomerate accumulations rest directly on basement.

Member 2 consists of a superimposed series of maroon-coloured, laterally extensive, erosion-surface bounded fining-up sequences, up to 5 m thick. These sequences typically begin with a poorly sorted massive or cross-bedded basal conglomerate of subangular to subround quartz, basement, or intraclasts set in a matrix of granular or very coarse sand grade feldspathic arenite or litharenite. This basal conglomerate rests on a scoured erosion surface and rapidly fines up into increasingly lateritised medium to fine feldspathis litharenites or sublitharenites, or silty mudstones, which are truncated by another erosion surface underlying the succeeding unit. Subordinate silcretes occur at various levels. In places the basal

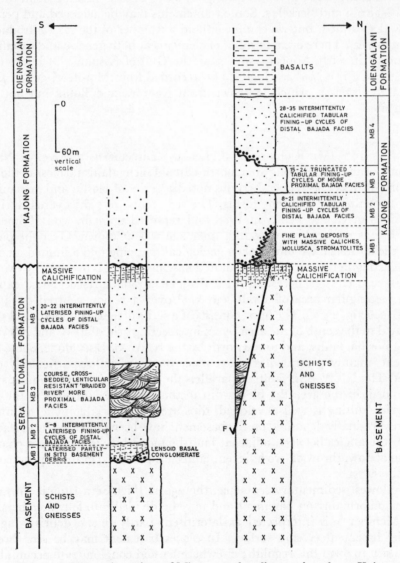

FIG. 23:2. Schematic sections of Miocene and earlier stratigraphy at Kajong.

conglomerates thicken into elongated lensoid bodies up to 8 m across which protrude below the general level of the lower surface of the unit and which appear to be the distributaries from which it was deposited. The basic sedimentary unit in member 2 seems similar to Bluck's Conglomerate 'A' (Bluck 1967) and is considered to be the result of sheet-flood deposition from a ramifying distributary system on a distal coalesced fan front or bajada.

Member 3 consists of a thick, greyish-yellow stacked and intercalated series of trough and planar cross-bedded fining up sandstone lenticles. Basally these lenticles may be very coarse, incorporating conglomerates of subangular to angular quartz, basement and intraclasts, set in a matrix of granular or very coarse feldspathic arenite or litharenite; the clasts may reach small cobble size. This member is generally coarser than member 2, and fines are very subordinate, although they may occur to the tops of certain lenses, to silty mudstone grade; otherwise the lenses may grade up to coarse or very coarse silty feldspathic arenites and sublitharenites. Occasionally member 2—type fining-up sequences may be intercalated in the succession. The sandstone lenticles may be up to 15 m across. The member seems similar to Bluck's conglomerate 'C' (Bluck 1967), and is interpreted as a braided river deposit laid down on a more proximal portion of the bajada than member 2, at a point where the braided system had not yet broken down to give the more isolated distributaries seen at the bases of the fining-up sequences in member 2. Member 3 grades rapidly into member 4, which is very similar to member 2. The clastics in these deposits appear to have been derived from a basement terrain, and volcanic clasts are absent.

Transport directions, based on cross-bedding, change in maximum clast size and on sphericity decline in the quartz 5–35 mm size grade for units 2–4 indicate that sediment was moving into the area from the south on a bearing of around 020°.

Members 2 and 3 are considered to record the building out of a coalesced fan-front or bajada into the area, which transgressed a low-relief, lateritised regolith-mantled basement erosion surface from the south. Member 4 apparently records the waning of this system, presumably due to levelling of its source.

The top of the Sera Iltomia Formation appears everywhere to have been removed by erosion and the formation may originally have been considerably thicker than the 250 m presently preserved.

The second sedimentary sequence in the area is the 210 m thick Kajong Formation, which is divisible into four members (Fig. 23:3). This sequence rests non- or disconformably on basement or on the Sera Iltomia Formation, both of which are massively calichified at their contact with the Kajong.

Member 1 of the Kajong Formation is seen only in the northern areas. It consists of a series of bright red, green and white featureless medium-fine feldspathic arenites, feldspathic sublitharenites and silty mudstones. Micrite cement is ubiquitous throughout the whole Kajong Formation. Its presence is attributed to pedogenesis, and several well marked cycles of calcrete formation can be discerned in member 1 and succeeding units. Typically, sandstones with some micrite cement and relatively minor solution of clastics pass up into beds with very embayed and dissolved grains dispersed in a micrite which displays growth

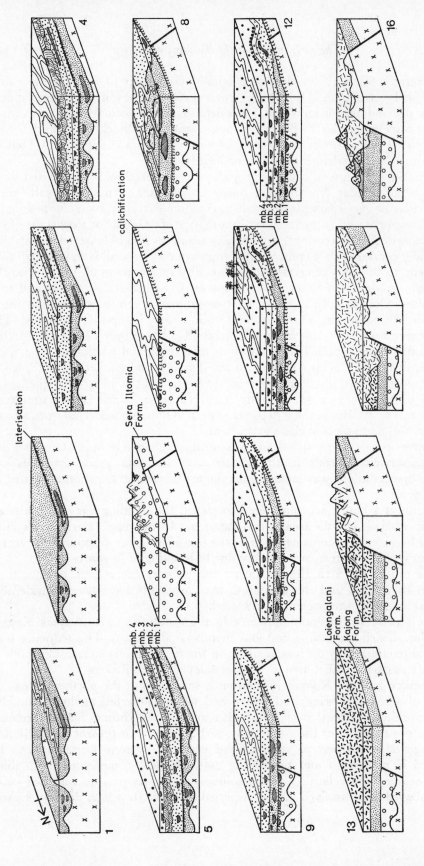

N

laterisation

calichification

Sera Illomia Form.

mb. 4
mb. 3
mb. 2
mb. 1

mb. 4
mb. 3
mb. 2
mb. 1

Loiengalani Form.

Kajong Form.

Fig. 23:3. Early development of the Kajong region, see legend on facing page.

zones. Higher in the succession true duricrusts up to 2 m thick may be developed with well marked micrite growth-zones and partial re-crystallisation of micrite to sparry calcite with few or no clast grains preserved. In addition to these calcite cycles member 1 shows veins of halite at certain localities. Lenses of large mode C type S–S oncolitic stromatolites (Logan *et al.* 1964) up to 0.3 m across occur in association with rich mollusc faunas (*Melanoides, Cerastua, Pseudobovaria, Mutela*), and vertebrate bones. Mollusca, bones and intraclasts may all form cores for the oncolites. Extensive tabular algal mats also occur, while *Etheria* reefs are common at certain locales. Although rarely in direct association with the stromatolite lenses, smashed *Etheria* often form cores for algal growth. Member 1 is interpreted as a playa deposit; areas of perennial standing and flowing water are suggested by certain of the mollusca. Locally turbulent flow is indicated by the *Etheria* reefs (Pilsbry & Bequaert 1927).

Member 2 consists of a superimposed series of laterally very extensive, erosion-surface bounded, fining-up sequences, up to 5 m thick. These sequences typically begin with a white, massive, coarse, poorly sorted conglomerate of subangular to rounded quartz and pumice clasts to large-pebble size, set in a matrix of granule or very coarse sandgrade feldspathic arenite or sublitharenite. Micrite cements are ubiquitous, and the apparent sorting in these sands has probably been improved by pedogenesis due to the solution of fines during the

LEGEND TO FIGURE 23:3

1. *Sera Iltomia Member 1*: Erosion surface cut in Basement is mantled by partly *in situ*, partly reworked Basement debris.

2. Basement debris lateritised.

3. *Sera Iltomia Member 2*: Fining-up sequences of distal bajada system transgressing from south in response to downwarping of depression floor.

4. *Sera Iltomia Member 3*: Braided systems of more proximal bajada transgress from south.

5. *Sera Iltomia Member 4*: Fining-up sequences of distal bajada again deposited as source levelled.

6. Depression undergoes uplift and NNW-SSE/E-W faulting. Sera Iltomia formation locally preserved by downfaulting.

7. Erosion surface cut in faulted terrain. Local accumulation of conglomerates and massive calichification on this surface.

8. As a result of downwarping of the depression, playa sediments of *Kajong Member 1* deposited to north, distal bajada fining-up sequences of *Kajong Member 2* to south.

9. *Kajong Member 2* distal bajada fining-up sequences transgress *Kajong Member 1* playa deposits to north.

10. Truncated fining-up sequences of *Kajong Member 3*, representing more proximal bajada deposition, succeed *Member 2*.

11. Local lateral movement on basement faults cause folding in overlying Kajong sediments in southern areas, leading to an angular unconformity between *Kajong Members 3 and 4*. Lag gravels and caliches develop on this surface, but in other areas *Members 3 and 4* are conformable. The finer fining-up sequences of *Kajong Member 4* indicate source levelling.

12. *Kajong Member 4* sediments bury folds in earlier Kajong sediments, and sedimentation is resumed over the whole of the area.

13. Volcanics of the *Loiengalani Formation* conformably succeed the Kajong deposits.

14. Turkana depression uplifted and faulted. Kajong and Loiengalani Formations locally preserved by downfaulting, largely on re-activated pre-existing fault zones.

15. Erosion surface cut in faulted terrain, and this landscape abruptly covered by newer volcanic series.

16. Continued erosion re-exhumes previous landscape. Erosion of all previous deposits in progress at present.

calichification process. This congolomerate rests on a scoured erosion surface and fines up rapidly into medium to coarse red feldspathic arenites. To the top of each fines sequence is a calichification sequence with discrete carbonate nodules which may pass up into a ramifying nodule framework and then a duricrust. Micrite cement and solution of fines is again ubiquitous in these finer sequences. A variable degree of removal of the top of the fining-up sequence by the erosion surface beneath the overlying sequence seems to be a major control on the degree to which the calichification cycle is apparently developed. The repeated development of palaeosurfaces in these deposits is a clear indication of the sporadic nature of their deposition. The basic sedimentary unit in member 2 is similar to Bluck's conglomerate 'A' (Bluck 1967), and is considered to have been deposited by sheet floods on a distal coalesced fan front or bajada. In the northern areas member 2 rests directly on member 1, but to the south member 2 rests directly on basement.

Member 3 is similar to member 2 but in general the erosion on the surfaces between successive fining up sequences was sufficient to remove entirely the fine red tops with their calichification cycles from each sequence. The member therefore consists simply of a stacked series of erosion-surface bounded coarse white conglomerates similar to, but coarser than, the basal conglomerates of the fining sequences in member 2. It is considered that the erosional surfaces between successive fining-up sequences in all these members are due to abrasion of the underlying sequence as the sheet-flood that deposited a later sequence scoured down into the underlying bed.

Member 3 was deposited by higher energy, more competent sheet-floods than member 2, as indicated by its much coarser basal conglomerates. These higher energy sheet floods incidentally removed far more of the underlying units than the lower energy floods which deposited the more complete but generally finer fining-up sequences in member 2. Member 3 is considered to have been deposited in a situation similar to, but more proximal than, that of member 2. The fining-up sequences of member 3 become finer and more complete towards the top of the member; a distinctive waterlain tuff marks a convenient boundary between members 3 and 4. Member 4 is lithologically similar to member 2, and must record similar depositional environments.

In the northern areas Kajong member 4 rests conformably on Kajong member 3, but in the southern areas there is a major angular unconformity between member 4 and all previous sediments. Lower deposits of the Kajong Formation, and the underlying Sera Iltomia Formation, have been thrown into a series of folds whose axes trend about 025°–205°. This folding is apparently due to limited lateral movements on the broadly E–W trending faults in the underlying basement of this region. The unconformity was cut in the crests of these folds, which must have formed local highs on the palaeosurface. Deposition apparently continued on the lows between folds, and sediments of Kajong member 4 can be seen to lap onto these folds, eventually burying them. The unconformity is marked by a sequence of lag-gravels, with wood and bone fragments, and well-developed calichification cycles. This sequence of caliches and gravels is naturally best developed on the fold-crests, but much less developed between folds, where a thin lag-gravel alone may be present. By upper member 4 times sedimentation had recommenced over the whole of the area. A thick volcanic and volcaniclastic sequence over 200 m thick, the Loiengalani Formation, conformably overlies the Kajong member 4 deposits. More recent volcanics lie above these.

On the basis of palaeocurrent directions from the very sporadic cross-bedding throughout Kajong Formation members 2–4, the direction of sphericity decline in the 5–35 mm size class of quartz clasts from the base of certain fining-up sequences and the decline in maximum clast size, sediment can be demonstrated to have travelled into the area from the south on a bearing of a about 025°. Members 2 and 3 are considered to record the building out of a coalesced alluvial fan front or bajada into the area and over the playa deposits of member 1 to the north. Member 4 is considered to represent the decline of this system, presumably due to the levelling of its source.

Certain distinctive fining-up sequences or groups of fining up sequences can be traced over much of the area of outcrop of the Kajong Formation; even beds traceable over 100 km² show no significant lithological or thickness change to the limits of outcrop. It seems likely that the depositional system responsible for the Kajong Formation was extremely extensive. In modern alluvial fan systems beds are commonly 5 to 20 times as extensive along a fan-radius as they are across its contour (Bull 1972). On this basis a minimum N–S length for the Kajong Formation would be from 50–170 km and it was almost certainly considerably larger than this.

The bajada deposits of both the Sera Iltomia and Kajong Formations were apparently derived from basement rocks and moved into the area from the south. A possible source for these deposits is the faulted terrain likely to result from differential movement between a downwarping Turkana depression and an upwarping or static domal area to the south. The E–W component of the fault system in the southern regions of the Kajong exposures, near Mount Porr, dies out to the north and south. No significant E–W faulting is known in the northern exposures; it is rare in the South Horr region (Dodson 1963), and absent at Baragoi still further south (Baker 1963); in the region south of Loiengalani and South Horr it is not well exposed. This zone of E–W/NNW–SSE faulting may mark the junction of the Kenyan dome and the Turkana depression as tectonic entities, and the exposures under discussion would then lie near this junction. It seems likely that the fan-systems of which the Kajong and Sera Iltomia bajada members form the distal extremities were derived from the northerly inclined faulted terrain likely to have been produced to the south of the Kajong region by differential movement between the Turkana depression and the Kenyan dome. Similar movements must have caused the Turkana and Uganda scarps on the western walls of the depression, which are the acknowledged sources for the coarse clastic sequences west of Lake Turkana.

Generalised sections through the Kajong and Sera Iltomia Formations shown in Fig. 23:2 indicate a basically cyclic pattern in the evolution of this area. It seems that sedimentation in the area commenced as bajada deposits of the 250 m thick Sera Iltomia Formation transgressed from the south onto a low relief laterised-regolith mantled basement erosion surface, in response to early differential movement between the depression and the Kenyan dome. Fining of the upper part of the formation could be due to source levelling and was followed by uplift and faulting of the area. The cutting of a new erosion surface in this uplifted terrain removed the Sera Iltomia deposits except where these had been preserved by the E–W and NNW–SSE faulting. The erosion surface cut into the Basement System and the Sera Iltomia Formation was then massively calichified. After an unknown interval of time the 210 m thick bajada deposits of the Kajong Formation transgressed from the south, due to renewed differential movement between the depression and the Kenyan dome, although playa deposition occurred initially in the northern areas until these were overrun by the bajada later in Kajong times. Minor movement on the fault system in the basement during this period is reflected in folding and the cutting of a local unconformity on these folds within the Kajong sediments in the southerly areas west of Kajong. The fining of the Kajong members could be due to source levelling. The deposition of the Kajong Formation was followed by a conformable volcanic sequence, the Loiengalani Formation, and an episode of major faulting and uplift then ensued. The Kajong and Loiengalani formations were locally preserved due to the movement, in part on the formerly active E–W and NNW–SSE trending fault system, and in part in the angles between the fault bounded blocks due to new NNW–SSE faults. Newer volcanics subsequently blanketed a landscape cut in the older volcanic and sedimentary series, and in the basement. The 2 major cycles of erosion and deposition seen at Kajong are clearly major, long-term events and are likely to be regional in scope, though now exposed in limited areas due to extensive later volcanic cover.

There are unfortunately no radiometric dates available to the authors on the age of the lavas in the Kajong area. The fauna of mammals so far identified is listed in Table 1. Most of the specimens are poorly preserved, but of considerable interest is the almost complete skull and mandible of a gomphothere; this specimen has complete tusks but unhappily the molars are poorly preserved. The flatness of the lower tusks is reminiscent of the *Platybelodon* from Loperot, though there is no sign of the dentine structures seen on the latter specimen. A specimen described by MacInnes (1942) from Maboko as *Trilophodon angustidens kisumuensis* has tusks of similar section.

Buluk

During 1973 in the course of mapping lavas on the Asille plateau between Lake Turkana and Lake Stefanie, Mr R. T. Watkins of Birkbeck College, University of London, found volcaniclastic sediments with a mammalian fauna. A brief account of the succession and the contained Miocene fauna was published by Harris & Watkins (1974) and the mammalian fauna is tabulated in Table 1. The authors hope to undertake detailed collecting in the area in the near future. The sediments rest on a weathered and eroded basalt near Buluk waterhole, and to the north-east they rest unconformably on Basement System; the basement exhibits large grid faults similar to those at Kajong. The sediments are around 50 m thick and comprise silty clays, sandstones and tuffs. Most of the mammals come from the red-brown clays, which also contain root casts, algal structures and burrows. The sediments are capped by a lava which has been dated at 17.3 ± 1.4 m.y. Later sediments up to 70 m thick are sandwiched between lavas with radiometrically determined dates of 16.2 and 12.3 m.y. (Fitch, Watkins & Miller 1975); the fauna of these sediments appears to include rhinocerotids, proboscideans, anthracotheres and primates.

Discussion

It is useful to consider other early deposits already known from the Turkana depression in order to see how far they reflect events at Kajong and how these events may be reconciled with current ideas on the early history of the rift in general and the Turkana depression in particular. It is clear from the previous discussion that Miocene and earlier sediments occur in one of two main settings and are of two main types:

(1) Coarse clastic fan/bajada sequences which may be associated with playa deposits, derived from basement, lying unconformably on basement

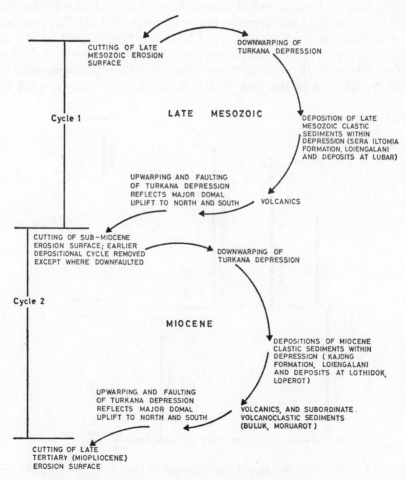

Fɪɢ. 23:4. Possible cyclic development of the early Turkana depression.

where the base is seen and often capped by volcanics (i.e. deposits at Lapurr, Loperot, Kajong).

(2) Lacustrine, alluvial, and playa deposits intercalated within volcanic sequences (i.e. Lothidok, Moruarot, Buluk).

Volcaniclastic deposits are either subordinate or absent in the first category of sediments, but dominate in the second, which may largely consist of reworked waterlain volcaniclastics.

The Kajong sites clearly indicate that at least two distinct episodes of sedimentation, possibly separated by a considerable period of time, are of the type in the first category above. The age of the second of these episodes of sedimentation (the Kajong Formation) is demonstrably Miocene at Kajong, from palaeontological evidence (see Table 1). A similar age is also indicated for certain fossiliferous coarse clastic sequences with subordinate volcanics to the west of the lake (e.g. Loperot). The earlier episode of sedimentation, the Sera Iltomia Formation, is undatable at present, having yielded only wood fragments. It must, however, predate the Kajong by a period of time sufficient for (i) the earlier formation to have been downfaulted into Basement in certain areas, and in other areas its entire thickness (at least 250 m) removed, and (ii) the resulting faulted surface bevelled and calichified to at least 8 m. In places a pure calcite duricrust 3 m thick overlies this surface. The only demonstrably non-Miocene coarse clastic sequence lying on Basement to the west of the lake is that in the Lapurr range, which recently

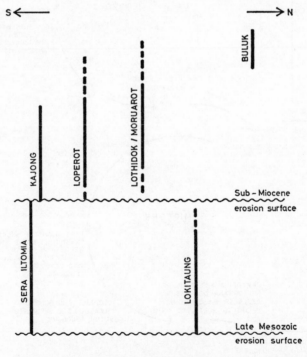

FIG. 23:5. Stratigraphic relationships of early fossiliferous deposits in the Turkana depression.

yielded a sauropod humerus and is presumably Mesozoic in age as noted above. Of the second major category of sediments, those intercalated in lava sequences, both Buluk and Moruarot have yielded Miocene faunas indistinguishable at present from those of the Miocene coarse clastic sequences in the area.

It is of course possible that there have been many distinct cycles of sedimentation and erosion during the early history of the Turkana depression, and that the Sera Iltomia and Kajong cycles are simply the only two of these that happen to have been preserved in the Kajong region.

However three distinct cycles of uplift and erosion have been generally recognised on the Ethiopian and Kenyan domes, late Mesozoic, early Tertiary (sub-Miocene) and end Tertiary in age. On this basis a very tentative correlation of Miocene and earlier sediments in the Turkana depression might be as shown in Fig. 23:5.

Summary

It seems that the early sedimentary history of the Turkana depression since the late Mesozoic has reflected periods of differential movement between the depression itself and the surrounding regions, which to the north and south are areas of intermittent domal uplift.

There were probably at least two complete early cycles of deposition within the depression, each initiated by differential movements which downwarped pre-existing erosion-bevels in the area and resulted in clastic sedimentation within the depression. A volcanic sequence or sequences was then deposited, in places with subordinate volcaniclastic sediments, and this succeeded by a period of uplift and faulting.

A new erosion surface was then cut in this terrain, removing the earlier cycle of deposits, except where this had been preserved by downfaulting. These major cycles of erosion and deposition within the Turkana depression are tentatively summarised in Fig. 23:4.

It is clear from Fig. 23:3 that each of these major erosion-deposition cycles involves two major episodes of movement separated by relative crustal stability. The first such episode downwarps the erosion-bevelled surface, then occupying the site of the depression, creating a faulted boundary from which fan/bajada systems invade the depression. Major volcanism then ensues, as a precursor to the second major episode of earth movements, which involves uplift and faulting of the floor of the depression. This uplifted, faulted terrain then undergoes erosion, and a new erosion-bevel is cut, removing the earlier cycle of deposits except where these had been preserved by down-faulting during the uplift phase. This second uplift episode of earth movement in each cycle seems likely to be a manifestation of these major episodes of domal uplift to the north and south of the Turkana depression which, in turn, terminated the late Mesozoic, and sub-Miocene erosion cycles on these domes.

This model therefore postulates downwarping of the Turkana depression somewhat prior to major uplift of the flanking domes, a succession of events apparently suggested for the Miocene phase of domal uplift by Baker *et al.* (1972).

It is quite possible that there are in fact more than the three main erosion surfaces presently recognised on the dome, and that deposits in the Turkana depression tentatively correlated in part on the two earliest of these surfaces may in fact represent a greater number of erosion-deposition cycles than are recognised here.

The Buluk and Loperot sediments underlie lavas with radiometrically determined dates of around 17 m.y. There are no dates for the Kajong Formation and those for the Lothidok and Moruarot sites do not seem reliable, having a very wide span.

The mammal faunas have many resemblances with the faunas from the Kisumu area, Rusinga in particular, which are dated at around 18 m.y. On all counts the Lake Turkana Basin faunas can be considered earlier than those of Fort Ternan, dated at 14 m.y. It is probable that they are of different ages and from different ecologies, but our knowledge at present is insufficient to be more precise.

However, this generalised and tentative model may serve as a framework for further enquiry into the early history of the Turkana depression.

An apparently longer and more complex history for the Turkana depression and certain of its major fault systems than has hitherto been inferred poses interesting questions with regard to the nature and timing of rifting in this area.

The early development of the Turkana depression may apparently be considered in terms of the cycles sketched in Fig. 23:4, but in the latest Tertiary this cyclicity is not apparent, presumably due to the apparent acceleration in the development of the rift at this time. The late Pliocene uplift of the Kenyan dome was far larger than any previous uplift, and was accompanied by extensive rift faulting as a major graben came into being for the first time (Baker *et al.* 1972).

Acknowledgements

Our thanks are due to the Natural Environment Research Council for a grant to undertake field studies in Kenya and to Mr R. E. F. Leakey, Director of the National Museums of Kenya, for facilitating our field work. We would like to acknowledge helpful discussions with Mrs Shirley C. Coryndon, who also helped in the field and in the identification of the faunas. Professor Bryan Patterson kindly allowed us to examine his Loperot collections in Harvard and Mr R. T. Watkins kindly conducted us through his mapping area.

References

ARAMBOURG, C. 1933a. Découverte d'un gisement de Mammifères burdigaliens dans le Bassin du Lac Rodolphe (Afrique Orientale). *C.R. Soc. geol. Fr.* **14**, 221–2.

—— 1933b. Les formations prétertiares de la bordure occidentale du Lac Rodolphe (Afrique Orientale). *C.R. Acad. Sci.* Paris **197**, 1663–5.

—— 1933c. Mammifères Miocènes du Turkana (Afrique Orientale). *Ann. de Paléo.* **22**, 123–46.

—— 1935. Esquisse géologique de la bordure occidentale du Lac Rodolphe. Mission Scient. Omo 1932–3, *Mus. natn. Hist. nat.* Paris **1** (1), 59 pp.

—— 1943. Contribution a l'étude géologique et paléontologique du Bassin du Lac Rodolphe et de la Basse Vallée de L'Omo. 1re Partie, Géologie *Mission Scient. Omo 1932–1933 Mus. natn. Hist. nat.* Paris, 230 pp.

—— 1947. Les mammifères du Turkana. *Mission Scient. Omo 1932–1933, Mus. natn. Hist. nat.* Paris **1** (3), 78–82.

—— & WOLFF, R. G. 1969. Nouvelles données paléontologique sur l'ages des 'gres du Lubur' (Turkana grits) a l'ouest du Lac Rodolphe. *C.R. Soc. geol. Fr.* **6**, 190–2.

BAKER, B. H. 1963. Geology of the Baragoi area. *Rep. geol. Surv. Kenya*, **53**, 74 pp.

——, WILLIAMS, L. A. J., MILLER, J. A. & FITCH, F. J. 1971. Sequence and Geochronology of the Kenya Rift Volcanics. *Tectonophysics* Amsterdam, **11**, 191–215.

——, MOHR, P. A. & WILLIAMS, L. A. J. 1972. Geology of the Eastern Rift System. *Geol. Soc. Am. Spec. Paper.* **136**, 66 pp.

BISHOP, W. W. 1967. The Later Tertiary in East Africa—Volcanics, sediments, and faunal inventory. In: *Background to Evolution in Africa.* W. W. Bishop and J. D. Clark (eds), Univ. Chicago Press, 31–56.

BLUCK, B. J. 1967. Deposition of some Upper Old Red Sandstone conglomerates in the Clyde area. A study in the significance of bedding. *Scott. J. Geol.* **3** (2), 139–67.

BULL, W. B. 1972, Recognition of alluvial-fan deposits in the stratigraphic record. In: Recognition of Ancient Sedimentary Environments. J. K. Rigby and W. K. Hamblin eds. *Soc. Econom. Palaeo. Min. Spec. Pub.* **16**, 63–83.

CHAMPION, A. M. 1937. Physiography of the region to the west and south-west of Lake Rudolf. *Geogrl. J.* **89**, 97–118.

CHURCHER, C. S. 1970. Two new Upper Miocene Giraffids from Fort Ternan, Kenya, East Africa: *Palaeotragus primaevus*, n.sp. and *Samotherium africanum* n.sp. In: *Fossil Vertebrates of Africa.* **2**, L. S. B. Leakey and R. J. G. Savage eds. Academic Press London 1–105.

CLARK, W. E. Le GROS & LEAKEY, L. S. B. 1951. The Miocene Hominoidea of East Africa. *Fossil Mammals of Africa.* **1**, 117 pp. B.M. (N.H.). Lond.

DAMME, D. VAN & GAUTIER, A. 1972. Some fossil Mollusca from Moruarot Hill (Turkana District, Kenya). *J. Conch.* **27**, 423–6.

DERANIYAGALA, P. E. P. 1951. A hornless rhinoceros from the Mio-Pliocene deposits of East Africa. *Spolia zeylan.* **26** (2), 133–5.

DIXEY, F. 1945. Miocene Sediments in South Turkana. *J.E. Afr. nat. Hist. Soc.* Nairobi **18**, 1 & 2 (81 & 82), 13–14.

DODSON, R. G. 1963. Geology of the South Horr area. *Rep. geol. Surv. Kenya.* **60**, 53 pp.

—— 1971. Geology of the area south of Lodwar. *Rep. geol. Surv. Kenya.* **87**, 36 pp.

FITCH, F. J., WATKINS, R. T. & MILLER, J. A. 1975. Age of a new carbonatite locality in north Kenya. *Nature, Lond.* **254**, 581–3.

FUCHS, V. E. 1934. The Geological Work of the Cambridge Expedition to the East African Lakes 1930–1. *Geol. Mag.* **71**, 97–112.

—— 1935. The Lake Rudolf Rift Valley Expedition, 1934. *Geogrl. J.* **86**, (2), 111–42.

—— 1939. The Geological History of the Lake Rudolf Basin, Kenya Colony. *Phil. Trans. R. Soc. Lond.* **229**, 219–74.

HARRIS, J. M. & WATKINS, R. 1974. New early Miocene vertebrate locality near Lake Rudolf, Kenya. *Nature, Lond.* **252**, 576–7.

HÖHNEL, L. R. VON, ROSIWAL, A., TOULA, F. & SUESS, E. 1891. Beiträge zur geologischen Kenntniss des Östlichen Afrika. *Denkschr. Akad. Wiss., Wien. Math-nat. Kl.* **58**, 447–584.

HOOIJER, D. A. 1966. Miocene Rhinoceroses of East Africa. *Bull. Br. Mus. nat. Hist. Lond. Geol.* **13** (2), 117–90.

—— 1968. A note on the mandible of *Aceratherium acutirostratum* (Deraniyagala) from Moruraret Hill, Turkana District, Kenya. *Zool. Med. Museum Leiden.* **42** (21), 231–5.

—— 1971. A new Rhinoceros from the Late Miocene of Loperot, Turkana District, Kenya. *Bull. Mus. Comp. zool.* **142**, 339–92.

—— 1973. Additional Miocene to Pleistocene Rhinoceroses of Africa. *Zool. Med. Museum Leiden.* **46** (11), 149–77.

JOUBERT, P. 1966. Geology of the Loperot area. *Rep. geol. Surv. Kenya.* **74**, 52 pp.

LOGAN, B. W., REZAK, R. & GINSBURG, R. N. 1964. Classification and environmental significance of algal stromatolites. *J. Geol.* **272**, 68–83.

MACINNES, D. G. 1942. Miocene and Post-Miocene Proboscidea from East Africa. *Trans. Zool. Soc.* **25** (2), 33–106.

MADDEN, C. T. 1972. Miocene Mammals, Stratigraphy and Environment of Muruarot Hill, Kenya. *Paleobios.* **14**, 1–12.

MAGLIO, V. J. 1969. A Shovel-tusked Gomphothere from the Miocene of Kenya. *Mus. Comp. Zool. Breviora.* **310**, 1–10.

MASON, P. & GIBSON, A. B. 1957. Geology of the Kalossia-Tiati area. *Rep. geol. Surv. Kenya.* **41**, 34 pp.

MEAD, J. G. 1975. A Fossil beaked whale (Cetacea: Ziphiidae) from the Miocene of Kenya. *Jl. Palaeo.* **49** (4), 745–51.

MURRAY-HUGHES, R. 1933. Notes on the Geological succession, tectonics and economic geology of the western half of Kenya Colony. *Rep. geol. Surv. Kenya.* **3**, 8 pp.

PILSBRY, H. A. & BEQUAERT, J. 1927. The aquatic molluscs of the Belgian Congo. *Bull. Am. Mus. nat. Hist.* **53** (2), 69–202.

REILLY, T. A., MUSSETT, A. E., RAJA, P. K. S., GRASTY, R. L. & WALSH, J. 1966. Age and polarity of the Turkana lavas, north-west Kenya. *Nature, Lond.* **210**, 1145–6.

SAVAGE, R. J. G. 1965. The Miocene Carnivora of East Africa. Fossil Mammals of Africa (19), *Bull. Br. Mus. nat. Hist.* **10** (8), 239–316.

VERDCOURT, B. 1963. The Miocene Non-Marine Mollusca of Rusinga Island, Lake Victoria and other localities in Kenya. *Palaeontographica.* Abt.A **121**, 1–37.

WALSH, J. & DODSON, R. G. 1969. Geology of Northern Turkana. *Rep. geol. Surv. Kenya.* **82**, 42 pp.

WHITWORTH, T. 1954. The Miocene Hyracoids of East Africa. *Fossil Mammals of Africa.* **7**, 58 pp. Brit. Mus. Nat. Hist. Lond.

—— 1958. Miocene Ruminants of East Africa. *Fossil Mammals of Africa.* **15**, 50 pp. Brit. Mus. Nat. Hist. Lond.

WILKINSON, A. F. 1976. The Lower Miocene Suidae of Africa. In: *Fossil Vertebrates of Africa*, R. J. G. Savage and S. C. Coryndon eds. **4**, Academic Press, London. 174–282.

24

CARL F. VONDRA & BRUCE E. BOWEN

Stratigraphy, sedimentary facies and paleoenvironments, East Lake Turkana, Kenya

The East Lake Turkana sequence, some 325 m in thickness, has been differentiated into four lithostratigraphic units (Bowen & Vondra 1973). Four major lithofacies have been recognized: (1) the laminated siltstone facies; (2) the arenaceous bioclastic carbonate facies; (3) the lenticular fine-grained sandstone and lenticular-bedded siltstone facies and (4) the lenticular conglomerate, sandstone and mudstone facies. These represent: (1) prodelta and shallow-shelf lacustrine; (2) littoral-lacustrine beach, barrier beach and lagoon; (3) delta plain-distributary channel and interdistributary flood basin and (4) fluvial channel and flood basin environments respectively. The East Lake Turkana basin was occupied by an embayment of the lake during the Plio-Pleistocene. This was filled by sediments provided by perennial and ephemeral streams and derived from older Cenozoic volcanics and Precambrian basement exposed along the basin margin and of the Stephanie Arch.

The Upper Cenozoic deposits in the Lake Turkana region have been the subject of intensive study during the past ten years as a result of several significant discoveries important to an understanding of the paleontology, anthropology and archaeology of East Africa (Patterson 1966; Patterson *et al.* 1970; Howell 1968; Leakey *et al.* 1970; Leakey 1971; 1972; 1973; Isaac *et al.* 1971; Maglio 1972; Robbins 1972 and Coppens 1972; 1973). The exposures along the north-eastern shore of the lake, here referred to as the East Lake Turkana Area, have yielded over one hundred hominid cranial and post-cranial specimens along with stone artifacts (Leakey *et al.* 1970; Leakey 1971; 1972; 1973 and Isaac *et al.* 1971) and abundant well-preserved vertebrate fossils (Maglio 1972).

This report provides a description of the stratigraphic units established and now formalized and a description and interpretation of the environment of deposition of the facies present.

The Plio-Pleistocene rocks of some 2500 square km of the East Lake Turkana Area have been studied. Fundamental rock units were established and 'marker beds' were delineated (Bowen & Vondra 1973). Both were mapped at a scale of 1:24 000 on aerial photographs. A composite generalized small-scale version of the resulting geologic maps is presented in Fig. 24:1. Over 120 critically located outcrops were measured, described and sampled in order to determine the lithology, major stratigraphic relationships and depositional history of the strata exposed.

Detailed laboratory analyses of the samples collected have been, and are still

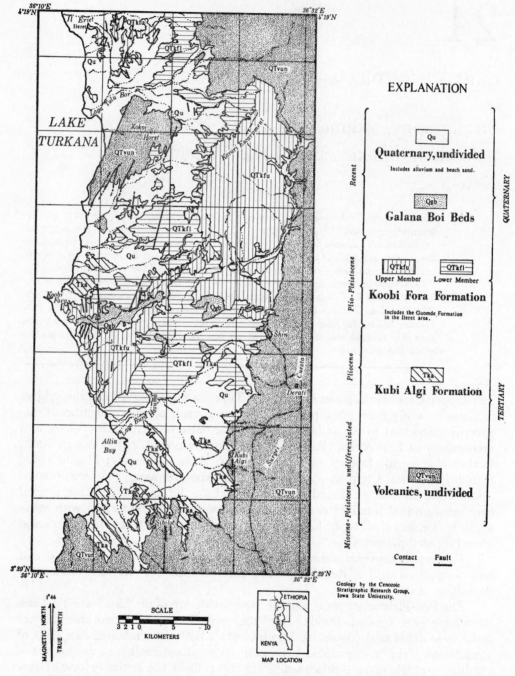

FIG. 24:1. Geologic map of the East Lake Turkana area, Kenya (modified after Vondra & Bowen 1976).

being carried out to supplement the field work and to provide additional support for paleoenvironmental interpretations. The petrology of the sediments will constitute a separate report which will be published at a later date.

Stratigraphy

The Upper Cenozoic deposits of the north-eastern portion of the Lake Turkana basin consist of a 325 m thick, inter-tongued complex of fluvial, deltaic, transitional-lacustrine and lacustrine sediments deposited disconformably on Upper Miocene and Pliocene volcanics. The sediments outcrop in a band 10 to 40 kilometres wide and 80 kilometres long which extends along the lake from Ethiopia southward to near Jarigole south of Allia Bay. Exposures are low lying, discontinuous and generally mantled by unconsolidated Holocene terrace, beach or aeolian sands. Except for occasional reversals due to local faulting, the sediments dip gently away from the Suregei Cuesta toward the lake and off the Kokoi horst.

During the early stages of geologic field work in the East Lake Turkana Area, the sedimentary exposures were separated into three areas because of the difficulty of physically correlating strata between them (Vondra *et al.* 1971). The exposures at Ileret, the northernmost area, are physically separated from those along the Koobi Fora ridge, the central area, by the Kokoi horst and a large Holocene alluvial plain complex to the east of the Kokoi. The Holocene floodplain deposits of the ephemeral, Laga Bura Hasuma separate the Allia Bay area, the southernmost, from the Koobi Fora area. Two separate sets of informal stratigraphic terminology were established. The sedimentary sequence in the Ileret area was subdivided into three unconformable units designated the lower, middle and upper units, while the sequence in the Koobi Fora and Allia Bay areas was subdivided into four units designated as Koobi Fora I, II, III and Galana Boi (Leakey *et al.* 1970; Vondra *et al.* 1971).

Continuous marker beds were subsequently recognized in the Upper Cenozoic sediments and were physically traced from the Ileret area to the Koobi Fora and Allia Bay areas. With the correlation established, the stratigraphic nomenclature now in use was developed and formalized (Bowen & Vondra 1973). This consists of, from oldest to youngest, the Kubi Algi, Koobi Fora and Guomde formations and the Galana Boi beds.

Kubi Algi Formation

The coarse-grained Pliocene strata in the East Lake Turkana Area which lie disconformably on Miocene or Pliocene volcanics or are in fault contact with the volcanics and are conformably overlain by fine-grained Plio-Pleistocene strata were named the Kubi Algi Formation (Bowen & Vondra 1973). Although these strata vary laterally, they retain certain features which give them unity for the purposes of mapping. These are (1) their position with respect to both older and younger rocks, (2) their graded nature, i.e. fining upward from pebble conglomerates with large cut and fill structures to very fine-grained sandstones and laminated claystones, and (3) their high admixture of volcanic rock fragments which are contained in the coarse sediments. The upper contact of the Kubi Algi Formation is the base of a complex of laminated bentonitic tuffs and claystones termed the Suregei Tuff Complex (Bowen & Vondra 1973). The type locality of the formation was designated as the midpoint (3°45′ N latitude and 36°19′ E longitude) of a

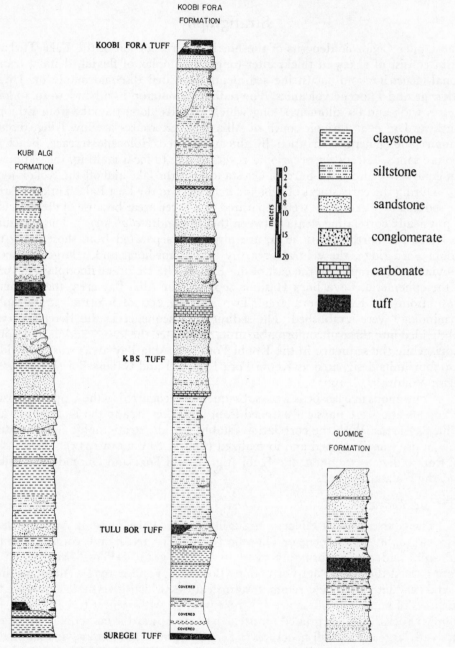

FIG. 24:2. Type-sections of the Upper Cenozoic deposits, East Lake Turkana (modified after Bowen & Vondra 1973 and Vondra & Bowen 1976).

traverse taken along a terrace trending toward Kubi Algi, a prominent butte located 20 km east of Allia Bay. The type section (Fig. 24:2) consists of a composite of the strata exposed in low lying outcrops along the terrace from a point 4 km south of Kubi Algi to a point near the Laga Bura Hasuma 4 km N 50° E of Allia Bay. The formation as proposed by Bowen & Vondra (1973) is equivalent to the lower portion of the informal Koobi Fora I (Vondra *et al.* 1971).

The Kubi Algi exhibits a considerable amount of lateral variation in thickness. Southward the sequence thickens from 98 m at the type locality to 153 m near Jarigole. It is the upper portion of the formation, measured from a key tuff horizon, which thickens from 32 m near Kubi Algi to 100 m near Jarigole while the lower portion does not vary significantly in thickness from north to south.

At its type locality the Kubi Algi Formation consists of a series of light coloured (pale yellow brown, 10YR6/2) basalt cobble conglomerates and conglomeratic sublitharenites, dark coloured (dark yellowish brown, 10YR4/2) fine-grained sublitharenites, drab (light olive gray, 5Y6/1, pale yellowish brown, 10YR6/2 and yellowish grey, 5Y7/2) siltstones and claystones and light gray (N–7) to pale yellowish brown (10YR7/2), fine-grained tuffs. The sandstones fine upward from coarse to fine-grained and tend to decrease in colour value from 10YR6/2 to 10YR4/2. Primary bedding structures also decrease in magnitude vertically from large-scale trough cross-bedding to ripple-lamination and eventually parallel lamination. Beds for the most part are lenticular and relatively thick (averaging 3.5 m) in the lower portion, but become noticeably thinner (averaging 1.7 m) upward. Laterally, they grade or interfinger with each other.

Fossils are relatively rare although locally, ostracods, gastropods and small bivalves are abundant. Abraded mammalian remains occur throughout the sequence but are only rarely encountered. Trace fossils, such as burrows and root casts are present in most of the siltstones and in some of the claystones and sandstones.

The lithologies of the clasts constituting the coarse sediments vary from north to south reflecting local differences in source rock. At the type locality cobble to pebble-sized basalt fragments are the dominant constituent of the conglomerates while northward there is a sizeable fraction of rhyolitic clasts. Toward Jarigole the conglomerates consist almost entirely of ignimbrite cobbles and pebbles.

The average particle size and the volume of coarse-grained units tend to increase toward Jarigole. Ignimbrite pebble and cobble conglomerates and very coarse-grained litharenites dominate the sequence. Unlike the type section, the units in the south are composed of fining upward cycles with conglomerates at the base and siltstones at the top of each. Tuffs increase in thickness southward but their particle-size tends to decrease.

Koobi Fora Formation

The Koobi Fora Formation (Bowen & Vondra 1973) consists of the deposits which lie between the basal contact of the Suregei Tuff Complex and the upper contact of either the prominent Chari Tuff (1.28 ± 0.23 m.y. BP) or the Karari Tuff (1.32 ± 0.10 m.y. BP) or the basal contact of the Holocene, grey diatomaceous, predominantly lacustrine siltstones where they rest unconformably on strata beneath the Chari or the Karari. The exposures located east of Koobi Fora spit at 3°56′ N latitude and 36°15′ E longitude (HBH 6136, East Africa Grid) best illustrate the lithology of the formation and have been designated as the type locality (Bowen & Vondra 1973). The type section of the Koobi Fora Formation is shown in Fig. 24:2. The unit includes the upper portion of the informal Koobi Fora I and Koobi Fora II and III units of Behrensmeyer (Leakey *et al.* 1970) and Vondra *et al.* (1971).

The Koobi Fora is a heterogeneous sequence of boulder to granule-sized conglomerates, coarse to fine-grained sandstones, variegated siltstones and claystones, bioclastic carbonates and tuff horizons. It varies in thickness from 135 m along the Karari escarpment near the basin margin to 175 m near the Koobi Fora spit and from 210 m at Ileret to 47 m near Derati. Most of the strata comprising the unit are highly lenticular and grade vertically or interfinger with each other laterally. However, certain tuffs proved to be good marker horizons. Their lateral continuity, character and stratigraphic position and the thickness and similarity of the interval separating them were the basis for establishing the correlations of low-lying exposures in the East Lake Turkana Area.

Although there are several tuff horizons in the Koobi Fora Formation, only a few are useful for correlation purposes and thus were given names (Bowen & Vondra 1973). The Suregei Tuff Complex occurring at the base of the formation and outcropping more or less continuously along the western backslope of the Suregei cuesta, was used to establish a correlation between the Ileret and Koobi Fora areas. The unique thinly laminated sequences of interbedded tuffs and claystones allowed the Suregei Tuff Complex to be easily identified and traced southward beyond the Laga Bura Hasuma.

The Tulu Bor Tuff (3.18 ± 0.09 m.y. BP) outcropping extensively between the Laga Tulu Bor and the southern margin of the Kokoi consists of two separate and distinct units. The lower unit is fine-grained and generally parallel laminated, while the upper contains calcareous concretions and lenses of pumice pebbles and sand. Locally the Tulu Bor is trough cross-bedded.

The artifact bearing KBS Tuff (2.61 ± 0.26 m.y. BP) is well exposed along the Karari Escarpment and the Koobi Fora ridge. It is highly varied in lithology, thickness and general appearance, thus only its stratigraphic position with respect to the Tulu Bor Tuff and an overlying sequence of coarse-grained large scale trough cross-bedded channel sandstones of variable thickness which occur at the base of the Ileret Member allows the KBS Tuff to be readily recognized in isolated outcrops at Ileret and between the Koobi Fora ridge and the mouth of the Laga Bura Hasuma.

Tuffs above the KBS Tuff are of local occurrence and cannot be traced laterally for any great distance. The Chari, a 2 to 3 m thick tuff, occurs along the top of the Ileret ridge, while the Karari Tuff, 0.5 to 1 m thick, caps the Karari escarpment. Although the Chari and Karari Tuffs cannot be traced laterally into one another radiometric dates obtained from sanidine crystals which occur in the pumice pebbles and cobbles of both (Fitch & Miller 1976) and the oxygen isotope ratios of the glass shards (Cerling *et al.* 1975) indicate that the two are correlative. The Koobi Fora Tuff Complex is a 12 m thick sequence of interbedded and intricately interfingering sandstone, siltstone and tuff units that are locally small-scale cross-bedded and contain gastropods. This complex is exposed along Koobi Fora ridge near Koobi Fora spit and cannot be physically correlated with any other tuff unit. Radiometric dates (1.57 ± 0.00 m.y. BP) indicate that the complex is slightly older than either the Chari or Karari (Fitch & Miller 1976) but oxygen isotope ratios permit correlation with the Chari and Karari Tuffs (Cerling 1973, and Cerling *et al.* 1975).

The Koobi Fora Formation was subdivided into two members in the Ileret area and in the Koobi Fora–Allia Bay areas (Bowen & Vondra 1973) on the basis of lithology and distribution. Although the change in lithology between the two occurs at the base of channel complexes stratigraphically above the KBS Tuff, the upper contact of the tuff was originally selected as the contact between the Lower Member and Ileret Member in the Ileret area and the Lower and Upper Members in the Koobi Fora–Allia Bay areas (Bowen & Vondra 1973). The easily mapped KBS Tuff was chosen as the contact rather than the base of the channels because of the inherent difficulty in mapping and correlating channels. In order to avoid confusion and to follow the code of stratigraphic nomenclature (1961) the contact between the two members is here redefined as the base of the conglomeratic channel sandstone complex which occurs stratigraphically immediately above the KBS Tuff in the Koobi Fora area. The separate member terminology for the Ileret and Koobi Fora areas was the result of the lack of secure correlations between the two areas. Correlations have now been established and the need for a separate nomenclature no longer exists, therefore the term Upper Member is here redefined to include the strata originally assigned to the Ileret Member in the Ileret area.

Lower Member. The Lower Member, as here defined, consists of the portion of the Koobi Fora Formation below the basal contact of the first conglomeratic channel sandstones occurring stratigraphically above the KBS Tuff. The unit outcrops extensively east of the Ileret ridge to the backslope of the Suregei cuesta and south of the Kokoi horst complex to Derati (Fig. 24:1). At the periphery of the basin east of Ileret the Lower Member consists of a series of thick (4 m) loosely consolidated molluscan subarkoses capped by thin (0.1 to 0.5 m) bioclastic carbonates with occasional lenses of thin conglomerates and siltstones. This sequence thins slightly toward the south near the Karari escarpment and interfingers with interbedded thick (3 m) laminated claystones and thin (0.1 to 0.3 m) bioclastic carbonates. Vertically these are truncated by 53 m of intercalated small-scale planar cross-bedded, medium-grained sublitharenites and lenticular-bedded, mud cracked siltstones which contain calcareous root casts. This coarse clastic sequence is followed by the Tulu Bor Tuff and overlying thinly laminated, limonitic siltstones which contain occasional discontinuous lenses of small-scale cross-bedded sandstone and isolated lenses of gastropods. The strata above the siltstones are characterized by an increase in average grain size. They consist of eight fining-upward cycles from conglomerate to siltstone which interfinger westward along the Karari escarpment with medium-grained ripple-laminated feldspathic litharenites, thin bioclastic carbonates which are occasionally capped by algal stromatolites, laminated to lenticular-bedded siltstones and intraformational, limonite clast conglomerates. This sequence contains numerous vertebrate fossils including hominid cranial and post-cranial remains. The average grain size tends to decrease laterally toward Koobi Fora spit. The sequence grades into and interfingers with 10 m thick sequences of laminated, limonitic siltstones capped by thin bioclastic carbonates with an occasional lens of fine-grained molluscan lithic subarkoses. South of Koobi Fora spit near the mouth of Laga Bura Hasuma, the Lower Member consists of a sequence of interbedded thin bioclastic carbonates and siltstones which grade vertically into 37 m of large-scale planar cross-bedded, medium-grained sandstones above the Tulu Bor Tuff. Thin algal stromatolite units and bioclastic carbonates interfinger with these sandstones toward the north and east. The thickness of the Lower Member varies from 123 m along the margin of the basin to 78 m near the Koobi Fora spit.

Upper Member. The term Upper Member as now proposed includes the beds which lie between the basal contact of the first channel sandstones occurring stratigraphically above the KBS Tuff and the top of the Chari or Karari tuffs which cap the Koobi Fora Formation.

The Upper Member in the Ileret area is restricted to a very small exposure and exhibits little lateral variation. Here the base of the member is marked by a sequence of small-scale trough cross-bedded fine-grained subarkoses containing vertebrate fossils. These overlie a series of interbedded thin bioclastic (ostracod) carbonates and fine-grained molluscan subarkoses which occur immediatly above the KBS Tuff. The basal strata coarsen upward and give way to lenticular conglomerates and trough cross-bedded, coarse-grained sublitharenites which are interbedded with tuffaceous, lenticular bedded siltstones and laminated to ripple-laminated tuffs. One exception to this coarsening upward sequence is a laterally continuous, very pale orange (10YR8/2) fine-grained subarkose which contains numerous disarticulated fish fossils and is capped by a thin carbonate unit. This occurs about 8 m below the Chari Tuff.

The entire sequence in the Ileret area contains mammalian fossils but they are most numerous in the upper portion. Hominid fossils and artifacts occur here as well.

Along the Karari escarpment and the eastern end of the Koobi Fora ridge, the Upper Member is characterized by large-scale trough cross-bedded sublitharenites; basalt cobble conglomerates; tuffaceous, lenticular-bedded siltstones; and thin tuffs which disconformably overlie the Lower Member. Here the upper portion of the member, which becomes quite tuffaceous and is locally interbedded with discontinuous lenticular tuffs, contains occasional vertebrate fossils and numerous artifacts. Individual beds generally decrease in thickness and average grain-size laterally toward Koobi Fora spit and interfinger with fine-grained sandstones which locally contain lenses of granule conglomerate, bioclastic carbonates and algal stromatolites and are interbedded with occasional laminated or lenticular-bedded siltstones and tuffs.

The disconformable relationship between the Upper and Lower Members along the Karari escarpment and the eastern portion of the Koobi Fora ridge is marked by a complex of channel sandstones and conglomerates. Locally, these may display as much as 8 m of downcutting. Each channel can be traced laterally into either a caliche (very arenaceous carbonate) or a laminated tuffaceous fine-grained sandstone both of which are apparently conformable. Near Koobi Fora spit the amount of downcutting decreases significantly and the disconformity appears to grade into conformity or paraconformity.

The Upper Member attains a maximum compiled thickness of 89 m in the Ileret area. There are no localities in the Ileret area where the entire member is exposed in a continuous sequence from bottom to top. In the Koobi Fora area, it thickens southwestward from 43 m along the Karari escarpment to 88 m near the Koobi Fora spit. It thins to 6 m southward toward Shin.

Guomde Formation

The strata overlying the Chari Tuff and underlying the sequence of Holocene, grey diatomaceous siltstones were named the Guomde Formation (Bowen & Vondra 1973). The term was taken from Kolum Guomde, a tributary of the Laga Tulu Bor which dissects the southern end of the Ileret ridge. The formation is restricted to the Ileret ridge and is best exposed at the southern and northern ends of the ridge. The exposures located at 4°18′ N latitude and 36°15′ E longitude (HBH 625752), best illustrate the lithology of the formation and were designated the type locality (Bowen & Vondra 1973). The Guomde comprises the middle unit at Ileret as originally defined by Vondra *et al.* (1971). The type section of this formation is shown in Fig. 24:2. and has been described elsewhere (Bowen & Vondra 1973).

The unifying features of this unit are the yellowish-grey (5Y6/2) to light olive brown (5Y6/6) laminated siltstones and the intercalated thin bioclastic carbonates. The northern exposures are transected by many faults the northern-most of which truncates the formation at the upper tuff. The base of the unit is faulted out in the southern exposures. The Guomde Formation thickens slightly toward the south along the ridge from 32 to more than 37 m.

Galana Boi Beds

The Holocene grey diatomaceous siltstones which cap the Guomde Formation in the Ileret area, the Koobi Fora Formation in the Koobi Fora area and the Kubi Algi Formation in the Allia Bay area have been designated the Galana Boi beds (Bowen & Vondra 1973). Although the lithology and stratigraphic position of the widely distributed strata assigned to this unit are similar, their exact lateral relationships are not yet fully known, thus the nomenclature has been kept informal.

The Galana Boi beds, which have been dated 9360 ± 135 m.y. BP (Vondra *et al.* 1971) at one locality in the Ileret area, occur as a thin discontinuous mantle over most of the older sediments within a 2 to 4 km wide band along the present lake shore. Locally the strata thicken to as much as 32 m at Koobi Fora and occur in isolated outcrops up to 18 km inland in the Allia Bay area. They are for the most part, horizontal but in at least one instance they have been faulted to 120 m above the present lake shore and occur in a cul-de-sac in the Kokoi horst area.

Artifacts, algal stromatolites, gastropods (*Melanoides*) and *Etheria* banks are common in this sequence of yellowish grey (5Y7/2) parallel laminated diatomaceous siltstones, tuffaceous mudstones and calcareous, trough cross-bedded to ripple-laminated, coarse-grained subarkoses. The units are thin, highly lenticular and grade vertically and laterally into each other.

Facies and environments of deposition

Four major lithofacies, each consisting of several subfacies, have been recognized in the Upper Cenozoic sediments in the East Rudolf area. These are (1) the laminated siltstone facies; (2) the arenaceous bioclastic carbonate facies; (3) the lenticular fine-grained sandstone and lenticular-bedded siltstone facies; and (4) the intertongued lenticular conglomerate, sandstone and mudstone facies (Vondra & Bowen 1976). These are characterized by properties indicative of four major depositional environments; (1) prodelta and shallow shelf lacustrine; (2) littoral lacustrine—beach and barrier beach and associated barrier lagoons; (3) delta plain—distributary channel, levee, and interdistributary flood basin; and (4) fluvial channel and flood plain (Vondra & Bowen 1976).

The facies are intricately interbedded and intertongued. Their position at any given time was related to the level of the lake which in turn was controlled by variations in climate and tectonic activity. In general, the four major facies occur in north-south trending belts which migrate to the west during the late Cenozoic recording a general regression of the lake. A preliminary examination of the facies and environments of deposition represented by the Upper Cenozoic sediments in the East Lake Turkana embayment has been presented by Vondra & Bowen (1976) and the following discussion is based on that presentation. Thus no further reference will be made.

Laminated siltstone facies

The stratigraphic occurrence of the facies is unique. Occurring 2 to 3 m above the Tulu Bor Tuff near the eastern end of the Karari escarpment, it thickens from 5 m to 42 m near Koobi Fora spit. Here it immediately overlies the Tulu Bor Tuff and continues to 4 m above the KBS Tuff.

The facies consists of sequences of thinly bedded to laminated, yellowish grey (5Y7/2), limonitic, argillaceous siltstones near the eastern end of the Karari escarpment. Toward the present lake margin this sequence becomes lenticular-bedded with isolated, discontinuous, flat lenses of greyish orange (10YR7/4), packed molluscan biosparudites and laminated light grey (N 7), bentonitic tuffs. The siltstones contain thin lenses of sandstone and conglomerates consisting of pebbles of bioclastic carbonates. Thin plate- or lens-shaped, limonite nodules occur parallel to laminae, and very thin (generally less than 5 millimetres in thickness) subvertical to horizontal veins of selenite are common.

The laminated siltstones generally range from 2 to 4 m in thickness, although units as thick as 18 m occur in exposures just east of the Koobi Fora spit near the present lake margin. Their basal contacts are usually sharp but non-erosional while their upper boundaries are gradational. Individual units, as well as the entire sequence, tend to coarsen upward to argillaceous and silty sandstones capped by packed molluscan biosparudites. The sandstones are ripple-laminated and occasionally bioturbated. Fossils in this facies are rare when compared with the facies with which they are interfingered. They usually consist of scattered thin lenses of whole, unabraded, lacustrine gastropods (*Cleopatra* sp., *Melanoides* sp.) in the siltstones and as broken and abraded molluscan fragments in the thin bio-sparudites.

The fine-grained and laminated nature of the siltstones indicates deposition under low energy conditions with the coarsening-upward and the lenticular bedding with isolated discontinuous lenses of sandstone suggesting periodic current activity. The isolated lenses formed as incomplete ripples of sand on a muddy bottom in an environment in which the sand supply was limited. This along with the sporadic occurrence of gastropods, the distribution and geometry of the facies and its relationship to interfingering facies suggests that the facies was deposited in a prodelta or shallow shelf environment (Donaldson *et al.* 1970; Allen 1970).

Arencaceous bioclastic carbonate facies

The geographic extent of this facies is difficult to define because it is so intimately interbedded with other facies. It is exposed throughout the basin and occurs intermittently in the entire Upper Cenozoic sedimentary sequence. Eastward it interfingers with and vertically passes into the lenticular, fine-grained sandstone and lenticular-bedded siltstone facies.

Lithologically, near the Koobi Fora spit the facies consists of dark yellowish orange (10YR6/6) to moderate yellowish brown (10YR5/4), very arenaceous, gastropod packed and/or ostracod biosparudite which may grade laterally into dark yellowish brown (10YR4/2) to yellowish grey (5Y7/2) very calcareous and fossiliferous, fine to medium-grained, lithic subarkose or into greyish orange (10YR7/4), biolithites possessing algal stromatolite structure. Individual beds vary in thickness from 5 cm to 3 m and all diminish in thickness and pinch out to the east. Vertically they display a distinct sequence of primary structures. The basal portion of each unit is a laminated, silty fine-grained sandstone, which is gradational with the underlying laminated siltstone. It contains vertical burrows and occasional internal moulds of the bivalve *Mutela*. This becomes a ripple-laminated and large-scale trough and planar cross-bedded, fine to medium-grained, very calcareous sandstone or arenaceous bioclastic carbonate in the middle portion, and a fine-grained sandstone which is structureless with the exception of occasional root casts and ripple marks in the upper portion.

The upper half of the middle portion is often packed with the shells of gastropods and/or ostracods. Disarticulated fish remains, fragments of algal mats or spheroids, occasional fragments of, and complete bivalves and relatively rare abraded mammalian bones may also occur. Laterally, as well as vertically, this sometimes gives way to thin algal biolithites. Three basic stromatolites—mats hemispheroids and spheroids—are dominant (Johnson 1974). Oncolites or concentrically stacked spheroids (SS–C) (Logan *et al.* 1964) and inverted stacked hemispheroids (SH–I) (Kendall & Skipwith 1968) are common forms (Johnson 1974). Although forms described by Kendall & Skipwith (1968), such as the cinder and polygon zones were noted, no attempt was made to relate them to any lateral sequences of environments.

Near Koobi Fora spit the carbonates and sandstones of this facies are often associated with 1 to 2 metre thick beds of massive, pale yellowish brown (10YR6/2) siltstone. They are characterized by gradational lower contacts and are sandy at the base becoming very argillaceous at the top. The siltstone contains calcareous concretions, calcareous root casts and thin subvertical veins of gypsum. An occasional thin limonite clast, pebble or cobble intraformational conglomerate with a very sharp erosional lower boundary is often intercalated with these siltstones.

At Ileret this facies consists of a series of 3 cm to 4 m thick pale yellowish brown (10YR6/2), very fine to very coarse-grained, with lithic subarkoses capped by thin (3 cm to 10 cm), dark yellowish brown (10YR2/2), packed gastropod biosparudites. Occasionally the biosparudites thicken laterally to 1.5 m and contain mud cracks at the base. Claystones and siltstones are rare in this sequence. They are generally moderate yellowish brown (10YR5/4) or pale yellowish brown (10YR6/2) and grade vertically from thin-bedded to massive. Vertically they become very sandy and extensively bioturbated.

The loose to friable, poorly sorted, lenticular sandstones show indistinct low angle, small scale, planar cross-beds. Their lower boundaries are usually gradational. They contain mud cracks, load casts, calcareous root casts and numerous vertebrate fossils. Although they do tend to become coarser-grained vertically, there is no detectable sequence of primary structures. Algal stromatolites are conspicuously absent.

Relatively thick tuffs (up to 4 m) may be locally extensive in this facies. Locally they contain rounded pumice cobbles up to 30 cm in diameter and fine to coarse-grained sand and are very argillaceous. Root casts often occur throughout the tuff units as well as in capping caliche horizons. Artifacts and vertebrate fossils occur locally.

Lenticular fine-grained sandstone and lenticular-bedded siltstone facies

This facies occurs throughout the basin and comprises the greatest volume of all the facies. It constitutes most of the upper portion of the Kubi Algi Formation and the basal and uppermost portions of the Guomde Formation. In the Koobi Fora Formation the base of the lenticular fine-grained sandstone and lenticular bedded siltstone facies climbs stratigraphically from about 30 m below the Tulu Bor Tuff along the margin of the basin near the Karari escarpment to well above the KBS Tuff near Koobi Fora spit. In the Ileret area this facies constitutes almost all of the Upper Member and about half of the Lower Member and is complexly intertongued with the arenaceous bioclastic carbonate facies from the Suregei Tuff Complex up to the KBS Tuff along the margin of the basin. The lenticular fine-grained sandstone and lenticular-bedded siltstone facies usually overlies the arenaceous, bioclastic carbonate facies but locally it rests unconformably on the laminated siltstone facies. It interfingers laterally with the lenticular conglomerate, sandstone and mudstone facies.

The facies consists of 1 to 25 m thick lenticular channels of greyish orange (10YR7/4) fine to medium-grained sandstone which grade into or interfinger with

1 or 2 m thick, pale yellowish brown (10YR6/2), lenticular-bedded siltstone. Occasionally a thin, dark yellowish orange (10YR6/6) limonite clast, intraformational conglomerate or a thin lenticular ripple-laminated to laminated, light grey (N 7) tuff interrupts this sequence. The sandstone bodies usually grade upward in grain size from the underlying siltstones and are predominantly small-scale trough (Nu) cross-bedded but often locally display a sequence of structural features ranging upwards from pebble conglomerates through large-scale trough (Pi) and planar (Alpha) cross-bedding to small-scale trough (Nu) cross-bedding and horizontal bedding. These features are grouped into sets usually no more than 15 cm thick and may be capped by a thin very argillaceous silt drape. Pebble and clay gall lenses, distorted bedding, slump structures, sand volcanoes, load casts and mud cracks are common. Convolute laminations occur less frequently in the silty sandstones. A distinct sequence of primary structures occurs in these sandstone units. The basal portion displays ripple-drift cross-lamination in which both grain size and ripple amplitude increase upward. The cross-laminae of the uppermost ripple are usually disturbed and overturned. This is followed by distorted bedding which is occasionally truncated by thin horizontal silt laminae. Locally, root casts are very abundant and occasionally the sandstones are capped by lenses of *Etheria*.

The thin, pale yellowish brown (10YR6/2) siltstones are very sandy and contain thin (1 to 3 cm) horizontal to subhorizontal lenses of fine-grained sandstone which extend laterally 1 to 10 cm before pinching out. Adjacent to the channel the siltstones contain thin (3 to 10 cm) intercalations of poorly sorted pebble conglomerates and coarse-grained sandstones. The siltstones become progressively more fine-grained and more argillaceous laterally with respect to the axis of the channel. They contain limonite concretions or laminae near the base, abundant mud cracks throughout, 10 cm thick lenses of claystone locally, and numerous calcareous root casts associated with capping caliche horizons. Slightly abraded and disarticulated vertebrate fossils are common but articulated fossils do occasionally occur in this facies.

Lenticular conglomerate, sandstone and mudstone facies

This facies is the most heterogeneous of the four facies described. It consists of a complex variety of subfacies which grade laterally and vertically into one another, wedge in, thicken and pinch out. Along the Karari escarpment the facies maintains a nearly constant thickness of about 40 m but thins noticeably to 20 m south and west along the Koobi Fora ridge. Westward along the Koobi Fora ridge it grades laterally into the lenticular, fine-grained sandstone and lenticular-bedded siltstone facies but south of the ridge the facies rests unconformably on the laminated siltstone facies.

Lenses of greyish orange (10YR7/4), granule to cobble conglomerate and fine to coarse-grained sandstone (arkose or feldspathic litharenite) occurring in sinuous depressions eroded into older deposits, and associated finer-grained, very pale orange (10YR8/2) to greyish orange (10YR7/4) siltstones, claystones, mudstones and light grey (N 7) tuffs comprise this facies. The conglomerates are most abundant to the east along the margin of the basin and in the axial portion of channels or in the basal portion of fining-upward lenses. They diminish in grain size toward the centre of the basin as well as vertically within individual channels. The sandstones display a variety of primary structures but two basic sequences predominate. The first occurs in sandstone lenses several metres in thickness with a basal surface eroded into older deposits. This begins with a thin veneer of conglomerate (channel lag) followed by a high angle (up to 25°), large-scale trough (Pi) cross-bedded, conglomeratic, coarse-grained sandstone with cosets that are usually uniform in thickness and up to 70 cm thick giving way vertically to small-scale trough (Nu) cross-bedded and to ripple-laminated (Kappa), fine to medium-grained sandstone. This is usually overlain by a very argillaceous sandstone or sandy siltstone that is parallel laminated and contains numerous calcareous root casts and concretions. This sequence which may be up to 8 m thick is frequently followed by several more. The coarse-grained basal sandstones contain clay galls and armoured mud balls, abraded vertebrate fossils and load casts. The ripple-laminated fine-grained argillaceous sandstones contain numerous calcareous root casts and an occasional small, lenticular, conglomerate channel which may be trough (Nu) cross-bedded. Lenses or banks of the freshwater oyster *Etheria* occasionally occur within these units. The second sequence is not consistent and is present in lenses which are usually no more than 2 m in thickness. The basal surface is usually erosional and is succeeded by either weakly imbricated conglomerates or by trough (Pi) cross-bedded, medium- to coarse-grained sandstones with cosets that are highly variable in thickness but are no more than 15 cm thick. The trough sets diminish in size (to about 3 cm in thickness) upward and are usually planed by erosion or are occasionally draped with a thin veneer of argillaceous silt. Both basal sequences are overlain by large-scale planar (Gamma) cross-bedded, fine- to medium-grained sandstone which contains alternating coarse-fine foresets vertically. These cross-beds generally dip at about 22° and the cosets although highly variable range up to 30 cm in thickness. This sequence may be capped with thin laminated siltstones or sandstones but is usually truncated and followed by another similar sequence. Calcareous root casts and concretions are numerous in these sandstones. Vertebrate fossils and artifacts are locally abundant. *Etheria* is absent in this second type of primary structure sequence.

Laterally the sandstones grade and interfinger with coarse-grained siltstones, mudstones and highly lenticular tuffs which are laminated and lenticular-bedded to massive. Convolute bedding is relatively rare in the sandstones but does occur locally in the siltstones and tuffs. These fine-grained sediments contain numerous sand and clay lenses. Calcareous root casts with accompanying concretions and/or caliche, mud cracks and fossil incipient soil horizons are locally contained in them.

The siltstones, mudstones and tuffs were deposited in a variety of minor flood plain environments ranging from those immediately proximal to the channel to the distal flood basin. Tuffs were deposited and preserved as usually lenticular beds in the ephemeral streams and in backwater and cutoff channel segments, swales and flood basin depressions associated with ephemeral as well as permanent streams.

Synthesis

Paleogeographic reconstructions of the northeastern part of the Lake Turkana Basin based on the four major facies, their areal distribution and the correlation of the tuffs can be made in a general way. Assuming that the correlation of the tuff units is correct and that individual tuffs were deposited everywhere at the same time, the tuffs can be used for time-stratigraphic control. By delineating the facies between successive tuffs a general but reasonable picture of the major depositional environments, their areal distribution, the approximate position of the shore line and the approximate location of major streams for a given interval of time can be reconstructed. The paleogeographic history of East Lake Turkana is summarized by the strandline maps, shown in Fig. 24:3a, b, c, d and interpretive cross-sections (Fig. 24:4a, b).

The development of the Lake Turkana Basin has been a continuous process since the late Miocene. Middle Pliocene arching of major domes to the east and faulting to the west in the Turkana depression resulted in a shallow asymmetrical westward tilted north-south trending trough or half-graben in which the Upper Cenozoic sedimentary wedge accumulated. The uplift was accompanied by igneous intrusion and the associated eruption of obsidian and ignimbrite followed by late Pliocene basaltic volcanism forming the major volcanic features along the margin of the basin prior to deposition of the basal units of the Kubi Algi Formation. Faults along the sediment-volcanic contact at Kubi Algi and Sibilot reflect the continuing development of the basin and suggest a much more subdued relief during the Plio-Pleistocene than in the present. Volcanism continued along the margin of the basin well into the late Pleistocene as documented by the numerous tuff horizons which occur throughout the Upper Cenozoic sedimentary sequence.

The northeastern part of the Lake Turkana Basin was occupied by a large embayment of the lake during much of the late Pliocene and Pleistocene. This developed first during the late Pliocene (about 4.0 m.y. BP) in the Allia Bay areas then extended northward shortly thereafter. At the time of the deposition of the Suregei Tuff Complex the lake was at its greatest extent along the Suregei cuesta east of Koobi Fora and Ileret, but had withdrawn to about the position of the present shoreline in the Allia Bay area. An interpretation of the depositional environments of the Suregei Tuff Complex indicates the approximate position of the shoreline as shown in Fig. 24:3a. The facies occurring below the Suregei Tuff Complex indicate that ephemeral braided streams drained the Suregei cuesta and the local volcanic uplands to the southeast. They formed small alluvial fans composed of boulders and cobbles along the edge of the basin in the Ileret and Koobi Fora areas and small deltas in the Allia Bay area. Alluvial fans formed locally along the margin of the lake and interfingered with muds accumulating on the shallow lake shelf and sands deposited on interfan beaches and barrier beaches. Thin algal growth mats formed in shallow littoral environments down current from the small deltas in the Allia Bay area.

The depositional environments of the Tulu Bor Tuff and the facies between the Suregei Tuff Complex and the Tulu Bor Tuff indicate a general westward regression of the lake in the Koobi Fora area (Fig. 24:3b). Primary structures of

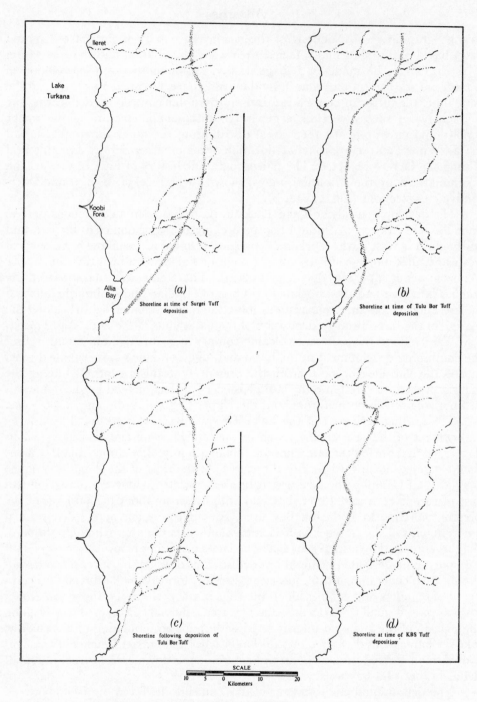

FIG. 24:3. Maps showing the approximate position of the strand line of lake during the Late Cenozoic.

the lenticular sandstone, lenticular-bedded siltstone facies and the mineralogy of the coarse-grained sediments indicates delta formation by a perennial stream heading to the east. Barrier beach and shelf lacustrine environments were dominant along the basin margin in the Ileret and Allia Bay areas while prodeltaic and shallow-shelf deposits accumulated toward the centre of the basin.

The widespread laminated siltstone facies, interpreted as pro-deltaic and shallow-shelf lacustrine deposits, overlying the Tulu Bor Tuff document major transgression and southward shifting of the deltaic complex shortly after deposition of the Tulu Bor Tuff (Fig. 24:3c). By the time the KBS Tuff was deposited, perennial streams of low sinuousity had developed prograding deltas in the Ileret

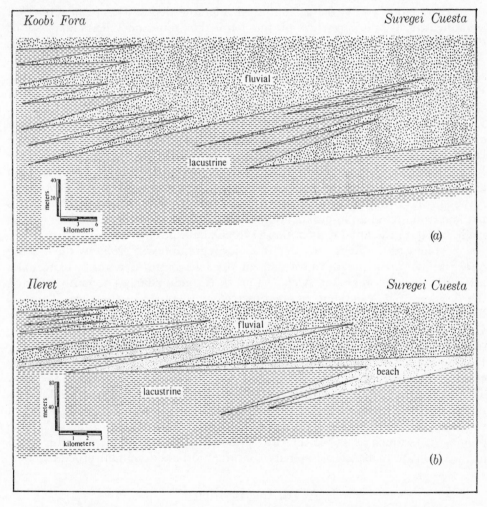

FIG. 24:4. Interpretative cross-sections of the Upper Cenozoic sediments in the East Lake Turkana.

　　a. Cross-section from the Suregei cuesta to the Koobi Fora spit.
　　b. Cross-section from the Suregei cuesta to Ileret.

and Koobi Fora areas (Fig. 24:3d). Thin lenses of littoral lacustrine deposits were intercalated with the point bars and other deltaic sediments documenting the short-lived lacustrine transgressions. Primary structures and the mineralogy of the coarse-grained sediments indicate that two major perennial streams entered the East Lake Turkana embayment. These headed to the east and north-east and derived a large portion of their sediment from Precambrian plutonic igneous and metamorphic rocks exposed on the flanks of the Stephanie Arch. The fluvial sediments of one of the stream complexes have been traced along the headwaters of the Laga Bura Hasuma toward the Stephanie Basin. This further suggests that the streams headed on the flanks of the Stephanie Arch and flowed through 'gaps' cut in the volcanics into the Lake Turkana Basin.

Progradation of the deltas continued after the deposition of the KBS Tuff as indicated by the presence of the lenticular sandstone, lenticular-bedded siltstone facies along the present shore line. At Ileret, major regression occurred. Alternation of the arenaceous bioclastic carbonate facies and the lenticular sandstone, lenticular-bedded siltstone facies in the upper portion of the Koobi Fora Formation documents a series of minor transgressions and regressions with a major regression occurring at the end of the deposition of the unit (Fig. 24:4, b). Fluvial conditions persisted in the Allia Bay and Koobi Fora areas until the advent of the major Holocene transgression which resulted in the deposition of the lacustrine Galana Boi strata (Vondra *et al.* 1971). In the Ileret area, however, the Late Pleistocene Guomde Formation represents a limited lacustrine transgression and regression prior to the Holocene transgression and Galana Boi deposition. Although the stratigraphy and lateral relationships of the Holocene sequence is not known in detail an approximation of the maximum extent of the lake can be made based on the areal distribution of the Galana Boi beds.

The impetus for delta growth and major regression was probably the arching of the source area and downwarping of the lake basin as a result of tectonic activity along the Stephanie Arch. Although climatic changes certainly affected the position of the shoreline, numerous tectonic features in the Lake Turkana Basin indicate that tectonic activity occurred throughout the deposition of the Upper Cenozoic sequence and very likely was the dominant process controlling sedimentation. Major faulting on the west side of the lake occurred in the early Pliocene and again in the Pleistocene (Walsh & Dodson 1969). This follows very closely the activity reported by Dodson (1963) for the southern end of the lake. The major faults of the Kokoi horst complex, the occurrence of Galana Boi sediments at elevations up to 120 m above the present shoreline in the eastern part of the Lake Turkana Basin and Galana Boi equivalents on the west side of the lake dipping at as much as 45° to the east (personal communication, J. Walsh 1973) further indicate that tectonic activity continues to be a dominant process.

Acknowledgements

This work was supported by Grants GA25684 and GS37814 to C. F. Vondra, and in part by a National Science Foundation grant to Dr G. L. Isaac and Mr R. E. F. Leakey, and National Geographic Society and L. S. B. Leakey Foundation grants to Mr. R. E. F. Leakey. The writers wish to express their deepest gratitude for the assistance and cooperation of the National Museums of Kenya and the Government of Kenya. Sincere thanks are given to Mr R. E. F. Leakey for his support and hospitality and to all members of the East Lake Turkana Research Group, particularly to A. K. Behrensmeyer, I. C. Findlater, F. J. Fitch, G. Ll. Isaac and G. D. Johnson. We especially thank and acknowledge H. N. Acuff, R. B. Bainbridge, D. R. Burggraf, T. E. Cerling, H. J. Frank and H. J. White of the Iowa State University Research Team. All have provided valuable assistance and contributions to every phase of the research. Finally, appreciation is extended to D. R. Burggraf who drafted all the figures in this report.

References

ALLEN, J. R. L. 1970. *Physical process of sedimentation. An introduction.* London. (G. Allen and Unwin.)

AMERICAN COMMISSION ON STRATIGRAPHIC NOMENCLATURE. 1961. Code of Stratigraphic Nomenclature. *Bull. Am. Assoc. Petrol. Geol.* **45**, 645–60.

BLACK, C. A. 1968. *Soil-plant relationships.* New York (John Wiley and Sons, inc.).

BOWEN, B. E. & VONDRA, C. F. 1973. Stratigraphical relationships of the Plio-Pleistocene deposits, East Rudolf, Kenya. *Nature, Lond.* **242**, 391–3.

CERLING, T. E. 1973. *Correlation of Plio-Pleistocene tuffs utilising O^{18}/O^{16} isotope ratios, East Rudolf Basin, Kenya.* Unpublished M.S. thesis. Library, Iowa State University of Science and Technology. Ames, Iowa.

——, BIGGS, D. L., VONDRA, C. F. & SVEC, H. J. 1975. Use of oxygen isotope ratios in correlation of tuffs, East Rudolf Basin, northern Kenya. *Earth and Planet. Sci. Letters.* **25**, 291–6.

COPPENS, Y. 1972. *Tentative de zonation du Pliocene et du Pleistocene d'Afrique par les grand mammiferes.* C. R. Acad. Sci. **274**, 181—4.

—— 1973. Les restes d'Hominides de formations Plio-Villafranchiennes de l'Omo en Ethiopie (recoltes 1970, 1971 et 1972). *C. R. Acad. Sci.* **276**, 1823–6 and 1981–4.

DODSON, R. G. 1963. Geology of the South Horr area. *Rep. geol. Surv. Kenya.* no. **60**.

FITCH, F. J. & MILLER, J. A. 1976. Conventional K–Ar and argon–40/argon 39 dating of volcanic rocks from East Rudoif. In Coppens, Y., Isaac, G. Ll. and Leakey, R. E. F. (Eds). *Earliest Man and Environments in the Lake Rudolf Basin.* Chicago (University of Chicago Press).

HOWELL, F. C. 1968. Omo Research Expedition. *Nature, Lond.* **219**, 567–72.

HOYT, J. H. 1962. High-angle beach stratification. Sapelo Island, Georgia, *J. sedim. Petrol.* **32**, 209–11.

ISAAC, G. LL., LEAKEY, R. E. F. & BEHRENSMEYER, A. K. 1971. Archaeological traces of early Hominid activities, east of Lake Rudolf, Kenya. *Science.* **173**, 1129–34.

KENDALL, C. G. ST. C. & SKIPWITH, P. A. 1968. Recent algal mats of a Persian Gulf lagoon. *J. sedim. Petrol.* **38**, 1040–58.

LEAKEY, R. E. F. 1971. Further evidence of Lower Pleistocene Hominids from East Rudolf, North Kenya. *Nature, Lond.* **231**, 241–5.

—— 1972. Further evidence of Lower Pleistocene Hominids from East Rudolf, North Kenya, 1972. *Nature, Lond.* **243**, 170–3.

—— 1973. Evidence for an advanced Plio-Pleistocene Hominid from East Rudolf, Kenya. *Nature, Lond.* **242**, 447–50.

——, BEHRENSMEYER, A. K., FITCH, F. J., MILLER, J. A. & LEAKEY, M. D. 1970. New Hominid remains and early artefacts from northern Kenya. *Nature, Lond.* **226**, 223–30.

LEBLANC, R. J. & HODGSON, W. D. 1959. Origin and development of the Texas shoreline. *Trans. Gulf Coast Assoc. geol. Socs.* **9**, 197–220.

LEOPOLD, L. B. & WOLMAN, M. W. 1957. River channel patterns; braided, meandering and straight. *U.S. Geol. Surv. Prof. Papers.* **282–B**, 39–85.

LOGAN, B. W., REZAK, R. & GINSBURG, R. M. 1964. Classification and environmental significance of algal stromatolites. *J. Geol.* **72**, 68–84.

MAGLIO, V. J. 1972. Vertebrate faunas and chronology of hominid-bearing sediments east of Lake Rudolf, Kenya. *Nature, Lond.* **239**, 379–85.

PATTERSON, B. 1966. A new locality for early Pleistocene fossils in north-west Kenya. *Nature, Lond.* **212**, 577–81.

——, BEHRENSMEYER, A. & SILL, W. D. 1970. Geology and fauna of a new Pliocene locality in north-western Kenya. *Nature, Lond.* **226**, 918–21.

ROBBINS, L. H. 1972. Archeology in the Turkana District, Kenya. *Science.* **176**, 359–66.

SMITH, N. D. 1972. Some sedimentological aspects of planar cross-stratification in a sandy braided river. *J. sedim. Petrol.* **42**, 624–34.

VONDRA, C. F. & BOWEN, B. E. 1976. Plio-Pleistocene deposits and environments East Rudolf, Kenya. In Y. Coppens, G. Ll. Isaac and R. E. F. Leakey (Ed). *Earliest man and Environments in the Lake Rudolf Basin.* Chicago (University of Chicago Press).

——, JOHNSON, G. D., BEHRENSMEYER, A. K. & BOWEN, B. E. 1971. Preliminary studies of the East Rudolf basin, Kenya. *Nature, Lond.* **231**, 245–8.

WALSH, J. & DODSON, R. G. 1969. Geology of northern Turkana. *Rep. geol. Surv. Kenya.* no. **82**.

25

IAN C. FINDLATER

Isochronous surfaces within the Plio-Pleistocene sediments east of Lake Turkana

The map accompanying this paper (Fig. 25:1) illustrates the outcrop of seven isochronous surfaces within the Plio-Pleistocene fluvio-lacustrine sediments east of Lake Turkana (formerly East Rudolf). The surfaces immediately underlie the seven major tuff horizons recognised in the area studied (Fig. 25:2). Bowen and Vondra 1973 summarised the stratigraphic relationships and introduced a stratigraphic nomenclature for the study area. The map represents a major elaboration of this earlier work.

Mapping the outcrop of these surfaces enables palaeontologists to assign relative ages to their fossil finds. Similarly the location in a relative time sequence of archaeological sites from widely separated localities is made possible. I am currently using the map as the basis for an account of the history of sedimentation in the study area and it will form a framework for the integration of microstratigraphic studies at important archaeological and palaeontological sites.

Identification of the isochronous surfaces in the field

Sedimentary units to be of practical use as chronostratigraphic marker horizons must be laterally extensive, have been deposited in a brief period and should be readily recognisable in the field. It is an advantage if the unit is of a known age, either by determinations on the unit itself or by inference from data derived from the enclosing sediments.

The tuff horizons are the obvious choice as markers in the study area. The sedimentary history of the area is of a deltaic complex prograding westward from the eastern basin margin. The resulting spatial distribution of the depositional environments is complex. Though the tuffs reflect environmental changes, their obvious lithologic contrast with the enclosing sediments makes them easily recognisable in the field. The sharply defined basal contact of the tuff is taken as the closest approximation to an isochronous surface. The upper surface of the tuff is usually gradational with the 'normal' terriginous sediment but may sometimes be partially removed by penecontemporaneous erosion. The eruption, transport and deposition of an individual tuff was probably very rapid. Inevitably there must have been some time required for transport into the basin and dispersal of the tuff within the basin. The time involved is insignificant when compared with the best resolution available from geophysical dating methods. Although basal contacts of individual tuffs are diachronous in detail they are considered to be, for practical purposes, time equivalent throughout the study area.

When an individual tuff unit is not present, due to local non-deposition or penecontemporaneous erosion, it may be possible to estimate the stratigraphic level at which it should have occurred. This requires the identification of a surface which is isochronous with the basal facies of the tuff. The reliability of the position of this surface in the absence of the tuff depends on the selection of time equivalent or approximately time equivalent marker beds. These marker beds, like the tuffs, must be laterally extensive and have been deposited or formed in a short period. If correctly interpreted these marker beds will allow a realistic estimate of the stratigraphic level at which the tuff would occur if it were present. Clearly the greater the distance between outcrops of tuff the less reliable the estimated position will be. The reliability of the estimated position is also significantly affected by the environments of deposition of the sediments through which the isochronous surface passes. In lacustrine and coastal plain situations reliability of the inferred position of the isochronous surface may be high as lithologies are often laterally extensive. Sand and sandy bioclastic sheets resulting from short term transgressive/regressive lake stages and deposits such as offshore bars, and delta front sand sheets may, if properly interpreted, be useful as marker horizons. In environments dominated by floodplain lithologies individual beds are not so extensive and it becomes more difficult to estimate the position of the isochronous surface. Soil profiles and aeolian sand sheets are two lithologies which may serve as marker horizons but often bioturbation and restricted lateral extent reduce their usefulness. When fluvial channel complexes are present it is often impossible to project an isochronous surface through the sediment body. However, it may be possible to identify the base of the channelling sequence which in the absence of other criteria may be of limited help. As a rough guide to the level of reliability of the outcrop position of the isochronous surfaces on the map, the straighter the outcrop boundary the lower the reliability.

Field character of the tuffs that immediately overlie the isochronous surfaces

In the field the tuff horizons have a very varied aspect. In thickness, continuity along the strike and internal structure they reflect the environments in which they were deposited. All the major tuffs show internal structures and have field relationships which suggest deposition from an aqueous medium.

Typically a tuff horizon has a well defined sharp base resting on channel sands, floodplain silts or lacustrine sediments. The basal facies may be very fine grain silt to clay grade often with many thin intercalations of clay grade terrigenous sediment. Internal structures, though small in scale, include cross stratification, ripple-and plain-laminations. This facies probably represents airfall volcanic dust blown or sheet washed from the landscape either near the source or possibly locally within the study area if the ash was wind borne for sufficient distances. Subsequently this volcanic dust was deposited in quiet conditions and mixed or intercalated with 'normal' terrigenous sediment. Locally these fine grained ashes may have altered to bentonite.

The bulk of the deposit comprises one or more layers of fragmentary volcanic glass with a limited suite of primary volcanic crystal fragments, sanidine, pyroxene

and ilmenite with occasional bipyramidal high quartz and zircon. Mixed with these primary volcanic elements are found varying proportions of other detritus including crystal and lithic fragments from the Miocene volcanic and Lower Palaeozoic Precambrian basement terrains. Often the initial deposits at the base of a tuff have a high degree of purity. It is probable that the source areas were heavily mantled by tephra. The newly deposited ash would be eroded and transported before 'normal' terrigenous sediment again became available. Grain size of the tuffs varies from silt to medium and coarse sand grade. Internal structures which may be present include cross stratification both solitary and grouped. Locally strong contorted bedding may occur as a result of dewatering on compaction or slumping. Root and burrow passages are common. Colour varies from chalk white through greys and blues to pale green and in tuffaceous sediments to orange-brown. Occasionally laterally extensive thin clay and silt horizons separate the tuff layers. This suggests that multiple layered deposits of fairly pure tuffs may have been deposited over a period recording consecutive flood stages within the drainage system. Increasing contamination by non primary volcanic detritus occurred in the later tuff layers.

When pumice clasts are present they are commonly associated with channel and beach littoral complexes, and are only locally abundant on the alluvial plains. Pumice is absent from the lacustrine environments. Pumice lumps range in size from granule gravel to boulder gravel. In the latter category lumps up to 25 cm long axis have been found. Though there is an apparent diminution in size from the basin margin toward the present lake shore, size variation at any locality is often great and there can be little doubt that floating was an important transport mechanism. Transport of a proportion of the smaller pumice lumps, up to very coarse pebble gravel size, as bed load accounts for the inclusion of pumice within layers of tuff. This occurs when the pumice become waterlogged. The pumice is then incorporated in the deposit rather than stranded in depressions on the top surface of a layer as is the case for the floated pumice. Pumice is not common in the early layers of multiple layered deposit and often occurs in concentrations on the upper surface only. Pumice stranded on the upper surface of the initial tuff deposits at the termination of flood conditions would dry out and if not partly buried would, during the next high water, float further downstream. Such pumice would be deposited on top of the new tuff layer with any new pumice. This mechanism would tend to concentrate pumice toward the top of a multiple-layered unit. The pumice contains phenocrysts of sanidine and pyroxene with these phases usually in complex intergrowth. Often associated with these inter-growths are crystals of ilmenite and in some pumice horizons bypyramidal high temperature quartz is an accessory. The pumices and their phenocrystic sanidine crystals have special significance in the study area as they offer the best available material for K–Ar dating studies.

The upper boundary between the tuff horizons and the 'normal' terrigenous sediment is gradational. Reworking of the upper surfaces by wind and bioturbation mixes tuff and overlying deposits resulting in a blurred contact. Sediment following the tuff is initially rich in glass and crystal debris but the availability of these clasts appears to decline quite rapidly in the source areas and 'normal'

sediment again becomes dominant. Contamination of the upper contact by local post depositional reworking occurs to a depth of a few inches. Tuffaceous sediment derived from the source areas may be 2 to 3 metres thick. Diagenetic modification appears to be very important locally. That there has been considerable post depositional mobility of carbonates is to be seen in the extensive concretionary

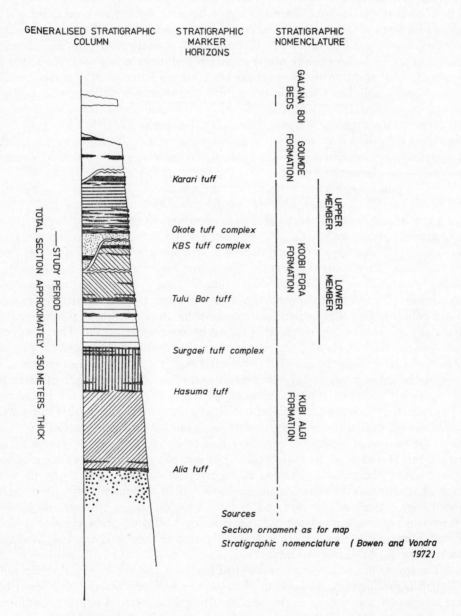

Fɪɢ. 25:2. Generalised stratigraphic column illustrating the seven tuffs whose lower boundaries define the isochronous surfaces.

calcrete (caliche) development throughout the succession. Within the tuffs carbonate is present but its distribution is easily modified by weathering on exhumation. This is well displayed by the exhumed pumice, which may be completely leached of soluble carbonate whereas partially or totally buried pumice lumps from the same locality may be solid with carbonate.

Zeolites are reported (G. L. Isaac pers. comm.) from the KBS Tuff in area 105, presumably resulting from the alteration of volcanic glass. The zeolites are effectively cementing the remaining glass. Carbonate is the major induration agent but it is characteristic of the tuffs that they are commonly unlithified.

This paper designates formal names to three marker beds, the Okote Tuff Complex, the Hasuma Tuff and the Alia Tuff. The stratigraphic positions of these three tuffs in relation to the tuffs formalised by Bowen & Vondra 1973 are shown in Fig. 25:2. The Okote Tuff Complex is synonymous with the previously informally named Lower/Middle Tuff Complex and the BBS Tuff Complex and now replaces these two informal terms. The Okote Tuff Complex outcrops extensively in the Okote area (HBH 8060). The Hasuma Tuff takes its name from the prominant hill in the south of the study area (HBH 7023) and outcrops extensively to the south of this feature. The Alia Tuff is named after the major lacustrine embayment, Alia Bay (HBH 6020) the tuff has restricted outcrop to the east of the bay. The upper and lower contacts and the lithology of the marker beds are described in the previous section.

Age of the Isochronous surfaces

In the study area correlation has relied exclusively on lithostratigraphic parameters to identify the time equivalent horizons. Biostratigraphic parameters were not used for correlation purposes as a framework independent of the fossil record is required. The temporal relationships of the fossils form a significant part of the palaeontological studies and their use for correlation purposes would introduce a significant measure of circularity into discussion of the palaeofaunas. Although the correlated framework gives a measure of the relative age of the fossil faunas it is necessary to have determinations of absolute age if any comparisons are to be made with other study areas. As it is usually the faunas which are to be compared it is unsatisfactory to use them to make the correlations. In any event the effect of faunal migration and the inherent inaccuracy of time assessment using zonal faunas blunts the potential for correlation on biostratigraphic grounds to an unacceptable degree. The geophysical dating methods potentially offer the best opportunity for determining the age of a particular stratigraphic level and for establishing correlations between sedimentary sequences occurring outside the study area. The chronology of the study area has been investigated by Fitch & Miller (1970) and Fitch *et al.* (1974) using the K–Ar irradiation technique, and by Curtis *et al.* (1975) and Curtis (1975) using the conventional K–Ar method. At present despite a great volume of work it has not proved possible to substantiate with reasonable certainty the age of any of the isochronous surfaces in the study area. The best fit ages for the Koobi Fora Formation tuffs are as follows:

TUFF HORIZON	AGE (m.y.)		
Chari	1.32 ± 0.10		
Okote	1.57 ± 0.00		
KBS	2.61 ± 0.26	1.82 ± 0.04 and 1.60 ± 0.05	
		Curtis *et al.* 1975	

Tulu Bor	3.18 ± 0.09
Surgaei	not dated

Source Fitch *et al.* 1974

Acknowledgements

My thanks are due to F. J. Fitch, for supervising this research and to the Natural Environment Research Council and Birkbeck College who funded the research; R. E. Leakey made the Koobi Fora Research Project a reality; G. L. Isaac, A. K. Behrensmeyer and B. E. Bowen provided helpful discussion in the field.

References

BOWEN, B. E. & VONDRA, C. F. 1973. Stratigraphical Relationships of the Plio-Pleistocene Deposits, East Rudolf, Kenya. *Nature, Lond.* **242**, 391.

FITCH, F. J. & MILLER, J. A. 1970. Radioisotopic age determinations of Lake Rudolf artefact site. *Nature, Lond.* **226**, 226–8.

——, FINDLATER, I. C., WATKINS, R. T. & MILLER, J. A. 1974. Dating of the rock succession containing fossil hominids at East Rudolf, Kenya. *Nature, Lond.* **251**, 213–15.

CURTIS, G. H. 1975. Improvements in potassium argon dating 1962–75. *World Archaeology*, **7**, 198–209.

——, DRAKE, R., CERLING, T. E. & HAMPEL, J. 1975. The age of the KBS tuff in the Koobi Fora Formation, East Rudolf, Kenya. *Nature, Lond.* **285**, 395–8.

26

ANNA K. BEHRENSMEYER

Correlation of Plio-Pleistocene sequences in the northern Lake Turkana Basin: a summary of evidence and issues

The northern part of the Lake Turkana (formerly Lake Rudolf) Basin offers unique opportunities for testing and cross-checking methods of correlation in continental sediments. Such methods include ^{40}K–^{39}Ar and ^{40}Ar–^{39}Ar dating, fission track dating, paleomagnetic reversal stratigraphy, faunal similarities and major and minor element analysis. Intensive research on deposits of the Omo Basin and the East Lake Turkana region have shown that the different methods of correlation indicate time equivalence between the upper parts of the Shungura Formation and the Koobi Fora Formation. However, correlations between the lower parts of the sequences differ, depending on which method is used. Faunal similarities can be used to show a correlation between Shungura F–G and Koobi Fora faunal zone 'a'. Radiometric dating and paleomagnetic reversal stratigraphy have indicated a correlation between Shungura units C–D in the Omo basin and the KBS Tuff in the East Lake Turkana area, which lies above faunal zone 'a'. Discrepancies in the correlation suggest either: (a) that the KBS Tuff is 0.5 to 0.7 m.y. younger than its radiometric age of about 2.6 m.y., (b) that Shungura F–G is 0.5 to 0.7 m.y. older than indicated by radiometric and paleomagnetic data, (c) that strong faunal similarities can be time-transgressive, (d) that dates and faunal correlations are all in error to some degree. Interest generated by the conflicting results has resulted in re-examination and further intensive research in all methods of correlation. Although the correlation in the northern part of the Lake Turkana Basin remains unresolved, issues relevant to problems in other areas and time periods have been clarified by this work.

Introduction

Interest in the geological evolution of the East African Rift System and the biological evolution of the sub-Saharan Neogene vertebrate fauna have resulted in high-resolution age control for many sedimentary and volcanic deposits in East Africa. Precise stratigraphic control on the fossils has been possible because geological mapping, stratigraphy and radiometric dating have been done prior to, or during, the fossil collecting. Thus, the East African Neogene is providing unique opportunities for testing and defining the consistency and limitations of the various methods of age-dating and correlation, including radiometric dating (^{40}K–^{40}Ar and ^{40}Ar –^{40}Ar), paleomagnetic reversal patterns, major and trace element analysis and faunal similarities. Nowhere can this be better demonstrated than in the northern part of the Lake Turkana (formerly Rudolf) Basin, where the Plio-Pleistocene deposits of two separate but adjacent areas, the Omo Basin and East Lake Turkana (formerly East Rudolf), are currently under intensive study. The

fact that correlations between the two areas have not been entirely in agreement
has created a controversy which should ultimately result in increased understanding of all methods used in stratigraphic correlation. This will have relevance to
geological problems extending beyond the Plio-Pleistocene and the geographical
limits of the Gregory Rift.

   Conflict between various methods of correlation in the northern Turkana
Basin was first brought into focus in September of 1973 during a workshop conference sponsored by the Wenner-Gren Foundation for Anthropological Research.
It became apparent during the conference that faunal correlations between certain
parts of the Omo and East Lake Turkana sequences were not consistent with
correlations based on radiometric dating of volcanic material interstratified with
the fossil-bearing deposits. A summary of the problem was presented by the author
at the 1973 Geological Society of America Symposium, 'Vertebrate Paleontology
as a Discipline in Geochronology'. Much research since September 1973 has been
directed toward resolving the correlation problem. The Geological Society of
London's 1975 Symposium (the basis for this volume) provided a platform where
much of the new evidence concerning correlations was aired and discussed. This
proved highly beneficial in clarifying the issues of controversy, although the
central problems were still unresolved. Several other papers included in this
volume present detailed evidence concerning the Omo-East Lake Turkana correlation. My aim is to provide a general and partially historical summary of the
many issues which have become important in discussions of the correlation
problems.

The Omo–East Lake Turkana correlation

A. *General setting and geological evidence*

   Geological and paleontological research near the northern end of Lake
Turkana has been carried out by international teams since 1967. The lake itself
lies in a basin which is tectonically part of the East African Rift System (Fig. 26:1).
The Omo sequence lies north of the lake within the axis of the north-south trough,
and represents accumulations of a major river (the ancestral Omo) plus some
lacustrine and alluvial deposits (de Heinzelin, *et al.* 1971). The cooperative French
and American Omo Research Expedition, under the leadership of Dr Y. Coppens
and Dr F. C. Howell, has been responsible for the research in this region. East
Lake Turkana is an area lying along the northeastern shore of Lake Turkana and
extending some 20–30 kilometres inland. The sedimentary deposits lie adjacent
to the rift axis, essentially on the shoulder of the downwarp which forms the
northern part of the lake. Facies are mixed deltaic, lacustrine and fluvial (Bowen
& Vondra 1973 and Vondra & Bowen this volume.). The Koobi Fora Research
Project (formerly 'East Rudolf Research Project'), consisting of Kenyan, British
and U.S. workers under the leadership of Mr R. E. Leakey and Dr G. L. Isaac,
has been active in the area since 1968.

   The Plio-Pleistocene deposits of the Omo and East Lake Turkana areas are
between 60 and 100 km apart. The intervening region east of the Omo River is
low-lying and devoid of outcrop. There is very little possibility that the two areas

can be linked by surface mapping, and there are no obvious lithological sequences which can be correlated stratigraphically. Sub-surface geology would be helpful, but no boreholes have yet been drilled in the area. Hence, correlation must fall back on secondary methods, which at present include faunal similarities, radio-

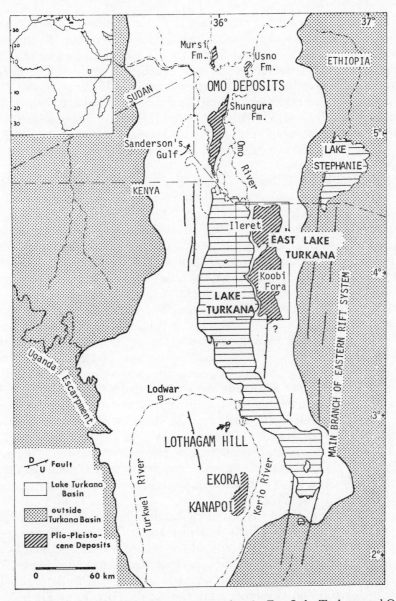

FIG. 26:1. Map of the Turkana Basin area showing the East Lake Turkana and Omo regions near the north end of Lake Turkana. Present area draining into Lake Turkana is shown in white. Sites shown south-west of the lake are discussed by Savage (this volume).

metric dates, major and trace element analysis, and the paleomagnetic reversal chronology.

The longest continuous section in the Omo Basin is represented by the Shungura Formation, which is about 700 metres thick. This has been divided into members using the prominent tuffaceous units; each major tuff and its overlying deposits are referred to as Member A, B, C, etc. or Shungura A, B, C, from the base of the section upward (de Heinzelin, *et al.* 1971; Butzer 1971 a & b). Shungura units A through to lower G are primarily fluvial in origin, the upper part of G is lacustrine, and H–I plus the Kalam units are again primarily fluvial (de Heinzelin *et al.* 1971; de Heinzelin *et al.* 1976). Feldspar-rich pumices from many of the tuffs have been dated using conventional K–Ar methods. The dates are consistent with the relative stratigraphic positions of the tuffs (Brown & Lajoie 1971; Brown & Nash 1976). This internal consistency, along with the careful initial work and re-dating of many of the tuffs, supports the validity of the dates, and they have been used as a standard for calibrating faunal evolution throughout East Africa.

Recent detailed paleomagnetic sampling and the resulting magnetostratigraphy for the Shungura Fm. give a 'best fit' with the standard reversal chronologies which agrees with the K–Ar dates (Shuey, Brown & Croes 1974). With the evidence from two relatively independent dating methods in agreement, the Shungura Fm. has been considered one of the best established high-resolution dated sequences for continental sediments from any region or time period.

The Plio-Pleistocene sequence east of Lake Turkana is represented primarily by the Koobi Fora Fm., which is approximately 150–200 metres thick. A widespread tuffaceous unit, the 'KBS Tuff', divides the Koobi Fora Fm. into upper and lower members. The Lower Member is primarily deltaic and lacustrine in origin, while the Upper Member includes mainly fluvial deposits with some lacustrine and deltaic units. The formation as a whole appears to be the result of a prograding deltaic complex with major stream components originating from the region of the modern Lake Stephanie (Bowen & Vondra 1973; Vondra & Bowen, this volume; Findlater, this volume). Although some of the streams may have been perennial, it is unlikely that any reached the scale or duration of flow of the ancestral Omo River. The fluvial cycles so evident in the Shungura Fm. are absent at East Lake Turkana. Instead, there is evidence for lacustrine transgression and regression, erosional episodes of some magnitude, and alluvial fan deposition within the Koobi Fora Fm. (Vondra & Bowen, this volume; Johnson & Raynolds 1976; Findlater, this volume).

The KBS Tuff provided the initial calibration for the Koobi Fora Fm. with its date of 2.61 ± 0.26 m.y. (Fitch & Miller 1970). This date, done using ^{39}Ar ^{40}–Ar age-spectra analysis, was considered the best of several obtained from ^{39}Ar ^{40}–Ar and conventional K–Ar techniques. Additional dates on other tuffs in the sequence agree with the stratigraphic succession and the KBS date, providing a relatively high-resolution chronology for the Koobi Fora Fm. (Fitch & Miller 1976 and this volume). The heat-step methods of Fitch & Miller give evidence of Argon loss events as well as the original age of crystal cooling through determinations of the ^{39}Ar $-^{40}$Ar ratios of isotopes released at successively higher temperatures.

Most of the tuffs dated so far at East Lake Turkana show evidence for argon loss (over-printing), interpreted by Fitch & Miller (1976) as perhaps due to diagenetic events ('metasomatism') which affected the feldspars and glass subsequent to deposition. Over-printing can result in a wide scatter of apparent dates from different samples of the same tuff, requiring the expertise of geochronologists familiar with the methodology of ^{39}Ar ^{40}–Ar dating for interpretation.

Paleomagnetic sampling of the Koobi Fora Fm. has provided a polarity reversal stratigraphy (Brock & Isaac 1974) which agrees well with the radiometric dates of Fitch & Miller. Like the Shungura Fm., the internal consistency of the two dating methods and the stratigraphy at East Lake Turkana gave a chronology for the Koobi Fora Fm. which has been considered reliable based on evidence available up to the end of 1974.

The correlation of the Shungura and Koobi Fora Formations by age determination is shown in Fig. 26:2. The two sequences span approximately the same time period between 3.2 and 1.0 m.y. BP. There is a marked time break indicated in the lower part of the Upper Member of the Koobi Fora Fm. This may correspond to evidence for a widespread unconformity and a marked change in sedimentary regime (Brock & Isaac 1976). The tuffaceous units do not show any marked correspondence between the two areas, except possibly for the KBS Tuff and Tuff D, and the Chari Tuff and Tuff L. Brown & Nash (1976) have discussed the possible correlation of Tuff D and the KBS, and show by major element analyses that the two may be separated on their Barium and Iron contents. Their conclusion is that the two are not likely to be from the same source. The element analyses for Tuff L and the Chari indicate close similarity, and Brown & Nash (1976) state that a correlation cannot be ruled out on this evidence, although it cannot be considered proved. The Chari Tuff and Tuff L are broadly similar in field aspect. A stratigraphic correlation between the two would indicate extremely widespread deposition of a massive tuffaceous unit (up to 3 m thick in the Koobi Fora Fm.) from a source common to both the East Lake Turkana and Omo paleo-drainage areas.

With the exception of the uppermost parts of the Koobi Fora and Shungura Formations, it is evident that source areas and depositional processes were probably markedly different throughout most of the Plio-Pleistocene history of the Omo Basin and East Lake Turkana. The greater thickness of the Shungura Fm. appears to reflect its tectonic setting along an axis of downwarping, where local subsidence permitted relatively uninterrupted sediment accumulation. East Lake Turkana apparently subsided less than the Omo region, and this is consistent with its position on the eastern margin of the westward dipping half-graben which forms the northern part of Lake Turkana.

B. *Faunal evidence*

(i) *Background*. Historically, paleontologists might have provided the only basis for correlating the Omo and East Lake Turkana sequences, using various criteria of faunal similarity. Such evidence remains important for strata that cannot be dated using radiometric chronometry, paleomagnetics or other methods. For vertebrates, as well as other fossil faunas, biostratigraphic correlations generally carry some sense of similarity in time; faunas may be said to be 'contemporaneous' if they meet certain criteria of similarity. But what is really meant by 'contemporaneous'? Biostratigraphic correlations in some cases may be ecological and potentially time transgressive, particularly when two faunas are correlated from widely separated localities. The Omo and East Lake Turkana faunas present a unique opportunity to test the precision of vertebrate faunal correlations in two adjacent areas against time correlations based on the best dating technology available at the present time.

Lines of evidence used to indicate a faunal correlation include:

1. General faunal similarity; numbers of species or genera present in both areas being sampled outnumber those unique to either area.

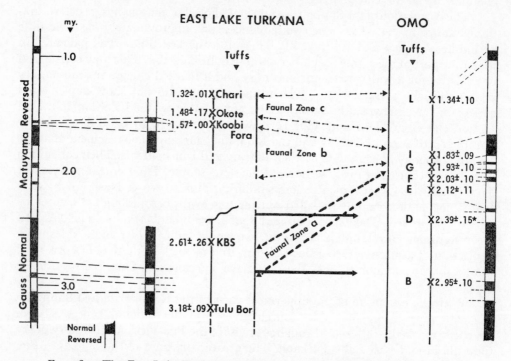

FIG. 26:2. The East Lake Turkana and Omo sequences are aligned on the figure according to the radiometric dates ($^{40}K-^{40}Ar$ and $^{39}Ar -^{40}Ar$) and paleomagnetic columns, which can be compared with the standard reversal chronology of Cox (1969) given on the left. The faunal zones of the Koobi Fora Fm., established by Maglio (1972) are indicated as 'a', 'b', and 'c' (see Table 1). The correlation problem between the two areas, which are 60–100 km apart, is represented by the oblique broken black lines connecting East Lake Turkana faunal zone 'a' and Shungura (Tuffs) F and G. Heavy, solid horizontal lines indicate the range of experimental error on the KBS Tuff date. Faunal comparisons indicate that these two faunal assemblages should be more similar in age than the radiometric and paleomagnetic dating suggest. Either the geochronological dates in one or both sequences are wrong for these assemblages, or the faunal similarity is a time-transgressive phenomenon possibly related to ecological factors. The lengths of the two columns represent time span, not relative thickness.

2. The occurrence of similar evolutionary stages of known lineages in the two areas.

3. The appearance in both areas of a new form, presumed to be a migrant or newly evolved morphotype, or the disappearance of a form common to both areas, through evolution or extinction.

4. Similarity in the relative numbers of the more common animals in the two areas.

The many aspects of the Omo–East Lake Turkana faunal correlations are complex and will not be treated in detail here. Instead, examples will be given of the kinds of faunal evidence used for correlation and the varying opinions concerning the contemporaneity of similar faunas from the two areas.

Some discussion of the faunal succession in the Koobi Fora and Shungura Formations is a necessary prerequisite to consideration of faunal correlations between the two. For East Lake Turkana, Maglio (1972) established three faunal zones in the Koobi Fora Formation. These are based on 'distinct evolutionary stages in the suid and proboscidian lineages' (Maglio 1972, p. 382). Evidence for rapid evolutionary change in the tooth and skull morphology of these two groups in the Plio-Pleistocene has been presented by Cooke & Maglio (1972). The evolutionary stages of *Mesochoerus limnetes* and *Elephas recki*, the two best-represented lineages, are based on the stratigraphic succession of the morphological forms in

TABLE 1: *The East Lake Turkana faunal zones established by Maglio (1972) for the Koobi Fora Formation. The relative age of these zones was based primarily on the evolutionary stages of* Mesochoerus *and* Elephas, *and was in general agreement with the stratigraphic succession and geochronological evidence as understood from 1972–4. Faunal zones are currently being revised by J. Harris.*

	Characteristic Forms
Zone 'c' or '*Loxodonta africana*' Collections from the Ileret Mb. of the Koobi Fora Formation.	*Elephas recki* (Stage III–IV) *Loxodonta africana* *Mesochoerus limnetes* (v. progr.) *Metridiochoerus andrewsi* (progr.) *Hippopotamus* cf. *amphibius* *Ceratotherium simum*
Zone 'b' or '*Metridiochoerus andrewsi*' Collections from the Upper Mb. of the Koobi Fora Formation, below and including the Koobi Fora Tuff.	*Elephas recki* (Stage III) *Mesochoerus limnetes* (progr.) *Metridiochoerus andrewsi* (typical) *Hippopotamus* sp. nov. *Ceratotherium simum* *Tragelaphus grandis*
Zone 'a' or '*Mesochoerus limnetes*' Collections from the Lower Mb. of the Koobi Fora Formation, from 10– 20 m below the KBS Tuff up to the KBS Tuff.	*Elephas recki* (Stage II) *Notochoerus scotti* *Mesochoerus limnetes* (typical) *Hippopotamus* sp. nov. (early) *Ceratotherium simum* (early) *Tragelaphus nakuae*

both the Shungura and Koobi Fora Formations, with supporting evidence from other localities such as Olduvai Gorge and the Lake Baringo region. The original Koobi Fora Formation faunal zones proposed by Maglio (1972) were named according to their most diagnostic component and characterized by other common forms. The zones corresponded to stratigraphic units designated as shown in Table 1. The assignment of a new faunal sample to a particular faunal zone at East Lake Turkana technically required a large enough sample of the critical lineages to establish that the morphological variation in this sample could be matched with one of the distinctive morphological stages.

Since the Omo fauna is accurately tied in with the tuff horizons of the Shungura Formation, discussions of correlation with other areas most often employ the lithostratigraphic units, *e.g.*, Shungura A, B, etc., in referring to faunal assemblages from these units. Formal zones have not been established for the Omo *per se*, although Coppens (1972) has proposed a zonation for East Africa as a whole which specifies Zone V as represented by Shungura A–D and Zone IV as Shungura D–G. These zones are defined by several diagnostic forms, rather than specific evolutionary stages of one or two lineages. In informal discussion, Shungura A–C, D–F and G–L have been grouped into three faunal 'zones' which appear to have significant differences. The apparent lack of large scale time breaks in the Shungura Formation makes clear-cut faunal zonation difficult, but the faunas from the well-defined lithostratigraphic units can be used instead of independently defined biostratigraphic zones in discussions of correlation.

(ii) *Faunal correlations.* Correlations between the Koobi Fora and Shungura Formations have been attempted using the methods listed at the beginning of this section, with varying emphasis on one or another:

(1a) *General faunal similarity.* Well over 150 distinct mammal taxa have been identified in the two areas. Micro-mammals have been well sampled so far only in the Shungura Formation and cannot be used in correlation. Of approximately 100 different species of large mammals, 55 are shared between the Koobi Fora and Shungura Formations (based on faunal lists available in early 1975). This results in an overall index of faunal resemblance (FRI) of 0.75 according to the method of Simpson (1960).[1] Species not shared between the two areas include primarily members of the Bovidae, Carnivora and Primates. The resemblance index is relatively free from possible discrepancies due to differing taxonomic schemes in the two areas since collaboration between specialists has been extensive, and in some cases the same paleontologist has studied both collections. According to calculated FRI's for recent faunas by Van Couvering & Van Couvering (1976), 0.75 is reasonable for spatially separate faunas from similar habitats. The fact that the FRI is not higher between the two adjacent areas indicates inadequate sampling and/or faunal differences between the Omo and East Lake Turkana regions which occurred throughout the 2 million year time span of the Plio-

---

[1] The percentage of faunal similarity is calculated using the method of Simpson (1960): % Similarity = $C/N_2 \times 100$, where C is the number of taxa common to both samples and N_2 is the total number of taxa in the smaller sample of the two (East Lake Turkana, $N_2 = 73$; Omo, $N_1 = 85$). This index tends to give misleadingly high percentages of similarity if samples (N_1, N_2) are widely different in size.

Pleistocene deposits. These could hold important paleoecological information for the history of the northern end of Lake Turkana.

(1b) *Faunal similarity between members.* Using Simpson's method of calculating faunal similarity (see Footnote 1), rather interesting patterns of correlation emerge, as shown in Table 2. The Koobi Fora Formation fauna from each of the three faunal zones is more similar to Shungura G–L than to A–C or D–F. Likewise, the Shungura fauna from each of these stratigraphic groupings is most similar to the *Loxodonta africana* zone (zone 'c') of East Lake Turkana. Using this measure of faunal similarity alone, the entire Shungura Formation could be correlated with the *Loxodonta* zone, or the Koobi Fora Formation could be matched with Members G–L of the Shungura Formation. The limitations of Simpson's faunal resemblance index as a method of chronological correlation are indicated by the fact that Shungura G–L is more similar to Koobi Fora 'a' and 'c' than to 'b', *i.e.*, faunal resemblance due to ecological (or other) factors may be important enough to swamp the effects of faunal change through time. This has also been pointed out recently by Van Couvering & Van Couvering (1976) with reference to Miocene faunas of East and Central Africa. Detailed analysis of East African faunal resemblances using Simpson's FRI is being undertaken by Shuey *et al.* (this volume). and this should help to isolate factors in faunal correlation which are due to time as opposed to ecology.

TABLE 2: *The number of species and genera in common and the percentage of similarity of Koobi Fora Formation faunal zones and Shungura Formation assemblages arranged in three informal groups. Percentage of similarity between faunas is calculated according to the method of Simpson (1960):* $\% = C/N_2 \times 100$, *where C = number of taxa in common and N_2 = the total number of taxa in the smaller of the two samples. Only macro-fauna have been used in the totals and percentages.*

Koobi Fora Formation Faunal Zones	Shungura Formation Members			Total No. Taxa Koobi Fora Formation
	G–L	D–F	A–C	
'a'	38 (79 %)	24 (50 %)	22 (46 %)	48
'b'	38 (64 %)	28 (56 %)	26 (62 %)	59
'c'	34 (76 %)	27 (60 %)	28 (62 %)	45
Total No. Taxa, Shungura Formation	63	50	54	

(2) *Correlation by evolutionary stage.* Large collections and detailed stratigraphic control have allowed the reconstruction of morphological change through time in *Elphas recki* and *Mesochoerus limnetes* samples from East Lake Turkana and the Omo Basin. Smaller samples of several other lineages, including *Tragelaphus*, *Menelikia*, *Kobus*, *Theropithecus* and *Hippopotamus*, also show evolutionary trends through the sequences (Gentry 1976, Coryndon 1976, M. E. Leakey 1976). The *Elephas* and *Mesochoerus* lineages show the most consistent and quantifiable evolutionary trends, and therefore are most useful for purposes of correlations. The rationale behind evolutionary stage correlation using suids and proboscidians

is given by Cooke & Maglio (1972). Similar evolutionary stages of *Mesochoerus limnetes*, as measured principally on molar length, occur in Shungura F–G and Koobi Fora 'a', Shungura G–H and Koobi Fora 'b' and 'c' (Cooke 1976). This indicates that the Koobi Fora Formation correlated best with Shungura F–L. Evolutionary stages of *Elephas recki* show closest similarities between Koobi Fora 'a' and Shungura D–F, Koobi Fora 'b' and Shungura G–H and Koobi Fora 'c' with Shungura H–L (Maglio 1972; Beden 1976). Evolutionary stages in the elephants are based on various characters of the molars, including crown height, length, number of plates and enamel thickness. The two examples of evolutionary stage correlation are not in exact agreement, since the elephants indicate a somewhat earlier correlation between the Shungura Formation and Koobi Fora 'a' than do the pigs. Both lines of evidence indicate contemporaneity between the upper part of the Shungura Formation and the Upper Member of the Koobi Fora Formation.

(3) *The appearance of new elements in the fauna.* The appearance of *Equus* in Shungura Member G provides a good example of how this evidence can be used for correlation. *Equus* is present throughout the Koobi Fora Formation faunal zones. Therefore, the obvious initial interpretation is that all of the East Lake Turkana faunal zones are equivalent in time with Omo G and above, since the invasion of *Equus* into East Africa is thought to have been a significant, widespread event which should have affected the Omo and East Lake Turkana areas simultaneously. However, the evidence is subject to various potential sampling problems. The Omo collection below Member G may not be large enough to provide specimens of *Equus*. *Hipparion* is abundant in all members up to and including G, where the ratio of *Hipparion* to *Equus* specimens is 3/2. If this ratio held for Member F, then there should be about 12 *Equus* specimens from F. If *Equus* was present in the Northern Turkana Basin in Shungura F time, and was four times rarer in the Omo area than in G, it might have escaped being sampled along with *Hipparion* in the Omo collections. This is possible if it was ecologically excluded from the Omo region prior to Member G. Alternatively, taphonomic factors could have hindered the preservation of *Equus* parts prior to G, but this seems unlikely since *Hipparion* is present throughout the sequence where *Equus* is absent, and similar taphonomic factors should affect both equids. Partial or complete ecological exclusion from the Omo region prior to Shungura G appears to be a possible alternative to the interpretation which regards the first appearance of *Equus* in the Shungura and Koobi Fora Formations as a synchronous event and hence a basis for time correlation.

(4) *Similarity in the relative numbers of common animals.* The Bovidae are the most abundant and diverse family in both areas. Since bovids were collected without bias for or against any particular form, the relative abundances of the various tribes in the collections can be used as a measure of their relative abundances in the taphocoenoses, and secondarily in the original biocoenoses. The currently available data (Gentry 1976; Harris 1976) show that there is closest similarity between the tribe abundances in Koobi Fora 'a' and 'b' and Shungura F–G (Table 3). There are consistent differences in the greater abundance of reduncines in the Koobi Fora Formation (except 'c') and the greater abundance of tragel-

aphines in the Shungura Formation. Alcelaphines are comparable in overall abundance (except in Shungura B–C) but differ generically between the two areas, with *Aepyceros* generally more common in the Omo and *Parmularius* at East Lake Turkana (Gentry 1976; Harris 1976). There is an added factor regarding the alcelaphines; most of the material below Shungura G consists of teeth, while more complete and varied skeletal parts are preserved in G. This may indicate that much of the alcelaphine material prior to G was transported and not locally derived. In contrast, relatively complete alcelaphine material is available throughout the section at East Lake Turkana.

TABLE 3: *Numbers of specimens and relative abundance comparisons of the four most common bovid tribes in the East Lake Turkana and Omo collections. Shungura Formation numbers are based on teeth only and are taken from Gentry (1976). Alcelaphines in the Omo collection consist mainly of* Aepyceros; *in the East Lake Turkana collection,* Parmularius *plus less abundant* Aepyceros. *Data for the Koobi Fora Formation are based on all identifiable material and were taken from Harris (1976).*

	Koobi Fora Formation Faunal Zones			Shungura Formation Members		
	'a'	'b'	'c'	B–C	D–E	F–G
Tragelaphini	17 (21 %)	59 (17 %)	23 (19 %)	162 (38 %)	180 (41 %)	343 (24 %)
Bovini	1 (1 %)	27 (8 %)	20 (17 %)	59 (14 %)	22 (5 %)	35 (2 %)
Reduncini	38 (46 %)	144 (41 %)	43 (36 %)	121 (28 %)	84 (19 %)	540 (38 %)
Alcelaphini	26 (32 %)	118 (34 %)	33 (28 %)	81 (19 %)	149 (34 %)	502 (35 %)
Total No.	82	348	119	423	435	1420

The bovid frequency data (Table 3) indicate possible ecological differences between the Omo Basin and East Lake Turkana, using the habitat preferences of modern analogues of the tribes represented in the two collections. The consistently high proportion of open-grassland forms (the Alcelaphini, excluding *Aepyceros*) in the Koobi Fora Formation (Harris 1976) indicates the presence of drier grassland proximal to depositional areas, whereas such grasslands may have been distal to the Omo river depositional environments until Shungura F–G (Gentry 1976). The high proportion of tragelaphines suggests that moderately dense thicket and scrub was more extensive in the Omo region that at East Lake Turkana. Greater abundance of reduncines in Koobi Fora 'a'–'b' indicates larger areas of grassy marshland than in Shungura B–F. In general, it is easier to relate observed differences in bovid abundances in the two areas to ecological factors than to time-dependent ones.

(iii) *Summary*. Much of the faunal evidence indicates correlation between Koobi Fora 'a'–'c' and Shungura F–L, implying a time similarity as well. However, as can be noted from Fig. 26:2, this correlation of the two sequences is not in agreement with the suggested interpretation of the radiometric and paleomagnetic evidence. Both geologists and paleontologists have been reluctant to stretch their ranges of error to accommodate this discrepancy. Thus, all workers have been forced to re-examine their evidence and basic assumptions. At present the problem remains unresolved, but the ongoing discussion should prove highly beneficial to the larger problem of vertebrate fossils as geochronological tools. Important points raised in regard to the Omo-East Lake Turkana correlation problem will be presented below, and they may provide insight into correlation problems in other regions and time periods.

Discussion: Geological issues

Given the conflict between faunal and radiometric dates, the faunal evidence was initially questioned more than the geochronological data. However, a number of points can be made concerning possible errors in the radiometric dating and paleomagnetic chronology. These are as follows:

A. *KBS Tuff: depositional factors.* The KBS Tuff may have been, at the time of deposition, considerably younger than the age of at least some of its included pumices (2.6 ± 0.26 m.y.). Therefore, the Lower Member of the Koobi Fora Formation would be younger than the indicated date, if these older pumices constitute the dated sample.

It is possible that some older pumices could have been included in the fluvial reworking of pyroclastic material which created the KBS Tuff, and this possible source of error cannot be ruled out. However, there is no geological evidence presently available which can be used to show that the eruption of a considerable volume of pyroclastic material and the final deposition of the KBS Tuff were not essentially synchronous. Instead, field relationships of the unit indicate that deposition followed closely upon the eruption and swamping of drainage sources with volcanic ash and pumice. The tuff is a distinct sedimentary unit in essentially non-tuffaceous deposits, it is widespread and relatively homogeneous, and there is some evidence for an initial air-fall layer at the base of the tuff in its type area (Area 105). Spatially distant localities have yielded similar ages using the ^{39}Ar– ^{40}Ar– method (Fitch & Miller 1976). If heterogeneity in the primary source material of the KBS Tuff is responsible for erroneous dates, then this must also be considered a likely source of error in other East Lake Turkana and Omo tuffs.

B. *KBS Tuff: dates and overprinting.* The spread of dates from the KBS Tuff is broad (3.63 ± 2.1 to 0.52 ± 0.33 m.y.) and poses the question of possible inherited argon as well as the proposed overprinting due to hydrothermal events subsequent to deposition (Fitch & Miller 1976).

The technical quality of the KBS Tuff dates has been discussed by Fitch & Miller (1976), who regard the problem of inherited argon as minor and easily corrected for. Although there is evidence for 'slight contamination and Argon-Loss errors' from the step-heating ^{11}Ar –^{11}Ar analysis, Fitch & Miller report that the oldest significant plateau in the age spectrum plots occurs at an average of 2.61 ± 0.26 m.y.—the most probable date of formation of the sanidine crystals used for dating. Conventional K–Ar dating gives an age of 2.37 ± 0.3 m.y. on the same crystal concentrate (Fitch & Miller 1970). Plateaus indicating younger ages were ascribed to 'thermal overprinting,' i.e., events which partially reset the K–Ar clocks in the parts of the sanidines where Ar was not tightly bound by the crystal structure (Fitch & Miller, Pers. Comm.). This theory of thermal over-printing has been questioned on two counts: (a) thermal events which create similar overprints in samples separated by tens of kilometers in relatively shallow deposits are difficult to explain in areas where highly localised volcanic activity seems to be the rule, (b) thermal events which affect sanidines but leave associated glass unaltered are geochemically unknown (R. L. Hay, Pers. Comm.) (Koobi Fora Formation tuffs and pumices include a considerable amount of unaltered glass). Non-thermal diagenetic loss of argon from sanidines is being explored as an alternate explanation for the young plateaus in the age spectra. At present, how-ever, no such mechanism is known which would not also decompose associated glass.

Given the general stratigraphic framework of the Koobi Fora Formation, the paleomagnetic stratigraphy, and dates on other tuff horizons, there has been little internal reason to question the general validity of the KBS Tuff date, even though interpretation of the age spectra from ^{39}Ar ^{40}–Ar dating can be questioned from theoretical grounds. However, dating of the KBS Tuff by other laboratories, and by other methods (*e.g.*, fission track), has been undertaken as a cross-check of the original dates.

   C. *Paleomagnetic alternatives for the KBS tuff.* The magnetostratigraphy of the Koobi Fora Formation can be matched moderately well with standard scales by extending the Lower Member into the base of the Matuyama Reversed Epoch (Brown & Shuey 1976). This would indicate that the KBS Tuff should lie within the Olduvai or Reunion events and have an age of 1.6–1.8 or 2.0–2.1 m.y. rather than 2.6 m.y.

Paleomagnetic sampling of the Koobi Fora Formation has not yet been done on the detailed scale of the Shungura Formation. However, there is evidence from two separate columns of samples that the Lower Member is consistently normal except for two reversed segments above the Tulu Bor Tuff (Brock & Isaac 1974 and 1976). Assigning the Lower Member to the top of the Gauss and the base of the Matuyama, as suggested by Brown & Shuey (1976) as an alternative fit, requires deposition of most of the Lower Member, including the KBS Tuff, during the 2–3×10^5 years encompassed by the Reunion and 'X' events of normal polarity, or during the normal polarity of the Olduvai/Gilsa event. This inter-pretation implies increased rates of sedimentation during the normal periods relative to the reversed ones, since most of the 40–50 m of the Lower Member is normal with narrow zones of reversed polarity. Brock & Isaac (1974, 1976) have considered this chronology for the Lower Member improbable, based on the re-ported degree of faunal change between faunal zones 'a' and 'b', on the stratigraphic evidence for relatively long periods of sedimentation during normal polarity and on the evidence for an erosion surface between 'a' and 'b'. However, they suggest means of testing the alternate fit, while maintaining the paleomagnetic chronology for the Koobi Fora Formation as shown in Fig. 26:2.

   Recent work at Stanford University by Cox, Hillhouse, Ndombi and Smith (Pers. Comm.) indicates that the Lower Member may include too much normally polarized sediment due to remagnetization during the Brunhes Epoch. The problem of 'poor magnetic memories' may be chronic in Koobi Fora Formation deposits; if so, then much of the Lower Member may actually be reversed 'overprinted' with normal polarity in sediments originally deposited during the Matuyama Epoch.

     D. *Omo tuffs and possible overprinting.* The Omo dates for Members E, F, G and I are clumped within about 300 000 years and may have been affected by overprinting events which would not be detected by conventional K–Ar dating techniques. Therefore, the dates on some, or all, of these members (and their faunal assemblages) may be too young.

    Overprinting seems to be part of the history of many of the East Lake Turkana tuffs (Fitch & Miller 1976) but has not been formally noted as a possible source of error in the Omo tuffs. The Ar^{39}–Ar^{40} age spectrum analysis used by Fitch and Miller on the East Lake Turkana tuffs has not yet been applied to the Shungura Formation, but this is planned as a cross-check on the dates and should help to resolve the question one way or the other. Older dates for Shungura E, F and G would help to resolve the faunal correlation problem. However, it is somewhat difficult to postulate hydrothermal events which affected only these tuffs and not the older ones. Pushing back the entire lower Shungura Formation would create many new problems.

     E. *Paleomagnetic alternatives for the Shungura Formation.* The Omo magnetostratigraphy can also be fitted to the standard scales so that Shungura E–F are older than the chronology indicated by the K–Ar dates and the paleomagnetic fit proposed by Shuey, Brown & Croes (1975). (See Fig. 26:2.)

    Given the 2 m spacing of the Omo paleo-polarity determinations and the long continuous section sampled, it is difficult to propose any alternative fit to the one which agrees with the K–Ar dates. The only reason for attempting another chronology is the discordance in the Omo-East Lake Turkana faunal correlation, which could be resolved if Shungura F and G are older than the dates indicate. As shown in the paleomagnetic columns of Fig. 26:2, F and G are characterized by a Normal-Reversed-Normal pattern which could be placed at the Gauss-Matuyama transition or across the Kaena Event in the Gauss Normal Epoch. The former seems more probable, but both fits create many more problems in matching the upper and lower parts of the sequence with the standard paleomagnetic scale.

    Questions raised above regarding the geochronological dating and correlation of the Koobi Fora Formation and the Shungura Formation are leading to additional research which will firmly establish time scales for the two sequences, within the limitations of the current methods and interpretations. As of the end of 1974, the radiometric geochronology did not provide any obvious reasons why the dating in either formation should be strongly doubted.

Discussion: Faunal issues

A. *Paleontologic assumptions*

    Many of the problems in fitting the faunas to the proposed K–Ar and paleomagnetic time scale arise because this requires certain bovid and suid species or stages to occur some 5–7 hundred thousand years earlier at East Lake Turkana than in the Omo. Considering that only about 100 km separate the two areas, and there were no major geographical barriers (except possibly Lake Turkana itself), these factors are difficult for the paleontologists to accept. They seem to base their concern on the following assumptions:

1. Spacially adjacent areas, through time, should show general faunal similarity (at least in the kinds of environments sampled in the fossil record).
2. Gene flow will be such that two populations of the same species in adjacent areas will not show significant morphological divergence over time, unless there is a substantial geographical or ecological barrier.
3. The appearance of a new morphological variant in one area will mean its nearly instantaneous appearance in adjacent areas.
4. Morphological variation in skeletal parts within a species today is a valid model for morphological variation on any given time plane for related species.
5. Past adaptations, as reflected in morphology, are similar to those of modern species.
6. Sampling of the original populations in the fossil record will tend to preserve the average morphological characters of those populations through time, and even small samples in collections will usually represent average characters.
7. The presence of a few, or even one, specimen of a given species or genus in a large sample is enough to allow some ecological interpretations and comparisons with other faunas.
8. The absence of a form from a large sample probably indicates absence from or extreme rarity in the original fauna.

While the above assumptions are not usually stated, they are fundamental to the way many modern vertebrate paleontologists approach correlation and match faunal zones and strata in time. In many cases, the degree of precision of their correlations is dependent on these assumptions, which tend to minimize possible ecological factors.

B. *Ecological hypothesis*

Radiometric and paleomagnetic evidence have indicated a time correlation between Shungura units C–D and the KBS tuff above Koobi Fora 'a'. However, the faunas from these levels are markedly different. A possible explanation, assuming that the radiometric dates are approximately correct, would be that the lack of faunal similarity is due to differing ecologies in the two areas. There is some evidence to support this hypothesis. Various lineages occur almost exclusively in either the Omo or East Lake Turkana deposits throughout their time span from about 3.0 to 1.0 m.y. BP. Ecological conditions of the two areas today are markedly different, with East Lake Turkana generally drier and subject to greater seasonal aridity, and it seems probable that similar climatic differences characterized the two areas in Plio-Pleistocene times. A larger proportion of reduncines and alcelaphines (excluding *Aepyceros*) in the Koobi Fora Formation indicates that marsh and grassland conditions were common in areas proximal to depositional environments, apparently not the case for the Omo until Shungura F–G. The geology of the Shungura Formation shows plainly a transition from fluvial to lacustrine conditions in G, while the Koobi Fora Formation is primarily lake margin in 'a', followed by mixed fluvial and lake margin facies in 'b' and 'c'. Micro-mammal and pollen evidence in the Shungura Formation suggests at least periodic ecological changes to drier conditions about 2.0 m.y. BP in E–G (Jaeger & Wesselman 1976; Bonnefille 1976).

The ecological hypothesis thus proposes that the Omo region was markedly wetter and less seasonally arid than East Lake Turkana from Shungura A–F. Bush and scrub, with a diverse riverine forest, dominated the areas proximal to the Omo depositional environments, while grasslands plus less extensive bush and forest characterized East Lake Turkana. In Shungura F or G, environmental conditions in the Omo changed and converged on those present at East Lake Turkana, with a rise in lake level, expanded areas of marshy grassland and seasonally grassy mudflats, and a decrease in the riverine forests due to drier local climatic conditions.[2] This change brought about the increase in faunal similarity.

The question remains, however, whether advanced forms of *Menelikia, Kobus* and *Mesochoerus* could have existed for perhaps 0.5 m.y. at East Lake Turkana without mixing with their presumed ancestral forms in the Omo. This morphological separation seems unlikely based on modern analogy, as these animals today are morphologically homogeneous over wide areas of East Africa. It may be valid, in this case, to question whether the present is really a reliable key to the past. Species diversity in some groups such as the reduncines, suids and large ungulates was considerably greater during the Plio-Pleistocene than at present. This would have required greater partitioning of the available resources. Habitats were probably more restricted and morphological variation less than is typical of modern forms. The presence of several contemporaneous species of *Hippopotamus* and crocodile in the Plio-Pleistocene fauna of the Lake Turkana Basin, while only one species of each survives today, exemplifies the difference in diversity. It follows that gene flow between adjacent populations or sub-species may have been more restricted than today, if animals were more closely adapted to particular food and water requirements. Whether or not genetic separation could have been maintained over 60–100 km for several hundred thousand years is the question of interest for evolutionists, and it should remain an open question in spite of intuitive feelings (based on our recent faunas) that such long-term separation is highly unlikely. The more 'advanced' forms may have been present earlier at East Lake Turkana because drier or more open marshland conditions existed there. These animals could have successfully invaded the Omo region when such habitats became available and the surviving relict populations of the 'primitive' forms

[2] It is worth noting in this context that the recent fluctuations of lake level in Lake Turkana appear to be related to rainfall over the Ethiopian Highlands far to the north and are virtually independent of climatic conditions close to the lake (Butzer 1972). Furthermore, fluctuations in lake salinity can have important effects on whether lake margin areas will be open grassland, woodland or forest, regardless of the local climate.

were displaced or selected against by the new conditions. Primitive and advanced are relative terms; if conditions had become more forested around the lake, the Omo forms might have displaced those in East Lake Turkana and would have become the lineages which continued to present times.

C. *Taphonomic factors*

It is also possible that taphonomic factors have contributed to the observed differences between *Mesochoerus* in the lower part of the Shungura Formation and Koobi Fora 'a'. Since tooth size is critical in correlating evolutionary stages in this lineage, slight sorting factors which biased the Omo sample toward preservation of the smaller end of the morphological range, and/or the East Lake Turkana sample toward the larger end of the range, might be enough to create the observed discrepancy (8 mm difference between the means) in correlation between East Lake Turkana 'a' and the Omo sample below Member F. This discrepancy is interpreted by various paleontologists (*e.g.* Cooke 1976) as evidence that zone 'a' is not the time equivalent of Shungura D–E. Although effects such as this seem unlikely, they are theoretically possible and deserve further research in the realm of experimental taphonomy.

A second possible taphonomic effect is differential seasonal sampling of the Turkana Basin Plio-Pleistocene faunas. If dry season mortality near the lake shore contributed significantly to the fossil fauna of East Lake Turkana, then water-dependent grazing herbivores may be more common in the fossil assemblages than they actually were in the living community. This would contribute to apparent ecological differences between the Omo and East Lake Turkana and might even bias samples of different sub-species in the two areas. It seems apparent from recent taphonomic observations that sampling of faunas from riverine as opposed to lake margin habitats is likely to result in different bone assemblages, but this has yet to be tested by detailed study.

Conclusion

If the paleontological correlation between Omo F–G and East Lake Turkana zone 'a' could be attributed in part to ecology rather than time, then this would provide an extremely important example for vertebrate paleontology and geochronology. Many supposed time correlations between faunas more spacially distant than the Omo and East Lake Turkana may, in fact, be ecological correlations. It would seem imperative to consider faunal similarity in context with sedimentary environments and general paleoecological frameworks before interpreting it as time dependent. If 'advanced' and 'primitive' species could coexist for some 0.5 m.y. (or even 0.2 m.y.) (?) at the Omo and East Lake Turkana, then 'advanced' and 'primitive' groups at various taxonomic levels might coexist for several million years in earlier Cenozoic time, in widely separated or ecologically distinct environments. The appearance of a new form may indicate no more than that a new ecological setting is being sampled.

In the case of the East Lake Turkana and Omo correlation problem, the ecological hypothesis is not yet proved, but only serves as a possible explanation for the observed differences in geochronological and faunal correlations between the two sequences. Research on this problem will probably continue for several years, and it may well be that some of the radiometric and/or paleomagnetic dates will be proved erroneous. It is also possible that the final analysis will show that none of the methods can provide reliability at the level of resolution required for detailed correlation of the Omo and East Lake Turkana sequences. However, this example of a correlation problem is useful in underlining that ecological factors could be more important than has been realized in the past. The increased resolution and credibility of chronostratigraphic correlations should be the inevitable result of test cases such as this.

Acknowledgements

The discussion of the East Lake Turkana–Omo correlation problem arose at the Wenner-Gren Foundation's workshop conference, 'Stratigraphy, Paleoecology and Evolution in the Lake Rudolf Basin,' in September of 1973. The Foundation and Mrs Lita Osmundsen, Director of Research, are to be commended for organizing this very important meeting. Nearly all of the participants at this conference contributed in some way to the ideas and discussion presented in this paper, and I am grateful for the privilege of relating some of their contributions to a wider and more general audience. However, I take full responsibility for the interpretations of the issues presented in this paper where these differ from the interpretations of principle researchers involved.

My special thanks go to Prof. W. W. Bishop, Prof. F. C. Howell and Prof. G. Ll. Isaac for their assistance, encouragement and moral support as I assembled, rethought and revised the contents of this paper. For specific advice and information during its preparation, I am grateful to Frank Brown, H. B. S. Cooke, Yves Coppens, Alan Cox, Ian Findlater, Frank Fitch, Alan Gentry, John Harris, Richard Hay, Richard E. Leakey, Jack Miller, Ralph Shuey and Judy and John Van Couvering. I thank those who have been generous with unpublished material, all of whom should receive full credit for work indicated as 'personal communication' or 'in press'. Ms. Beagle Brown and Ms. Mayme Matsumoto have earned thanks for their speed and consideration in typing the manuscript.

References

BEDEN, M. 1976. Concerning the Plio-Pleistocene Proboscidians from the Omo Group Formations. In: '*Stratigraphy, Paleoecology and Evolution in the Lake Rudolf Basin*', Coppens, Y. *et al.*, eds, Univ. of Chicago Press. 193–208.

BONNEFILLE, R. 1976. Palynological evidence for an important change in the vegation of the Omo Basin between 2.5 and 2 million years. In: '*Stratigraphy, Paleoecology and Evolution in the Lake Rudolf Basin*', Y. Coppens, *et al.*, eds, Univ. of Chicago Press, 421–31.

BOWEN, B. & VONDRA, C. 1973. Stratigraphical relationships of the Plio-Pleistocene deposits, East Rudolf, Kenya. *Nature, Lond.* **242**, 391–3.

BROCK, A. & ISAAC, G. 1974. Paleomagnetic stratigraphy and chronology of hominid-bearing sediments east of Lake Rudolf, Kenya. *Nature, Lond.* **247**, 344–8.

—— & ISAAC, G. 1976. Reversal stratigraphy and its application at East Rudolf. In '*Strati-*

graphy, Paleoecology and Evolution in the Lake Rudolf Basin', Y. Coppens, *et al.*, eds, Univ. of Chicago Press, 148–62.

BROWN, F. & LAJOIE, K. 1971. Radiometric age determination of Pliocene/Pleistocene formations in the Lower Omo Basin, Ethiopia. *Nature, Lond.* **229**, 483–5.

—— & NASH, W. 1976. Radiometric dating and tuff mineralogy of Omo Group deposits. In: *'Stratigraphy, Paleoecology and Evolution in the Lake Rudolf Basin'*, Y. Coppens, *et al.*, eds, Univ. of Chicago Press, 50–63.

—— & SHUEY, R. 1976. Preliminary magnetostratigraphy of the Lower Omo Valley, Ethiopia. In: *'Stratigraphy, Paleoecology and Evolution in the Lake Rudolf Basin'*, Y. Coppens, *et al.*, eds, Univ. of Chicago Press, 64–78.

BUTZER, K. W. 1971a. *Recent History of an Ethiopian Delta*, Res. Pap. No. 136, Dept. Geogr. Chicago. The University of Chicago Press. 184 pp.

—— 1971b. The Lower Omo Basin, geology, fauna and hominids of Plio-Pleistocene formations. *Naturwissenschaften*. **58**, 7–16.

COOKE, H. B. S. 1976. Suidae from Plio-Pleistocene strata of the Rudolf Basin. In: *'Stratigraphy, Paleoecology and Evolution in the Lake Rudolf Basin,'* Y. Coppens, *et al.*, eds, Univ. of Chicago Press, 251–63.

—— & MAGLIO, V. 1972. Plio-Pleistocene stratigraphy in East Africa in relation to proboscidean and suid evolution. In: *'Calibration of Hominoid Evolution'*, W. Bishop and J. Miller, eds, Edinburgh Scottish Academic Press, 303–29.

COPPENS, Y. 1972. Tentative de zonation du Pliocene et du Pleistocene d'Afrique par les grands Mammiferes. *C. R. Acad. Sci. Paris.* **274**, 181–4.

CORYNDON, S. C. 1976. Fossil Hippopotamidae from Pliocene/Pleistocene successions of the Rudolf Basin. In: *'Stratigraphy, Paleoecology and Evolution in the Lake Rudolf Basin'*. Y. Coppens, *et al.*, eds, Univ. of Chicago Press, 238–50.

COX, A. 1969. Geomagnetic reversals. *Science*, **163**, 237–46.

FITCH, F. J. & MILLER, J. A. 1970. Radioisotopic age determinations of Lake Rudolf artefact site. *Nature, Lond.* **226**, 226–8.

—— & ——. 1976. Conventional K–Ar and Argon-40/Argon-39 dating of volcanic rocks from East Rudolf. In: *'Stratigraphy, Paleoecology and Evolution in the Lake Rudolf Basin,'* Y. Coppens, *et al.*, eds, Univ. of Chicago Press, 123–47.

GENTRY, A. 1976. Bovidae of the Omo Group deposits. In: *'Stratigraphy, Paleoecology and Evolution in the Lake Rudolf Basin,'* Y. Coppens, *et al.*, eds, Univ. of Chicago Press, 275–92.

HARRIS, J. M. 1976. Giraffidae, Rhinocerotidae and Bovidae of the East Rudolf succession. In: *'Stratigraphy, Paleoecology and Evolution in the Lake Rudolf Basin'*, Y. Coppens, *et al.*, eds, Univ. of Chicago Press, 293–301.

DE HEINZELIN, J., BROWN, F. & HOWELL, F. C. 1971. Pliocene/Pleistocene formations in the lower Omo Basin, southern Ethiopia. *Quaternaria.* **13**, 247–68.

——, HAESAERTS, P. & HOWELL, F. C. 1976. Depositional history of the Shungura Formation. In: *'Stratigraphy, Paleoecologyy and Evolution in the Lake Rudolf Basin,'* Coppens, Y. *et al.*, eds, Univ. of Chicago Press, 24–49.

JAEGER, J. & WESSELMAN, H. 1976. Fossil remains of micromammals from the Omo Group deposits. In: *'Stratigraphy, Paleoecology and Evolution in the Lake Rudolf Basin,'* Y. Coppens, *et al.*, eds, Univ. of Chicago Press.

JOHNSON, G. D. & REYNOLDS, R. G. H. 1976. Late Cenozoic environments of the Koobi Fora Formation. In: *'Stratigraphy, Paleoecology and Evolution in the Lake Rudolf Basin,'* Y. Coppens, *et al.*, eds, Univ. of Chicago Press, 115–22.

LEAKEY, M. G. 1976. Carnivora and Cercopithecoidea of the East Rudolf succession. In: *'Stratigraphy, Paleocology and Evolution in the Lake Rudolf Basin'*, Y. Coppens, *et al.*, eds, Univ. of Chicago Press, 302–13.

MAGLIO, V. J. 1972. Vertebrate faunas and chronology of hominid-bearing sediments east of Lake Rudolf, Kenya. *Nature, Lond.* **239**, 379–85.

SHUEY, R. T., BROWN, F. H. & CROES, M. K. 1974. Magnetostratigraphy of the Shungura Formation, South-western Ethiopia: Fine Structure of the Lower Matuyama Polarity Epoch. *Earth and Plan. Sci. Let.* **23**, 249–60.

SIMPSON, G. G. 1960. Notes on the measurement of faunal resemblance. *Am. Jour. Sci.*, **258-A**, 300–11.

VAN COUVERING, J. & VAN COUVERING, J. 1976. Early Miocene mammal faunas from East Africa: aspects of geology, faunistics and paleoecology. In: '*Human Origins*,' G. Isaac and E. McCown, eds, W. A. Benjamin, Inc., Menlo Park, Calif. 155–207.

27

FRANK J. FITCH, PAUL J. HOOKER
& JOHN A. MILLER

Geochronological problems and radioisotopic dating in the Gregory Rift Valley

Geochronology is the science of geological time and has a wider connotation than radioisotopic dating. The two major tools of geochronology: stratigraphical palaeontology and radioisotopic dating are equally fallible and are best used in conjunction. Basic large-scale geological mapping and the erection of local rock-stratigraphies are necessary pre-requisites for the detailed geochronological analysis of an area. In the Gregory Rift Valley these requirements are beginning to be met at such places as Olduvai, Baringo, East Lake Turkana and Omo. Conventional K–Ar age determination of volcanic rocks and minerals has been used widely in the dating of East African Cenozoic successions. It is important that the theoretical and practical limitations of this otherwise very successful technique are fully appreciated. K–Ar and $^{40}Ar/^{39}Ar$ isochron dating are developments of the K–Ar technique that can be used to resolve some of the difficulties inherent in the conventional method. Isochron dating of rocks from Olduvai, the plateau phonolites of Kenya, East Lake Turkana and the Omo illustrates this point. Arguments are presented for rejecting discrepant apparent ages of less than 2.0 m.y. from the KBS Tuff, East Lake Turkana. It is suggested that fission track dating of volcanic zircons will resolve this controversy and provide independent time-scales for many East African and other volcanic successions in future.

Introduction

Geochronology is the science of geological time and, as such, has a much longer history and a wider connotation than *radioisotopic dating*. Since its innovation by Rutherford 70 years ago radioisotopic geochronometry has become the most prestigious branch of the subject: nevertheless, it is salutary to remind ourselves that radioisotopic dating is a geochronological tool and not an end in itself. Geochronometry must be judged by its success in solving geological problems and not by the analytical perfection of its techniques. In the Gregory Rift Valley, as elsewhere, the geochronologist is concerned to derive the true sequence and timing of geological events from the only primary matrix of evidence available, i.e. that which is contained within the actual exposed rocks of the area *and nowhere else* (Miller 1965). Major aspects of any geochronological study are: (i) *basic stratigraphical geochronology*: the description and classification of local rock stratigraphical sequences and the correlation of these local rock successions into individual sub-basin and basin-wide stratigraphies, followed by correlation between separated basins by comparison of their stratigraphies against the regional and world-wide standard stratigraphic scales; (ii) *dating specific events*: for example, the burial of a hominid fossil, the eruption of a particular tuff or lava flow or the movement on a specific fault. A high degree of precision and accuracy is required if the dating of individual events of geological, palaeontological or archaeological interest is to be of value; (iii) *dating the time-range over which certain processes were operative*: for example, episodes of erosion, upwarping, sedimentation, faulting, volcanism, geothermal activity, mineralization or faunal provincialization; (iv) *investigation of rates*: for example, rates of organic evolution, sedimentation, diagenesis, magmatism, magmatic differentiation or diastrophism.

Before satisfactory geochronological studies can proceed in an area, basic stratigraphical analysis and geological mapping must be undertaken. Geochronology is equally dependent upon the data provided by petrology, palaeontology, geomorphology, structural studies and other aspects of geology. Evidence of two main time-related processes of progressive change are to be found in rocks: organic evolution and radioactive decay. Study of these processes provides the principal tools of geochronology. Less universal, but equally important in certain instances are the small fragments of time-related processes seen in such things as varved clays and the evidence of geomagnetic polarity reversals obtainable from some rock successions. The combination of rock-stratigraphy and stratigraphical palaeontology gives us the relative geological time-scale. It is generally accepted that the accuracy of our correlation of local sequences with each other and with the regional or world-wide standard stratigraphical scales by this method depends upon the completeness of the rock record, the abundance and state of preservation of the fossil faunas and lastly, *but not least*, upon our interpretation of the faunal evidence. A study of radioactive decay processes can enable the cooling of igneous and metamorphic rock-forming minerals subsequent to crystallization or re-heating and the formation of certain sedimentary rocks and minerals to be dated *in years*. It is important to realize that the accuracy of the ages obtained by this means is, *likewise*, dependent upon the completeness and state of preservation of the radioisotopic record in the rocks and upon the correctness or otherwise of our interpretation of the results of the radioisotopic dating experiments. Thus, the two main tools of geochronology are equally fallible and are best used in conjunction rather than in rivalry. When there appears to be a discrepancy between the correlations suggested by faunal and radioisotopic considerations, it is imperative that *both* lines of evidence are subjected to an unflagging scrutiny until the cause of the discrepancy is eliminated.

Cenozoic geochronology in the Gregory Rift Valley

(A) STRATIGRAPHICAL GEOCHRONOLOGY

Whilst it is true that a great deal is known regarding the regional geology and geochronology of East Africa, only certain rather restricted areas have been examined in any detail. During the past ten years, the majority of geologists working on the Cenozoic rocks of the Gregory Rift Valley have been engaged almost everywhere upon the basic large-scale geological mapping and erection of local rock-stratigraphies that are essential pre-requisites for stratigraphical geo-chronology. Only now, in such places as Olduvai, Baringo, East Lake Turkana and the Omo is there beginning to be a sufficient body of basic geological data available for detailed geochronological analysis. As large-scale geological mapping has progressed in each of these areas, correlation of local rock successions into sub-basin and basin-wide stratigraphies has been possible. Correlation at this level does not necessarily require the assistance of radioisotopic dating: the palae-ontological, geomorphological, petrological or structural evidence contained in the rocks is normally sufficient to enable high probability correlation to be made within areas of continuous or virtually continuous rock exposure. Faunal com-parison then enables correlation of each individual area succession with one another and with standard stratigraphic scales. For the solution of correlation difficulties which inevitably arise wherever the palaeontological and other geo-logical evidence is sparse and in order to increase the resolution of both local, regional and world-wide correlation of individual successions, it is, however, necessary to utilize the radioisotopic dating tool. Magnetostratigraphic correla-tion, either between local sequences or with the world geomagnetic polarity reversal scale may be of assistance in certain instances, but it must be remembered that both the Geomagnetic Reversal Time-Scale and the world-wide Phanerozoic Time-Scale are themselves calibrated by K-Ar radioisotopic dating of rocks.

It is obvious that the value of the radioisotopic dating tool to stratigraphical geochronology is entirely dependent upon the accuracy and precision with which specific events of known stratigraphical age can be dated. In this respect, the results of the first seventy years of geochronometry have been distinctly disap-pointing. Firstly, we have, as a major inherent difficulty, the fact that the vast majority of the events and horizons of known stratigraphical age for which dates are needed by the stratigraphical geochronologist occur in rocks not suitable for radioisotopic dating (Fitch 1972; Fitch *et al.* 1974a; 1975a). Secondly, and equally disappointing, is the obvious geological inaccuracy and controversial nature of many otherwise analytically precise radioisotopic age determinations. Over the last decade a great proliferation of dating laboratories has provided geology with a very welcome flood of analytically precise apparent ages derived from all kinds of rocks throughout the world by a great variety of methods. Many previously intractable correlation and timing problems have been solved, but, at the same time it has become increasingly clear that all of the dating methods currently in use suffer from severe limitations which make the numerical results obtained widely erroneous in certain instances. Before practical geologists can regain an overall confidence in the application of radioisotopic geochronometry to stratigraphy, it is necessary that these uncertainties be resolved. Progress in strati-graphical geochronology requires that dating experts settle their internal con-troversies and develop methods and techniques which can demonstrably separate geologically accurate radioisotopic age estimations from those which are prob-lematical or otherwise of much less value to stratigraphy.

In this paper we indicate how geochronologists using potassium-argon dating methods might achieve this objective. Potassium-argon geochronometry is the most used and most useful method in rocks within the age range 0.25 m.y.– 25 m.y.

(Upper Pleistocene—Lower Miocene) with which we are most concerned in the Gregory Rift Valley. The data available and the problems which arise in the application of carbon-14 dating and other methods to rocks younger than 0.25 m.y. in the area are too voluminous to discuss here. U-Th-Pb and Rb-Sr isochron dating methods, which are mainly applicable to the older rocks of the East African basement, will likewise be ignored. Some mention must be made, however, of fission track dating, for this technique may be applied to a wide range of East African rocks in the future.

(B) DATING SPECIFIC EVENTS BY POTASSIUM-ARGON GEOCHRONOMETRY

Potassium-argon geochronometry can be undertaken on virtually any potassium-bearing rock or mineral of sufficient age for a measurable amount of radiogenic argon to have accumulated from the natural radioactive decay of the isotope ^{40}K. Dependent upon the potassium-content and size of the dating sample available, the practical lower limit of the method lies within the age range 0.01 to 0.5 m.y. Normal detrital sediments only rarely reveal dates of stratigraphical significance, but in the Gregory Rift Valley there are numerous lavas, ignimbrites and air-fall tuffs ideally suitable for K–Ar dating interstratified with the fossiliferous sediments of Pleistocene, Pliocene and Miocene age. There is already a considerable literature recording the results of K–Ar dating studies on these rocks. (See Bagdasaryan *et al.* 1973; Baker *et al.* 1971; Bishop *et al.* 1969, 1971; Brown 1972; 1975; Brown & Lajoie 1971; Carney *et al.* 1970; Curtis 1975; Curtis & Hay 1972; Curtis *et al.* 1975; Evans *et al.* 1971; Evernden & Curtis 1965; Fairhead *et al.* 1972; Fitch & Miller 1969, 1970; Fitch *et al.* 1974(b), 1975(b); Macintyre *et al.* 1974; Reilly *et al.* 1966; Van Couvering & Miller 1969 and the further references quoted therein.)

The geochronological problems that have been encountered in the Cenozoic of East Africa include both stratigraphical difficulties and problems inherent in the K–Ar dating method.

Over the past 25 m.y. continental sediments and volcanics have accumulated intermittently within and adjacent to the Gregory Rift Valley, continuity of deposition being broken by successive cycles of erosion initiated by tectonic events and base-level changes. Volcanism and volcanic rocks have been a dominant factor in the environment at all times. Flood lava eruptions, central volcanoes of many types, ash-flow and air-fall pyroclastics, sub-volcanic intrusives and a great variety of re-worked volcanic tuffs and volcanoclastic sediments alternate and interdigitate with aeolian, colluvial, fluvial, deltaic and lacustrine sediments, themselves often largely or in part derived from the weathering and erosion of volcanic rocks. During intervals of relative stability major rivers built large deltas into rift- and volcano-trapped lakes and excavated extensive valley systems in the bordering highlands. Many of the stratigraphically isolated sedimentary basins within the Gregory Rift Valley contain important fossil faunas or other items of particular scientific interest. Hominid fossils and archaeological sites are just two such items for which really accurate and precise dating is required. Some of the geochronological problems that may arise during the K–Ar dating of an isolated fossil hominid cranium, an artefact and a water-lain volcanic tuff are discussed here as examples of the difficulties that may be encountered.

1. *Dating a fossil hominid cranium.*

As it is normally not possible to date the substance of a fossil such as a hominid cranium by physical methods, an estimation of its age must be obtained from the dating of the sedimentary layer within which it was fossilized. Location of the enclosing sedimentary stratum is usually clear and unambiguous in excavation or in cliffs or other steep rock exposures, but may raise problems in the case of surface finds. When the exact location of a fossil is not certain, its stratigraphical age should be quoted as limited by the bracketing horizons between which it must have been derived and its most probable horizon of derivation indicated with an assessment of the degree of certainty available at the site. Of course, fossils are given only *minimum ages* by the age of their enclosing sedimentary stratum; a further study of the environment of deposition is required in order that some assessment of the time-lag between death and final burial and fossilization can be made. The most serious of the possible errors that may arise occurs when the fossil concerned has been re-worked from a previous sediment.

If the enclosing stratigraphical horizon from which it is certain that a hominid fossil was derived is itself datable, e.g. is a primary volcanic tuff, a lava flow or a datable soil horizon, then the dating problem is relatively straightforward and a (minimum, as derived above) apparent age for the fossil can be obtained by direct radioisotopic dating of the enclosed rock. The accuracy or otherwise of this apparent age estimate will depend upon the inherent accuracy of the isotopic method employed and upon the experimental precision of the individual age determinations, alone. If, however, the enclosing sedimentary horizon is not itself datable, then its age must be quoted as bracketed between the apparent ages of overlying and underlying datable horizons. The correct placing of a given sedimentary horizon between such enclosing brackets requires that basic geological mapping and section measurement must have been undertaken previously in the area. In some instances, a sedimentary sequence containing datable bracketing horizons may not be present. If only one datable horizon is available locally, then the apparent age of this horizon must be quoted as giving either a minimum or a maximum apparent age for the fossiliferous horizon. A further magnitude of uncertainty arises when no datable horizons occur in the local stratigraphical sequence and its relative age range must be assessed by geological, palaeontological and/or geomagnetic polarity comparison and correlation with sequences that do contain datable rocks.

Another form of uncertainty can arise when the sedimentary horizons concerned do contain or consist of datable materials, but are not of primary formation. Radioisotopic dating of detrital volcanic minerals and rock fragments in a sediment can provide no more than a maximum estimate for the age of sedimentation of the horizon in which they are found. In all instances where the horizon dated may consist, even in part, of reworked material, a thorough geological/petrological assessment of the possible sedimentary discrepancy of the apparent age obtained must be undertaken. Consideration of this further possible source of error is of major importance in East African geochronology because of the prevalence of water-lain (i.e. reworked) volcanic tuffs in some of the fossiliferous successions (see below). In general, the failure of K–Ar radioisotopic geochronometry to satisfactorily date a particular apparently datable horizon is usually due to a combination of 'geo-logical errors' (Fitch 1972) which include both the presence of extraneous argon (inherited or introduced) and the possible loss of radiogenic argon from the system arising from its involvement in subsequent geological events.

2. *Dating an artefact.*

All that has been said above regarding the dating of hominid fossils applies to the dating of artefacts. Initially, the dating of either individual artefacts or of a related suite of artefacts requires that their location within the local stratigraphical sequence be known with as much precision as is possible in the circumstances and that the possibility of a time-lag between the age of manufacture and use and the date of their inclusion within the enclosing sediment be considered. Similarly, bracketing minimum and maximum apparent ages must be defined in those instances where the enclosing horizon is not itself composed of datable materials—and all dates quoted must be regarded as subject to the other uncertainties outlined above. Artefacts, however, are often themselves made of datable rock and this fact adds a further dimension to their geochronological analysis. The apparent age of the substance of an artefact made of lava, for example, can be derived either from the artefact itself or from a matched sample from the outcrop of the source rock. Locating and matching the source rock of an artefact can be done only after geological mapping, petrological and geochemical analysis of the source region have reached an advanced stage, but when it is possible, the apparent age of the rock from which an artefact is made will provide a maximum age for the manufac-ture of the artefact. In favourable circumstances, say where it is possible to show that the raw materials were derived from a contemporaneous lava flow or from a local boulder deposit of the right composition, matching of source rocks may enable very close maximum apparent ages to be assigned to some artefact suites.

3. *Dating water-lain volcanic tuffs.*

Water-lain volcanic tuffs are common in many East African fossiliferous successions. Where intercalations of datable primary lava flows or tuffs are available, these are obviously preferable horizons for stratigraphical geochronological analysis, but it often occurs that the only datable horizons in a given sequence are water-lain, 're-worked' volcanics. Even when, as in the Omo and East Lake Turkana outcrops, there are cogent geological grounds for believing that each tuff flood seen in the local stratigraphical column represents a virtually immediate downstream response to major volcanic eruptions in the headwater regions of their respective drainage basins, some doubts regarding the extent of the possible time-lag between eruption and deposition must remain. At East Lake Turkana, for example, the abruptness with which each of the major tuffs appears within the sediments, instantaneously overwhelming a whole series of different contemporary sedimentary environments, the apparently sediment-free nature of the lower portions of individual pulses of tuff that then show an accelerating increase of detrital contamination upwards, combined with the virtual absence of primary unworn volcanic debris in the intervening sediments, all provide good reasons for supposing that eruption, erosion, transport and deposition of each tuff occurred as a continuous process of short duration. Thus, whilst some time-lag intervened between, say, the crystallization of volcanic sanidine in the source vent and its deposition within a water-lain basin tuff, it would appear to be a reasonable assumption in circumstances such as are found at East Lake Turkana to regard this time-lag as insignificant when compared with the experimental imprecision of K–Ar age determination on volcanic sanidines.

Contamination is a major problem in all tuff dating. In water-lain tuffs, contaminating fragments of older rocks and minerals can be present (a) as partially assimilated xenoliths and xenocrysts incorporated in the magma (b) as angular fragments torn from the walls of the vent during explosive eruption (accidental debris) and (c) as sedimentary detritus picked up during erosion and transportation of the tuff from near the vent to its final resting place. It is often very difficult to separate the primary, juvenile volcanic components of a tuff from the accompanying accidental and detrital contamination. All dates obtained from tuff samples in which contamination is suspected must be regarded as discrepantly high. Pumice is a very unstable rock and exposed to the atmosphere under normal surface conditions it has a very short life. Thus, pumice concentrations in most water-lain tuffs of the type described above from East Lake Turkana are usually found to consist of one generation only, derived from the same volcanic eruption as their enclosing tuff. Nevertheless, the possible presence of xenolithic pumice lumps can never be totally disregarded in any pumice collection. A further possibility of contamination that must be avoided when using mineral concentrates from pumice is the likely inclusion of extraneous detritus in the vesicules and tubules of the glass froth during transportation and sedimentation. It is recommended that mineral concentrates for dating from water-lain pumice be made by careful hand-picking beneath a stereo POL microscope and that only euhedral unworn mineral grains *actually suspended by attached glass fibres* be selected. Primary volcanic minerals selected from the inner cores of pumice lumps in this way should be free from all sources of contamination other than the very rare presence of cognate or accessary xenocrysts of the same mineral species, derived from solidified earlier batches of the same magma, which may or may not have been completely outgassed during the current eruption. Fortunately, this latter possibility appears to amount to no more than a minor, occasional, source of error in the K–Ar dating of primary volcanic minerals derived from pumice. In general, volcanic glass is not a favourable material for K–Ar dating as it is extremely prone to argon loss discrepancy resulting from dehydration, devitrification and alteration. In unusually favourable circumstances the glass fraction of vitric tuffs and pumice can be used for K–Ar dating, but, if a potash-rich primary mineral phase is present, the mineral sample is always to be preferred.

K–Ar isochron geochronometry

The successes, failures and problems of K–Ar dating can best be illustrated by isotope ratio or correlation diagrams in which, for each analytical sample, the ratio $^{40}Ar/^{36}Ar$ is plotted against the ratio $^{40}K/^{36}Ar$ (or, when neutron activation techniques are employed to measure the potassium-content, the equivalent $^{39}Ar/^{36}Ar$ ratio). In conventional total fusion K–Ar dating, two major simplifying assumptions have to be made: (1) that no extraneous argon is present in the sample

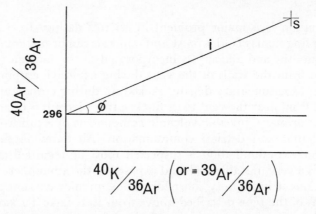

FIG. 27:1. The 'ideal' K–Ar dating situation illustrated on an isotope ratio or correlation diagram. The two basic simplifying assumptions are made (i) the $^{40}Ar/^{36}Ar$ ratio of all extraneous (i.e. non-radiogenic) argon present is assumed to be that of the modern atmosphere (296) and (ii) it is assumed that the sample has remained a closed system for potassium and radiogenic argon since its time of formation. In these ideal circumstances the apparent age of the sample represented by the spot datum point s is a function of the slope of the line i such that

$$\tan \varnothing = \frac{^{40}Ar/^{36}Ar - 296}{^{40}K/^{36}Ar} = \frac{rad\ ^{40}Ar}{^{40}K}$$

FIG. 27:2.

Discrepancy in the conventionally calculated K–Ar age of a total fusion sample (assumed to be a function of the slope of line i_2 when it should be a function of the slope of line i_1) resulting from (i) a small argon loss which would move the spot datum point from s_1 to s_2 and (ii) the incorrect assumption that the $^{40}Ar/^{36}Ar$ ratio of the extraneous argon present was 296.

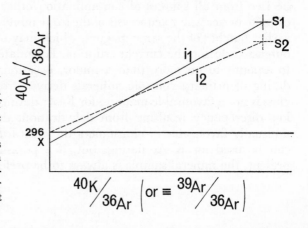

other than that derived from the modern atmosphere and (2) that the sample has remained a closed system for potassium and radiogenic argon since its time of formation. When these assumptions are made, the 'ideal' K–Ar dating situation is as is illustrated in Fig. 27:1. Then the apparent K–Ar age of the sample producing the single 'spot' datum point s, by conventional or $^{40}Ar/^{39}Ar$ total fusion analysis, is a function of the slope of the line i joining the datum point s to that point on the $^{40}Ar/^{36}Ar$ axis which represents the modern atmospheric ratio (296).

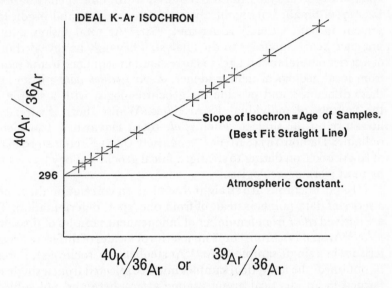

FIG. 27:3. An ideal K–Ar isochron plot, obtained from a suite of samples free from any extraneous argon or argon loss discrepancies.

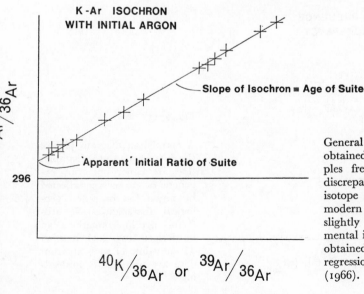

FIG. 27:4.

General case of a K–Ar isochron plot obtained from a cogenetic suite of samples free from significant argon loss discrepancy but formed in an argon isotope atmosphere unlike that of the modern atmosphere. The data points are slightly scattered as a result of experimental imprecision and the 'isochron' is obtained by a 'best fit' least-squares regression analysis modified from York (1966).

It is a generally accepted, but often largely ignored, fact that neither of the two basic simplifying assumptions of K–Ar dating are ever fully met in practice. Even small deviations from the ideal situation can be very significant in the dating of young volcanic rocks because of the small volumes of radiogenic ^{40}Ar involved. In these circumstances, a discrepant slope value accepted for the line i, resulting from quite small losses of radiogenic argon or as a result of the assumption of an incorrect extraneous argon intercept on the $^{40}Ar/^{36}Ar$ axis, can make the quoted apparent age very far from the required truth. One such situation is illustrated in Fig. 27:2. Small, but significant, extraneous argon and argon loss errors can be present in conventional K–Ar and $^{40}Ar/^{39}Ar$ total fusion dating without their presence being apparent to the analyst. They may be revealed only by the failure of a series of total fusion ages to correspond in sequence to the sequence established from local geological field evidence. *K–Ar isochron dating* is an attempt to resolve these difficulties and provide the geochronologist with a simple analytical test of the validity of individual K–Ar dates. Whilst there is already a considerable literature on the subject (Fitch *et al.* 1969; Hayatsu & Carmichael 1970; Shafiqullah & Damon 1974; Mellor & Mussett 1975; Fitch *et al.* 1976a) the importance of K–Ar isochron dating to stratigraphical geochronology has yet to be appreciated by most working geologists.

K–Ar isochrons are straight line plots on correlation diagrams obtained from a series of data points instead of from one 'spot' determination. These data points are derived *either* from a number of independent samples of the same rock (K–Ar or $^{40}Ar/^{39}Ar$ spot isochrons) or from a series of locales of different argon release characteristics in a single sample ($^{40}Ar/^{39}Ar$ stepheating isochrons). For a true isochron to be obtained, the analytical samples must be selected from a single rock or rock-suite formed in an identical argon isotope atmosphere and not subject to any partial argon losses. (Totally overprinted rock-suites, i.e. those that have suffered com-

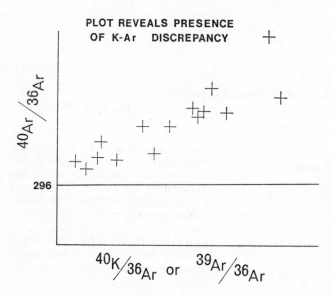

FIG. 27:5.

A correlation diagram plot of the data from a suite of samples that are either not truly cogenetic or are severely affected by argon loss or other 'geological' discrepancies. No satisfactory age interpretations can be derived from such a suite. Mean values of such aberrant data are particularly misleading.

plete loss of previously accumulated radiogenic argon during a subsequent geological event, will also produce K–Ar isochrons.) An ideal K–Ar isochron, constructed from samples in which no extraneous argon is present other than that derived from the modern atmosphere and from which no partial loss of radiogenic argon has occurred, is illustrated in Fig. 27:3. Fig. 27:4 illustrates the more general case of a suite of samples formed in an argon isotope atmosphere significantly different from that of the modern atmosphere. It also illustrates a more realistic practical situation, where the analytically determined data points are scattered by small experimental errors and have been interpreted as falling on a straight line least-squares 'best fit' isochron using a modified regression analysis after York (1966). Fig. 27:5 illustrates the situation that can arise when the suite of samples is either not genuinely co-genetic and/or variable argon loss errors are present. When significant extraneous argon or argon loss discrepancy is present in samples subjected to K–Ar or ^{40}Ar/^{39}Ar total fusion dating, repeat analyses or the averaging of results from many samples cannot remove the discrepancy. Thus it can be seen that K–Ar isochron age determinations are superior to single (spot) total fusion age estimates in four ways (i) because they can produce more precise ages, (ii) because this technique clearly separates discrepant from non-discrepant ages, (iii) because when using this technique it is not necessary to make any assumption regarding the ^{40}Ar/^{36}Ar ratio of any extraneous argon present—its value for a suite of samples from which an isochron is obtained is measured by the intercept of the isochron on the ^{40}Ar/^{36}Ar axis (an intercept value of 296 becoming a special case) and (iv) because the closeness with which the data points fall on a straight line provides an internal check on the validity of the assumptions made (but see certain special case reservations discussed by Shafiqullah & Damon 1974). Of the utmost importance are the two special cases in which apparent K–Ar ages obtained from total fusion spot-dating will be identical to those obtained by isochron experiments. The first of these is when no extraneous argon, other than a small unfractionated contaminant derived from the modern atmosphere, is present during the dating of a cogenetic suite of samples and when it also happens that none of the samples have been subjected to any subsequent argon loss overprinting during their geological lifetime. This is the *ideal* K–Ar dating situation and samples of this type, identified as a result of the coincidence of spot and isochron dates, are those of the greatest value to stratigraphical geochronology. A similar, but slightly different situation can occur when total overprinting (say, as a result of metamorphism or metasomatism) has completely outgassed a suite of samples so that their K–Ar 'clocks' are re-set at the time of the overprinting event. In these circumstances, whilst it is likely occasionally that the isochron intercept on the ^{40}Ar/^{36}Ar axis will be 296, often it will not be close to this value, and then, therefore, the spot and isochron ages obtained from the suite will not agree. When this occurs, however, the individual spot ages will be seen to be systematically in error as a result of the incorrect use of the ratio 296 in the calculation of their conventional K–Ar ages.

Applications of K–Ar isochron dating in East Africa

Potassium-argon isochron dating has yet to be applied systematically to East African rocks. The examples which follow are the results of pioneer work in this field of geochronology and have been selected to give some idea of the scope and promise of the method rather than to resolve specific geological problems.

(A) OLDUVAI BASALT MEMBER

Obviously discrepant K–Ar total fusion apparent ages ranging from 1.3 m.y. to 4.4 m.y. have been obtained from samples of the basalt member (Hay 1963; 1967; 1973) at Olduvai (Von Koenigswald *et al.* 1961; Leakey *et al.* 1962; Evernden &

FIG. 27:6.

K–Ar spot isochron plot of a sample FM7013B from the Olduvai basalt member. Apparent K–Ar isochron age 1.88 ± 0.02 m.y. with an intercept value of 296 ± 0.3.

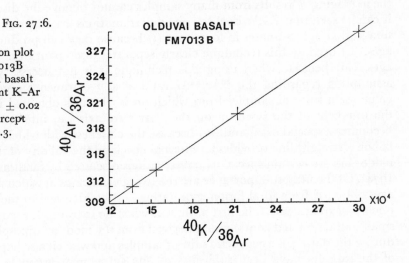

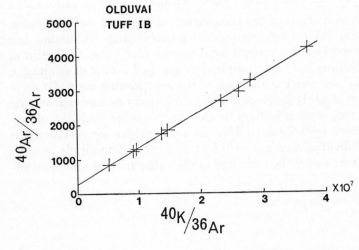

FIG. 27:7

K–Ar spot isochron plot of the data on samples from Olduvai Tuff IB quoted in Curtis & Hay (1972). Apparent K–Ar isochron age 1.81 ± 0.02 m.y. with an intercept value of 276 ± 29.

Curtis 1965). After careful review of the data Evernden and Curtis concluded that the best date from this basalt member was 1.92 m.y. (obtained from sample KA 1100). Curtis and Hay re-examined and recalculated the data obtained from three hydrofluoric acid-treated samples of this basalt member in 1972. As a result they suggested a best average K–Ar age of 1.96 ± 0.093 m.y. for this horizon. Fig. 27:6 is a K–Ar spot isochron obtained from the analysis of four sub-samples of Olduvai basalt FM 7013B carried out in Cambridge. These sub-samples were not subjected to hydrofluoric acid-treatment but were pre-heated in the argon line before analysis. The average conventional K–Ar apparent age of this sample is 1.86 ± 0.01 m.y. The apparent K–Ar isochron age is identical within error at 1.88 ± 0.02 m.y. with an intercept value of 296 ± 0.3. From this we conclude that sample FM 7013B was an ideal K–Ar dating sample and that the best apparent age for this rock is the K–Ar isochron age of 1.88 ± 0.02 m.y.

(B) OLDUVAI TUFF IB

This horizon is described as a nuée ardente with accompanying ash-fall by Curtis & Hay (1972). From numerous clean mineral dating samples separated from Tuff IB pumice, Curtis and Hay quote a scatter of data with an average conventional K–Ar apparent age of 1.79 ± 0.03 m.y. They also point out that the mean of 35 tuff dates from Bed I at Olduvai below Tuff IF is 1.82 m.y. A fission track date of 2.0 ± 0.28 m.y. was obtained from Tuff IB pumice glass by Fleischer *et al.* (1965). Fig. 27:7 is a K–Ar spot isochron obtained by re-computing the data from Tuff IB available in Curtis & Hay (1972). The apparent K–Ar isochron age obtained is 1.81 ± 0.02 m.y. with an intercept value of 276 ± 29. Again, we regard the isochron age as the best estimate of the date of this tuff horizon currently available.

(C) KENYA PLATEAU PHONOLITES

Fifteen conventional K–Ar total fusion age determinations on Kenya plateau phonolites are quoted in Bishop *et al.* (1969). The conventionally calculated apparent ages range from 11.8 ± 0.3 m.y. to 13.6 ± 0.6 m.y. Interpretation of this range as either an experimental scatter or the real duration of volcanism was not possible in 1969. In addition, however, one ^{40}Ar/^{39}Ar total fusion age determination and one six-point ^{40}Ar/^{39}Ar step heating isochron age of 12.6 ± 0.7 m.y. and 12.5 ± 0.4 m.y. respectively, were quoted for the phonolite lava MB/16 which overlies the main fossiliferous sequence at Fort Ternan, whereas the conventional K–Ar apparent ages from this same phonolite were 11.8 ± 0.3 and 11.8 ± 0.2 m.y. Fig. 27:8 is a K–Ar spot isochron obtained from the re-computation of the total fusion K–Ar data available in Bishop *et al.* (without MB/8, in which they report the presence of xenoliths). The K–Ar isochron age of 12.50 ± 0.15 m.y. with an intercept value of 295 ± 5.6 obtained clearly resolves the apparent inconsistencies of the earlier work and suggests that these phonolites belong to a very short eruptive cycle.

(D) EAST LAKE TURKANA MIOCENE VOLCANICS

Stepheating ^{40}Ar/^{39}Ar isochron ages of 16.22 ± 0.10 m.y. and 16.14 ± 0.13 m.y. for sanidine concentrates separated from Miocene ignimbrites of the Gum Dura Formation are quoted by Fitch *et al.* (1975b). One of these virtually identical single sample stepheating isochrons is illustrated in Fig. 27:9. It can be regarded as indicating a very precise age for this episode of ignimbrite volcanism.

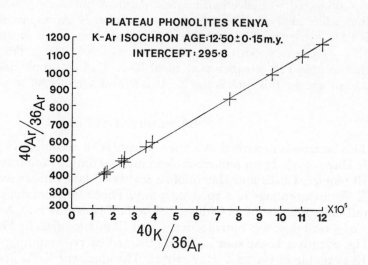

FIG. 27:8. K–Ar spot isochron plot of the data from samples of Kenya plateau phonolites quoted in Bishop *et al.* (1969). Apparent K–Ar isochron age 12.50 ± 0.15 m.y. with an intercept value of 295.8 ± 5.6. Conventional K–Ar spot ages ranged between 11.8 and 13.6 m.y. The ^{40}Ar/^{39}Ar stepheating isochron age of one of these phonolites was reported as 12.5 ± 0.4 m.y. in Bishop *et al.*

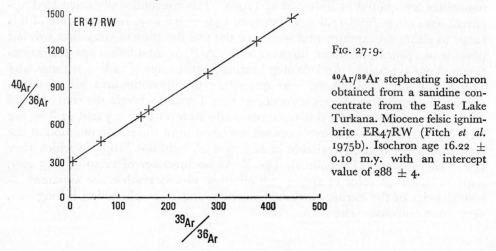

FIG. 27:9.

^{40}Ar/^{39}Ar stepheating isochron obtained from a sanidine concentrate from the East Lake Turkana. Miocene felsic ignimbrite ER47RW (Fitch *et al.* 1975b). Isochron age 16.22 ± 0.10 m.y. with an intercept value of 288 ± 4.

(E) EAST LAKE TURKANA KBS TUFF, AREA 105
(LEAKEY I)

A ^{40}Ar/^{39}Ar stepheating plateau age of 2.61 ± 0.26 m.y. for a sanidine concentrate (Leakey I) from pumice contained within the KBS Tuff Area 105, East Lake Turkana, was quoted by Fitch & Miller (1970). These early ^{40}Ar/^{39}Ar stepheating data have been recomputed and the four data point plateau is plotted as a correlation diagram in Fig. 27:10. The corrected apparent age derived from this correlation diagram is 2.42 ± 0.01 m.y. (Fitch *et al.* 1976b). This corrected apparent age is within the error bracket of the age quoted in 1970 but is more precise. As total degassing K–Ar and ^{40}Ar/^{39}Ar ages of sample Leakey I are identical within error, it can be regarded as a near ideal K–Ar dating sample. Thus, the ages obtained from this sample can still be regarded as good estimates for the age of emplacement of the KBS Tuff.

(F) EAST LAKE TURKANA TULU BOR TUFF, AREA 116

A new isochron age of 3.19 ± 0.08 m.y. obtained from a sanidine concentrate (FMA 301) from pumice contained within the Tulu Bor Tuff, Area 116, East

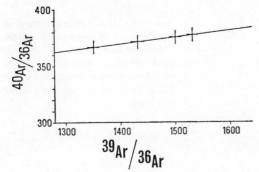

FIG. 27:10.

Recomputation of ^{40}Ar/^{39}Ar stepheating isochron obtained by Fitch & Miller (1970) from a sanidine concentrate (Leakey I) from pumice in the KBS Tuff, Area 105, East Lake Turkana. The four data points relate to the age spectrum plateau and give a revised age of 2.42 ± 0.01 m.y. with an intercept value 293.9 ± 0.5: (Fitch *et al.* 1976b).

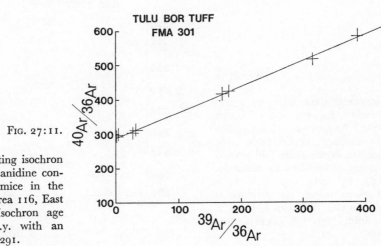

FIG. 27:11.

^{40}Ar/^{39}Ar stepheating isochron obtained from a sanidine concentrate from pumice in the Tulu Bor Tuff, Area 116, East Lake Turkana. Isochron age 3.19 ± 0.08 m.y. with an intercept value of 291.

Lake Turkana is illustrated in Fig. 27:11. Because total degassing ages from this sample are co-incident, it can be regarded as a good date from an ideal dating sample.

(G) EAST LAKE TURKANA KBS TUFF, AREA 105 (FMA 294)

The $^{40}Ar/^{39}Ar$ stepheating data obtained from a sanidine concentrate (FMA 294) from pumice collected from the KBS, Area 105 East Lake Turkana are illustrated in Fig. 27:12. These data cannot be interpreted as an isochron. It is certain, therefore, that K–Ar age discrepancy is present in the sample.

(H) EAST LAKE TURKANA KBS TUFF, AREAS 129, 131, 10 & 105

The data obtained from eleven total fusion $^{40}Ar/^{39}Ar$ age determinations on samples of sanidine concentrated from KBS pumice at a wide variety of sites at

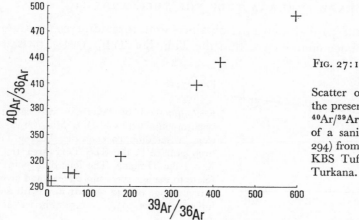

FIG. 27:12.

Scatter of data points revealing the presence of discrepancy in the $^{40}Ar/^{39}Ar$ stepheating age analysis of a sanidine concentrate (FMA 294) from pumice contained in the KBS Tuff, Area 105, East Lake Turkana.

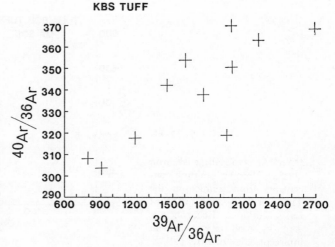

FIG. 27:13.

Scatter of data points on a correlation diagram plot of eleven $^{40}Ar/^{39}Ar$ spot age determinations on sanidine concentrates from pumice in the KBS Tuff from a variety of localities in Areas 10. 129, 131 and 105, East Lake Turkana reveals the presence of widespread discrepancy in KBS sanidines ($^{39}Ar/^{36}Ar$ values normalised)

East Lake Turkana are illustrated in Fig. 27:13. The scatter of data points makes it certain that variable K–Ar age discrepancy is present in some sanidine samples from the KBS Tuff (Fitch *et al.* 1976b).

(I) OMO TUFF D

Conventional K–Ar total fusion data on sanidine samples from Tuff D of the Omo succession are quoted in Brown (1975). Fig. 27:14 is a correlation diagram plot of the data and reveals the presence of K–Ar discrepancy.

(J) K–AR TIME-SCALE EAST LAKE TURKANA

The principles and techniques outlined in this paper have been applied to all of the K–Ar and $^{40}Ar/^{39}Ar$ total fusion and $^{40}Ar/^{39}Ar$ stepheating age data currently available from East Turkana (including numerous conventional K–Ar dates undertaken by Curtis and his associates at Berkeley and very kindly made available

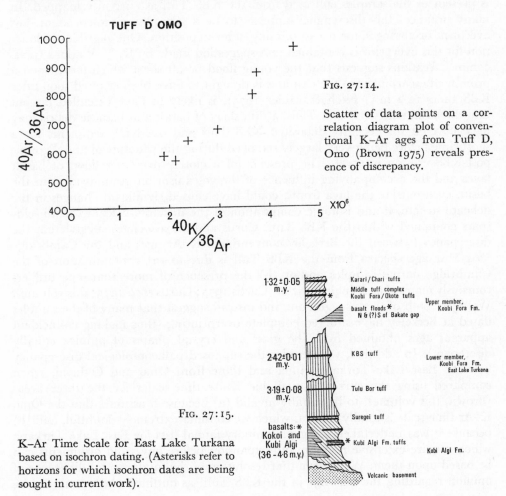

TUFF 'D' OMO

FIG. 27:14.

Scatter of data points on a correlation diagram plot of conventional K–Ar ages from Tuff D, Omo (Brown 1975) reveals presence of discrepancy.

FIG. 27:15.

K–Ar Time Scale for East Lake Turkana based on isochron dating. (Asterisks refer to horizons for which isochron dates are being sought in current work).

1·32±0·05 m.y. — Karari/Chari tuffs, Middle tuff complex, Koobi Fora/Okote tuffs — Upper member, Koobi Fora Fm.
basalt floods *
N & (?)S of Bakate gap

2·42±0·01 m.y. — KBS tuff — Lower member, Koobi Fora Fm. East Lake Turkana

3·19±0·08 m.y. — Tulu Bor tuff

Suregei tuff

basalts: *
Kokoi and Kubi Algi
(3·6 - 4·6 m.y.) — * Kubi Algi Fm. tuffs — Kubi Algi Fm.

Volcanic basement

to us prior to their publications in this volume and elsewhere). Full details of our analysis of this data is too voluminous for presentation here, but our conclusions are summarized in Fig. 27:15. Briefly, we find that a number of ideal K–Ar dating samples can be obtained by the careful extraction of juvenile sanidine from the pumice concentrates within East Lake Turkana tuffs. Co-incidence of conventional K–Ar and isochron determinations from the Karari/Chari Tuff level suggest an age of 1.32 ± 0.05 m.y. for this horizon, an age of 2.42 ± 0.01 m.y. for the KBS Tuff in Area 105 and an age of 3.19 ± 0.08 m.y. for the Tulu Bor Tuff level. Older tuffs and basalt lavas exposed below the Suregei Tuff have apparent ages between 3.6 and 4.6 m.y. Isochron age analysis of these and other relevant rocks from the East Lake Turkana area is in progress. When this work is complete further refinement of the East Lake Turkana K–Ar time-scale will be possible. The only horizon where any real difficulty is being experienced is that of the KBS Tuff. Numerous unsuccessful attempts at K–Ar isochron analysis suggests that over much of Areas 10, 129, 131 and part of Area 105 severe discrepancy is present in the samples collected from the KBS Tuff as currently mapped. In many instances this discrepancy appears to be a virtually complete argon loss overprint occurring some 0.4 to 0.8 m.y. after deposition. One possible explanation for this overprint is contained in a suggestion made by R. T. Watkins (pers. comm.). Watkins suggests that the young flood basalt sheet which forms such a prominent escarpment to the east and is thought to have been erupted soon after KBS times (2.2 m.y. Fitch & Miller 1975), is likely to have extended far out into the basin when erupted. Thus, a thin sheet of basalt and basaltic debris may have covered the recently deposited KBS Tuff and overlying sediments for a short time before being very largely removed during the widespread and vigorous post-KBS erosion interval. The presence of a closely overlying flood of basalt lavas and the accompanying influence of the volcanism on groundwaters in the basin, even outside the lava fronts, could have caused significant changes in the delicate structural and isotopic composition of the unstable volcanic alkali feldspars contained within the KBS Tuff. Curtis and his associates suggest that the discrepancy between the Berkeley conventional K–Ar ages and the Cambridge ^{40}Ar/^{39}Ar age spectra from the KBS Tuff is due to (a) contamination of the Cambridge dating samples and/or (b) the presence of more than one tuff erroneously mapped as a single horizon (Curtis 1975; Curtis *et al.* 1975; this volume). We reject both of these suggestions and instead suggest that many of the samples dated at Berkeley have suffered complete overprinting (thus making co-incident apparent ages obtained from the glass and crystal phases of pumice equally discrepant). In addition, we consider the supposed palaeontological discrepancy between East Lake Turkana fauna and those from Omo and Olduvai, when compared using the currently available K–Ar time-scales for the three areas (Brown, this volume), to be equally invalid (a) because it assumes that the Omo K–Ar time-scale is totally correct, which we consider extremely doubtful, and (b) because it was undertaken using preliminary faunal lists for East Lake Turkana which require extensive revision and expansion before any valid arguments can be based upon them. We believe that resolution of the fundamental difference of opinion regarding the true age of the KBS Tuff, as outlined above, will not be obtained readily from further K–Ar dating of the few samples available: it requires the application of an independent radioisotopic dating technique to the problem.

Exploratory work in London, Denver and Melbourne (Hurford 1974; Hurford, Gleadow & Naeser 1976 and further work in progress) suggests that fission track dating of juvenile accessory zircon crystals from young volcanic pumice can be successfully undertaken and will provide entirely independent fission track dating time-scales for the East Lake Turkana succession and elsewhere. Preliminary fission track dating results obtained from the KBS Tuff suggest that it has an age of 2.42 m.y. (Hurford, Gleadow & Naeser 1976).

Conclusions and discussion

The principal conclusion to be reached from this work is that K–Ar and ^{40}Ar/^{39}Ar isochron geochronometry will make a significant contribution to East African Cenozoic geochronology. The application of this technique to a variety of rocks, including basalt and phonolite lavas and a variety of tuffs has been shown to result in a considerable increase in dating precision and confidence. Whilst the apparent alignment of data points as a straight line plot on a correlation diagram cannot be said to *prove conclusively* that the samples from which they were derived are truly isochronous, it is undoubtedly certain that if a scatter of data points is obtained, it does indicate the presence of K–Ar age discrepancy.

When an ideal K–Ar dating sample or a set of ideal cogenetic samples are analysed, concordance is obtained between total fusion and isochron dates. Even under non-ideal conditions, as long as argon loss overprinting is either insignificant or complete, isochron dating can still produce geologically correct K–Ar ages in situations where total fusion dating is hopelessly discrepant. The character of a correlation diagram plot is, in fact, a sensitive test for K–Ar discrepancy. We believe that all of the above conclusions are valid and will be of considerable practical use to working geochronologists—regardless of whether our earlier and more controversial conclusions regarding the value and interpretation of ^{40}Ar/^{39}Ar age spectra are accepted (Fitch *et al.* 1969; 1974a; Fitch & Miller 1975).

Acknowledgements

Geochronological research undertaken by the Fitch-Miller group in East Africa is supported by the Natural Environment Research Council, the National Museums of Kenya and the Universities of London and Cambridge.

References

BAGDASARYAN, G. P., GERASIMOVSKIY, V. I., POLYAKOV, A. I. & GUYASYAN, R.KH. 1973. Age of volcanic rocks in the rift zones of East Africa. *Geochemistry International* 1973. 66–71. *Trans. from Geokhimiya.* No. **1**. 83–90. 1083.

BAKER, B. H., WILLIAMS, L. A. J., MILLER, J. A. & FITCH, F. J. 1971. Sequence and geochronology of the Kenya rift volcanics. *Tectonophysics.* **11**, 191–215.

BISHOP, W. W., MILLER, J. A. & FITCH, F. J. 1969. New potassium-argon age determinations relevant to the Miocene fossil mammal sequence in East Africa. *Amer. J. Sci.* **267**, 669–99.

——, CHAPMAN, G. R., HILL, A. & MILLER, J. A. 1971. Succession of Cainozoic vertebrate Assemblages from the Northern Kenya Rift Valley. *Nature, Lond.* **233**, 389–94.

BROWN, F. H. 1972. Radiometric dating of sedimentary formations in the lower Omo Valley, Ethiopia. In: W. W. Bishop & J. A. Miller (eds). *Calibration of Hominoid Evolution.* 271–87. Edinburgh, Scottish Academic Press.

—— 1975. Radiometric dating and paleomagnetic studies of Omo Group deposits. In: Y. Coppens (eds). *Earliest Man and Environment in the Lake Rudolf Basin: Stratigraphy, Paleoecology and Evolution.* Chicago. Univ. of Chicago Press. 50–78

——, & LAJOIE, K. R. 1971. K–Ar ages of the Omo group and fossil localities of the Shungura formation, south-west Ethiopia. *Nature, Lond.* **229**, 483–5.

CARNEY, J., HILL, A., MILLER, J. A. & WALKER, A. 1971. A late Australopithecine from the Baringo district, Kenya. *Nature, Lond.* **230**, 509–14.

CURTIS, G. H. 1975. Improvements in potassium-argon dating. 1962–75. *World Arch.* **7**, 198–209.

——, & HAY, R. L. 1972. Further geological studies and potassium-argon dating at Olduvai Gorge and Ngorongoro Crater. In: W. W. Bishop & J. A. Miller (eds). *Calibration of Hominoid Evolution.* 289–301. Edinburgh, Scottish Academic Press.

——, DRAKE, R., CERLING, T. & HAMPEL, J. (sic). 1975. Age of KBS Tuff in Koobi Fora Formation, East Rudolf, Kenya. *Nature, Lond.* **258**, 395–8.

EVANS, A. L., FAIRHEAD, J. D. & MITCHELL, J. G. 1971. Potassium-argon ages from the volcanic province of Northern Tanzania. *Nature, Lond.* **229**, 19–20.

EVERNDEN, J. F. & CURTIS, G. H. 1965. Potassium-argon dating of late Cenozoic rocks in East Africa and Italy. *Current Anthropology.* **6**, 343–85.

FAIRHEAD, J. D., MITCHELL, J. G. & WILLIAMS, L. A. J. 1972. New potassium-argon determinations on Rift volcanics of South Kenya and the bearing on age of rift faulting. *Nature, Lond.* **238**, 66–9.

FITCH, F. J. 1972. Selection of suitable material for dating and the assessment of geological error in potassium-argon age determination. In: W. W. Bishop & J. A. Miller (eds). *Calibration of Hominoid Evolution.* 77–91. Edinburgh, Scottish Academic Press.

——, & MILLER, J. A. 1969. Age determinations on feldspar from the Lower Omo Basin. *Nature, Lond.* **222**, 1143.

—— & ——. 1970. Radioisotopic age determination on Lake Rudolf Artefact Site. *Nature, Lond.* **226**, 226–8.

—— & ——. 1975. Conventional K–Ar and argon–40/argon–39 dating of volcanic rocks from East Rudolf. In: Y. Coppens, *et al.*, (eds). *Earliest Man and Environments in the Lake Rudolf basin: Stratigraphy, Paleoecology and Evolution.* Chicago. The University of Chicago Press. 123–47

——, —— & MITCHELL, J. G. 1969. A new approach to isotopic dating in orogenic belts. In: P. E. Kent, *et al.*, (eds). *Time and Place in Orogeny.* 157–96. London (Geological Society).

——, FORSTER, S. C. & MILLER, J. A. 1974a. Geological Time Scale. *Rep. Prog. Phys.* **37**, 1433–1496.

——, FINDLATER, I. C., WATKINS, R. T. & MILLER, J. A. 1974b. Dating of the rock succession containing Fossil Hominids at East Rudolf, Kenya. *Nature, Lond.* **251**, 213–14.

——, FORSTER, S. C. & MILLER, J. A. 1975a. The dating of the Ordovician. In: M. G. Basset (ed.). *The Ordovician System; Proceedings of a Palaeontological Association Symposium,* Birmingham, September 1974. pp. 15–27. University of Wales Press and National Museum of Wales Press, Cardiff.

——, WATKINS, R. T. & MILLER, J. A. 1975b. Age of new carbonatite locality in northern Kenya. *Nature, Lond.* **254**, 581.

——, MILLER, J. A. & HOOKER, P. J. 1976a. Single whole rock K–Ar isochrons. *Geol. Mag.* **113**, 1–10.

—— 1976b. Argon–40/argon–39 dating of the KBS Tuff in Koobi Fora Formation, East Rudolf, Kenya. *Nature, Lond.* (**263**, 740–4).

FLEISCHER, R. L., PRICE, P. B., WALKER, R. M. & LEAKEY, L. S. B. 1965. Fission-track dating of Bed I, Olduvai Gorge. *Science*, **148**, 72–4.

HAY, R. L. 1963. Stratigraphy of Beds I through IV, Olduvai Gorge, Tanganyika. *Science*. **139**, 829–33.

—— 1967. Revised stratigraphy of Olduvai Gorge. In: W. W. Bishop & J. D. Clark (eds). *Background to Evolution in Africa*, 221–8. Univ. Chicago Press.

—— 1973. Geologic background to Beds I and II, Olduvai Gorge. Stratigraphic summary. In: M. D. Leakey, *Olduvai Gorge Vol.* **3**, 9–18, Cambridge University Press.

HAYATSU, A. & CARMICHAEL, C. M. 1970. K–Ar isochron method and initial argon ratios. *Earth. Planet. Sci. Lett.* **8**, 109–17.

HURFORD, A. J. 1974. A fission track dating of a vitric tuff from East Rudolf, N. Kenya. *Nature, Lond.* **249**, 236–7.

——, GLEADOW, A. J. W. & NAESER, C. W. 1976. Fission track dating of pumice from the KBS Tuff, East Rudolf, Kenya. *Nature, Lond.* **263**, 738–40.

LEAKEY, L. S. B., CURTIS, G. H. & EVERNDEN, J. F. 1962. Age of basalt underlying Bed I, Olduvai Gorge, Tanganyika. *Nature, Lond.* **191**, 478–9.

MACINTYRE, R. M., MITCHELL, J. G. & DAWSON, J. B. 1974. Age of fault movements in Tanzanian sector of East African Rift System. *Nature, Lond.* **247**, 354–6.

MELLOR, D. W. & MUSSETT, A. E. 1975. Evidence for initial argon–36 in volcanic rocks, and some implications. *Earth & Planet. Sci. Lett.* **26**, 312–18.

MILLER, T. G. 1965. Time in stratigraphy. *Palaeontology*, **8**, 113–31.

REILLY, T. A., MUSSETT, A. E., RAJA, P. R. S., GRASTY, R. L. & WALSH, J. 1966. Age and polarity of the Turkana lavas, north-west Kenya. *Nature, Lond.* **210**, 1145–6.

SHAFIQULLAH, M. & DAMON, P. E. 1974. Evaluation of K–Ar isochron methods. *Geochimica Cosmochim. Acta.* **38**, 1341–58.

VAN-COUVERING, J. A. & MILLER, J. A. 1969. Miocene stratigraphy and age determinations, Rusinga Island, Kenya. *Nature, Lond.* **221**, 628–32.

VON KOENIGSWALD, G. H. R., GENTNER, W. & LIPPOLT, H. J. 1961. Age of the basalt flow at Olduvai, East Africa. *Nature, Lond.* **192**, 720–1.

YORK, D. 1966. Least-squares fitting of a straight line. *Can. J. Phys.* **44**, 1079.

28

G. H. CURTIS, R. E. DRAKE, T. E. CERLING,
B. W. CERLING & J. H. HAMPEL

Age of KBS Tuff in Koobi Fora
Formation, East Lake, Turkana, Kenya

Conventional K–Ar dates on pumice from the KBS Tuff horizon in the Koobi Fora Formation at East Turkana, Kenya, distinguish two tuff units. Samples from area 10 and area 105 gave an age of 1.60 ± 0.05 m.y. whereas those from area 131 give 1.82 ± 0.04 m.y. Both ages differ from the previously published age of 2.61 ± 0.26 m.y. obtained by $^{40}Ar/^{39}Ar$ technique.

Plio-Pleistocene lacustrine and fluvial strata exposed along the east side of Lake Turkana, Kenya, now known as 'East Turkana', have been under study since 1969 when Behrensmeyer began geological investigations, which led to the identification of the KBS Tuff in area 105 (Fig. 28:1)[1-4]. These strata, resting uncomformably on late Miocene and Pliocene volcanics, are notable for their extraordinarily rich assemblages of fossil vertebrates. which include a large number of hominid remains. Artefacts also occur in some places[5-8].

Interspersed in the stratigraphic sequence are several rhyolitic tuffs, mostly reworked, which have served as marker beds and for calibration of the fossil vertebrates and hominid artefacts by potassium argon dating[9-11] and in one instance by fission-track dating.[12] Dates on tuffs within the Koobi Fora Formation range from 3.18 m.y. for the Tulu Bor Tuff in the Lower Member, to 1.22 m.y. for the Chari Tuff and the correlated Karari Tuff in the upper part of the Formation. The KBS Tuff from the type locality in area 105 was assigned an age of 2.61 m.y., although subsequent age determinations resulted in a broad scatter of dates, both older and younger than this value.[10, 11]. Contamination by ancient bed rock material during the reworking of the tuffs was suggested to account for the anomalously old dates, whereas subsequent alteration, 'over-printing', of the pumice fragments used for dating, by alkaline-rich and possibly heated ground water may explain the anomalously young dates by partial loss of radiogenic argon.

Using the incremental heating technique of $^{40}Ar/^{39}Ar$ dating, Fitch & Miller believed they could resolve these overprinting events by analysis and interpretation of age spectra, and arrived at the 'best apparent age' for the KBS Tuff of 2.61 ± 0.26 m.y. (ref. 10).

When some palaeontologists compared fauna associated with the KBS Tuff in East Rudolf with those of other, supposedly well calibrated localities, the reliability of the date of 2.61 m.y. for the KBS was questioned. Although Maglio found that the morphology of elephant fossils fit with a 2.5 m.y. date,[7] Cooke & Maglio, in 1972, pointed out that fossil pigs from below the KBS Tuff horizon at East Rudolf seemed to correlate best with those from beds dating close to 2 m.y. in the Omo river area to the north in Ethiopia.[13-16] Because of the great interest in the dating and the question of reliability raised by the scatter of results from the KBS Tuff, we repeated the dating using the conventional method of K–Ar dating.

It is argued that the conventional K–Ar method is unable to evaluate results obtained by the $^{40}Ar/^{39}Ar$ method, because it assumes that all non-radiogenic argon in the sample has the same relative abundance as in the atmosphere, an assumption demonstrably wrong in some cases. Although this may be true for a single determination, multiple K–Ar determinations made on several samples of a tuff, and on more than one type of material can be plotted on an isochron diagram to show the isotopic composition of the non-radiogenic argon in the tuff.

With respect to the $^{40}Ar/^{39}Ar$ dating method, Dalrymple and Lanphere have pointed out (personal communication) that even after 5 yr of research with this method they occasionally obtain anomalous ages for geologically young rocks (late Tertiary and younger) which they are unable to explain. Furthermore, the interpretation of the $^{40}Ar/^{39}Ar$ isochrons that yield multiple apparent ages from a single phase, such as Fitch and Miller have obtained on KBS sanidines, is uncertain. Their interpretations rely on oversimplified and unproved models for the diffusion of argon in solids, both in nature and in the laboratory procedure necessary to make an age determination. For this reason we feel that their results, even when they are reproducible to high precision, may be an artefact of experimental procedure, and thus not geologically meaningful.

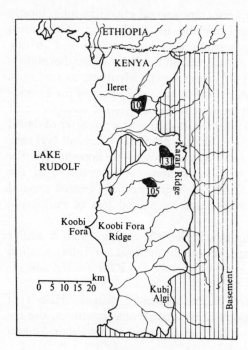

FIG. 28:1.

Map of the East Turkana region, showing the sampling areas.

Analytical procedure

During the summer of 1974 samples of pumice were collected from outcrops now classified as the KBS Tuff in three localities. One additional pumice sample from the eastern part of area 105 had been collected by Lucas previously for geochemical analysis during palaeomagnetic sampling in the summer of 1973. These samples were subsequently prepared for dating, using selected pumice fragments which were first examined both by eye and in thin section. Most sections showed the principal minerals, soda sanidine and pyroxene, to be fresh, although the enclosing vesicular glass ranged from highly altered in most specimens to fresh in only three. Calcite filled most vesicles, although zeolite was present in some.

Sanidine samples were separated from the glass with bromo form after the pumice had been cleaned, crushed and sieved. The concentrates were then treated with 1 N HCl to remove calcite, and 5 per cent HF for 1 min to remove adhering clay. Splits of these samples were then analysed for argon, using procedures described by Dalrymple and Lanphere,[17] and for potassium on a Zeiss PF5 flame photometer following dissolution of samples in HF and neutralisation.

The first dates (KA 2816, KA 2817, KA 2818, and later KA 2851) obtained from samples prepared in the manner described showed no agreement (see Table 3). Microscopic examination of the remaining concentrated crystals revealed numerous yellow grains of rounded and altered K-spar, clearly detrital in origin and probably derived from ancient bedrock. These grains had been tightly cemented on the convoluted surfaces of the rounded pumice, and had escaped detection in spite of careful cleaning. Subsequent work proved that it was virtually impossible to exclude these detrital sand grains completely when mineral separations were made using heavy liquids. To avoid this problem, we carefully crushed individual pumices in a hand mortar, extracting the large phenocrysts with tweezers under a

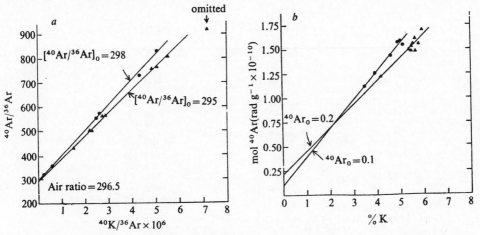

FIG. 28:2. Isochrons for KBS Tuff. *a*, $^{40}Ar/^{36}Ar$ against $^{40}K/^{36}Ar$; *b*, ^{40}Ar against %K.
●, Data from area 131; ▲, areas 105 and 10. Plots obtained by least-squares fitting. The ratio of $^{40}Ar/^{36}Ar$ in air is 296.0.

magnifying glass or binocular microscope. The phenocrysts were then treated with HF, crushed again, and put through a magnetic separator to remove included and adhering pyroxene grains.

Only three specimens 131–0001, 131–0002, and the Lucas sample from area 105 had glass fresh enough for both sanidine and glass age determinations to test for possible effects of overprinting. Since sanidines and glass have markedly different physical and chemical properties, it is highly unlikely that any substantial overprinting event would produce an exactly proportional loss of argon, such that their apparent ages would remain the same. It is equally unlikely that several sanidines from widely separated samples would yield identical apparent ages, had overprinting occurred, as the intensity and duration of these events are not everywhere uniform. Thus, when concordant dates are obtained on two different phases from the same sample which also closely agree with single dates on other samples the probability becomes increasingly large that these dates are an accurate estimate of the true age of their formation.

Results

Three different pumice clasts from the KBS Tuff at area 105 and one from area 10 (Ileret) give an average age of 1.60 ± 0.05 m.y. for nine analyses, while two separate pumice clasts from the tuff in area 131 (Karari Ridge), hitherto correlated with the type KBS in area 105, give an average age of 1.82 ± 0.04 m.y. on six analyses (Table 1).

The narrow range of dates and high precision obtained on all samples, once sufficient care was taken in sample preparation, we believe distinguish the tuff in area 131 from the KBS in area 105 and area 10. This same distinction is seen by the comparison of average K contents for the uncontaminated sanidine separates used for age determination. The sanidines from area 105 and area 10 average

TABLE 1: *KBS Tuff ages (m.y.)*

Area 105		Area 10	Area 131	
1.65 (S)	0001	1.54 (S)	1.84 (G)	
1.50 (S)		1.60 (S)	1.83 (S)	0001
			1.81 (S)	
1.62 (S)		Mean of 2 = 1.57		
1.56 (S)	0002		1.83 (G)	
1.62 (S)			1.85 (S)	0002
			1.73 (S)	
1.68 (G)	Lucas			
1.61 (S)			Mean (± s.d.) of	
			6 = 1.82 ± 0.04 Myr	
Mean of 7 = 1.61				
Mean (± s.d.) of 9 from 105 and				
10 = 1.60 ± 0.05 Myr				

G, glass: S, sanidine.
Samples with detrital contamination have been omitted

5.6 per cent with a range of 5.36–5.94 per cent, whereas those from area 131 average 4.90 per cent with a range of 4.60–5.11 per cent.

X-ray fluorescence analyses[18] show the distinctive Nb, Zr, Y, and Rb contents of pumice glass from the two tuff units (Table 2). Sample 0001 from area 131 and those from area 105 are, however, similar in appearance, being composed of dark brown glass with open vesicles enclosing large euhedral sanidine phenocrysts, whereas sample 0002 from area 131 is composed of a grey glass with elongated tubular vesicles enclosing less abundant sanidine phenocrysts. These pumices with fresh glass are rare compared with those which are completely altered to clay minerals and zeolites.

Although the difference between KBS–105 and KBS–131 is clear, the data for the samples from area 131 show that two different pumices have been sampled. Even though they are of the same age, they may be derived from different eruptive centres. This introduces the possibility that older pumices may also be present in the KBS Tuff horizon which could account for the 2.61 m.y. date reported by Fitch & Miller. For this reason, all dates obtained on pumices should be considered as maximum ages for the tuff itself.

The high precision and concordancy of dates obtained on all samples, in spite of the fact they ranged from fresh to highly altered, indicates to us that none has lost significant argon after deposition.

This conclusion is further supported by two isochron plots (Fig. 2) (and see also ref. 19). The $^{40}Ar/^{39}Ar$ against $^{40}K/^{33}Ar$ isochron shows that the non-radiogenic argon component in both age groups is nearly the assumed atmospheric composition.

The ^{40}Ar against per cent K isochron indicates a slight excess ^{40}Ar component for each group, which would make the apparent ages too old. The narrow range of potassium in this isochron, however, makes the error in extrapolation large; but the fact that a positive rather than negative ^{40}Ar intercept result argues against argon loss for either sample group.

We thank R. E. Leakey for help and F. J. Fitch and J. A. Miller for assistance in obtaining the samples. The work was supported by the NSF.

TABLE 2: *Concentrations of niobium, zirconium, yttrium and rubidium in pumice*

| Locality | Sample number | Concentrations (p.p.m.) | | | |
		Nb	Zr	Y	Rb
	ER72-105-0004	500	757	144	98
KBS 105	Lucus–TC–A	502	746	144	101
	Lucus–TC–B	503	735	139	103
KBS 131	ER74-131-0001	333	1,357	121	87
	ER74-131-0002	926	1,664	303	146

Nb and Y were analysed using BCR-1 as standard. Zr and Rb were analysed using G-2 as standard.

TABLE 3: *KBS analytical data*

KA no.	Sample no.	Material dated	%K	% atmospheric ^{40}Ar	mol ^{40}Ar rad g^{-1} l^{-1} x10^{-11}	Age x 10^6 yr	Remarks
2816	ER74-010-0002	Sanidine	5,167±0.013	17.8	6.203	6.90^{+}0.05	detrital contamination
2817	ER74-105-0002	Sanidine	5,035	37.8	1,755	2.01±0.03	detrital contamination
2818	ER74-105-0002	Sanidine	5.006	41.2	2.089	2.40±0.03	detrital contamination
2823	ER74-131-0001	Sanidine	4.808	40.6	1.807	2.16±0.03	detrital contamination
2836	ER74-131-0001	Sanidine	4.993±0.019	35.7	1.584	1.83±0.02	hand picked crystals,HF 2 min
2837	ER74-131-0001	Glass	3.922±0.020	92.5	1.252	1.84±0.07	heavy liquid separation, HF 2min
2841	ER74-131-0002	Sanidine	4.896±0.031	51.6	1.569	1.85±0.03	hand picked crystals, HF 2 min
2842	ER74-131-0002	Sanidine	5.110±0.045	40.6	1.539	1.73±0.03	hand picked crystals, coarser fraction than KA2841
2843	ER74-131-0002	Glass	3.484±0.0002	82.9	1.105	1.83±0.03	heavy liquid separation, HF 2min
2850	ER74-105-0001	Sanidine	5.665±0.040	32.2	1.479	1.50±0.02	hand picked crystals, HF 2 min
2851	ER74-105-0001	Sanidine	5.94	31.0	2.605	2.53±0.02	detrital contamination
2851R	ER74-105-0001	Sanidine	5.936±0.007	39.1	1.698	1.65±0.02	same as KA2851 with detrital grains picked out by hand
2852	ER74-010-0002	Sanidine	5.785±0.004	68.9	1.545	1.54±0.02	hand picked crystals, HF 2 min
2854	Lucas-105-East	Glass	4.171±0.0001	52.7	1.217	1.68±0.01	heavy liquid separation, HF 10min to remove zeolite
2855	ER74-105-0002	Sanidine	5.468±0.024	58.9	1.481	1.56±0.02	hand picked crystals, HF 5 min
2856	Lucas-105-East	Sanidine	5.487±0.020	36.7	1.529	1.61±0.02	hand picked crystals, HF 5 min
2857	ER74-105-0001	Sanidine	5.543±0.021	52.5	1.558	1.62±0.02	hand picked crystals, HF 5 min
2859	ER74-105-0002	Sanidine	5.643±0.034	58.7	1.613	1.65±0.02	hand picked crystals, HF 5 min
2861	ER74-010-0002	Sanidine	5.359±0.0009	38.8	1.485	1.60±0.01	hand picked crystals, HF 5 min
2869	ER74-131-0001	Sanidine	4.598±0.0002	53.4	1.441	1.81±0.02	hand picked crystals, HF 5 min

$^{40}O_K/K = 1.18 \times 10^{-4}$; $^{40}O_{K_\lambda} = 5.480 \times 10^{-10}$ yr^{-1}; $^{40}O_{K_{\lambda\beta}} = 4.905 \times 10^{-10}$ yr^{-1}; $^{40}O_{K_{\lambda e}} = 0.575 \times 10^{-10}$ yr^{-1}.

References

1 Behrensmeyer, A. K. 1975. *Earliest Man and Environments in the Rudolf Basin: Stratigraphy, Palaeoecology, and Evolution* (edit. by Coppens, Y., Howell, F. C., Isaac, G. L., and Leakey, R. E.), (University of Chicago Press, Chicago.)

2 Leakey, R. E. F., Behrensmeyer, A. K., Fitch, F. J., Miller, J. A. & Leakey, M. D. 1970. *Nature*, **226**, 223–30.

3 Vondra, C. F., Johnson, G. D., Behrensmeyer, A. K. & Bowen, B. E. 1971. *Nature*, **231**, 245–8.

4 Vondra, C. F. & Bowen, B. E. 1975. *Earliest Man and Environments in the Rudolf Basin: Stratigraphy, Palaeoecology, and Evolution* (edit. by Coppens, Y., Howell, F. C., Isaac, G. L. and Leakey, R. E. F.), (University of Chicago Press, Chicago.)

5 Leakey, R. E. F., Mungai, J. M. & Walker, A. C. 1971. *Am. J. phys. Anthrop.*, **35**, 175–86.

6 Leakey, R. E. F. 1972. *Nature*, **242**, 170–3.

7 Maglio, V. I. 1972. *Nature*, **239**, 379–85.

8 Isaac, G. L., Leakey, R. E. F. & Behrensmeyer, A. K. 1971. *Science*, **173**, 1129–34.

9 Fitch, F. J. & Miller, J. A. 1970. *Nature*, **226**, 226–8.

10 Fitch, F. J., Findlater, I. C., Watkins, R. T. & Miller, J. A. 1974. *Nature*, **251**, 213–15.

11 Fitch, F. J. & Miller, J. A. 1975. *Earliest Man and Environment in the Rudolf Basin: Stratigraphy, Palaeoecology, and Evolution* (edit. by Coppens, Y., Howell, F. C., Isaac, G. L., and Leakey, R.E. F.), (University of Chicago Press, Chicago.)

12 Hurford. A. J. 1974. *Nature*, **249**, 236–7.

13 Cooke, H. B. S. & Maglio, V. J. 1972. *Calibration of Hominoid Evolution* (edit. by Bishop, W. W. and Miller, J. A.), 303–68 (Scottish Academic, Edinburgh).

14 Brown, F. H. 1969. *Quaternaria*, **11**, 7–14.

15 Brown, F. H. & LAJOIE, K. R. 1971. *Nature*, **229**, 483–5.

16 Maglio, V. J. 1972. *Nature*, **239**, 379–85.

17 Dalrymple, G. B. & Lanphere, M. A. 1969. *Potassium–Argon Dating: Principles, Techniques, and Applications to Geochronology* (Freeman, San Francisco).

18 Jack, R. N. & Carmichael, I. S. E. 1969. *The Chemical 'Fingerprinting' of Acid Volcanic Rocks*, Calif. Div. Mines and Geology Short Contr. SR 100, 17–30.

19 Shafiqullah, M. & Damon, P. E. 1974. *Geochim. cosmochim. Acta*, **38**, 1341–58.

29

ANDREW BROCK

Magneto-stratigraphy east of Lake Turkana and at Olduvai Gorge: a brief summary

At the Geological Society Symposium Professor Brock gave an informal account of the work of the palaeomagnetic laboratory at the University of Nairobi in relation to the magneto-stratigraphy of the East Lake Turkana area. Results from material collected up to the end of the 1972 season are already published, and do not warrant a further paper. The main results are given in the paper by Brock & Isaac entitled 'Palaeomagnetic stratigraphy and chronology of hominid-bearing sediments east of Lake Rudolph, Kenya' published in *Nature* in 1974 (Vol. 247, pp. 344–8). A more detailed account with more descriptive material and a greater emphasis on the basic principles of magneto-stratigraphy is given in the chapter by Brock & Isaac entitled 'Reversal stratigraphy at East Rudolf' in the Wenner-Gren Symposium volume, *Earliest Man and Environments in the Rudolph Basin: Stratigraphy, Paleoecology and Evolution* (edited by Y. Coppens, F. C. Howell, G. Ll. Isaac & R. E. F. Leakey, Chicago University Press, Chicago, 1976). The results given in the above two papers are in broad general agreement with the isotopic dates obtained by the Fitch-Miller group.

More recently, debate has arisen about the age of the KBS Tuff (see elsewhere in this volume), and palaeomagnetic collections made in the 1973 and 1974 seasons obviously bear upon this. These collections are being measured in the Stanford laboratory of Professor Alan Cox, and it is already clear that some of the samples have suffered magnetic overprinting, and interpretation will be less simple than was first thought. Preliminary results are given in a paper in *Nature* by Hillhouse, Ndombi, Cox & Brock (1977: Vol. 265, 411-415).

It is worth noting that the Nairobi laboratory has also done some work on samples from Olduvai in which the stratigraphic limits of the Olduvai event in the Olduvai Gorge section have been fairly closely defined. A short note on this (by Brock & Hay) has appeared in *Earth and Planetary Science Letters* (1976: Vol. 29, 126-30).

30

F. H. BROWN, F. CLARK HOWELL
& G. G. ECK

Observations on problems of correlation of late Cenozoic hominid-bearing formations in the North Lake Turkana Basin

In the lower Omo valley (Ethiopia), north of Lake Turkana (formerly Lake Rudolf) a long succession of fossiliferous sediments and pyroclastics is exposed, the age of which has been determined by conventional K/Ar and paleomagnetic methods to fall between 0.8 m.y. and 3.2 m.y. Still older sediments and pyroclastics occur at the northernmost margin of the basin and have ages both greater and just less than 4 m.y. Another succession of fossiliferous sediments and pyroclastics is exposed over a large area in Kenya east of Lake Turkana. This succession has also been subjected to K/Ar and paleomagnetic analyses, which have yielded ages between about 1.2 and 4.5 m.y. (Fig. 30:1).

Large collections of vertebrate fossils have been recovered from both areas. In the analysis of these assemblages, various workers have suggested that the geophysical age determinations of one or both of these successions is incorrect. In particular, there has been substantial antechamber speculation about whether or not the age of 2.61 m.y. assigned to the KBS Tuff in the East Lake Turkana area by Fitch & Miller (1970, 1976) is correct. That tuff separates the Lower Member from the Upper Member of the Koobi Fora Formation. This controversy arose not from the morphology of the now well-publicized ER–1470 skull (Leakey 1973a, b)—recovered 38 m below the KBS Tuff in Karari Area 131—as some participants in the discussion have suggested. It arose instead from an apparent discrepancy between mammalian taxa derived from pre-KBS Tuff sediments when compared with those taxa from Member C of the Shungura Formation in the Omo valley. Using K/Ar and paleomagnetic data, these two sets of deposits are supposed to have similar ages (lower Member of the Koobi Fora Formation, 2.61 m.y.; Member C, Shungura Formation, 2.4–2.7 m.y.); however, faunal correlations indicate that the deposits of the lower Member of the Koobi Fora Formation are considerably younger than those of Member C.

The problem is thus one of correlation of two proximate sedimentary sequences, situated on the northern and north-eastern margins of the same sedimentary basin. Unfortunately, as a consequence of the vicissitudes of tectonic events in this segment of the African Rift System, and their separation by national boundaries, it has been almost impossible to link one succession directly with another (see below). One correlation has been suggested by radiometric age data, in part supplemented by paleomagnetic data; another, very different correlation has been suggested by faunal comparison, that is, mammalian biostratigraphy (Behrensmeyer 1976 in this volume).

Our purpose here is to review in a concerted fashion the paleogeography, stratigraphy, tuff chemistry, K/Ar age determinations, magnetostratigraphy, and vertebrate faunal evidence relevant to the correlation of these areas of the North Lake Turkana Basin.

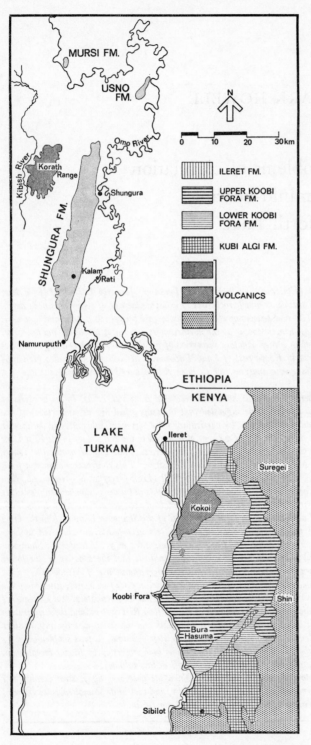

FIG. 30:1.

Map of North Lake Turkana area and distribution of principal late Cenozoic lithological units. In legend: note that the Lower Koobi Fora, Upper Koobi Fora and Ileret units are three members of the Koobi Fora Formation.

Paleogeographic aspects

The paleogeography of the Lake Turkana basin prior to about 4 m.y. is not directly relevant to the purpose of this paper. However, by about 4 m.y. ago the major features of the basin existed. The Omo and Usno rivers had established a generally north to south course, approximately along the 36° E meridian, and were depositing their sedimentary load in the lower reaches of the Omo valley, where they debouched into the northern end of an ancestral Lake Turkana, the size and conformation of which is still ill-defined. Over this time, sediments were also being deposited along the eastern margin of that lake by one or several perennial streams with headwaters in the highlands north of the Lake Stephanie (Chew Bahir) basin of the Ethiopian rift. The known Omo succession is presumed to represent sediments accumulated in the axis of the trough, and the sediments eastward of the lake to represent an accumulation adjacent to the rift axis on the margin of a downwarp (see Fitch & Vondra 1976).

The paleogeographic setting was not then fundamentally unlike that of the present. However, young features like the Korath Range (Brown & Carmichael, 1969) did not exist at that time, and the vast plain north of Lake Turkana had not yet been severely fractured by faulting. The streams (how many is still uncertain) which brought sediments into the area east of the lake had not as yet had their headwaters truncated by faulting between that area and the Stefanie basin, but were through flowing, debouching into the Lake Turkana basin. Some distance to the south and west of these rivers and streams, a lake existed as a common sump, and received sediments predominantly derived from the north and the east. This appears on present evidence to have been the paleogeographic setting for the interval 4 to 1 m.y. ago (cf. Behrensmeyer 1974).

Bodies of water in closed basins, such as those of the early and present day Lake Turkana basin, respond to changes in precipitation and/or evaporation rates through simultaneous changes in water level around their margins. We think, therefore, that the major episodes of lake transgression recorded in the sediments both in the Omo valley and at East Lake Turkana may provide a means of stratigraphic correlation between the two areas. Past differential tectonic activity in the two areas may cause problems for this method of correlation. The fact that tectonic activity in the northern Lake Turkana basin was apparently relatively quiescent during the period of deposition of the Shungura and Koobi Fora Formations (see below), however, leads us to think that this complicating factor is of little importance.

Differential drainage gradients between the two areas and/or differences in the distances between the outcrops of the various formations and the centre of the basin may also complicate the picture. The drainage gradients are now dissimilar on the northern and eastern margins of Lake Turkana. The slope of the Omo valley is on the order of 0.0003 and that of the eastern margin of the lake on the order of 0.003. Thus, for each metre rise in lake level, the shoreline should advance approximately ten times as far to the north as to the east. Given slope values of this approximate magnitude in the past, it is expected that lacustrine sediments would have far greater extent in the Omo valley than in the East Lake Turkana area.

On the other hand, the outcrops of the early members of the Shungura Formation are situated between 40 and 60 kilometres north of the present lake margin, while those of the Koobi Fora Formation crop out essentially adjacent to the lake margin and extend up to 40 km to the east (Vondra & Bowen 1976). Thus, small rises in lake level, on the order of 15 m or less, might not be documented in the more distal outcrops of the Shungura Formation. Lacustrine sediments of the Koobi Fora Formation have been recorded as much as 25 km east of the present shoreline of Lake Turkana (Bowen 1974). These occurrences imply a rise in lake level of 45 m, if a slope of 0.003 is assumed. Even if the paleoslope were on the order of 0.001, a rise of 25 m in lake level would have been necessary to produce deposits at this distance. A transgression of this magnitude would have shifted the lake shoreline sufficiently northward to have had such an event documented in the Shungura Formation exposures. It would appear then that the lower drainage gradient of the Omo valley 'compensates' for the more distal location of its exposures, and the more proximal location of the outcrops at East Lake Turkana 'compensates' for the greater drainage gradient found in this area.

Thus, it would appear to us that records of major transgressive/regressive events in the sediments at East Lake Turkana and in the Omo valley provide a means of stratigraphic correlation between the two areas. The data and arguments for such a correlation are presented below.

Stratigraphic aspects

The stratigraphy of these north Lake Turkana sedimentary formations is set out briefly here. For details the reader should consult de Heinzelin & Brown (1969), de Heinzelin, Brown & Howell (1970), de Heinzelin (1971), and de Heinzelin, Haesaerts & Howell (1976) in regard to the lower Omo basin, and Behrensmeyer (1970), Vondra *et al.* (1971), Bowen & Vondra (1973), Bowen (1974) and Vondra & Bowen (1976) in respect to East Lake Turkana. Two formations exposed in the lower Omo valley, the Usno and Shungura Formations, are relevant to this discussion. The oldest, most distant transgressions are recorded in the Mursi Formation and the Nkalabong Formation, which are older and just younger than 4 m.y., respectively, and which crop out 100–20 km north of the present lake at the foot of Nkalabong mountain (Butzer 1976). It is unknown if the transgressive deposits recorded at Kanapoi, southwest of the present basin, and of about the same age on radiometric and faunal grounds was accumulated in a part of early Lake Turkana, or in another isolated, restricted basin (Behrensmeyer 1974). A schematic section of the Omo succession in which fluvial and lacustrine sediments are indicated along with the major associated tuffs is depicted in Fig. 30:2. Only these major depositional units are indicated because they are likely to afford the best basic data for purposes of correlation. A similar section is also presented for the sedimentary sequence east of Lake Turkana in the same figure.

The Shungura Formation has a total thickness of about 775 m. Within this largely fluviatile succession of deposits four episodes of lacustrine sedimentation are recorded—in the Basal Member, in (mid) Member C, in (upper) Member G, and in Member L. All four depositional units afford molluscan assemblages (Van

Damme & Gautier 1971; Gautier 1976). The lacustrine sediments of the Basal Member have been correlated with lacustrine sediments (unit U–3) of the Usno Formation on the basis of comparable molluscan assemblages (Gautier 1976), sedimentary-stratigraphic grounds (de Heinzelin, Haesaerts & Howell 1976), and magnetostratigraphic evidence (Brown & Shuey 1976).

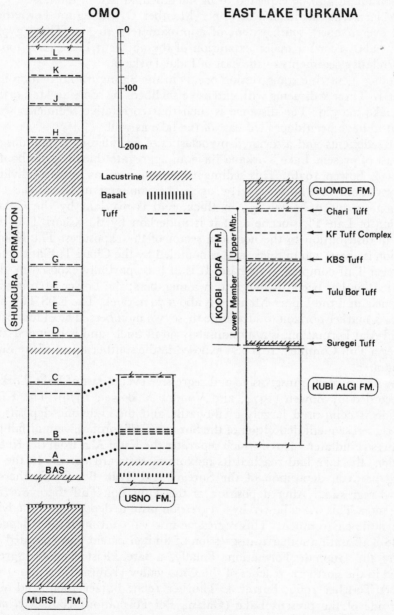

FIG. 30:2. Schematic diagram (to scale) of formations of the Omo Group and the East Lake Turkana area.

The Usno Formation exposures crop out some 20 km north of the Shungura Formation exposures and approximately 80 km north of the present lake margin. Thus, the lacustrine sediments at the Usno Formation localities record a major transgression of the lake. Correlative sediments should be found, if exposed, to the east of Lake Turkana. A brief lacustrine incursion in Member C, Shungura Formation might also be expected to be documented east of the lake.

The lacustrine sediments of upper Member G, Shungura Formation, are exposed over a north-south extent of approximately 40 km. These gypsiferous sediments also record a major expansion of the adjacent lake. They too should have correlative sediments to the east of Lake Turkana.

Another lacustrine transgression occurs in the Shungura Formation in upper Member L. These sediments with extensive shellbeds lie some 40 km north of the present lake margin. The distance is such that correlative sediments would be expected to have been deposited east of the lake as well.

The sediments and associated pyroclastics of fossiliferous formations exposed to the east of present Lake Turkana have an aggregate thickness of about 325 m (Vondra & Bowen 1976). This sedimentary sequence has been subdivided into three formations of Plio/Pleistocene age. The youngest of these, the Guomde Formation, is approximately 35 m thick, and is overlain by the Galana Boi Formation of Late Pleistocene age; it is underlain by the Chari Tuff. It is restricted in distribution to the northern sector of the exposures. The Koobi Fora Formation is about 230 m thick, and is bounded by the Chari Tuff at the top, and the Suregei Tuff complex at its base. It is at least partially exposed in all areas. This formation has been divided into two members; the Lower Member is about 160 m thick, and the Upper Member is about 70 m thick. The KBS Tuff has been chosen as a marker horizon to subdivide these two members. The oldest formation, the Kubi Algi Formation, is approximately 90 m thick, and is bounded above by the Suregei Tuff Complex. It is best exposed in the southern part of the East Lake Turkana area.

The sequence of transgressive and regressive events east of Lake Turkana has been recorded by Bowen (1974; also Vondra & Bowen 1976). The Kubi Algi Formation is comprised largely of fluviatile and fluvio-deltaic deposits. At the time of the subsequent deposition of the Suregei Tuff (probably an airfall tuff into lake waters (Findlater 1976)), which separates the Kubi Algi from the Koobi Fora Formation, the lake had reached its maximum eastward extent. In the ensuing interval, between deposition of the Suregei and Tulu Bor Tuffs, there was a westward regression. After deposition of the Tulu Bor Tuff there was a major transgression. This was followed by a regression prior to deposition of the KBS Tuff in a fluviatile environment. This regressive interval continued after deposition of the KBS Tuff until another transgression of limited extent was recorded by sediments of the Guomde Formation. Finally, a late Pleistocene transgression is recorded in the northern reaches of the Omo valley (Kibish Formation) (Butzer, Brown & Thurber 1969; Butzer & Thurber 1969; Butzer 1969) and along the eastern side of the present Lake (Galana Boi Formation) (Butzer *et al.* 1972; Vondra & Bowen 1976).

At least three major ancient transgressions are thus well documented in the stratigraphic record both to the north and to the east of present Lake Turkana. Interareal correlation might be sought first by equating these major transgressive episodes in their relative stratigraphic order—lacustrine sediments below the Suregei Tuff with those of the Basal Member, Shungura Formation; lacustrine sediments of the Lower Member, Koobi Fora Formation with those of upper Member G, Shungura Formation; and Guomde Formation lacustrine sediments with those of upper Member L, Shungura Formation. On the basis of conventional K/Ar radiometry and paleomagnetic evidence the Basal Member of the Shungura Formation has been determined to be about 3.2–3.3 m.y. in age, and to correspond with that part of the Gauss Normal Epoch preceding the Mammoth (Reversed) Event.

Correlation on this basis would be appropriate only if no major episodes of tectonic movements occurred during the accumulation of these sedimentary bodies. None is documented within the main Omo succession, and there is no recorded incidence within the relevant East Lake Turkana section. Tentative correlation lines in Fig. 30:4 are drawn so as to indicate these possible temporal correspondences. Admittedly this is only one kind of evidence. So far as is known, the suggested correlation is not incompatible with what is known of the molluscan faunas from these respective horizons. Van Damme & Gautier (1971) and Gautier (1976) have distinguished lower and upper molluscan faunal zones in the Omo Group succession. The assemblages from the two later transgressions recorded in East Lake Turkana appear to correspond with the upper zone (Members H and L) of the Shungura Formation (A. Gautier, pers. comm.). Shellbeds are recorded below the Tulu Bor Tuff, but to our knowledge that assemblage appears to differ in some species from that of the lower zone of the Omo succession. Other kinds of evidence must be examined to determine whether such a hypothetical correlation is reasonable.

Chemistry of tuffs in the two areas

Tuffs occur extensively in both successions and are very useful stratigraphically as they represent isochronous or near-isochronous horizons. Unfortunately their source is still unknown, although many possibilities have been eliminated on petrological grounds. At present the most feasible source of the tuffs appears to lie in the upper Omo basin, perhaps near Mount Damota. What is known of volcanic rocks from the Damota area (Brown & Nash 1976) suggests that they are fundamentally similar in petrology to the tuffs of the Omo and East Lake Turkana sedimentary sequences although they differ in minor details. The Damota area and Gughe highlands lie along the divide between the Omo river and the Galana Sagan, which now flows into Lake Stephanie. The proto-Galana Sagan is postulated to have carried much of the sediment deposited in East Lake Turkana. Therefore, if the Damota region (including the Gughe highlands) is the general source region of the tuffs in the Omo basin, it is likely that some correlative tuffs will also be found at East Lake Turkana.

Brown & Nash (1976) have suggested a possible correlation between the uppermost or Chari Tuff in East Lake Turkana with Tuff L of the Shungura Formation on the basis of similar chemistry of microphenocrysts in their pumices.

Equation of these two tuffs is consistent with the correlation proposed on the basis of successive lacustrine transgressions as set out before; both tuffs occur slightly below the uppermost lacustrine sediments in each area. In East Lake Turkana an unconformity occurs between the Chari Tuff and the overlying Guomde Formation; an unconformity is also inferred in submember L2 of the Shungura Formation on paleomagnetic and radiometric grounds which separates Tuff L from the overlying lacustrine sediments.

Brown & Nash (1976) also sought to correlate the KBS Tuff with Tuff D of the Shungura Formation. The attempt was made because of the apparent similarity in age of the tuffs on the basis of K/Ar determinations available at that time (1973). No other correlations were suggested as the dating of both areas seemed securely established. However, these tuffs have been found to differ chemically, notably with respect to the concentration of barium and iron in their feldspars.

If correlation is based initially on successive lacustrine transgressions common to both areas, the question then arises as to whether or not another tuff within the upper Shungura Formation between upper Member G and Member L, might be correlative with the KBS Tuff of East Lake Turkana. Examination of the chemical analyses of feldspars presented in Brown & Nash (1976) reveals that the only feldspars analysed from a tuff of the Shungura Formation which resemble feldspar from a tuff identified as the KBS (ERL–70–4) are those of Tuffs E, I2 (now termed H2) and I4 (now termed H4). The best fit of the three is with Tuff H2; both Tuffs E and H4 contain more Fe_2O_3 in their feldspars. Analyses of clinopyroxenes and ilmenites separated from pumice of Tuff H2 also correspond closely in composition to those from the KBS Tuff as identified in Area 131 (Karari). These pyroxenes are quite distinct in minor element content from those derived from Tuff L and the Chari Tuff. Hence, correlation of the KBS Tuff with Tuff H2 of the Shungura Formation is distinctly possible on chemical grounds. This correlation is consistent with that suggested by the equation of the lacustrine transgressions. The evidence is not conclusive, but it is suggestive.

Unfortunately, the Chari and KBS Tuffs are the only ones from East Lake Turkana for which sufficient chemical data are available for such a comparison. Other correlations of tuffs between the two areas might be sought if comparable chemical data were available.

Magnetostratigraphy

The magnetostratigraphy of the Shungura Formation has been published by Shuey, Brown & Croes (1974; also, Brown & Shuey 1976). It has since been confirmed in all essential respects by more intensive resampling (Brown, Croes & Shuey, in preparation). Brock & Isaac (1974; also 1976) have published a magnetostratigraphy of the Koobi Fora Formation. Additional sampling has failed to confirm some aspects of that work (Brock *et al.*, 1974). In particular, the presumed dominant normal polarity reported for the pre-KBS sediments of the lower Member of the Koobi Fora Formation may be largely reversed (editor's note in Brock & Isaac 1976). The magnetic polarity logs of each succession are shown in Fig. 30:3.

The magnetostratigraphic logs may be used in several ways. Each of the logs could be correlated with the 'standard' polarity scale (Cox 1969), after which named events could be matched. The logs might also be matched directly, and for purposes of regional correlation the latter method is preferable, although it is hampered by dissimilar sedimentation rates in these two areas. The following points are particularly worthy of note in this regard.

A polarity transition from reverse to normal occurs at the end of the lacustrine succession below the KBS Tuff in East Lake Turkana, and at the top of Member G of the Shungura Formation. The transition is followed by a long sequence of normal sediments, after which there is a long sequence of sediments with dominantly reversed polarity. It therefore could be maintained that the polarity data are consistent with correlations proposed on the basis of lacustrine transgressions common to the two areas, at least in the upper parts of their respective sections.

In the East Lake Turkana succession, however, there is a reversed-normal pair of

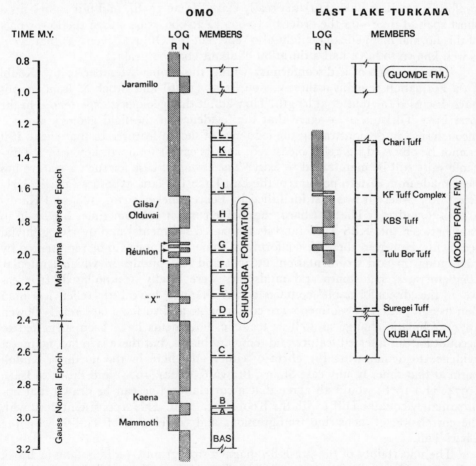

FIG. 30:3. Paleomagnetic logs from the Shungura Formation (Omo basin) and the Koobi Fora Formation, East Lake Turkana.

magnetozones in the magnetostratigraphic log near the upper boundary of the long normal interval which is not apparent in the Omo succession. Two explanations might account for this seeming discrepancy; either this event was missed in sampling the Shungura Formation, or the normal event is spurious at East Lake Turkana. The former alternative appears unlikely in view of the closely spaced samples (ca. 1–2 m) taken from the Shungura Formation. However examination of Fig. 3 of Shuey, Brown & Croes (1974) reveals such a structure to be present in Member H–7, although these authors considered the uppermost normal samples of Member H to be spurious. It should also be noted that Brock & Isaac (1974) have correlated the upper part of this normal magnetozone with the Olduvai normal event. However, they did not so correlate the lower part because a disconformity was inferred in that part of the Koobi Fora succession not far above the KBS Tuff (cf. Behrensmeyer 1970). Shuey, Brown & Croes (1974) correlate the entire normal sequence from the upper part of Member G through to the top of Member H, Shungura Formation, with the Olduvai normal event. If this disconformity at East Lake Turkana really exists (Isaac 1976), and if it represents a short span of time—on the order of 10 000 to 20 000 years—then the lower part of this normal magnetozone might also represent the Olduvai event, as that event is well known to have had a duration of about 200 000 years.

The existence of a disconformity within the Upper Member of the Koobi Fora Formation remains a thorny issue. Isaac (1976) and Brock & Isaac (1976) have discussed the matter at length. They admit that geologists experienced in the East Lake Turkana area agree that the 'evidence of the field geology alone is inconclusive' in demonstrating the existence of such a feature. In that event, if it cannot be observed, its relief measured, and its extent mapped, how can the disconformity still be maintained to exist? The strongest case for such a feature has been made from eastern portions of the East Lake Turkana exposures, particularly Areas 105 (where it was first identified by Behrensmeyer 1970), 108, and Karari areas 130 and 131. There a sharp, mappable contact has sometimes been said to exist between the 'KBS Tuff' (and its associated sediments) and overlying fluvial sediments. However, these 'disconformities' generally appear to be represented by coarsening upward sedimentation and cut and fill episodes in which lenticular conglomerates, sandstones and mudstones were locally accumulated. In areas where the 'disconformable' contacts occur, channels have little relief, less than 8 m usually; laterally sediments are conformable and surfaces lack relief (Bowen 1974). Climatic change as well as tectonic movements have been suggested to account for this inferred feature and temporal hiatus, but there is in fact no direct evidence to demonstrate the effect of either anywhere in the northern Rudolf basin at that time. In any case Shuey, Brown & Croes (1974) and Brock & Isaac (1974; also 1976) would all agree that a correlation line can be drawn that approximately equates Tuff J with the Koobi Fora Tuff. This is consistent both with the correlation of lacustrine transgressions and correlation of Tuff L with the Chari Tuff.

The uncertainty of the pre-KBS magnetostratigraphy (editors note in Brock & Isaac 1976) now limits any possibility of correlation of that portion of the East Rudolf succession on the basis of magnetostratigraphy. Brock & Isaac (1974; also

1976) originally assigned the part of the section below the KBS Tuff to the Gauss Normal Epoch, and the two short reversed intervals thought to be recorded there were attributed to the Kaena and Mammoth events. Brock & Isaac, however, have not recognized the existence of the Reunion events. These events, which are now increasingly well documented (see also Curtis & Hay 1972) may well be of importance in any attempted correlation with the 'standard' magnetic polarity scale. This is especially so now that less of the part of the East Rudolf section below the KBS Tuff is accepted as being of normal polarity.

Potassium-argon ages:

Potassium-argon age determinations on tuffs of the Shungura and Usno Formations are summarized in Brown & Nash (1976; also Brown & Lajoie, 1971; Brown 1972). Fitch *et al.* (1974) and Fitch & Miller (1976 and this volume) provide age determinations on tuffs in the East Lake Turkana succession. The principal results are set out in Table 1.

TABLE 1: *K/Ar ages of principal tuffs in the North Lake Turkana Basin.*

Shungura Tuff	Age (m.y.)	East Rudolf Tuff	Age (m.y.)
L	1.34 ± 0.10	Chari	1.26 ± 0.20
I$_2$ (= H2)	1.83 ± 0.09	Karari	1.32 ± 0.01
G	1.93 ± 0.10	Lower Tuff, Ileret	1.48 ± 0.02
F	2.03 ± 0.10	Koobi Fora	1.57 ± 0.00
D4	2.12 ± 0.11	BBS	1.56 ± 0.02
D	2.39 ± 0.15	KBS	2.61 ± 0.26
B10	2.95 ± 0.10	Tulu Bor	3.18 ± 0.09
Usno Tuff	2.97 ± 0.3		
Usno Basalt	3.31 ± 0.42		

Bowen (1974; Bowen & Vondra 1973, 1976) has correlated the Chari Tuff (of the Ileret area) with the Karari Tuff (of the Karari area). Brown & Nash (1976) proposed a possible correlation of the Chari Tuff with Shungura Formation Tuff L based on microphenocryst chemistry. G. H. Curtis and associates (pers. comm.) have recently ascertained ages (on the basis of total degassing method) of ca. 1.37 for the Karari Tuff, and ca. 1.35 m.y. for the Chari Tuff (values comparable to those reported by Fitch & Miller 1976). The K/Ar ages for all three tuffs are essentially the same within the limits of error, and hence confirm our proposed correlations, as set out above.

Oxygen isotope ratio analyses (Cerling 1976) support the correlation of the Chari & Karari Tuffs. They suggest, furthermore, that these tuffs might well be correlative with the Koobi Fora Tuff (of the Koobi Fora area). A correlation of the Koobi Fora Tuff with the Chari and Karari Tuffs and thus with Tuff L of the Shungura Formation is inconsistent with the magnetostratigraphic correlation proposed above and the K/Ar age assigned to the Koobi Fora Tuff (Table 1). Because the Koobi Fora Tuff cannot be distinguished from the Chari and Karari

Tuffs, or from the Ileret and BBS (now Okote) Tuff complexes, using $^{18}O/^{16}O$ analyses, we think that correlation of the Koobi Fora Tuff with Tuff J of the Shungura Formation is more accurate.

In the preceding section correlation based on magnetostratigraphy was proposed for the Koobi Fora Tuff and Tuff J (or a proximal tuff in the strati-graphic section), Shungura Formation. Each is just slightly younger than the top of the Olduvai normal event. The top of the Olduvai event is about 1.65 m.y. in age. Thus this correlation appears to be confirmed as the K/Ar age of the Koobi Fora Tuff is ca. 1.57 m.y. (Fitch & Miller 1976).

The proposed correlation of Tuff H2, Shungura Formation, with the KBS Tuff, Koobi Fora Formation, might appear to be problematical. The ages assigned to these tuffs are 1.83 m.y. (Brown & Nash 1976) and 2.61 m.y. (Fitch & Miller 1970, 1976; also Fitch *et al.* 1974), respectively. Clearly these figures do not coincide. The mid-point ages are different, and the dates do not even overlap within the limits of error. The three ages obtained by the conventional K/Ar method on Tuff H2, from which the mean is derived, are 1.81, 1.81, and 1.87 m.y.; all determinations are in good agreement. Apparent ages on the KBS Tuff obtained by the $^{40}Ar/^{39}Ar$ step-heating method, and the $^{40}Ar/^{39}Ar$ total degassing method range from 0.52 to 2.64 m.y. (Fitch & Miller 1976). Recently, Curtis *et al.* (1975) have dated both sanidine and glass from pumices from what has been identified as the KBS Tuff from three different areas of East Lake Turkana, and found those tuffs to be of dissimilar ages. The tuff from the type occurrence (Area 105) afforded an age of 1.60 ± 0.05 m.y.; that in Ileret Area 10 had an average age of 1.57 m.y.; whereas that in Area 131 afforded an age of 1.82 ± 0.04 m.y. If these data are correct then the correlation of one 'KBS Tuff' with Tuff H2 is not only possible, but highly probable. Given this age, it is mandatory that the older KBS Tuff be placed within the Olduvai Normal Event; the age of this has been determined by the same methodology, and not that employed by Fitch & Miller to arrive at their age assessment. The close correspondence of the younger set of ages obtained by Curtis *et al.* (1975) from the occurrences in Area 10 and Area 105 with the ages of the Koobi Fora and BBS (now termed Okote) Tuffs obtained by Fitch & Miller (1976) immediately causes one to suspect that the two 'KBS Tuffs' are not identical, and that an error has been made in stratigraphic attribution of these tephra.

Two further remarks are worthwhile in regard to the potassium-argon age data. Fitch & Miller (1976) report a date (FM 7053) on a basalt lava which either underlies or is intercalated with sediments of the Kubi Algi Formation. The age of this basalt is 3.8 m.y. Fitch & Miller (1976) are of the opinion that the ages obtained on the lavas are generally geologically correct. Thus the maximum age of the Kubi Algi sequence above this basalt is 3.8 m.y. The other dates of 3.9 m.y. and ca. 4.5 m.y. tentatively assigned to tuffs of that formation are internally inconsistent with that maximum age. The date of 3.8 m.y. on the basalt is accepted here as a maximum age for the entire East Lake Turkana sedimentary succession. Maximum age refers only to the age of the base upon which the deposits rest, and means that the oldest sediments must be younger than 3.8 m.y. The mammal fauna of the Kubi Algi Formation has been considered by Maglio (1970, 1972) and others to be comparable to, and hence of broadly the same age as that of the

Mursi Formation, the oldest unit within the Omo Group succession. A basalt capping the latter formation has afforded ages of 4.05 (Brown 1972; Brown & Lajoie 1971) and 4.4–4.1 m.y. (Fitch & Miller 1976). The small number of taxa known from the Mursi Formation results in a rather large uncertainty in faunal correlation of these two sequences.

Fitch & Miller (1976) have also published a 'best apparent age' of 3.18 m.y. (FMA 255) for the Tulu Bor Tuff. They note that two age components are present in the Tulu Bor Tuff at medium to high heating steps, and are thought to be due to the presence of two generations of alkali feldspar. These feldspars are supposed to differ in their argon diffusion characteristics perhaps because they differ slightly in composition or structural state. By sector analysis of the age spectrum, apparent ages of 4.04 m.y. and 3.22 m.y. were suggested for the ages of the older and younger components respectively, although the possibility of overlap of the age components is noted. Since the diffusion characteristics of the two feldspars are unknown, the possibility that there is considerable overlap of the two age components cannot be discounted. Thus the 'best apparent age' of 3.18 m.y. may be a composite age derived from feldspars of two different ages. Each component would yield some fraction of the argon finally analysed in proportion to its diffusion characteristics. For example, one component with an $^{40}Ar/^{39}Ar$ ratio corresponding to an age of 4.18 m.y. might yield half the argon while another component with an $^{40}Ar/^{39}Ar$ ratio corresponding to an age of 2.18 might yield the other half—the result would be a composite age of 3.18 m.y. The figures used in this example are meant to be illustrative only, and we make no suggestion as to the age of the Tulu Bor Tuff's youngest age component. The age sought (2.18 m.y. in our example) might never be revealed. If a situation such as this is the case for the Tulu Bor Tuff, then 3.18 m.y. is the maximum possible age of the younger feldspar component, and does not necessarily represent its real age.

Lacustrine sediments directly overlie the Tulu Bor Tuff, and it is the top of these lacustrine sediments which we have tentatively correlated with the lacustrine sediments of Upper Member G, Shungura Formation. If this proposed correlation is correct, then an unconformity must be inferred to exist within the lacustrine sediments between the Tulu Bor Tuff and the KBS Tuff, if one accepts an age of 3.18 m.y. for the Tulu Bor Tuff. If such an unconformity exists, the lower part of these lacustrine sediments would correlate best with the lacustrine sediments of the Basal Member of the Shungura Formation. If such an unconformity does not exist, and none has as yet been demonstrated to the best of our knowledge (cf. Vondra & Bowen 1976; Bowen 1974), then it must be assumed that the age of 3.18 m.y. on the Tulu Bor Tuff is in error.

Oxygen isotope analyses of this tuff are ambiguous (Cerling 1976) in that material collected from what was said to be the same tuff yielded several unexpectedly low values, considering its presumably early age. In other samples that tuff has been shown to differ markedly in $^{18}O/^{16}O$ composition not only from tuffs of the upper Koobi Fora Formation, but also from the KBS Tuff.

Thus far four lines of evidence have been considered—stratigraphy and sedimentation, tuff chemistry, magnetostratigraphy, and potassium-argon dating. The following set of correlations is proposed on the basis of these lines of evidence (Fig.

30:4). These correlations are felt to be reasonably secure; correlations with the lower part of the East Lake Turkana sequence fall more in the realm of speculation, and some of our speculations are presented in a later section of this paper.

Evolutionary grade of mammalian taxa

The phyletic evolution of African suids (Cooke & Maglio 1972a, b; Cooke 1976) and proboscideans (Maglio 1970, 1973; Cooke & Maglio 1972; Beden 1976) has

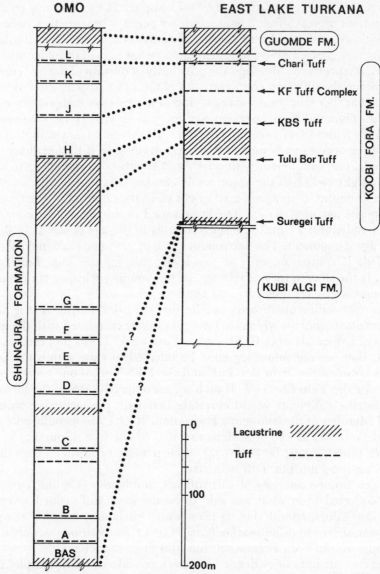

FIG. 30:4. Proposed correlation of late Cenozoic successions in the lower Omo basin and the East Lake Turkana area.

been intensively studied in recent years. Morphological trends in the evolution of particular suid and proboscidean taxa have been shown by these authors to occur progressively through the African late Cenozoic. These researches, in particular, and those of Gentry (1976, 1977) on Bovidae, as well as of S. Coryndon (1976) on Hippopotamidae, have more than any other single set of data, led to the controversy over the age assignment of 2.6 m.y. to the KBS Tuff of the East Lake Turkana succession, or alternatively, to an overall denunciation, implicit or explicit, of K/Ar determinations on Omo Group formations, and indeed, therefore, to all K/Ar results obtained by conventional, total-degassing methods.

Maglio (1972) suggested that the 'best fit' age of three faunal zones recorded by him in the East Lake Turkana area, with K/Ar dated horizons within the Omo Group succession were:

	'Best fit' Age (m.y.)	Maximum Age (m.y.)	Minimum Age (m.y.)
Loxodonta africana zone	1.3	1.6	1.0
Metridiochoerus andrewsi zone	1.7	1.9	1.5
Mesochoerus limnetes zone	2.3	3.0	2.0

The mammalian faunal assemblage representative of the *L. africana* zone derives only from the northern, Ileret area, and essentially only from the Lower/Middle Tuff Complex. According to the previous discussion it should be correlative with Members K and L, Shungura Formation, with an age between 1.0 and, at most, 1.5 m.y. This assignment is in good agreement with radiometric and paleomagnetic determinations.

The mammalian faunal assemblage taken as representative of an *M. andrewsi* zone has been collected from a more diverse series of localities and subjacent horizons in the Upper Member, Koobi Fora Formation (Maglio 1972; A. K. Behrensmeyer, pers. comm.). In the Koobi Fora area these collections were assembled from Areas 105 (above the 'disconformity'), 104 (above the 'disconformity,' and some 13–20 m above that 'KBS Tuff'), and 103 (where a 'disconformity' is ill-distinguished, but from some 10 to as much as 50 m above that 'KBS Tuff'). In the northern, Ileret area collections were assembled from above that 'KBS Tuff', and generally some 20 m or so above it (where no 'disconformity' has been suspected to exist). [It should be noted that according to Curtis *et al.* (1975) that the 'KBS Tuff' in Area 105, and presumably adjacent areas to the west, has an apparent age of 1.6 m.y. and that the 'KBS Tuff' in the Ileret area (10) has an apparent age of 1.57 m.y., hence these collections should be generally representative of a spatially diverse, but temporally comparable time range.] This faunal zone should be broadly correlative with Members H to J of the Shungura Formation. An age of ∼ 1.6 m.y. is reasonable. This is also in good agreement with correlations proposed previously on other grounds.

The mammalian faunal assemblage of the *M. limnetes* zone has been collected from the Lower Member, Koobi Fora Formation. Apparently all of the collections attributed to this assemblage by Maglio (1972) have been assembled from the upper part of that member, from a few metres to 10–12 m below the 'KBS Tuff,' that is, they derive from no more than the upper 1/4 to 1/5 of that sedimentary accumulation. However, Harris (1976) states that the *M. limnetes* zone extends to levels 35 m below the 'KBS Tuff'.

There is no well fixed point in the Koobi Fora section, unless one chooses to rely on the correlation of the lacustrine sequence at the base of the Shungura Formation, with the lacustrine sequence represented at the top (Suregai Tuff) of the Kubi Algi Formation. In that case ~ 3.2 m.y. would be a maximum allowable age. As the mammalian fauna taken as characteristic of the *M. limnetes* zone derives from levels only a limited distance below that 'KBS Tuff', it is to be expected that that assemblage should be closer to 1.8 m.y. than to 3.2 m.y. in age. Thus, Maglio's (1972) mid-point age estimate of 2.3 m.y. does not seem to be unreasonable. However, the maximum estimate (3.0 m.y.) might well appear to be excessive—perhaps 1.8–2.5 m.y. would be a more appropriate approximation.

Cooke (1976) has noted that 'the recent revision of the Omo time-scale has the effect of making ages below Member G somewhat younger than they were considered before and the new dates would serve to reduce the lower limits now appropriate to Maglio's estimates'. Using more data than was available heretofore he would place the *M. limnetes* zone at 'about the boundary between Members F and G of the Shungura Formation'. The best fit of the *M. andrewsi* zone is still with Member H (or above). Cooke (1976) supposes that the 'KBS Tuff' should be about the same age as the top of Member F at Omo, that is 1.95 m.y. His supposition is in far better agreement with an age of 1.8 m.y. for certain 'KBS Tuff' occurrences obtained by Curtis *et al.* (1975) than the value of 2.61 m.y. proposed by Fitch & Miller (1976). As the mammalian assemblage taken to be characteristic of the *M. limnetes* zone is collected from below a 'KBS Tuff' the inferred age is a maximum, and 1.8 m.y. is *not* inconsistent with the faunal data.

Additional faunal evidence relevant to relative age has been adduced from Bovidae (Gentry 1976), Equidae (Hooijer 1976; Eisenmann 1976), Hippopotamidae (Coryndon 1976), and some other mammalian taxa. Again these studies indicate that the *Mesochoerus limnetes* Zone fauna is best correlated with the fauna from upper Member F or lower Member G rather than with that of Member C of the Shungura Formation.

Correlation of mammal faunal assemblages

Shuey *et al.* (1976, in this volume) have considered a statistical correlation of the entire mammal megafauna recorded from the East Lake Turkana faunal zones with those recorded from Omo Group formations. R. E. Leakey kindly read that paper in manuscript. He informed us that as the published mammalian faunal lists for East Lake Turkana were under revision, comparative analysis of this sort was meaningless. That may indeed be so, and one must then re-examine the matter when revisions of previous identifications are forthcoming. The most notable additions to the mammal species lists would appear to be fuller documentation of the fauna below the Tulu Bor, Suregai and Kubi Algi Tuffs. In a letter to one of us (F.C.H.) J. M. Harris (20/1/75) has stated that he does not 'think that the new material changes the picture that was emerging by late 1973 in any way. . . '. However, the conclusions reached by Shuey *et al.* (1976) merit consideration, bearing in mind that the age assignments derived from the current East Lake Turkana list may be either too old or too young. Employing this method of analysis the *Loxodonta africana* Zone fauna equates best with the fauna of Members H–J of the Shungura Formation. The *Metridiochoerus andrewsi* Zone fauna equates best with the fauna of mid- to upper Member G. The *Mesochoerus limnetes* Zone fauna equates best with the fauna of Member F or lower Member G.

These correlations derived from statistical comparisons of mammalian faunas are at variance with correlations proposed above on other grounds, all appearing slightly older. It could be assumed, however, that this might be due to the relatively poor faunal assemblages known from Members J and K of the Shungura Formation. Hence one could postulate that all of the age assignments derived from this method of comparison may err on the old side. This possibility is discussed by Shuey *et al.* (1976). Even if the East Lake Turkana faunal zones are as old as indicated by this method, the age of the *Mesochoerus limnetes* Zone and the overlying 'KBS Tuffs' are much younger than that based on the K/Ar dates proposed by Fitch & Miller (1976).

Enviromental setting

It has been suggested to us (cf. Behrensmeyer 1976) that ecological barriers could prevent the use of normal methods of faunal comparison and correlation on the basis of faunal assemblages. This sector of the East African Rift has been suggested to constitute an intermediate zone between several distinct faunal provinces where different ecological conditions now prevail, and would have also in the past. A number of lines of evidence suggest that environmental differences are not as marked as some workers believe to be the case; these are discussed below. Because of this evidence, we feel that faunal data can be useful in establishing stratigraphic correlations between the two areas.

It is of course well known that the Sudanese area to the north-west, the Ethiopian highlands to the north-east and the East African plateaux and plains to the south represent different aspects of the Ethiopian faunal region (Bigalke 1972; Davis 1962). However this sector of the rift is conjunctive with adjacent areas in its natural plant and animal communities rather than being markedly distinctive (cf. Moreau 1952; Davis 1962; Bigalke 1972).

In its overall vegetation this sector of the rift is within the zone of wooded to subdesertic steppe as defined for example, in Keay (1958), Rattray (1960), Trapnell & Langdale Brown (1962) and Lind & Morrison (1974). The zone is characterized overall by low, widely spaced woody vegetation, usually *Acacia* and *Commiphora* species and (50–90 per cent) annual grasses, largely *Chrysopogon* and *Aristida* species. The annual rainfall (often less than 250 mm) falls in the spring (April–May) and autumn (October–November) for the most part with four to six intervening dry months.

The floristic affinities of the area around the Lake Turkana basin, including the Omo valley, are predominately East African (in fact the highland plants of southern Ethiopia are of East African rather than north Ethiopian affinities). In the lower Omo valley, where Carr (1976) carried out extensive studies of natural communities, floristic variability is quite low (Carr has recorded 193 genera and 321 species of 76 families). The principal prevailing vegetation classifications of the area are xerophyllous *Acacia* savanna, savanna and xerophile scrub, wooded steppe with *Acacia* and *Commiphora*, subdesert steppe, and Sudanese park steppe. However, Carr has stressed the markedly mosaic pattern of the plant communities there. Mesic environments, including riverine woodland and forest are well developed in association with the perennial Omo river, and to a lesser extent with some seasonal streams.

The natural communities around the eastern and western margins of the Rudolf basin remain, so far as we are aware from literature and varied individual experiences over this area, largely unstudied. However, major differences from the lower Omo basin situation are hardly to be expected other than those which might be associated primarily with the perennial Omo river. Furthermore, in the past, one or possibly several perennial rivers and other seasonal streams existed to the east of the lake at the time in question (Vondra & Bowen 1976). Moreover, the initial results of palynological research in that part of the basin (by R. Bonnefille, pers. comm.) suggest wooded grassland, and hence a less xeric environment than that of the present during a part of Upper Koobi Fora Member times.

From the descriptions presented above, we conclude that vegetational differences between the two areas are not marked, and were no more different in the past necessarily.

Any argument based on biostratigraphy and taxonomic diversity should, if possible, have a basis in modern vertebrate distributions and local occurrences. Throughout the broad vegetational zone described above, the mammalian fauna is distinctive (cf. Grimwood, 1963; Bigalke 1972). The zone extends from the horn of Africa through southern Ethiopia, northern Kenya (toward the Uaso Nyiro and Tana rivers), the southern Sudan, and northern Uganda. Its southern boundary in Kenya is sometimes referred to as Sclater's line. The north Lake Turkana area is one of the more remote in Africa, and its fauna is unfortunately very poorly documented even now. However some information relevant to Mammalia, the class primarily important in late Cenozoic biostratigraphic analyses, is available, particularly from the lower Omo basin and, to a lesser extent, from the area north-east of the present lake. For the former there is very inadequate data on 4 mammalian groups (Chiroptera, Insectivora, Hyracoidea, and lesser carnivores), and good to excellent data on another 12. Duff-MacKay (1967), Houin & Rhodain (1967–73), and Hubert (1973) have all kindly communicated the results of their findings to us; earlier records are afforded by Old-field Thomas (1900), Neumann (1901, 1902) and Beaux (1943). Large mammal records are based on these sources, Rhoads (1897), Urban & Brown (1968), and on repetitive, confirmed sightings by researchers associated with the Omo Research Expedition between 1966–74. For the area east of the present lake data for the poorly known groups is equally bad, however some data are available for the other 12 groups. See Stewart & Stewart (1963) and Stewart (1963) for distributional data east of Lake Rudolf. Further data has been obtained from collections made for taphonomic purposes (Univ. California, Berkeley) and the sightings of various members of the East Rudolf Research Project. Potential expected distributions may be inferred from maps and comments in Kingdon (1971, 1974). As the taxa of primary interest to us are large mammals, we confine our remarks to these.

Known occurrences of large mammals were tabulated by species, and the species combined into families for both the lower Omo valley, and the East Lake Turkana area. These data are presented in Table 2; the third column in Table 2 records the number of species in common between the two areas for each family.

It is apparent from the data presented in Table 2, based on admittedly incomplete observations, that mammalian species occurrence and distributional continuity is demonstrably very substantial over the North Lake Turkana area. With adequate documentation the occasional lack of overlap might well prove to be even more negligible than it now appears. At any rate the congruence in mammalian distribution is overall of a very high order.

TABLE 2: *Documented present occurrences of large mammals in the North Lake Turkana basin.*

Family	Omo Valley	East Rudolf	Common
Orycteropodidae	1	1	1
Manidae	1	1	1
Leporidae	1	1	1
Cercopithecidae	5	1	1
Elephantidae	1	1*	1
Equidae	2	2	2
Rhinocerotidae	1	1	1
Hippopotamidae	1	1	1
Suidae	2	1	1
Giraffidae	1	1	1
Bovidae	13	12	11
Canidae	5	3	3
Hyaenidae	3	3	3
Felidae	5	4	4
Totals	42	33	32

* No longer occurs east of the lake, but known to have been exterminated there prior to the turn of the century. They occur just north of the Nkalabong Range and northward to the Mwi River area.

It might be argued that small mammal species reflect environmental differences more sensitively. Unfortunately, groups such as insectivores, bats, rodents and lesser carnivores are often less well known. However, 2 chiropteran and 2 shrew species documented in the Omo would be expected on distributional evidence to occur also east of the lake; the same is true in the case of most of the 4 species of lesser carnivores documented in the Omo. At least 20 species of 7 families of rodents are documented in the recent Omo fauna, only 3 of which seem to be restricted to the present riverine forest area. Of these 20 species 1 is already known east of the lake, and practically all of the remainder would be expected to occur there on distributional grounds (although there are as yet no trapping records available from that area).

Because of the congruence in distribution of these modern mammals, we feel that the biostratigraphic arguments in the preceding section rest on a fairly firm foundation. Before we can abandon these arguments, we must find unequivocal evidence that environments in the past differed markedly in these two areas. Some have argued, *in camera*, that such evidence exists in the lack of congruence between the faunas of the *Mesochoerus limnetes* zone at East Lake Turkana and Member C of the Shungura Formation. If the date of 2.61 m.y. for the 'KBS Tuff' is correct, these two faunas should show strong similarities, and as shown above, they do not. However, if one assigns dates of less than 2 m.y. to these tuffs, as proposed above, and then correlates the *M. limnetes* Zone fauna with that of upper Member F or lower Member G, the lack of congruence between the faunas of the two areas disappears and so does the basis for the argument that the environments of the two areas differed greatly. Why assume that past environments were different without hard evidence to support such an assumption?

One further point should be made in regard to the 'ecological question'. The fossil assemblages with which we are dealing in correlation of these two areas do not represent the animals living in either area at a particular point in time. Rather they represent a collection of remains preserved over a period of several tens of thousands of years. Walker & Bambach (1971) in writing about marine communities in the fossil record have observed that 'recent communities appear enormously complex because of fluctuations in population size, population patchiness and other factors. Despite complexities in instantaneous communities, repetitive fossil communities indicate that environments operate to produce predictable "equilibrium" associations or time-averaged "climax" assemblages which display regular trophic structure.' We see no reason that this concept of time-averaged assemblages cannot be extended to mammalian faunas. Surely there were climatic fluctuations in the past which would have affected vegetational and faunal distributions, but we feel that since we are in effect averaging each faunal assemblage over a considerable period of time, these fluctuations would tend to average out. The resulting assemblages will represent not one, but many environments which existed during the time the sediments accumulated.

Speculations on correlation of lower parts of the East Lake Turkana section

If the correlations proposed in the preceding section are correct for the upper part of the section, we see immediately that sedimentation in the lower Omo valley occurred roughly in proportion to that at East Lake Turkana, albeit more rapidly. The proportionality factor is roughly 4:1 to 2:1, i.e. for every metre of net accumulation of sediment at East Lake Turkana, two to four metres of sediment accumulated in the lower Omo valley.

Vondra & Bowen (1976) and Fitch & Vondra (1976) postulate that the principal source of East Lake Turkana sediments was a perennial stream (or streams) draining through what is now the Stephanie basin with headwaters still farther to the north. This drainage basin has an area of about 22 000 km², whereas that of the Omo river is about 73 000 km² (Butzer 1971). This difference in catchment area might be the explanation for the differing sedimentation rates. If the basins were being denuded at approximately the same rate, three times as much material would be available to the ancient Omo river, everything else being equal. However, in reality the depositional plain of the lower Omo valley is somewhat larger than that of the plain lying east of Lake Turkana. On the other hand the Omo river drains a considerably larger area of high mountains where weathering and erosion would presumably have been more rapid.

The sedimentation rates as computed between various pairs of magnetic boundaries in the Shungura Formation are tabulated in Table 3. Sedimentation rates computed between pairs of tuffs, assuming that our correlations are correct, are tabulated for the East Lake Turkana sequence in Table 4.

TABLE 3: *Sedimentation rates for the Shungura Formation*

Boundaries	Thickness (m)	Approximate Duration (years)	Rate (cm/1000 yr)
Bottom Mammoth–Top Kaena	100	240 000	41
Top Kaena–Top Gauss	132	400 000	33
Top Gauss–Bottom Reunion Events	104	350 000	30
Bottom Reunion Events–Top Reunion Events	146	150 000	91
Top Reunion–Bottom Olduvai	77	50 000	154
Bottom Olduvai–Top Olduvai	82	200 000	41
Top Olduvai–Tuff L	95	350 000	27
Bottom Jaramillo–Top Jaramillo	26	60 000	43

TABLE 4: *Sedimentation rates for the Koobi Fora Formation*

Boundaries	Thickness (m)	Approximate Duration (years)	Rate (cm/1000 yr)
Chari Tuff–Koobi Fora Tuff	36	350 000	10
Koobi Fora Tuff–KBS Tuff	30	240 000	13
KBS Tuff–Bottom of lacustrine sediments below KBS Tuff	34	60 000	57
Tulu Bor Tuff–Suregei Tuff	50	1 200 000	4

Sedimentation rates for the fluvial parts of the Shungura Formation fall generally in the range from 30–40 cm/1000 years. The rate calculated for the lacustrine part of the section in Member G is distinctly higher. The same relation holds between fluvial and lacustrine rates of deposition in the Koobi Fora Formation, although rates there are about one-third those of the Shungura Formation.

Only one value in Tables 3 and 4 appears clearly anomalous. This is the value of 4 cm/1000 years computed for the sediments between the Tulu Bor Tuff and the Suregei Tuff. There are several possible reasons for this discrepancy:

1. The proposed correlation of the lacustrine sediments below the Suregei Tuff with those of the Basal Member, Shungura Formation is incorrect. In this case we would have to assume that the Suregei Tuff should be correlated with some higher part of the Shungura Formation which fails to record a *major* lacustrine sequence.

2. There is a hiatus in the sedimentary sequence above the Tulu Bor Tuff prior to deposition of the lacustrine sediments above it. Such a hiatus has not to our knowledge been proposed.

3. The correlation is correct, but there are hiatuses in the section between the Tulu Bor and Suregei Tuffs. Again we know of no disconformities having been proposed.

If one uses the sedimentation rates arrived at for the upper part of the East Lake Turkana sequence (10–13 cm/1000 years) to predict the age of the Suregei Tuff, one finds that it should be about 380 000 to 500 000 years older than the base of the lacustrine sediments which lie above the Tulu Bor Tuff. If the base of these sediments is assumed to be the same age as the base of the lacustrine sediments in upper Member G, Shungura Formation (ca. 1.9 m.y.), then an age of about 2.4 m.y. would be derived for the Suregei Tuff. It is perhaps not coincidental that a *minor* lacustrine sequence appears in upper Member C of the Shungura Formation (de Heinzelin, Haesaerts & Howell 1976), which is about 2.5 m.y. in age. The possibility of correlation of this lacustrine sequence with the lacustrine sequence below the Suregei Tuff might bear consideration when more data are available on radiometric ages, magnetostratigraphy, or tuff chemistry from the lower part of the East Lake Turkana sequence. Magnetic polarity data known at this time do not make such a correlation impossible.

If sedimentation rates of about 10 cm/1000 years prevailed during deposition of the Kubi Algi Formation, the maximum age of this formation would be about 3.3 m.y. This is not inconsistent with the date of 3.8 m.y. on the basalt near the base of the Kubi Algi Formation discussed above.

We wish to stress that the arguments in this section of the paper are highly speculative, and may have to be modified when further data regarding the radiometric age, magnetostratigraphy, and tuff chemistry of the lower portion of the East Lake Turkana sequence have been obtained.

Conclusions

Our conclusions are based on data available to us up to mid-1975

1. Ages of about 1.35 m.y. for the Chari and Karari Tuffs and 1.57 m.y. for the Koobi Fora Tuff are consistent with both geological and faunal data.

2. The age of 2.61 m.y. assigned to the 'KBS Tuff' by Fitch & Miller (1976) is inconsistent with other geological data and with the faunal data. The age of about 1.60 m.y. for the younger 'KBS Tuff' and the age of about 1.82 m.y. for the older 'KBS Tuff' proposed by Curtis *et al.* (1975) are more consistent with other geological data and the faunal data.

3. The maximum apparent age of the Tulu Bor Tuff is 3.18 m.y., and its correct geological age may be considerably less than this.

4. The age of the Suregei Tuff Complex may be either about 2.5 m.y. or just greater than 3.0 m.y. depending on which of the possible interpretations of the magnetostratigraphy, sedimentology, and sedimentation rates is correct.

5. The maximum age of the sedimentary sequence at East Lake Turkana can be taken as 3.8 m.y. If sedimentation rates were relatively constant throughout the time of deposition of this sequence, the maximum age may be only about 3.3 m.y.

6. Any fauna collected from the Guomde Formation must be less than 900 000 years in age.

7. The fauna of the *Loxodonta africana* Zone at East Lake Turkana is 1.0 to 1.5 m.y. in age.

8. The fauna of the *Metridiochoerus andrewsi* zone at East Lake Turkana lies between 1.5 and 1.8 m.y. in age.

9. The fauna assigned to the *Mesochoerus limnetes* Zone at East Lake Turkana lies between 1.8 and 2.0 m.y. in age.

10. Environments in the East Lake Turkana area and the lower Omo valley, while distinct, were not so different that mammalian biostratigraphy should fail to provide reasonable data for correlation of sediments deposited in these two areas in the past.

References

BEAUX, O. DE. 1943. Mammalia. Missione Biologica Sagan-Omo, vol. 7. Zoologia 1. Mammalia-Aves-Reptilia-Amphibia-Pisces. *Reale Accademia d'Italia, Centro Studi per l'Africa Orientale Italiana*, no. 6, Roma.

BEDEN, M. 1976. Proboscideans from Omo Group formations. In: *Earliest Man and Environments in the Lake Rudolf Basin: Stratigraphy, Paleoecology and Evolution.* Y. Coppens, F. C. Howell, G. Ll. Isaac, and R. E. F. Leakey, eds, University of Chicago Press, Chicago.

BEHRENSMEYER, A. K. 1970. Preliminary geological interpretation of a new hominid site in the Lake Rudolf basin. *Nature, Lond.* **226**, 225–6.

—— 1974. Late Cenozoic sedimentation in the Lake Rudolf basin, Kenya. *Annals of the Geological Survey of Egypt*, **4**, 387–406.

—— 1976. Correlation of Plio-Pleistocene sequences of the northern Lake Turkana basin: a summary of evidence and issues. In: *Geological Background to Fossil Man.* W. W. Bishop, ed., Geological Society of London, London.

BIGALKE, R. C. 1972. The contemporary mammal fauna of Africa. In: *Evolution, Mammals and Southern Continents.* A. Keast, F. C. Erk and B. Glass, eds, State University of New York Press, Albany, 141–94.

BOWEN, B. E. 1974. The geology of the Upper Cenozoic sediments in the East Rudolf embayment of the Lake Rudolf basin, Kenya. Ph.D. Dissertation, Iowa State University, Ames.

—— & VONDRA, C. F. 1973. Stratigraphical relationships of the Plio-Pleistocene deposits, East Rudolf, Kenya. *Nature, Lond.* **242**, 391–3.

BROCK, A., HILLHOUSE, J., COX, A. & NDOMBI, J. 1974. The paleomagnetism of the Koobi Fora Formation, Lake Rudolf, Kenya. *Transactions of the American Geophysical Union*, **56**, 1109 (Abstract GP 15).

—— & ISAAC, G. LL. 1974. Paleomagnetic stratigraphy and chronology of hominid-bearing sediments east of Lake Rudolf, Kenya. *Nature, Lond.* **247**, 344–8.

—— & ISAAC, G. LL. 1976. Reversal stratigraphy and its application at East Rudolf. In: *Earliest Man and Environments in the Lake Rudolf Basin: Stratigraphy, Paleoecology and Evolution.* Y. Coppens, F. C. Howell, G. Ll. Isaac and R. E. F. Leakey, eds, University of Chicago Press, Chicago.

BROWN, F. H. 1972. Radiometric dating of sedimentary formations in the lower Omo Valley, southern Ethiopia. In: *Calibration of Hominoid Evolution*. W. W. Bishop and J. A. Miller, eds, Scottish Academic Press, Edinburgh, 273–87.

—— & CARMICHAEL, I. S. E. 1969. Quaternary volcanoes of the Lake Rudolf region. 1. The basanite-tephrite series of the Korath range. *Lithos*, **2**, 239–60.

—— & LAJOIE, K. R. 1971. Radiometric age determinations on Pliocene/Pleistocene formations in the lower Omo basin, southern Ethiopia. *Nature, Lond.* **229**, 483–5.

—— & NASH, W. P. 1976. Radiometric dating and tuff mineralogy of Omo Group deposits. In: *Earliest Man and Environments in the Lake Rudolf Basin: Stratigraphy, Paleoecology and Evolution*. Y. Coppens, F. C. Howell, G. Ll. Isaac and R. E. F. Leakey, eds, University of Chicago Press, Chicago.

—— & SHUEY, R. T. 1976. Magnetostratigraphy of the Shungura and Usno Formations, lower Omo valley, Ethiopia. In: *Earliest Man and Environments in the Lake Rudolf Basin: Stratigraphy, Paleoecology and Evolution*. Y. Coppens, F. C. Howell, G. Ll. Isaac and R. E. F. Leakey, eds, University of Chicago Press, Chicago.

BUTZER, K. W. 1969. Geological interpretation of two Pleistocene hominid sites in the lower Omo basin. *Nature, Lond.* **222**, 1138–40.

—— 1971. Recent history of an Ethiopian delta. Research Papers, Department of Geography, University of Chicago, Chicago, no. 136.

—— 1976. The Mursi, Nkalabong and Kibish Formations, lower Omo basin (Ethiopia). In: *Earliest Man and Environments in the Lake Rudolf Basin: Stratigraphy, Paleoecology and Evolution*. Y. Coppens, F. C. Howell, G. Ll. Isaac and R. E. F. Leakey, eds, University of Chicago Press, Chicago.

——, BROWN, F. H. & THURBER, D. L. 1969. Horizontal sediments of the lower Omo valley: the Kibish Formation. *Quaternaria*, **11**, 15–29.

——, ISAAC, G. LL., RICHARDSON, J. L. & WASHBOURN-KAMAU, C. 1972. Radio-carbon dating of East African lake levels. *Science*, **175**, 1069–76.

—— & THURBER, D. L. 1969. Some late Cenozoic sedimentary formations of the lower Omo basin. *Nature, Lond.* **222**, 1132–7.

CARR, C. J. 1976. Natural communities and their diversity in the lower Omo basin. In: *Earliest Man and Environments in the Lake Rudolf Basin: Stratigraphy, Paleoecology and Evolution*. Y. Coppens, F. C. Howell, G. Ll. Isaac and R. E. F. Leakey, eds, University of Chicago Press, Chicago.

CERLING, T. E. 1975. Oxygen-isotope studies of the East Rudolf volcanoclastics. In: *Earliest Man and Environments in the Lake Rudolf Basin: Stratigraphy, Paleoecology and Evolution*. Y. Coppens, F. C. Howell, G. Ll. Isaac and R. E. F. Leakey, eds, University of Chicago Press, Chicago.

COOKE, H. B. S. 1976. Suidae from Pliocene/Pleistocene successions of the Rudolf basin. In: *Earliest Man and Environments in the Lake Rudolf Basin: Stratigraphy, Paleoecology and Evolution*. Y. Coppens, F. C. Howell, G. Ll. Isaac and R. E. F. Leakey, eds, University of Chicago Press, Chicago.

—— & MAGLIO, V. J. 1972a. Plio/Pleistocene stratigraphy in eastern Africa in relation to proboscidean and suid evolution. In: *Calibration of Hominoid Evolution*. W. W. Bishop and J. A. Miller, eds, Scottish Academic Press, Edinburgh, 303–29.

—— & MAGLIO, V. J. 1972b. Recent Pliocene-Pleistocene discoveries in East Africa. *Palaeoecology of Africa, the Surrounding Islands and Antarctica*, **6**, 163–7.

CORYNDON, S. C. 1976. Fossil Hippopotamidae from Pliocene-Pleistocene successions of the Rudolf basin. In: *Earliest Man and Environments in the Lake Rudolf Basin: Stratigraphy, Paleoecology and Evolution*. Y. Coppens, F. C. Howell, G. Ll. Isaac and R. E. F. Leakey, eds, University of Chicago Press, Chicago.

COX, A. 1969. Geomagnetic reversals. *Science*, **163**, 237–45.

CURTIS, G. H., DRAKE, R. L., CERLING, T. & HAMPEL, J. 1975. Age of the KBS Tuff in the Koobi Fora Formation, East Rudolf, Kenya. *Nature, Lond.* **258**, 395–8.

—— & HAY, R. L. 1972. Further geological studies and potassium-argon dating at Olduvai Gorge and Ngorongoro Crater. In: *Calibration of Hominoid Evolution*. W. W. Bishop and J. A. Miller, eds, Scottish Academic Press, Edinburgh, 289–301.

DAVIS, D. H. 1962. Distribution patterns of southern African Muridae, with notes on some of their fossil antecedents. *Ann. Cape Prov. Mus.* **2**, 56–76.

EISENMANN, V. 1976. Equidae from the Shungura Formation. In: *Earliest Man and Environments in the Lake Rudolf Basin: Stratigraphy, Paleoecology and Evolution*. Y. Coppens, F. C. Howell, G. Ll. Isaac and R. E. F. Leakey, eds, University of Chicago Press, Chicago.

FINDLATER, I. C. 1976. Tuffs and the recognition of isochronous mapping units in the Rudolf succession. In: *Earliest Man and Environments in the Lake Rudolf Basin: Stratigraphy, Paleoecology and Evolution*. Y. Coppens, F. C. Howell, G. Ll. Isaac and R. E. F. Leakey, eds, University of Chicago Press, Chicago.

FITCH, F. J., FINDLATER, I. C., WATKINS, R. T. & MILLER, J. A. 1974. Dating of the rock succession containing fossil hominids at East Rudolf, Kenya. *Nature, Lond.* **251**, 213–15.

—— & MILLER, J. A. 1970. Radioisotopic age determinations of Lake Rudolf artefact site. *Nature, Lond.* **226**, 226–8.

—— & MILLER, J. A. 1976. Conventional Potassium-Argon and Argon–40/Argon–39 dating of volcanic rocks from East Rudolf. In: *Earliest Man and Environments in the Lake Rudolf Basin: Stratigraphy, Paleoecology and Evolution*. Y. Coppens, F. C. Howell, G. Ll. Isaac and R. E. F. Leakey, eds, University of Chicago Press, Chicago.

—— & VONDRA, C. F. 1976. Tectonic and stratigraphic framework. In: *Earliest Man and Environments in the Lake Rudolf Basin: Stratigraphy, Paleoecology and Evolution*. Y. Coppens, F. C. Howell, G. Ll. Isaac and R. E. F. Leakey, eds, University of Chicago Press, Chicago.

GAUTIER, A. 1976. Assemblages of fossil freshwater mollusks from the Omo Group and related deposits in the Lake Rudolf basin. In: *Earliest Man and Environments in the Lake Rudolf Basin: Stratigraphy, Paleoecology and Evolution*. Y. Coppens, F. C. Howell, G. Ll. Isaac and R. E. F. Leakey, eds, University of Chicago Press, Chicago.

GENTRY, A. W. 1976. Bovidae of the Omo Group deposits. In: *Earliest Man and Environments in the Lake Rudolf Basin: Stratigraphy, Paleoecology and Evolution*. Y. Coppens, F. C. Howell, G. Ll. Isaac and R. E. F. Leakey, eds, University of Chicago Press, Chicago.

—— 1977. Order Artiodactyla. In: *Mammalian Evolution in Africa*. H. B. S. Cooke & V. J. Maglio, eds, Harvard University Press, Cambridge, Mass.

GRIMWOOD, I. R. 1963. The fauna and flora of East Africa. In: *Conservation of Nature and Natural Resources in Modern African States*. IUCN Publications New Series, no. 1, Morges, Switzerland, 179–88.

HARRIS, J. M. 1976. Rhinocerotidae from the East Rudolf succession. In: *Earliest Man and Environments in the Lake Rudolf Basin: Stratigraphy, Paleoecology and Evolution*. Y. Coppens, F. C. Howell, G. Ll. Isaac and R. E. F. Leakey, eds, University of Chicago Press, Chicago.

DE HEINZELIN, J. 1971. Observations sur la formation de Shungura (Vallée de l'Omo, Ethiopie). *C.R.A.S.*, Paris, **272**, 2409–11.

—— & BROWN, F. H. 1969. Some early Pleistocene deposits of the lower Omo valley: the Usno Formation. *Quaternaria*, **11**, 31–46.

——, —— & HOWELL, F. C. 1970. Pliocene/Pleistocene formations in the lower Omo basin, southern Ethiopia. *Quaternaria*, **13**, 247–68.

——, HAESAERTS, P. & HOWELL, F. C. 1976. Plio-Pleistocene formations of the lower Omo basin, with particular reference to the Shungura Formation. In: *Earliest Man and Environments in the Lake Rudolf Basin: Stratigraphy, Paleoecology and Evolution*. Y. Coppens, F. C. Howell, G. Ll. Isaac and R. E. F. Leakey, eds, University of Chicago Press, Chicago.

HOOIJER, D. A. 1976. Evolution of the Perissodactyla of the Omo Group deposits. In: *Earliest Man and Environments in the Lake Rudolf Basin: Stratigraphy, Paleoecology and Evolution*. Y. Coppens, F. C. Howell, G. Ll. Isaac and R. E. F. Leakey, eds, University of Chicago Press, Chicago.

ISAAC, G. LL. 1976. Introduction to Part I. Geology and Geochronology. A commentary on discussion at the symposium. In: *Earliest Man and Environments in the Lake Rudolf Basin: Stratigraphy, Paleoecology and Evolution*. Y. Coppens, F. C. Howell, G. Ll. Isaac and R. E. F. Leakey, eds, University of Chicago Press, Chicago.

KEAY, R. W. J. 1959. *Vegetation Map of Africa South of the Tropic of Cancer*. Oxford University Press, London.

KINGDON, J. 1971, 1974. *East African Mammals. An Atlas of Evolution in Africa*, vol. 1, 11A, 11B. Academic Press, New York, London.

LEAKEY, R. E. 1973a. Evidence for an advanced Plio-Pleistocene hominid from East Rudolf, Kenya. *Nature, Lond.* **242**, 447–50.

—— 1973b. Skull 1470. *National Geographic Magazine,* **143**, 818–29.

LIND, E. M. & MORRISON, M. E. S. 1974. *East African Vegetation.* Longman, London.

MAGLIO, V. J. 1970. Early Elephantidae of Africa and a tentative correlation of African Plio-Pleistocene deposits. *Nature, Lond.* **225**, 328–32.

—— 1972. Vertebrate faunas and chronology of hominid-bearing sediments east of Lake Rudolf, Kenya. *Nature, Lond.* **239**, 379–85.

—— 1973. Origin and evolution of the Elephantidae. *Trans. Am. Philos. Soc.,* n.s. **63**, 1–149.

MOREAU, R. E. 1952. Africa since the Mesozoic: with particular reference to certain biological problems. *Proc. Zool. Soc. London,* **121**, 869–913.

NEUMANN, O. 1901. Uber Hyraciden. *Sitzungsber d. Gesellschaft naturforschender Freunde zu Berlin,* **9**, 238–44.

—— 1902. From the Somali coast through southern Ethiopia to the Sudan. *Geog. J.* **19**, 1–29.

RATTRAY, J. M. 1961. *The Grass Cover of Africa.* F.A.O., Rome.

RHOADS, S. N. 1897. Mammals collected by Dr A. Donaldson Smith during his expedition to Lake Rudolf, Africa. *Proc. Acad. Nat. Sci., Philadelphia,* 1896, 517–46.

SHUEY, R. T., BROWN, F. H. & CROES, M. K. 1974. Magnetostratigraphy of the Shungura Formation, south-western Ethiopia: fine structure of the lower Matuyama polarity epoch. *Earth and Planetary Science Letters,* **23**, 249–60.

——, BROWN, F. H., ECK, G. G. & HOWELL, F. C. 1976. A statistical approach to temporal biostratigraphy. In: *Geological Background to Fossil Man.* W. W. Bishop, ed., Geological Society of London, London.

STEWART, D. R. M. 1963. Wildlife census—Lake Rudolf. *East African Wildlife Journal,* **1**, 121.

—— & J. STEWART. 1963. The distribution of some large mammals in Kenya. *Journal of East Africa Natural History Society and Coryndon Museum,* **24**, 1–52.

THOMAS, O. 1897–8. On the mammals collected during Captain Bottego's last expedition to Lake Rudolf and the upper Sobat. *Annali del Museo Civico de Storia Naturale di Genova,* series 2, no. 18, 676–9.

TRAPNELL, C. G. & LANGDALE BROWN, I. 1962. The natural vegetation of East Africa. In: *The Natural Resources of Africa.* E. W. Russell, ed. East African Literature Bureau, Nairobi, Kenya, 92–102.

URBAN, E. K. & BROWN, L. 1968. Wildlife in an Ethiopian valley. *Oryx,* **9**, 342–53.

VAN DAMME, D. & GAUTIER, A. 1971. Molluscan assemblages from the later Cenozoic of the lower Omo basin, Ethiopia. *Quaternary Research,* **2**, 25–37.

VONDRA, C. F. & BOWEN, B. E. 1976. Plio-Pleistocene deposits and environments, East Rudolf, Kenya. In: *Earliest Man and Environments in the Lake Rudolf Basin: Stratigraphy, Paleoecology and Evolution.* Y. Coppens, F. C. Howell, G. Ll. Isaac and R. E. F. Leakey, eds, University of Chicago Press, Chicago.

——, JOHNSON, G. D., BOWEN, B. E. & BEHRENSMEYER, A. K. 1971. Preliminary stratigraphical studies of the East Rudolf basin, Kenya. *Nature, Lond.* **231**, 245–8.

WALKER, K. R. & BAMBACH, R. K. 1971. The significance of fossil assemblages from fine grained sediments; time-averaged communities. *Geol. Soc. Amer.* Abstracts with Programs (1971).

31

YVES COPPENS[1]

Evolution of the hominids and of their environment during the Plio-Pleistocene in the lower Omo Valley, Ethiopia

In this paper I have concentrated on the lower Omo valley in Ethiopia, to the north of the Lake Turkana Basin. I have extended the study to include not only mammals, but all aspects of palaeontology related to the reconstruction of palaeoenvironments.

Geologically, the Omo Group is composed of four formations: The Mursi Formation, 4·5 to 4 millions years old: N'Kalabong Formation and Shungura Formation, including the Usno Formation, 3·5 to 1 million years old, and the much younger, Kibish Formation, possibly 100 000 years old. The most important is the Shungura Formation, which is divided into members by tuffs widely distributed throughout the area.

The group of formations is over 1000 metres thick, well exposed, fossiliferous and so well dated by radiometry, magnetostratigraphy and biostratigraphy, that it has developed into a type of standard scale. Unfortunately, the preservation of fossil vertebrates is not good as most of them were transported by streams. This makes palaeoecological studies more difficult.

Nevertheless, for its length and continuity of sequence, the amount of fossils collected and the range of dates obtained, the Omo succession is unique. As several Hominids are present in these strata as well as artifacts, it is particularly interesting to see how the environment changed during the period when the Hominids were evolving.

Geology

The Mursi Formation, 143 m thick, is composed of deltaic and fluvio-lacustrine deposits. The Shungura Formation consists of a long series of fluviatile cyclic units, indicating flood plain conditions varying from sandy channels to strong current flow with changing stream courses. These sedimentary units are usually capped by immature soil profiles corresponding to short periods of emersion and are interstratified with increments of volcanic ash, deposited either in water (e.g. tuff A or tuff C), or on emerged land surfaces (e.g. tuff E).

A tendency to prodeltaic conditions occurs in the upper part of member F but, in member G 11 and later, the rhythmic character of the sedimentation disappears and the conditions become estuarine and lacustrine.

Qualitative palaeontology of large vertebrates

Perissodactyla

Equids: From the base of the sequence to the top, there is an increase in

[1] Various sections of this paper owe a great deal to the work of my colleagues whose contributions are noted in the Acknowledgement.

hypsodonty of the incisors and of the cheek teeth of three species of *Hipparion*. These are successively *Hipparion turkanense*, *Hipparion cf. albertense*, *Hipparion aethiopicum*. Through the sequence the development of an extra-stylid, the ectostylid, on the lower cheek teeth can also be observed. True *Equus* appears in member G together with *Hipparion aethiopicum*. These developments are interpreted as indicating a trend towards more grass in the diet.

Rhinocerotids: Two genera of rhinoceroses, *Ceratotherium praecox* (developing into *C. simum*) and *Diceros bicornis*, both exhibit, like the Equids, a trend towards higher crowned teeth.

Artiodactyla

Bovids are numerous with 41 species representing 22 genera and 8 tribes. The abundance of Tragelaphini, Reduncini, Bovini in early levels and the great number of *Aepyceros* (nearly one-third of the total number of Bovids) suggest a savanna with scrub, acacias and nearby water. Above tuff G, Alcelaphini like *Megalotragus*, *Beatragus* and *Parmularius* arrive in the spectrum with *Tragelaphus strepsiceros* and *Antidorcas recki* (Antilopini) which are all adapted to open grassland.

Suids. The general tendency of almost all the Suids is to evolve increasingly reduced premolars and longer, more hypsodont molars with more pillars. This can be observed in the *Mesochoerus limnetes—Mesochoerus olduvaiensis* lineage, the *Notochoerus euilus—Notochoerus scotti* lineage and the *Metridiochoerus jacksoni-Metridiochoerus andrewsi* lineage. The evolution of the length of the last molar accelerates above tuff G. It is also good evidence of change to more grass in the diet. *Stylochoerus*, which appears to live exclusively in open savanna, has not been found below member H.

Hippopotamids. Hippopotamus protamphibius which is hexaprotodont up to tuff C and then tetraprotodont, is a middle size species with relatively slender limb bones, more adapted to walking on land than in mud. In contrast, in member G, the enormous tetraprotodont *Hippopotamus gorgops* with high crowned molar teeth, small premolars and short and robust limb bones, appears to be adapted to a lacustrine environment. Together with *Hippopotamus gorgops*, a very rare new pygmy Hippo, *Hippopotamus aethiopicus*, occurs. It has lophodont teeth and slender limb bones, and was possibly riverine rather than lacustrine. The Hippos also appear to suggest development of lacustrine conditions and of grassland since tuff G.

Proboscidea

Seven species of Proboscidians occur in the Omo beds, the Gomphothere, *Anancus kenyensis* was only found in strata older than 4 million years: the Stegodont, *Stegodon kaisensis*, below tuff C; *Loxodonta adaurora* possibly below tuff E; and two other *Loxodonta* represented by 3 tooth fragments. The Deinothere, *Deinotherium bozasi*, was found everywhere, and also the genus *Elephas* (*Elephas recki*). Two of these seven species, *Deinotherium bozasi* and *Elephas recki*, represent 98 per cent of the specimens.

Elephas is interesting because it is seen to evolve from the basal member to member L; the teeth become bigger and longer, the number of plates and the quantity of cement increases. The enamel thickness decreases and its folding becomes more complex. All these changes take place at different rates and allow four chronosubspecies to be recognized. *Elephas recki* I occurs in members A, B and possibly C; *Elephas recki* II A in members A, B, C, D, E and possibly F; *Elephas recki* II B in members F, G and possibly H and *Elephas recki* III in H, J, K and L.

These evolutionary trends lead to increase in the volume of the tooth by increasing its size and its height; increase in the number of plates by reduction of the thickness of the enamel, which in turn increases its surface by infolding the enamel; increase in amount of cement. They all indicate adaptation of the diet to more and more grass. The acceleration of the process starts in member F.

Primates

Cercopithecoids are very numerous but difficult to assign to definite taxonomic position. Very interesting preliminary work done by G. Eck distinguishes four Colobines and five Cercopithecines. Some of these nine Primates are very characteristic of the older levels (*Colobine* A. *Papionini* A); another one (*Colobus* sp.) is characteristic of recent levels. The six others were found almost everywhere.

It is too early to explain their distribution. However, it is interesting to note the clear decrease of Cercopithecines in G while Colobines were slowly increasing. At this time they are represented by smaller-sized animals, close to the living ones. This may be an indication of less precipitation and lower humidity.

Carnivora

Carnivore fossils are less good indicators of ecology than many herbivores and specimens are too few to reveal evolutionary trends. However, they seem to indicate woodland in lower members of the Shungura Formation (*Helogale* sp. from member B, for instance, is close to *Helogale varia* of the equatorial forest) and possibly a landscape with less shade in medium and upper members (*Helogale* sp. from member F, for instance, is closer to *Helogale hirtula* or to *Helogale parvula* of arid or semi arid areas).

Quantitative palaeontology of large vertebrates

Following the discovery of *Zinjanthropus*, by Mary Leakey, and the announcement of its geological age by Louis Leakey as 1.75 m.y., palaeontologists all over the world, understood for the first time that man had really lived in East Africa and was very ancient there. There was then a 'Hominid rush' and like locusts, palaeontologists began to arrive in Tanzania, Kenya and Ethiopia. Over the last 10 years, a total of 200 scientists have been investigating the 2000 kilometres of the Rift Valley from the Serengeti plains to the Red Sea.

This has resulted in the collection of about 200 000 specimens. It is obvious that such a figure must lead to a new style of research in laboratories, and that it now is possible to apply quantitative palaeontological techniques. Instead of using a limited number of specimens, we tried to investigate the frequencies of groups of animals relative to each other using a computer programme.

For instance the ratio between the number of specimens of a particular **group** from several strata and the total number of specimens collected in these levels shows an increasing number of Bovids in G while Primates, Hippos, Proboscidians and Suids all decrease.

Microvertebrate palaeontology

Micromammals, and especially rodents, are very good indicators of climate and environment because they are adapted to specific ecological niches. Unfortunately the geological levels which contain microvertebrates are rare and Jaeger and Wesselmann had to screen and wash 30 tons of sediment to collect 38 species of small mammals (23 Rodents, 5 Chiroptera, 3 Insectivora, 1 Lagomorpha, 1 Carnivora (Viverrid), 3 Primates (Lorisid) and 2 Hyracoids). Two major sites were discovered, one in member B (B 10, locality 1) and one in member F (F 1-2, locality 28).

Comparison of the two collections reveals surprising differences. In B 10, six genera of Murids are present and among them 85 specimens of *Mastomys*, which lives in tall grass or thick bush near water, and 13 specimens of *Pelomys* which also lives in tall grass. In F1, only three genera of Murids are present and 5 specimens of *Mastomys* while *Pelomys* is absent. However 18 specimens of *Aethomys* and 6 of *Thallomys* occur which are inhabitants of dry or open savanna. The Lorisidae, the Chiroptera *Eidolon* and *Taphozous* signify a well developed forest and the Thryonomyid Rodents, *Thryonomys*, which live in tall grasses, are present in B 10 and absent in F 1. Chiroptera *Coleura*, Cricetidae *Gerbillurus*, Dipodidae *Jaculus*, Bathyergidae *Heterocephalus* and Lagomorpha *Lepus*, adapted to open savanna and even some to semi-desertic country, were collected in F 1 but not found in B 10.

These microvertebrates, although scarce and difficult to find, suggest a landscape of riverine forest and woody savanna with tall grasses for member B, becoming more and more open savanna with scattered acacias on a carpet of low grass in member F.

Assemblages

Since the beginning of the field work at Omo in 1967, I have developed a scheme of biozonation which now has four zones:

— Omo zero, or *Hippopotamus protamphibius* hexaprotodon zone, composed of 12 genera and 13 species of mammals is the fauna of the Mursi Formation, characterized by the association of *Anancus kenyensis*, *Ceratotherium praecox*, *Hipparion turkanense*, *Tragelaphus* sp. ancestor to *Tragelaphus gaudryi*, *Aepyceros* sp. (ancestor to *Aepyceros* sp. of the Shungura Formation), *Nyanzachoerus jaegeri*, *Nyanzachoerus pattersoni*, *Loxodonta adaurora* and *Deinotherium bozasi*.

— Omo 1, or *Elephas recki I* zone, is composed of 74 genera and 113 species of mammals and characterized by *Stegodon kaisensis*, *Loxodonta adaurora*, *Nyanzachoerus pattersoni*, *Nyanzachoerus jaegeri*, *Notochoerus euilus*, *Kobus patulicornis*, *Hippopotamus protamphibius* hexaprotodon; it is the fauna of the Usno Formation and of the basal member and members A, B and C of the Shungura Formation.

— Omo 2, or *Hippopotamus protamphibius* tetraprotodon zone, is composed of 76 genera and 116 species of mammals and characterized by *Elephas recki II*, *Mesochoerus limnetes*, *Metridiochoerus jacksoni*, *Menelikia* (ancestor to *Menelikia lyrocera*) and *Menelikia lyrocera*, *Giraffa gracilis*. *Loxodonta adaurora*, *Notochoerus euilus* and *Nyanzachoerus pattersoni* disappear during the lower part of this zone; *Equus* and *Hippopotamus gorgops* appear in the topmost part. It is the fauna of the members C, D, E, F and G of the Shungura Formation.

— Omo 3, or *Hippopotamus gorgops* zone, is composed of 52 genera and 84 species of mammals and characterized by the association of *Elephas recki III*, *Equus*, *Mesochoerus olduvaiensis*, *Stylochoerus nicoli*, *Kobus ellipsiprymnus*, *Megalotragus kattwinkeli*, *Hystrix cf. cristata*, a progressive subspecies of *Menelikia lyrocera*, *Parmularius altidens*, *Beatragus antiquus*.

The first impression is of great general stability or very gentle transformation resulting in progressive evolution. There are no real faunal breaks but a gentle change. This situation obliges us to establish the notion of transitory periods like those represented by members C and G, belonging at the same time to the preceding and the following zone. Changes are a reality but, in a relatively stable climate, they take more time to become obvious.

The second trend is in the form of evidence from the bottom to the top of the sequence of assemblages of mammals favouring less shady habitats. Through 2 million years, *Anancus* and *Stegodon* disappear while *Nyanzachoerus*, *Elephas recki*, *Ceratotherium*, *Mesochoerus*, *Notochoerus*, *Metridiochoerus* and *Menelikia* evolve towards more hypsodont species or subspecies; *Equus*, *Phacochoerus*, *Hippopotamus gorgops*, *Stylochoerus* and the Alcelaphini favouring more open country and lake shores appear in increasing numbers.

Invertebrate palaeontology

The analysis of the molluscan fauna by A. Gautier, at first glance, gives the same general impression of faunal stability in the lower Omo valley, through a period of 3 million years. Levels with molluscs are rare but it is nevertheless possible to collect them in the Mursi Formation, the Usno Formation, in the basal member and members H to L of the Shungura Formation.

Gautier distinguishes two zones, a lower one (Mursi fauna, Usno fauna and fauna from basal member of the Shungura Formation) with *Cleopatra arambourgi arambourgi*, *Melanoides tuberculata* and *Pleiodon* sp.; and an upper one (members H to L) with *Cleopatra arambourgi* nov. subsp., descendant from *Cleopatra arambourgi arambourgi*, *Melanoides* nov. sp., descendant from *Melanoides tuberculata* and *Mutela* nov. sp. instead of *Pleiodon* sp. There are also numerous levels with *Etheria elliptica* through the sequence (B 2, B 11, C 4, C 5, C 6, E 4, G 8, H 4) which indicate a fast stream probably very similar to the present Omo river.

Palaeobotany—Palynology

One of the best ways to reconstruct the former landscapes is to use the evidence of plants. A good assemblage of fossil wood has been collected and is under study. As early as the second field season in 1968 I invited Raymonde Bonnefille to search for pollen. She has worked since that year on this difficult and often disappointing task. Only four samples have yielded pollen but they have proved very interesting. A poor assemblage was obtained from B 10 (locality 1). Two rich samples were obtained from member C (C 7/8 and C 9, Omo 18) and a fourth one from member E (E 4, Omo 57).

Comparison of the spectra from the rich samples reveals very interesting

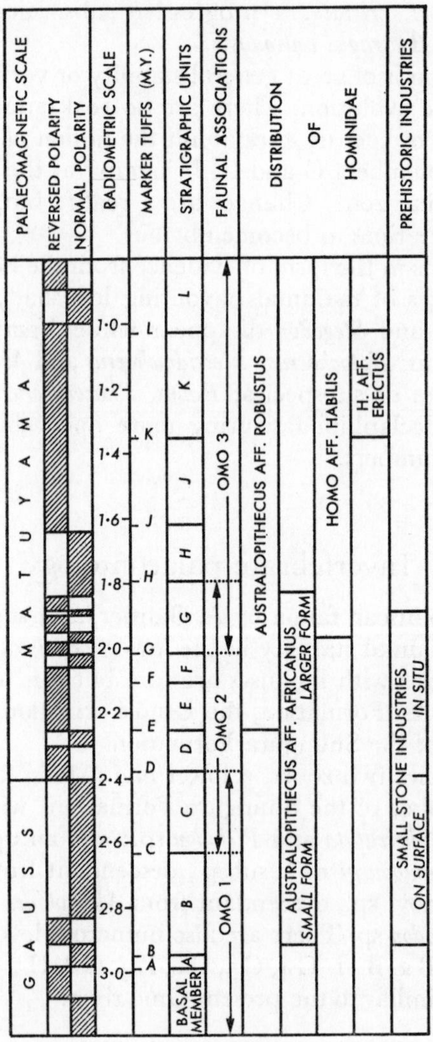

FIG. 31:1. Relationships between stratigraphic units, hominids and fauna, lower Omo valley.

differences: flora from C 7/8 and C 9 are very similar; but they are both very different from flora from E 4. For instance, it is possible to recognize 22 taxa of trees in C 7/8, 24 taxa in C 9, only 11 in E 4. The ratio between the number of pollens from trees and from grass shows a similar change: 0,4 in C 7/8, 0,15 in C 9, 0,01 in E 4. *Celtis* and *Acalypha*, indicating local humidity, are present in the two sites from member C but absent in E. Likewise, *Olea* requiring very high humidity, represents 23 % of tree pollens in C 7/8, 12 % in C 9 and 3,2 % in E 4. In contrast, *Myrica* is absent in C 7/8 and C 9 but present in E 4. All these differences indicate a drying up of the area and at the same time a diminution of numbers of taxa from C to E probably due to less rain.

Palaeoanthropology and prehistory

Two genera of Hominids, *Australopithecus* and *Homo* are present representing 4 species.
— from B 2 or Usno to G 13, *Australopithecus aff. africanus* occurs, possibly divided in two subspecies, a small one from B 2 to tuff C and a larger one above;
— a robust Australopithecine, *Australopithecus aff. robustus*, occurs from E to L.
— an ancient species of *Homo*, very close to what is called *Homo habilis*, occurs from G 5 to L.
— cranial fragments of *Homo aff. erectus* occur in member K.

Since 1969, artifacts have been known to occur *in situ*; they were found at possible occupation sites in member F (Omo 71, Omo 57, Omo 123, locality 204, locality 208, locality 396) where the most important excavations were carried out by J. Chavaillon and H. Merrick. Similar artifacts were found by J. de Heinzelin on the surface from B 2, and in sediments and tuffs in member D, in channel situations in member C and some isolated ones in lower member G.

Conclusion

There are differences between the various localities which are still being studied. It is possible, however, to make four concluding comments:

(1) All the different approaches show the same thing, the evolution of the climate from wet to less wet, and of the ecological setting from bush and scrub and tall grass to more open savanna and much shorter grass. Thus it is probable that a real event is recorded in the sedimentary and fossil record.

(2) The Hominids seem to be affected by the same climatic events as the other mammals. When the first assemblage (Omo 1) evolves into the second (Omo 2) about tuff C, *Australopithecus aff. africanus* evolves into another subspecies of increased size. When a minor change is indicated within the Omo 2 assemblage, in member E, *Australopithecus aff. robustus* appears. When a major change is in progress in the lower part of member G (change from Omo 2 zone to Omo 3 zone), *Homo aff. habilis* appears.

(3) *Homo aff. habilis* first occurs with *Equus* and *Hippopotamus gorgop* ; it appears when *Mesochoerus limnetes* is evolving into *Mesochoerus olduvaiensis*, when *Elephas recki* reaches its II B stage, when *Theropithecus brumpti* is decreasing and when the Alcelaphini appear. *Homo* arrived when the savanna was becoming less shaded and may well have been an animal of very open areas.

(4) From B 2 to G 5, over a range of more than a million years, we encounter artifacts and remains of only one Hominid, *Australopithecus*, and from B 2 to E, artifacts and remains of only one species of *Australopithecus*, *Australopithecus aff. africanus*.

Up-to-date reports by different specialists on all these subjects will be found in the volume published in 1976 by the University of Chicago Press under the title 'Earliest man and environments in the lake Rudolf Basin' edited by Y. Coppens, F. C. Howell, G. L. Isaac and R. E. F. Leakey.

Acknowledgements

I thank the following colleagues for their assistance in both field and laboratory: F. M. Brown, J. Chavaillon, P. Haesaerts, J. de Heinzelin (Geology); M. Beden, H. B. S. Cooke, S. C. Coryndon, G. Eck, V. Eisenmann, A. Gautier, A. Gentry, C. Guerin, D. Hooijer, F. C. Howell, J. J. Jaeger, G. Petter, H. Wesselmann, (Palaeontology); R. Bonnefille (Palynology); J. Chavaillon, H. Merrick (Prehistory).

32

PETER G. WILLIAMSON

Evidence for the major features and development of Rift Palaeolakes in the Neogene of East Africa from certain aspects of lacustrine mollusc assemblages

A simple conceptual model is proposed which relates certain rift lake basin features to the operation of five major processes, which in turn produce five major categories of East African lacustrine mollusca. Various lake basin types encourage the operation of characteristic combinations of processes which result in faunal category combinations broadly diagnostic for basin type. The model constitutes a useful, though at present generalised, diagnostic tool, whereby the major features and developmental histories of rift palaeolakes may be investigated.

Analysis of well-documented Plio-Pleistocene mollusc faunas, from the northern portions of both eastern and western rifts in East Africa, by this model, indicates a major climatic event in the area, dateable at 2.6 m.y. in the eastern rift.

Introduction

There has been increasing interest in recent years in the palaeontology and geology of the Neogene deposits of the East African rift system. At present, as in the past, actively sedimenting areas in and around the rift are often associated with lakes. Although major palaeontological interest usually focuses on the terrestrial fauna, including hominoids, it is useful to understand something of the nature of the lakes around which this fauna lived.

Of the sediments laid down at any one time during the history of a basin, only a relatively small fraction are usually available at outcrop, and many major features of the palaeolake are often obscure. Conventional palaeoecological analysis of fossil lacustrine mollusc assemblages gives valuable data about local palaeoenvironments over a given area, but it is usually hard to assess such large-scale features as the overall size and broad environmental character of the palaeo-lake, and its likely degree of hydrographic connection with other basins.

A simple model for the analysis of Rift Palaeolake features using mollusc assemblages

A simple Process-Product model which is of some use in the analysis of the broad nature of rift palaeolakes may be constructed following the recognition of five major categories of East African lacustrine mollusca.

Cosmopolitan mollusca may be defined for the purposes of this model as those evolutionarily conservative, geographically widespread species that appear to be the 'root stocks' of East African freshwater mollusc faunas. Their earliest Neogene representatives are often conchologically referable to modern species, and they probably represent the generalised pre-rift fauna of East Africa. A typical Cosmopolitan species-group by this definition would be the geographically widespread form *Melanoides tuberculata*, known from the Miocene to the Recent. Certain taxa which by the criteria of geological longevity and wide geographic distribution are apparently 'Cosmopolitan' earlier in the Neogene have subsequently undergone a major restriction of range in East Africa. They are now confined to small areas and we may term them *Relict* forms. A good example of such a taxon might be *Pleidon spekei*.

The peculiar nature of rift lacustrine environments, particularly the very long life-times of certain lakes, has favoured the periodic long-term ecological or geographical isolation of various Cosmopolitan forms, and endemic versions of these have resulted in various basins during the Neogene. The environmental diversity of large stable lakes, in which many environments and niches within environments are open to mollusca, can result in the evolution of what we may term *Radiative Endemics*. An adaptive radiation of a Cosmopolitan taxon results in one or more endemic species or genera; normal Cosmopolitan species characteristically accompany their endemic derivatives, and this feature we may take as the definition of Radiative Endemism. Examples appear to be furnished by the modern Melaniidae of Lake Tanganyika (Moore 1899, 1901, 1903; Gunther 1894; Pelseneer 1906; Yonge 1938; Bourguignat 1890; Smith 1904; Cunnington 1920; Fuchs 1936), and the Bellamyid radiation of the Plio-Pleistocene 'Kaiso fauna' of the Edward-Albert basin (Adam 1957; Fuchs 1936; Gautier 1967, 1970).

Conversely, interruption of gene flow into a basin due to its geographic isolation may also result in endemism, as the various Cosmopolitan species present in the basin become increasingly endemic by the effects of genetic drift or other mechanisms leading to allopatric speciation. Mild degrees of interruption of gene flow will lead to the formation of geographic races or subspecies of certain Cosmopolitan forms, as commonly occurs in many modern lakes, but a more rigorous genetic isolation might be expected to lead to a fauna consisting solely of fully endemic species descended from Cosmopolitan taxa, but reproductively isolate from them. Unlike Radiative Endemism, in this case normal Cosmopolitan taxa would not be expected to accompany their endemic derivatives, which we may term *Phyletic Endemics*. Another important difference between Phyletic and Radiative endemism is that in the latter case it is normal for only one or two families or genera to exhibit endemism, whereas in the former case all thoroughly aquatic taxa will ultimately exhibit some degree of endemism.

The well documented tectonic and climatic instability of the African rift has on occasion fostered the immigration of forms into regions to which they are known to be normally foreign. Such forms may be termed *Exotics*. An example are the primarily Holarctic taxa which periodically invaded the northern portion of the rift during the Plio-Pleistocene and Holocene (Brown 1973; Williamson, unpublished manuscript). It is clear that a given species could be allocated to a variety of the above categories at different times in its geological history. The correctness of the allocation of a given species to a given category at any point in time is therefore consequent on the extent to which its spatial distribution and evolutionary relationships are known or can be surmised. Although such allocations will clearly become less certain with increasing geological age of faunas, circumstantial evidence can usually be adduced even from the relatively incomplete record presently available. For example, *Pseudobovaria* as a genus is apparently extinct. However, closely related *Pseudobovaria* species are known for much of the Neogene from both eastern (Lake Turkana basin) and western (Lake Edward-Albert basin) rifts, from localities up to 500 km apart, and this species group might reasonably be regarded as Cosmopolitan for much of this period in East Africa.

Summarising the observations above it appears that one or more of five main processes may operate within a basin to produce one or more of five main faunal categories within that basin (Table 1, columns 2 and 3). However, certain broad features of the basin itself inevitably govern the extent to which each of these processes is operative within the basin, and hence determine the presence, absence and relative abundance of the five major faunal categories. Using qualitative proportionality arguments, the following broad generalisations can be made.

Immigration and consequent gene flow into a basin will be encouraged by hydrographic connection with other basins, and by a large lake surface area (encouraging aerial introductions by birds and aquatic insects) (Kew 1893). On both counts larger lakes are more likely to experience large-scale, regular immigration,

i.e. Immigration $\propto$ A, where A = Lake Surface area.

In addition, the successful immigration of a species moving into a basin for the first time requires that the species not only reach the lake but also establish a breeding population within it. This establishment is contingent on the existence within the lake of an environmental niche suitable for the immigrant species. In tropical lakes mollusca are frequently concentrated in shallower inshore waters and a longer absolute length of shoreline (SL) will tend to increase the likelihood of a large range of environments being available to immigrant mollusca, in one of which they may settle.

As (Environmental Range) $\propto$ SL,
and SL $\propto$ A

from purely geometric considerations, it follows that:

(Environmental range) $\propto$ A.

In addition, the characteristically more complex ecologies of larger lakes may mean that environments within such lakes are more likely to offer niches exploitable by immigrant forms, i.e.:

Niche availability $\propto$ A.

On these various grounds we might expect that the regularity and richness of immigration of Cosmopolitan forms into a given lake would be generally proportional to the area of that lake.

The immigration of Exotic forms into a basin will be conditioned by the features discussed above, and also by various extrabasinal features ('E') such as regional climatic change, i.e.:

$$\text{(Immigration of Exotics)} \propto \text{E.}$$

TABLE 1: *Probable relations between nature of lake basin and the prevalent molluscan evolutionary/migration processes, and their products.*

LAKE CONDITIONS FAVOURING PROCESS	PROCESS	PRODUCT (Faunal category produced in basin by process).
Good hydrographic connections with other areas. Large lake surface area. Large range of environments, niches.	IMMIGRATION OF COSMOPOLITAN TAXA i.e. immigration of evolutionarily conservative, geologically long-ranging taxa not restricted to a specific basin or area and characteristically occurring in both lacustrine and fluviatile situations.	COSMOPOLITANS
As above, other relevant factors extrabasinal (e.g. regional climatic change).	IMMIGRATION OF EXOTIC TAXA i.e. immigration of taxa normally foreign to area of basin.	EXOTICS
Large range of environments, niches. High shoreline development. Long-term lake stability. Meromictic conditions leading to resource poor/ resource unstable regimes.	RADIATIVE ENDEMISM i.e. the evolution of endemic forms within a basin by the adaptive radiation of COSMOPOLITAN or EXOTIC forms already present within the basin.	RADIATIVE ENDEMICS
Small range of environments, niches. Long-term lake stability. Poor hydrographic connections with other areas. Small lake surface area.	PHYLETIC ENDEMISM i.e. the evolution of endemic forms within a basin by allopatric speciation of genetically isolated Cosmopolitan faunas within the basin.	PHYLETIC ENDEMICS
Large range of environments, niches.	PERSISTENCE OF RELICTS i.e. local persistence of formerly widespread taxa.	RELICTS

Radiative Endemism will clearly be encouraged by a large environmental range and high niche availability, i.e.:

$$\text{(Radiative Endemism)} \propto A.$$

Radiative Endemism will also be encouraged by the possibility of periodic geographic subdivision of the lake basin by the closing off of bay mouths etc., as this will favour intrabasinal allopatric speciation (Brooks 1950). Such subdivision is most likely in lakes which have long and irregular shorelines relative to their areas. A measure of this feature is the quantity 'Shoreline Development' (SD) (Reeves 1968), where

$$\text{SD} = \text{SL}/2 \sqrt{\pi A}.$$

We might expect: (Radiative Endemism) $\propto$ SD.

Major restriction of gene flow may result in Phyletic Endemism, i.e.

$$\text{(Phyletic Endemism)} \propto \text{(1/Rate of Gene Flow)}$$

As   (Rate of Gene Flow) $\propto$ (Rate of Immigration

$$\text{(1/Rate of Gene Flow)} = \text{(1/Rate of Immigration)}.$$

However,   (Rate of Immigration) $\propto$ A (from above)

and   (1/Rate of Immigration) $\propto$ 1/A.

Therefore   (1/Rate of Gene Flow) $\propto$ 1/A

and   (Phyletic Endemism) $\propto$ 1/A.

True Phyletic Endemism, as defined, will be inhibited by the intra-basinal allopatric speciation likely to be consequent on high degrees of shoreline development, i.e.

$$\text{Phyletic Endemism} \propto 1/\text{SD}.$$

If intertaxon competition has contributed to the range restriction of forms now regarded as Relict, persistence of such forms is most likely where a wide range of environments and abundance of niches reduces selection pressures, i.e.:

$$\text{(Persistence of Relicts)} \propto A.$$

Most of the processes discussed above and summarised in Table 1 require a certain amount of time to produce their corresponding faunal category within a basin. Thus although a given process may be operative within a basin, the actual production of the appropriate faunal category may be inhibited by environmental instability within the lake, which persistently 'resets the clock' as far as the operating process is concerned. For example, a lake which persistently dries out or develops inimical water chemistries, or which shows rapid environmental shifts in time, is unlikely to ever develop an endemic fauna even though other features of the basin apparently favour this development. A convenient but obviously very general guide to the likely long-term environmental stability of a lake may be its maximum depth (MD), as the deeper a lake, the less likely is it to experience frequent development of unfavourable water chemistries, drying out, or rapid major environmental shifts (large-scale transgression/regression). Long-term lake stability is of greatest importance to those time-demanding processes concerned with the evolution of endemics and the persistence of relicts. There is evidence, however, that if the immigration of Cosmopolitan forms occurs at all it tends to occur quite rapidly in geological terms.

In addition to the likely connection between long-term lake stability (STB) and maximum depth of lake (MD), there is another major argument that might lead us to expect the quantity MD to influence certain of the processes discussed above. Large deep tropical lakes are known to develop thermoclines, with the consequence that nutrients important to the primary productivity of the lake become periodically 'locked-up' to a certain extent in the anoxic hypolimnion. Such meromictic lakes apparently have, on average, somewhat lower primary productivities than shallower lakes, which are unstratified (holomict/polymict) and do not develop thermoclines, due largely to wind-induced turbulence (Beadle 1974). The consequent regular circulation of nutrients and relatively higher primary productivities generally present in holomict/polymict lakes may well directly affect mollusca, even though these tend to be concentrated in inshore waters, as the total primary productivity of the lake must have some bearing on the resources available to inshore benthos. We might expect deeper meromictic lakes to be relatively resource-poor as far as mollusca are concerned, and holomict/polymict lakes to be relatively resource-rich.

Most tropical lakes appear to be essentially resource unstable, in the long term. Shallow lakes may experience such calm conditions that the cessation of turbulence results in the sudden onset of anoxic conditions, and ensuing drastic reductions in the carrying capacity of the lake for its various populations. Deeper lakes may experience storms violent enough to break down their stratification and result in massive release of nutrients from the hypolimnion and corresponding rapid increase in primary productivity.

Valentine (1974) has argued convincingly that in marine environments resource stability and availability directly affect the diversity and degree of specialisation of a population. In resource rich/resource unstable environments, selection is apparently in favour of high reproductive potentials and adaptive flexibility in populations, in order that they may best exploit the fluctuating but often rich resource supplies in an uncrowded environment. Resource utilisation may be inefficient in this case, and morphologically generalised populations are likely. This is the 'r-selection' of MacArthur and Wilson (1967). Conversely, in resource poor/resource stable situations, selection will be towards small, non-fluctuating populations exhibiting a high degree of specialisation and diversity in order to best exploit small resource supplies in a crowded environment. The high degree of specialisation necessary to exploit small resource supplies efficiently is clearly likely to be expressed by speciation and high taxonomic diversity in populations in this resource regime. This is the 'k-selection' of MacArthur and Wilson.

In resource rich/resource stable and resource poor/resource unstable regimes selection will be a compromise between r- and k-selection. It follows that in general the resource poor/resource unstable conditions of deep meromictic rift lakes should result in populations that have high reproductive potentials but which nonetheless may be expected to be more specialised and taxonomically diverse, due to a measure of k-selection, than populations from resource rich/resource unstable shallow holomict/polymict lakes where r-selection alone is of predominant importance. In our context, the taxonomic diveristy and speciation consequent on a measure of k-selection will be reflected by the presence of large Cosmopolitan faunas and Radiative Endemics derived from these.

In general, from resource considerations, we would expect rich Cosmopolitan faunas and specialised Radiative Endemic derivatives of these to occur in deep, and hence meromictic, lakes. Shallower holomict and polymict lakes are more likely to possess more generalised Cosmopolitan forms alone, or Phyletic Endemic forms derived from these by allopatric speciation.

The basin features which are probably of greatest importance in governing the operation of the various processes are summarised in Table 1, column 1. To summarise the discussion above, for a given basin we might expect:

(1) Immigration of Cosmopolitans $\propto A$
(2) Immigration of Exotics $\propto A + E$
(3) Radiative Endemism $\propto A + SD + MD + STB$
(4) Phyletic Endemism $\propto 1/A + 1/SD + STB$
(5) Persistence of Relicts $\propto A + STB$

Where A = Lake Surface Area.
       E = Extrabasinal factors promoting immigration of Exotics.
     SD = Shoreline Development.
    MD = Maximum depth of lake.
    STB = Lake Stability (length of time fauna in lake enjoys
                 uninterrupted biological and evolutionary continuity).

Of the factors above, STB will often, but not always, be proportional to MD, as discussed previously. In addition, STB will generally be proportional to A, as areally small lakes are not usually of great depth, and consequently tend to be environmentally unstable and geologically short-lived.

Although Phyletic Endemic mollusc assemblages are apparently known from the Plio-Pleistocene, Phyletic Endemism is rare in mollusc faunas, and no good examples are known from modern rift lakes. This rarity is probably due to the fact that to maximise the proportionality expression for Phyletic Endemism we require to maximise the values for both $1/A$ and STB. As discussed above, STB is likely to be proportional to A, and a high value for $1/A$ would normally imply a low value for STB, and vice-versa. The rarity of Phyletic Endemic faunas therefore probably relates to the rarity of those lakes likely to produce them, i.e. very continuous and persistent yet areally small bodies of water for which the values of both $1/A$ and STB are high.

Similarly, modern rift lake faunas do not include Exotic forms at present, though such Exotics have often occurred in the past. However, modern lacustrine faunas including the other molluscan categories do occur, and parameters such as A, SD, and MD (generally proportional to STB) can be determined for a range of modern rift lakes whose faunas are at least generally known. By substitution in the proportionality expressions given above we can rank these modern lakes in order of their theoretical likelihood to possess a given faunal category. Comparison of the predicted faunal categories from a lake with those actually known from it is a useful test of the model. Fig. 32:1 indicates that, within the limitations of the data presently available, the proportionality expressions proposed rank the series of modern rift lakes broadly as we might predict. The only major exception appears to be Lake Mweru, which has a considerable endemic fauna, but comparatively low values for the parameters MD and A. However, it is clear that this lake is at present rapidly disappearing, due partly to the sedimenting activities of its influent rivers, and partly to the fact that one of these (the Luvua) is steadily lowering the altitude of its outlet from the lake. The present topography indicates a recently much larger lake; as a consequence of the shrinkage of this body of water several endemic taxa are already rare in the live state (Pilsbry and Bequaert 1927). The endemic elements in this fauna may reasonably be viewed as a heritage from a recently very different aquatic regime.

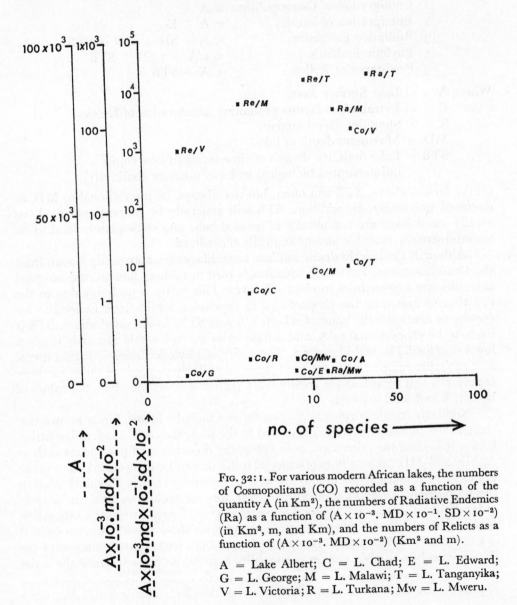

Fig. 32:1. For various modern African lakes, the numbers of Cosmopolitans (CO) recorded as a function of the quantity A (in Km²), the numbers of Radiative Endemics (Ra) as a function of $(A \times 10^{-3}. MD \times 10^{-1}. SD \times 10^{-2})$ (in Km², m, and Km), and the numbers of Relicts as a function of $(A \times 10^{-3}. MD \times 10^{-2})$ (Km² and m).

A = Lake Albert; C = L. Chad; E = L. Edward; G = L. George; M = L. Malawi; T = L. Tanganyika; V = L. Victoria; R = L. Turkana; Mw = L. Mweru.

This generalised model assumes five major faunal categories produced by five main processes. In a given basin it would therefore appear at first sight that any one of the 31 possible combinations of faunal categories could be produced by any one of the 31 possible combinations of processes, depending on the nature of the basin in question. However, the operation of several of the processes is mutually exclusive in a general sense due to the proportionality relationships already discussed. For example, Phyletic Endemism is theoretically proportional to $1/A$, and Radiative Endemism is proportional to A. We might therefore expect that these processes would tend to be mutually exclusive in nature, and would not expect to see lacustrine faunas which included both the faunal categories produced by these processes; i.e. the faunal categories Phyletic and Radiative Endemics should also be mutually exclusive. Consideration of the evolutionary origins of these categories reinforces this view. Certain other processes might similarly be expected to be contingent, again from consideration of the proportionality relationships. In this case we would not expect a certain process to operate without another process simultaneously operating. The predicted combinations of processes which are mutually exclusive and contingent are shown in Table 2. The majority of possible process combinations, and the faunal category combinations produced by these, are logically prohibited due to their failure to meet these conditions of mutual exclusivity or contingency. All possible process combinations are shown in Table 3; permitted process combinations, and resulting faunal category combinations, are underlined. A major argument in favour of this simple model is that none of the 24 prohibited faunal category combinations seems to be known from Neogene or recent rift lacustrine mollusc assemblages. On the other hand, of the seven faunal category combinations permitted by the model, six are known from various basins in the Neogene and Recent of the rift.

TABLE 2: *Mutually exclusive and contingent processes (see text for explanation)*

1) MUTUALLY EXCLUSIVE PROCESSES   (both processes unlikely to
be in operation simultaneously
in same basin).

   a) PHYLETIC ENDEMISM + RADIATIVE ENDEMISM

   b) PHYLETIC ENDEMISM + IMMIGRATION OF COSMOPOLITANS

   c) PHYLETIC ENDEMISM + IMMIGRATION OF EXOTICS

   d) PHYLETIC ENDEMISM + PERSISTENCE OF RELICTS

2) CONTINGENT PROCESSES   (first process unlikely to be operative
in basin without second process also
being operative).

   a) RADIATIVE ENDEMISM + IMMIGRATION OF COSMOPOLITANS

   b) IMMIGRATION OF EXOTICS + IMMIGRATION OF COSMOPOLITANS

   c) PERSISTENCE OF RELICTS + IMMIGRATION OF COSMOPOLITANS

   d) PERSISTENCE OF RELICTS + RADIATIVE ENDEMISM

Because the model predicts the broad nature of the basins in which the various permitted process combinations are likely to operate and produce their characteristic faunal category combinations, it provides a useful generalised interpretative tool. Certain broad features of rift palaeolakes which are not readily determinable from conventional palaeoecological analysis of faunas from areally limited exposures may be indicated by analysis of the faunal category combinations present; use of the model to determine basin conditions favouring the corresponding process combinations then gives some indication of the broad nature of the palaeolake.

TABLE 3: *Permitted and prohibited process, and resulting faunal category, combinations*

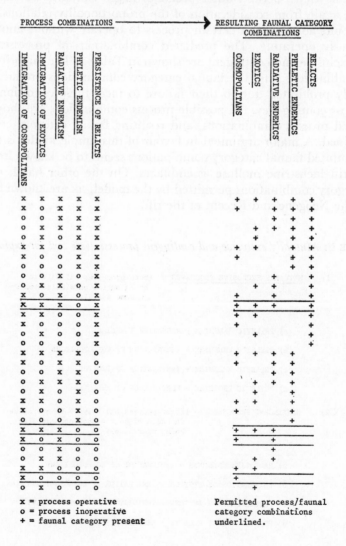

x = process operative
o = process inoperative
+ = faunal category present

Permitted process/faunal category combinations underlined.

Use of the model to trace the development of Rift Palaeolakes

The model gives information about the development of a lake through time because changes in the faunal category combinations in passing up a stratigraphic section will be the result of the history of the basin in which the succession was deposited. The changes in faunal category combinations which might be predicted with various patterns of basin history are summarised in Fig. 32 :2.

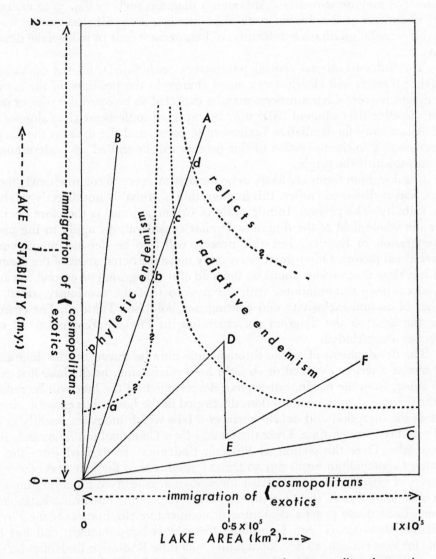

FIG. 32:2. Various patterns of lake development, and corresponding changes in faunal category combination present in the basin. See text for further explanation.

The ordinate in this diagram is the quantity 'Lake stability' (STB), defined previously. The abscissa is lake surface area (A). All the processes considered previously are either proportional, inversely proportional or non-proportional to lake stability and/or area; these parameters appear to be the main governors of the faunal category combination encountered in a given basin. The zones on the field of the diagram represent the broad range of values of STB and A for which the various processes are operative. These values are somewhat speculative at present, but probably broadly correct in the light of faunal category combinations known from modern lakes and the timing of events in radiometrically dated Plio-Pleistocene mollusc sequences. Although a diagram such as Fig. 32 :2 cannot be fully quantified at the present state of knowledge, it nonetheless can be used to provide a useful qualitative indication of long-term trends in palaeolake development.

The influence of less crucial parameters, such as SD, on the operation of certain processes will clearly cause some change in the positions of the areas on the diagram over which a process may be expected to be operative; for example, high shoreline development (SD) may be expected to decrease the values of STB and A necessary for Radiative Endemism to occur, and the area on the diagram corresponding to the operation of this process will be shifted an undeterminable amount towards the origin.

Cosmopolitan forms are likely to be the first category to colonise newly formed lakes, and as discussed earlier, this immigration is probably not directly related to lake stability. The process 'Immigration of Cosmopolitans' is therefore operative over the whole field of the diagram. Similar considerations apply to the process 'Immigration of Exotics', but this process will also be dependant on various extrabasinal factors. Other processes occupy more restricted areas of the diagram, and it is clear that various points on the field of the diagram correspond to various faunal category combinations, with the proviso that the previously stated conditions of mutual exclusivity and contingency hold (see Table 2). For example, over the area on the diagram corresponding to Phyletic Endemism all other processes are excluded.

The development of a lake through time may be traced on the diagram by any one of a series of curved or straight lines originating, as the lake first comes into being, from the origin. Subsequent development of the lake will be reflected by the faunal category combinations developed in the basin. For example line *OA* records the inception and development of a lake which increases steadily in area and stability through time. Over the section *Oa* a Cosmopolitan fauna is built up in the lake. Over the section *ab* Phyletic Endemics are derived from the pre-existing Cosmopolitan forms due to the still small size of the lake and consequent low rate of gene flow into it. Further increase in area leads to mounting immigration of Cosmopolitans, with or without Exotics, depending on various extrabasinal factors. Consequent genetic swamping or competitive elimination of the Phyletic Endemics then occurs (*bc*). Eventually the lake is large enough, and has been stable for long enough, to develop a fauna including Radiative Endemics (*cd*) and ultimately Relict forms (*dA*). Conversely, the line *OB* records the development of a lake which remains small but is very persistent. The initially Cosmopolitan fauna develops into a Phyletic Endemic fauna due to genetic isolation, but the persistently small lake area ensures that the fauna develops no further. The line *OC* records the development of a lake which although ultimately large is shallow and has a persistently low value for STB; it hence only ever develops a fauna of Cosmopolitan forms, with or without Exotics. If the fauna in the lake is eliminated by the development of inimical water chemistry, or drying, the line of lake development becomes disjunct, either returning to the origin (in the case of drying) or to a point on the diagram corresponding to 'STB = O' and the lake's new area, if this has changed. Line *ODEF* records the development, elimination and re-establishment of a lacustrine fauna, where the elimination is due to inimical water chemistry.

Examples of the use of the model

(a) East Lake Turkana Basin

The Plio-Pleistocene deposits in the eastern part of the Lake Turkana Basin in north Kenya have yielded an extremely important series of hominids, other terrestrial vertebrates, and early archaeological sites. In addition they have also yielded the most abundant and well-preserved series of Cenozoic lacustrine mollusc faunas in Africa, which prompted the development of the model proposed in this paper. The faunas range in age from about 4.0 to 1.5 m.y. (Williamson unpublished manuscript). Younger faunas at around 0.8 m.y. and of Holocene age are also known, but the older faunas alone will be considered here. The stratigraphic ranges of the various taxa are shown in a general way in Fig. 32:3, and this faunal succession may be used as the basis for a biostratigraphy, as many of the zonal boundaries can be demonstrated to maintain their stratigraphic positions over wide areas, and the mollusc zones shown to be non-interdigitating.

In the absence of a conceptual model such as that proposed here the underlying reasons for the succession of faunas seen in this and similar sections recorded from East Africa are obscure. In addition, although conventional palaeoecological analysis yields useful data about palaeoenvironmental features of the lake at various points in time, many broad features of the lake are necessarily obscure, at least in part due to the small areal extent of exposure relative to the original area of the sedimenting basin.

By grouping the taxa recorded from this section into various faunal categories proposed in this paper (Fig. 32:3) the picture becomes clearer. The sequence of faunal category combinations through time are seen to fall into a reasonable series corresponding to the line of lake development shown in Fig. 32:5. Generalised conclusions about the major features of the evolving palaeolake are summarised in Fig. 32:4. It appears that an initially small palaeolake persisted in the Rudolf basin from around 4.0 to around ?3.5 m.y. (Kubi Algi Formation times). The apparent development of the Zone 2 Phyletic Endemic from a normal Zone 1 Cosmopolitan fauna over this period must reflect both a persistent yet relatively small and shallow genetically isolated body of water.

At Suregei Tuff times, possibly around 3.5 m.y., a major transgression is

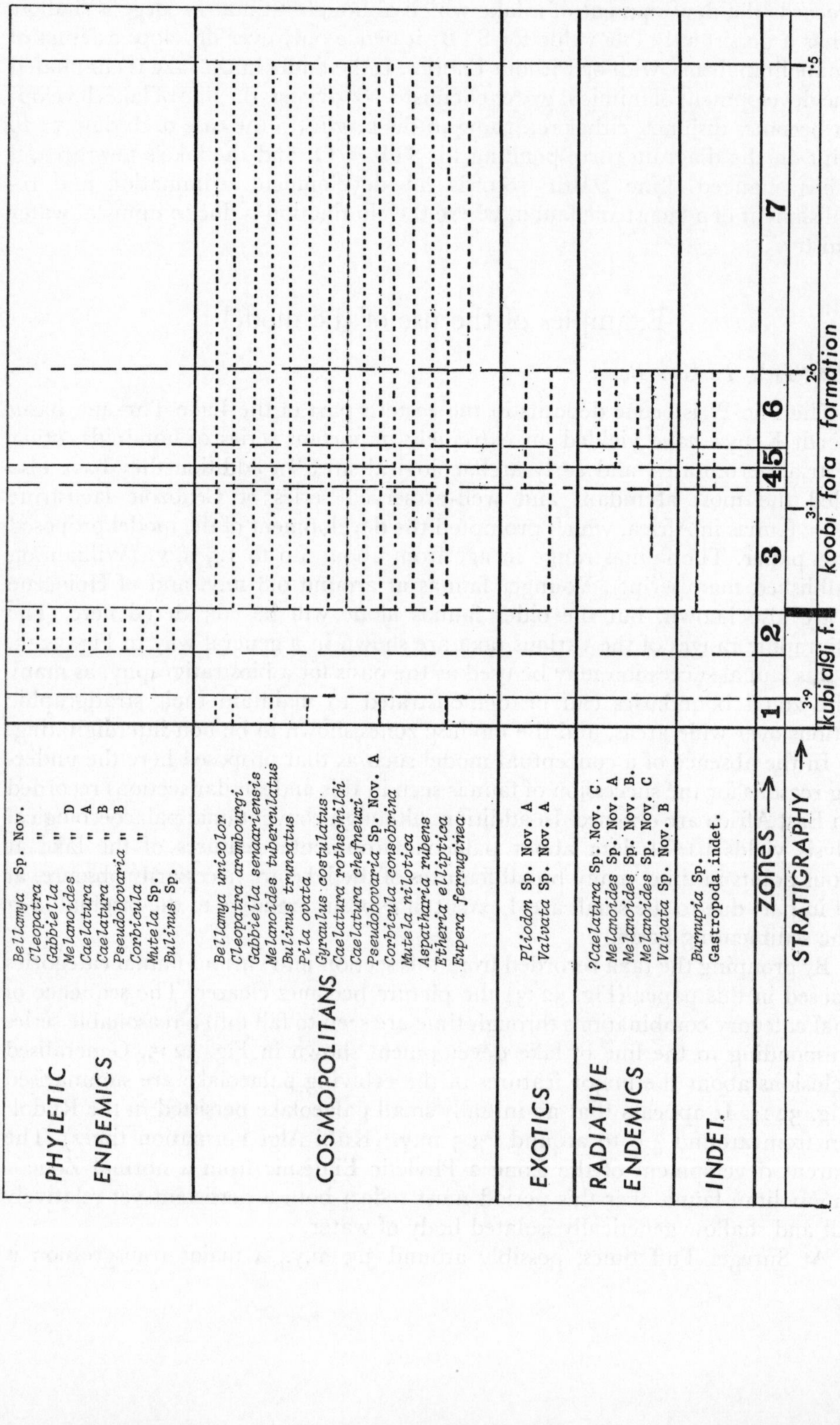

Fig. 32:3. Stratigraphic ranges of mollusca in the Plio-Pleistocene at East Lake Turkana, and a biostratigraphy based on these ranges.

recorded in exposures at East Rudolf (Pers Comm. I. Findlater). At about this time a normal Cosmopolitan (Zone 3) fauna rapidly replaces the Phyletic Endemic faunas of Zone 2. A genuine and major increase in the size of the lake at this time is indicated, which would inevitably lead to a rapid increase in the rate of immigration of Cosmopolitan forms and resulting competitive elimination and replacement of the pre-existing Phyletic Endemic faunas. The progressive development of Radiative Endemics from elements of both this Cosmopolitan fauna, and also from elements of the gradually accumulated Exotic forms, throughout the Lower Member of the Koobi Fora Formation (Zones 4–6), is in line with an expanded, probably meromictic, environmentally stable Lake Turkana over this period. High niche availability, large range of environments, and uninterrupted evolution with a measure of k-selection concomitant with a resource poor/resource

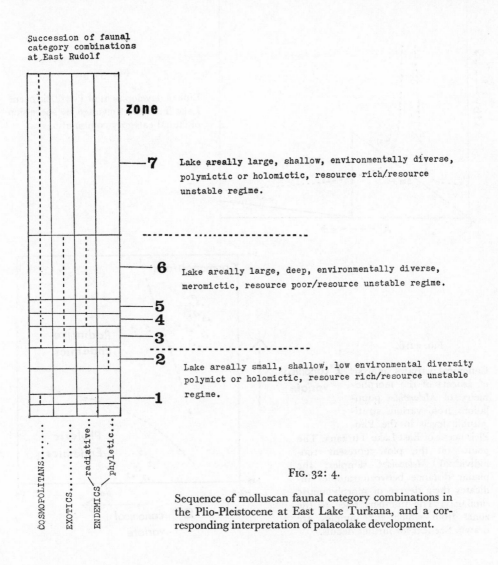

FIG. 32: 4.

Sequence of molluscan faunal category combinations in the Plio-Pleistocene at East Lake Turkana, and a corresponding interpretation of palaeolake development.

unstable regime are all features which would be associated with a larger, deeper lake and which would inevitably lead to the very diverse faunas of the Lower Member.

At about 2.6 m.y. a major extinction event apparently eliminated the fauna within the basin, probably due to the development of inimical water chemistry consequent on the lowering of lake level indicated at about this time in present-day inland (palaeocoastal) exposures (I. Findlater pers. comm.). Rapid subsequent immigration of all Cosmopolitan forms existing in the lake prior to the extinction event indicates little major permanent reduction in lake area. The persistently

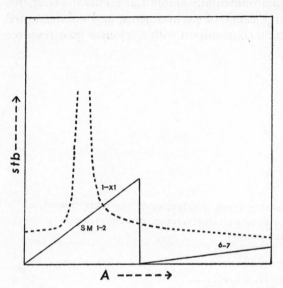

FIG. 32:5.

Line of development of Plio-Pleistocene Lake Turkana suggested by succession of faunal category combinations.

FIG. 32:6.

Canonical variates analysis of aspects of the morphometry of *Melanoides* populations from various stratigraphic levels in the Plio-Pleistocene of East Lake Turkana. The points on the plot represent 100-individual *Melanoides* samples; the planar distance between samples indicates their degree of morphometric similarity. The numbers refer to the zones from which the samples are drawn. See text for further details.

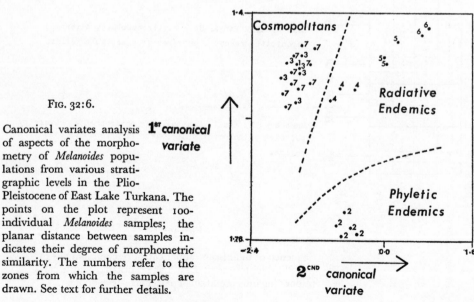

Cosmopolitan fauna of the Upper Koobi Fora Formation indicates, however, that the post-extinction event lake became shallow, probably holomictic or polymictic, experiencing a resource rich/resource unstable regime and environmental instability.

Many of the features of lake history mentioned above are reflected in the detailed palaeoecologies of the various zones (Williamson, unpublished manuscript). Available stratigraphic evidence (I. Findlater pers. comm.) and geochemical data (T. Cerling pers. comm.) appears supportive of, or reconcilable with, this putative history of lake development.

The proposed pattern of mollusc evolution, extinction and migration events is reflected by the changing morphometries of all the longer-ranging taxa. Fig. 32 :6 shows the results of a canonical variates analysis of 30 samples of *Melanoides* populations, each of 100 individuals, from various stratigraphic levels. Measurements recorded for each individual include the pleural angle, the height of the last whorl, total height of shell, and number of horizontal ribs on each whorl. The linear measurements were expressed as a ratio. Fig. 32 :6 is a plot of the first canonical variate against the second for these samples, and the planar distance between the points representing the various samples indicates the degree of

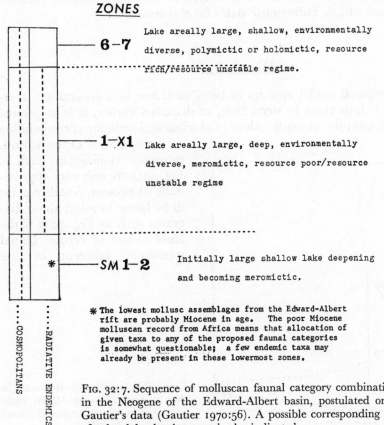

ZONES

6–7 Lake areally large, shallow, environmentally diverse, polymictic or holomictic, resource rich/resource unstable regime.

1–X1 Lake areally large, deep, environmentally diverse, meromictic, resource poor/resource unstable regime

SM 1–2 Initially large shallow lake deepening and becoming meromictic.

COSMOPOLITANS

RADIATIVE ENDEMICS

❋ The lowest mollusc assemblages from the Edward–Albert rift are probably Miocene in age. The poor Miocene molluscan record from Africa means that allocation of given taxa to any of the proposed faunal categories is somewhat questionable; a few endemic taxa may already be present in these lowermost zones.

FIG. 32:7. Sequence of molluscan faunal category combinations occurring in the Neogene of the Edward-Albert basin, postulated on the basis of Gautier's data (Gautier 1970:56). A possible corresponding interpretation of palaeolake development is also indicated.

morphometric similarity between populations of this genus from various points in the section. Populations from Zones 1, 3 and 7 cluster together—these populations consist purely of the generalised Cosmopolitan form *Melanoides tuberculata* (O. F. Müller). The Phyletic Endemic derivative of *M. tuberculata* from Zone 2 is morphometrically quite distinct from the parent stock, and these populations cluster separately. *Melanoides* populations from Zones 4 to 6 show a steady divergence from the 'Cosmopolitan cluster' as two Radiative Endemic species of *Melanoides* evolve and become of increasing numerical importance in *Melanoides* populations. All long-ranging taxa so far examined in this way exhibit similar 'Cosmopolitan' and 'Phyletic Endemic' clusters, with or without the development of Radiative Endemic forms in Zones 4 to 6.

(b) Edward-Albert rift

Analysis by the proposed model of the mollusc assemblages recorded by Gautier (1970) from the Neogene deposits of the Edward Albert rift is summarised in Fig. 32:7. A history of lake development corresponding to that in Figs. 32:7 and 8 is suggested. The material from these sites is not particularly well preserved and it seems likely that many of the gaps in the time ranges of various taxa recorded in Gautier (1970:56) are due simply to collection failure. Another problem is the lack of any radiometric dates for this sequence.

Discussion

Although the proposed model appears to be of some use as a generalised interpretative tool it is important to stress that, as discussed earlier, it is at present essentially qualitative; the absolute values of abscissa and ordinate corresponding to the operation of the various processes on diagrams such as Fig. 32:2 are only very approximate at present. Another point to be borne in mind about diagrams such as Fig. 32:2 is the ambivalence of certain faunal transitions inherent in the

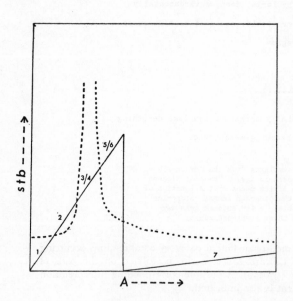

Fig. 32: 8.

Line of palaeolake development in the Neogene Edward-Albert basin suggested by the succession of faunal category combinations.

simplicity of the model. For example, a transition from a fauna of Phyletic Endemics to one of Cosmopolitan forms may occur in one of two ways according to Fig. 32 :2. It may result from a steady increase in lake size of hydrographic connections with other basins to a point where immigration of Cosmopolitan forms is such that competitive elimination of Phyletic Endemics occurs. On the other hand, Fig. 32 :2 also indicates that such a transition may result from a lake drying out or developing inimical water chemistry, eliminating the Phyletic Endemics living within the basin, and followed by the re-establishment of the lake as a viable molluscan habitat, with consequent Cosmopolitan immigration. As in the East Lake Turkana example, stratigraphic evidence may assist in deciding whether in a particular case the extirpation of Phyletic Endemics is due to growth in lake area or decrease in environmental stability.

Similarly, long-term persistence of 'Cosmopolitan-only' faunas is subject to two possible interpretations, according to Fig. 32 :2. This may be due to the persistent environmental instability of the lake, or to the fact that a lake is environmentally stable but of persistently intermediate size such that it is too small to develop Radiative Endemics but too large to be effectively genetically isolate and develop Phyletic Endemics. However, this latter situation seems unlikely to occur, as most rift lakes seem to have fluctuated in level periodically due to the climatic and tectonic instability of the area. Most cases of persistent 'Cosmopolitan-only' faunas are likely to be due to persistent environmental instability within the lake. Stratigraphical and sedimentological data may again be of use here.

Conversely, short-term Cosmopolitan-only faunas, from which endemic forms ultimately develop, indicate biological continuity and environmental stability within the lake. For example, in Zones 5 and 6 at East Turkana Radiative Endemics appear, derived from Cosmopolitan and Exotic forms present earlier in Zones 3 and 4. As the evolution of Endemics clearly requires time, this implies long-term lake stability over the interval Zone 3–Zone 6. We would therefore draw an important distinction between the implications of the relatively short-lived Zone 3/4 Cosmopolitan (and Exotic) faunas, from which Radiative Endemics were ultimately developed, and the much longer-lasting Cosmopolitan-only faunas of Zone 7, from which no endemics were derived. As discussed earlier, the latter probably indicate an environmentally unstable situation, at least in the long term, with frequent extinction events followed by reimmigration.

Care is therefore needed in interpreting Cosmopolitan-only faunas ($\pm$ Exotics). There significance can usually be determined by a consideration of their persistence, in absolute time terms, up a stratigraphic section, and by the nature of the preceding and succeeding faunal category combinations.

Conclusions

Five main categories of East African freshwater mollusca may be defined, and the presence of any one of these in a given lake basin related to the operation of a specific process. The operation of a given process is governed by certain broad basin features; various types of basin have characteristic process combinations and hence diagnostic molluscan category combinations. The molluscan category combinations encountered up a stratigraphic section may be used to indicate major basin features not immediately obvious from areally limited exposure, and to chart the qualitative development of a given lake through time.

Two of the best-documented Neogene mollusc sequences in the East African rift occur in the Edward-Albert Basin (Western rift) and in the Lake Turkana Basin (Eastern rift). These two sections are known from vertebrate evidence to sample a largely overlapping segment of Plio-Pleistocene time. It is therefore of considerable interest to note that the analysis of their respective sequences of mollusc assemblages by this model reveals a strikingly similar pattern of lake development in both basins over this period of time. Both sections apparently record lakes which grow in depth, area and stability during the Plio-Pleistocene, with consequent development of rich and in part Radiative Endemic faunas, until a sudden extinction event terminates these earlier faunas and the lakes remain shallower, environmentally unstable and relatively faunally depauperate thereafter. In the absence of radiometric dates for the Edward-Albert succession it is not possible to demonstrate the extent to which the major extinction event in both basins is synchronous, but it seems likely that it occurred at broadly the same point in Plio-Pleistocene time (c. 2.6 m.y. at East Lake Turkana).

The similarity in development history of the Turkana and Edward-Albert palaeolakes, which were over 500 km apart, apparently indicates that the fluctuations in the areas and depths of these lakes have been regulated by suprabasinal factors, presumably tectonic and/or climatic. The extinction event in both basins appears likely to have been due to the development of inimical water chemistry, an effect similar to the rise in alkalinity which has greatly reduced the rich Holocene mollusc faunas of Lake Turkana in under three thousand years. Such effects are clearly more likely when lakes become shallow, with reduction in total water volume within the basin. The persistently Cosmopolitan-only fauna which in both basins follows the extinction event apparently indicates that these lakes did indeed become shallower and increasingly environmentally unstable.

The similarity in longitude of both successions and the considerable geographic distance between them might incline us to favour a climatic basis for the broadly synchronous reduction in water levels implied by the faunal events in both basins. Such a climatic event would be accomplished, for example, by a southerly movement of the southern boundary of the North African desert belt (0–250 mm mean annual rainfall), which at present lies some 10° of longitude to the north of these basins. Whatever the precise mechanism involved, the suggestion of a major climatic change in this region at about 2.6 m.y. is interesting in the light of the profound climatic changes known to be occurring at about this time in the temperate regions.

Although more sections, and future work, will probably clarify the issue, at present it seems likely that large-scale climatic and tectonic features of rift evolution will prove to be directly mirrored by evolutionary events in rift-lacustrine mollusc faunas. Should this be the case, it would have interesting implications for the future biostratigraphic use of such faunas for inter as well as intra-basinal correlations in the African rift.

Acknowledgements

Thanks are due to Mr R. E. F. Leakey, Dr R. J. G. Savage, and Dr Glynn Ll. Isaac for the opportunity of fieldwork in northern Kenya. Thanks are also due to Dr R. J. G. Savage, Mr I. Findlater and Prof. J. W. Murray for useful discussion and criticism of the manuscript. This work was undertaken while in receipt of an NERC studentship.

References

ADAM, W. 1957. Mollusques quaternaires de la région du Lac Édouard. *I.P.N.C.B., Expl. Parc. Nat. Albert,* Mission J. de Heinzelin, fasc. **3**, 1–172.

BEADLE, L. C. 1974. *The Inland Waters of Tropical Africa.* 365 pp. Longman, London.

BROOKS, J. L. 1950. Speciation in Ancient Lakes. *Quart. Rev. Biol.* xxv, 30-60 (Part 1) and 131-176 (Part 2)

BROWN, D. S. 1973. The Palaearctic element in late Quaternary lake faunas of Southern Ethiopia. *J. Conch., Lond.,* **28**, 79–80.

FUCHS, V. E. 1936. Extinct Pleistocene Mollusca from Lake Edward, Uganda, and their bearing on the Tanganyika problem. *J. Linn. Soc. (Zoology),* **40**, 93–106.

GAUTIER, A. 1967. *New observations on the later Tertiary and early Quaternary in the Western Rift: the Stratigraphic and palaeontological evidence.* In *Background to Evolution in Africa,* Eds. Bishop and Clark, 73–87, University of Chicago Press.

—— 1970. Fossil Freshwater Mollusca of the Lake Albert–Lake Edward Rift (Uganda). *Annls. Mus. r. Afr. cent. Sér. 8VO. Sciences geologignes,* **67**, 1–144.

KEW, H. W. 1893. *The Dispersal of Shells.* 180 pp. International Science Series.

MACARTHUR, R. H., and WILSON, E. O. 1967. *The Theory of Island Biogeography.* Princeton University Press, Princeton.

MOORE, J. E. S. 1899a. The Mollusca of the Great African Lakes. III *Tanganyikia Rufofilosa* and the Genus *Spekia. Q. Jl microsc. Sci.,* **42**, 155–85.

—— 1899b. The Mollusca of the Great African Lakes. IV *Nassopsis* and *Bythoceras. Q. Jl microsc. Sci.,* **42**, 187–201.

—— 1903. *The Tanganyika Problem.* London.

PILSBURY, H. A., and BEQUAERT, J. 1927. The Aquatic Mollusks of the Belgian Congo. With a Geographical and Ecological Account of Congo Molacology. *Bull. Am. Mins. Nat. Hist.,* liii, 69-659.

REEVES, C. C. 1968. *Introduction to Palaeolimnology.* 228 pp. Elsevier.

SMITH, E. A. 1904. Some remarks on the Mollusca of Lake Tanganyika. *Proc. malac. Soc. Lond.,* **6**, 77–104.

VALENTINE, J. W. 1971. Resource supply and species diversity patterns. *Lethaia,* **4**, 51–61.

YONGE, C. M. 1938. The Prosobranchs of Lake Tanganyika. *Nature, Lond.* **142**, 464–6.

33

JACK W. K. HARRIS & INGRID HERBICH

Aspects of early Pleistocene hominid behaviour east of Lake Turkana, Kenya

Archaeological survey and excavations in the Upper Member of the Koobi Fora Formation provide evidence of hominid activities during the early Pleistocene which date between 1·2 and 1·6 million years ago. The analysis of lithic remains from a complex of 16 excavated localities reveals a distinctive local series of stone assemblages which we propose to call the Karari Industry. A series of heavy-duty core scrapers comprises the major component of this industry but choppers, discoids and polyhedrons are also common. Early Acheulian-type bifaces have been recovered in small numbers from the same beds.

The excavation strategy has been to sample traces of hominid activities in different sedimentary contexts. Geologic and micro-stratigraphic studies show that the archaeological sites were located in or near stream channels and in river floodplain contexts. The location of these sites provides a key to understanding the early Pleistocene man/land relationships represented in the area.

Archaeological sites with associated hominid remains and hominid finds in various parts of the paleolandscape show the sympatric coexistence of two hominid taxa: Homo sp. and Australopithecus c.f. boisei.

Introduction

The large expanse of sedimentary exposures to the east of Lake Turkana in Northern Kenya contains a unique fossil and artefact record that documents the existence of early hominids during the Pliocene/early Pleistocene and in addition provides evidence of their activities. Initial exploration and geologic reconnaissance of this region was begun in 1968 by Richard Leakey with a small interdisciplinary research team (Leakey *et al* 1970; Vondra *et al* 1971). In subsequent years, further survey by an expanded research group has led to the recovery of over 140 fossil hominid specimens and the equally important discovery of an extremely rich archaeological record revealing two major archaeological phases within the time range of the Pliocene/early Pleistocene (Isaac *et al.* 1971, 1976; Leakey 1971; 1972; 1973; 1974).

Investigations of the early archaeological phase have been in progress under the direction of G. Isaac, since 1970. Sites of this early phase, to which the name KBS Industry has been given, occur in outcrops of the KBS Tuff complex at the top of the Lower Member of the Koobi Fora Formation and have been dated at 2.61 ± 0.26 million years (Brock & Isaac 1974; Fitch & Miller 1970; 1975; Fitch *et al.* 1974; Isaac 1972a; 1972b; 1975; 1976). Sites of the later archaeological phase, which we propose to call the Karari Industry, occur in outcrops of the Upper Member of the Koobi Fora Formation and have been dated between 1.2 and 1.6 million years (Brock & Isaac 1974; Fitch & Miller 1970; 1975; Fitch *et al.* 1974; Harris & Isaac 1976; Isaac 1975).

This paper first describes archaeological sites and assemblages of the Karari Industry. The discovery of surface occurrences of artefacts in exposures of the Upper Member was made by G. Isaac and R. Leakey in 1971. During the succeeding field seasons 1972, 1973 and 1974, systematic archaeological survey and a large-scale excavation programme were carried out in exposures along the Karari Escarpment and at Ileret, two contiguous sub-regions of the East Lake Turkana Basin but separated by the Kokoi Horst. Fig. 33:1 shows the relationship of these archaeological study areas to the present-day topography. The location of known sites is shown on Figs 33:1 and 33:2.

One of the most important aspects of the archaeological investigations has been recognition of the significance of the relationship between the location of archaeological sites and their paleogeographic setting in understanding early Pleistocene man/land relationships. Detailed stratigraphic work made widespread correlation possible, as well as the reconstruction of major features of the early Pleistocene landscape (Behrensmeyer 1974; Bowen & Vondra 1973; Findlater 1976; Vondra & Bowen 1976). We will treat in discussion the combination of a set of archaeological sites in a variety of palaeogeographic and palaeoecologic

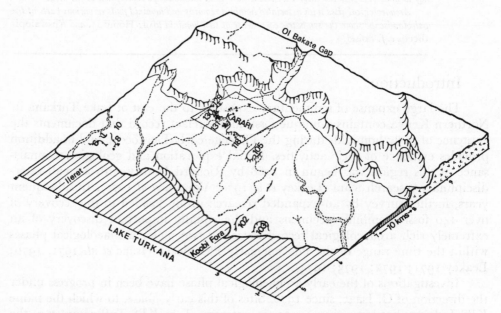

FIG. 33:1. An isometric view of the location of Karari sites along the Karari Escarpment and at Ileret in relation to present-day topography. Numbers 1–131 refer to the Expedition's Area numbers. Rectangle delimits the area shown in Fig. 33:2. ● = unexcavated site ▲ = excavated site (modified after G. Ll. Isaac *et al.* 1976).

settings as a basis for a general assessment of past behaviours and activities at one time horizon.

Stratigraphy and dating

To the east of Lake Turkana is the Suregei Cuesta, a range of low hills at the rim of the East Lake Turkana Basin composed of uplifted and faulted Miocene volcanics (Bowen & Vondra 1973; Vondra & Bowen 1976). A low ridge, the Karari Escarpment, extends from the base of the Surgei and extends laterally south for 20 to 25 km to Area 105 (Fig. 33:2). In the Karari area, the Upper Member disconformably overlies the KBS Tuff and comprises a suite of conglomerates and coarse fluvial deposits which interfinger with tuffaceous floodplain silts. These Upper Member sediments reach a maximum thickness of 50 m and the sequence is capped by the Karari Tuff (Bowen & Vondra 1973; Vondra & Bowen 1976).

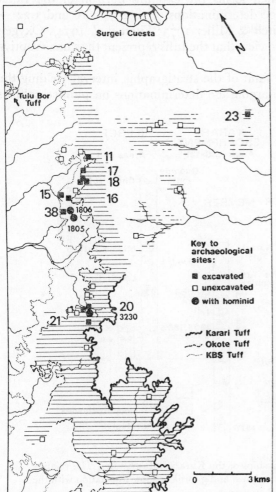

FIG. 33:2.

Map of the location of sites along the Karari Escarpment. The geology is based on data from I. Findlater. All 2 digit numbers are Kenya Museum (KNMER) catalogue numbers for hominid fossils (modified after Harris & Isaac 1976).

Archaeological sites were found stratified in exposures of beds representing both channel and floodplain conditions. They occur mainly along the eroded western face of the Karari Escarpment (Fig. 33:2), but several localities are also known at Ileret where they occur in low-energy floodplains and delta floodplain deposits between the Lower and Middle Tuff complex. It is probable that sites pertaining to the Karari Industry have a much greater areal distribution throughout the East Lake Turkana area and further survey is planned to the north and south of the areas already mentioned.

In the Karari area itself, sites belonging to the Karari Industry occur in the stratigraphic interval between the KBS Tuff and the Karari Tuff. Fig. 33:3 illustrates the stratigraphic relationships of the Karari sites; sites are particularly numerous in beds of the Okote Tuff complex, formerly the BBS Tuff complex. There now appear to be good grounds for correlating the Upper Member sequence in the Karari area with the Ileret Member (Vondra & Bowen, this volume). The Karari Tuff and Chari Tuff, which forms the capping horizon of the Ileret Member, have potassium-argon age determinations of 1.32 ± 0.10 and 1.22 ± 0.01 million years, respectively (Fitch & Miller 1975; Fitch *et al.* 1974). Oxygen isotope ratios are consistent with the view that the tuffs represent the same eruptive event (Cerling *et al.* 1975).

The KBS Tuff sets the lower limit of the stratigraphic interval yielding the Karari sites. A series of potassium-argon age determinations on pumice cobbles

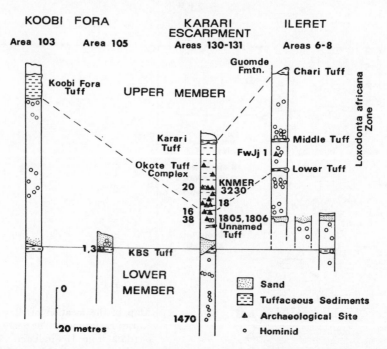

FIG. 33:3. The stratigraphic relationships of the Karari Industry sites. The relative position of groups of hominid fossils is shown and 3 fossils in close association with the sites are identified by their KNMER numbers.

recovered from the tuff gave a date of 2.61 ± 0.26 million years (Fitch & Miller 1970; 1975; Fitch *et al.* 1974). Another series of potassium-argon age determinations, however, gave a date of 1.8 million years (Curtis *et al.* 1975, and this volume). Although the question of the absolute age of the KBS Tuff is still under review, it does not affect the dating resolution of the Karari sites, as the majority of these are in fact found stratified in the Okote Tuff complex and its correlative at Ileret, the Lower and Middle Tuff complex. The potassium-argon age determinations for these tuff horizons are 1.56 ± 0.02 million years and 1.48 ± 0.17 million years, respectively (Fitch *et al.* 1974; Fitch & Miller 1976).

The palaeomagnetic polarity data collected from sediments of the Upper Member and the Ileret Member show reversed polarity (Brock & Isaac 1974; 1976). The Karari sites would thereby be correlated with the Matuyama Reversed Epoch. An age in the order of 1.2 to 1.6 million years, i.e. Upper Matuyama, is consistent with the potassium-argon age determinations.

The Ileret Member, and by correlation the Upper Member in the Karari area, fall within the *Loxodonta africana* zone (Maglio 1972). The 'best fit' age for this faunal zone is 1.3 million years with limits of 1.6 to 1.0 million years.

The isotopic dates, palaeomagnetic ages and palaeontological evidence, therefore, all suggest an age of 1.2 to 1.6 million years for the sediments containing sites of the Karari Industry.

Character of artifact assemblages

General description

An extremely rich and diverse archaeological record of early hominid activities has been recovered from the known localities along the Karari Escarpment and at Ileret. Excavation to date has been carried out at 16 sites which have yielded large samples of stone artifacts. Although more detailed quantitative studies remain to be completed, it is possible to outline the major features of the Karari Industry.

The typology devised by Mary Leakey for the analysis of stone artefacts recovered from Olduvai Gorge Bed I and Bed II (Leakey, M.D. 1971) was used

---

TABLE 1: *Table of potassium-argon dates for tuffaceous horizons within the Upper Member sedimentary sequence (source:* Fitch & Miller 1970; 1975).

Ileret	Karari Escarpment	Koobi Fora
Chari Tuff 1.22 ± 0.01 my (FMA 280)	Karari Tuff 1.32 ± 0.10 my (FMA 290)	
Lower and Middle Tuff Complex 1.48 ± 0.17 my (FMA 278)	Okote Tuff Complex 1.56 ± 0.02 my (FMA 266)	Koobi Fora Tuff 1.57 ± 0.0 my (FMA 270)

——————————————K.B.S. Tuff————2.61 ± 0.26 my——————————

in the classification of stone artefacts from the Karari sites, although several additional categories were established during the course of our analysis.

Two distinct classes of stone artifacts comprise the assemblages recovered from each site: (1) boulders, cobbles and blocks of stone, which exhibit to a greater or lesser degree some evidence of trimming, retouch and/or utilisation; and (2) small stone fragments which resulted from percussion flaking. Table 2 gives the percentage frequency of stone artifacts classified as tools, modified material, utilised material and debitage. The term manuport is used in circumstances when cobbles were introduced to a site by human agency (Leakey, M.D. 1967).

A large proportion of each assemblage is comprised of debitage, with percentage values ranging from 54.5 per cent to 98.4 per cent. An important feature of the assemblage composition from sites in floodplain contexts is the much larger proportion of whole flakes and flake fragments (Table 2). In every case, these flakes are found associated with bone debris. One interpretation for the association of flakes and small stone fragments with bones is that they functioned as cutting and scraping implements in the butchering and processing of meat (Clark 1970; Clark & Haynes 1969; Isaac 1971).

In contrast to the floodplain sites, those from the channel contexts have yielded a higher proportion of formal tools (Table 2). Many such sites are located at or in close proximity to channel-deposited basalt boulder and cobble conglomer-

TABLE 2: *Percentage frequency of Karari stone artifact assemblages.*

Site	N	% Tools	% Modified	% Utilized	% Debitage	% Manuports
FLOOD PLAIN SILTS						
FxJj 11	337	3.6	–	0.6	95.8	4.0
17	247	0.4	0.4	0.4	98.8	0.4
18IHS	3267	3.5	–	0.3	96.2	0.1
20M	2510	1.7	–	0.8	97.5	0.7
20E	1211	3.0	–	0.7	96.3	2.2
20AB	3462	0.7	–	0.6	98.7	0.3
CHANNEL SANDS AND GRAVELS						
FxJj 16	173	26.0	4.1	12.1	57.8	Not applicable since often manuports cannot be distinguished
18GS/L	1556	10.9	0.3	1.7	87.1	
18GS/U	214	10.7	0.5	1.9	86.9	
18NS	993	5.9	0.5	2.6	91.0	
21 Surf.	54	11.1	–	3.7	85.2	
23	73	1.4	–	1.4	97.2	
38	115	34.8	–	7.0	58.2	
ILERET: LAKE MARGIN FLOOD PLAIN						
FwJj 1	318	0.3	–	0.3	99.4	0.3

ates which provided an excellent source of raw material for stone artifact manufacture. While many of the heavy-duty tools such as choppers, scrapers, discoids, etc., had functional importance as implements, it must also be remembered that they could have been important as a source of flakes. Large boulder cores as well as a high proportion of cortical flakes are also a feature of Karari assemblages found in these channel contexts, adding support to the notion that early hominids visited these localities to manufacture stone tools. In this sense, the channel localities can be classified as 'factory' or 'workshop' sites.

The majority of the artifacts were made of lava, especially a rather distinctive trachy basalt; also used were feldspathic ignimbrite and rhyolite (F. J. Fitch, pers. comm.). The original source of the trachybasalt was the volcanic hills at the basin rim. Cobbles and boulders composed of these materials were transported by streams that drained these areas of high relief to the alluvial plain below, where they were deposited as conglomerates in the beds of river and stream courses. These boulders and cobbles were readily available as a source of raw material and had particularly good flaking qualities. Less than 5 per cent of the artifacts were made of siliceous materials such as chert and chalcedony.

TABLE 3: *Percentage composition of tool category of Karari stone artefact assemblages.*

	N	CHOPPERS	PROTO-BIFACES	BIFACES	DISCOIDS	POLYHEDRONS	CORE SCRAPERS	L/D SCRAPERS	R.C./C.F.	SUNDRY TOOLS	OTHER TOOLS	TOTAL SCRAPERS
KARARI INDUSTRY: FLOOD PLAIN SITES												
FxJj 11	12	33.3	-	-	8.3	16.7	-	-	16.7	-	25.0	-
17	1	-	-	-	100.0	-	-	-	-	-	-	-
18IHS	114	1.7	2.6	-	25.4	21.1	21.1	16.7	5.3	6.1	-	37.8
20M	42	11.9	4.7	-	19.1	7.1	2.4	31.0	21.4	-	2.4	33.4
20E	36	30.6	-	-	22.2	8.3	8.3	25.0	2.8	-	2.8	33.3
20AB	25	16.0	-	-	4.0	8.0	20.0	20.0	32.0	-	-	40.0
KARARI INDUSTRY: SITES IN SANDS AND GRAVELS												
FxJj 16	45	26.7	4.4	-	8.9	15.5	17.8	17.8	8.9	-	-	35.6
18GS/L	169	18.9	2.4	-	15.4	18.9	19.5	17.2	4.7	3.0	-	36.7
18GS/U	23	17.4	-	-	4.4	17.4	21.7	30.4	-	8.7	-	52.1
18NS	58	10.3	3.4	-	13.8	19.0	13.8	29.3	5.2	-	5.2	43.1
21	6	16.7	-	66.6	-	-	-	-	-	-	16.7	-
23	1	100.0	-	-	-	-	-	-	-	-	-	-
38	40	72.5	-	-	5.0	-	-	2.5	17.5	2.5	-	2.5
KARARI INDUSTRY: LAKE MARGIN FLOOD PLAIN AT ILERET												
FwJj 1	1	100.0	-	-	-	-	-	-	-	-	-	-

Tools

Within the tool category, a series of heavy-duty core scrapers and light-duty scrapers form a distinctive set (Table 3). The core scrapers are made on cobbles and split cobbles and have one flat striking platform from which flakes have been struck. Intensive flake removal has undercut the dorsal face to form a steep edge. A characteristic feature is that much of the perimeter of the scraper is trimmed. A large number of core scrapers are elongated with a high keeled or rounded cortical back (Fig. 33:5).

The light-duty scrapers are made on flakes and flake and core fragments; they are thin in cross-section in comparison to core scrapers. These specimens exhibit rudimentary features of design. The most characteristic form is elongated with a denticulate edge (Fig. 33:5).

Other tool forms recovered by excavation such as choppers, proto-bifaces, discoids, and polyhedrons occur with less frequency, A small number of bifaces and cleavers have been recovered from surface localities in circumstances that suggest their derivation from the Upper Member beds. No specimens have been recovered from excavated contexts.

Retouched core and cobble fragments characterised by trimming on the

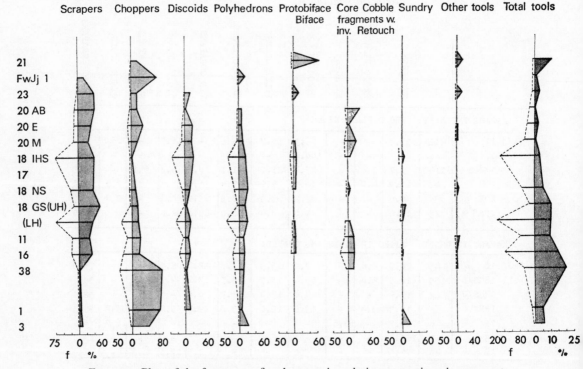

FIG. 33:4. Plots of the frequency of tool categories relative to stratigraphy amongst the available samples of assemblages. Sites are arranged in stratigraphic sequence. Broken line indicates the net frequency based on actual counts, solid line shows percentages. (*Note*: 1 and 3 are KBS Industry site located in the Lower Member.)

ventral face could not be accommodated within the type list proposed by Mary
Leakey; therefore, they have been included as a new category: core/cobble
fragments with inverse retouch. Various unstandardised trimmed forms have been

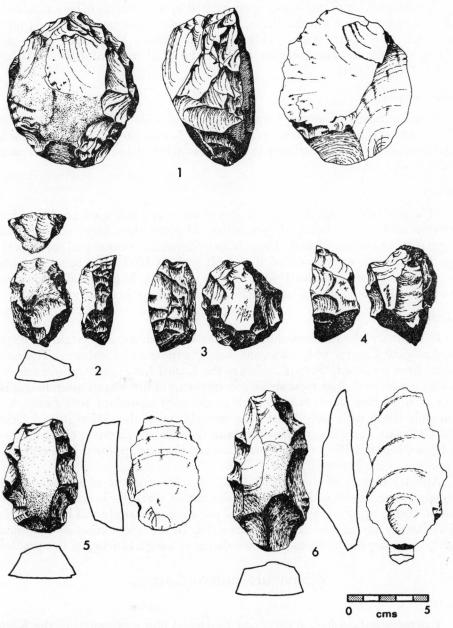

FIG. 33:5. 1, core scraper, FxJj 11; 2, core scraper, FxJj 20 Main; 3 4, core scrapers,
FxJj 20 AB; 5, core scraper, FxJj 18 IHS; 6, light-duty scraper on flake, FxJj 18 NS.
Made on lava. Drawings by Barbara Isaac.

classified as sundry tools while small tools such as awls and burins are relatively scarce.

Modified material

A small number of specimens characterised by a minimum of trimming or the lack of any patterned shaping have been placed in this category, so as not to distort the 'tool' percentages' This procedure follows suggestions made at a meeting of African archaeologists held at the University of Illinois, Urbana (Keller & Isaac 1971).

Utilised material

Small numbers of anvils, hammerstones and utilised flakes, and core and cobble fragments have been recovered from the Karari sites. These specimens are found in much greater abundance in channel contexts than in sites located on the floodplains.

Debitage

Characteristic morphological features of the whole flakes are small restricted platforms and diffuse bulbs of percussion. At some sites, large chunky flakes (larger than 10 cms) are found. There is no evidence from excavated contexts that large flake blanks were modified into tools such as bifaces. In some instances, particularly at sites associated with stream channels, large boulder cores and battered cobbles are found from which large flakes have been struck.

Summary

In summary, the stone assemblages of the Karari sites share certain distinctive morphological features which warrant their grouping as an industry. The name that we have proposed, 'Karari', refers to the Karari Escarpment, where many of the sites are located. The most striking component of the Karari assemblages is a series of distinctive core scrapers, the single most abundant tool form. As is commonly found with early Pleistocene assemblages of the Oldowan and Developed Oldowan, choppers, polyhedrons and discoids are common components of the Karari Industry (Fig. 33:5). Hand-axes and cleavers appear to be a very small component of the Karari Industry.

Another feature of the Karari Industry, characteristic of stone industries throughout the Lower and Middle Pleistocene, is variation in assemblage composition (Isaac 1972c). One of the aims of our long-term study will be to examine the variability among sites of the Karari series. In the next section we shall make some preliminary observations of factors that may have a bearing on this problem.

Behavioural interpretations

General

Extensive archaeological survey at Ileret and along exposures of the Karari Escarpment located more than 50 sites (Figs 33:1 and 33:2). Sixteen of these sites were excavated.

Since sites are components of a single settlement system, the distribution pattern clearly shows traces of hominid activities at:

   (i) low-lying channel and floodplain localities close to the lake margin;

   (ii) stream channel and floodplain localities on the alluvial plain, some 15 to 25 kilometres inland from the shoreline;

   (iii) alluvial fan localities at the foot of the low hills at the rim of the East Lake Turkana Basin, which may indicate that early hominids exploited resources in the valleys and areas of high relief.

The distribution of sites indicates the exploitation of a wide range of micro-habitats, which is consistent with the idea that early hominids were mobile, opportunistic hunters and gatherers who used a wide range of resources.

Several variables emerge from a preliminary analysis of the spatial distribution of sites, their palaeogeographic contexts and associated remains found at individual sites:

   (i) Proximity of sites to stream and river channels;

   (ii) Accessibility and availability of a source of raw material;

   (iii) Variability in stone assemblages and presence or absence of associated remains;

   (iv) Seasonal occupation of sites.

The effect of these variables on the observed patterns of site distribution may shed light on the adaptive behavioural patterns of early hominids.

Proximity of sites to stream and river channels

The distribution of known sites indicates that the locus of hominid activities was at or in close proximity to a water source such as stream or river channels. Even those sites located on the broad floodplains were never more than a few

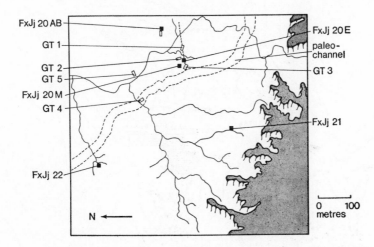

FIG. 33:6. Map of FxJj 20 site complex, FxJj 21 and FxJj 22, Area 131, Karari Escarpment, in relation to the present-day physiography. Hominid KNMER 3230 was recovered *in situ* from FxJj 20 East. Stipled area is Karari Escarpment. ■ = archaeological site; GT = geological trench.

hundred metres from a stream course. Fig. 33:6 shows excavated sites of the FxJj 20 site complex, located in floodplain silts, in relation to the paleo-channel.

This choice of location may be attributed to the physiological requirement for water by early hominids. Moreover, the lack of containers or receptacles for carrying or storing water may have determined the specific location of sites at or adjacent to stream channels.

Proximity to exploitable food sources must have been another critical factor in the choice of this location. Tree and bush lined streams and rivers would have provided food sources such as nuts and berries (Isaac 1969; 1971; 1972a). The congregation of animals at watering places along stream banks may have provided opportunities for hunting in such situations, as well as being likely localities for scavenging the remains of animals killed by carnivores.

Accessibility and availability of raw materials

The accessibility and availability of raw materials appears to have been another critical factor in the decision to locate sites close to stream courses. Sites cluster on the floodplains and alluvial fan deposits close to the basin margin, where

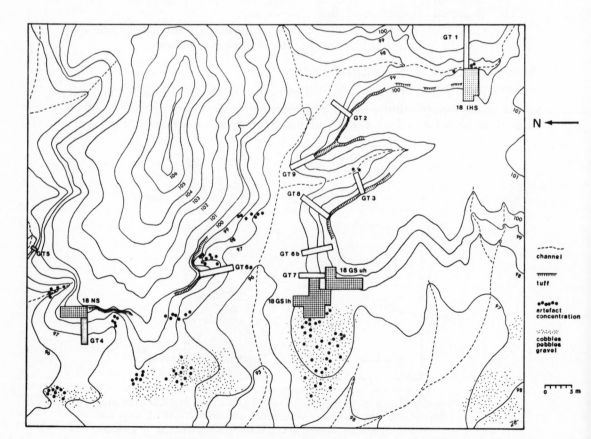

FIG. 33:7. A contoured plan of FxJj 18 site complex, Area 130, Karari Escarpment.

basalt cobbles and boulders were readily available in conglomerates in streams and rivers that drained the volcanic hills (Fig. 33:1). One example is site FxJj 18 GS. Here the evidence clearly indicates that stone artefact manufacture was carried out on the gravel-strewn sand bar of a stream channel (Fig. 33:7). This site also provides evidence for reoccupation of an area over a span in time: approximately one metre above the lower occurrence (18 GS Lower Horizon) was another stratified archaeological horizon (18 GS Upper Horizon). Conversely, there is only a thin scatter of sites of low artefact density close to the lake margin, where sources of raw material were scarce.

The dependence upon or the regularity in the use of stone artefacts in the behavioural repertoire of early man at this stage of his evolution may account for sites clustering on the alluvial plain, close to an available source of raw material. In general, Karari sites show a greater diversity of tool types than is present in assemblages of the KBS Industry (Fig. 33:4). A comparable shift is observed between Bed I and middle Bed II at Olduvai Gorge (Leakey, M. D. 1971). This greater diversity in stone tool forms is accompanied by an increase in the density of stone artefact assemblages on the Karari sites. Considered together, these factors seem to be indicative of a greater propensity by early hominids toward the utilisation of stone artefacts.

Variability of stone assemblage composition and the presence or absence of associated remains

The scope of the archaeological investigations has encompassed the reconstruction and survey of a local sedimentary basin and the location and excavation of sites in a mosaic of diverse habitats. We suggest that there is a pattern to the distribution of sites which may account for the variety of stone artefacts and associated remains found at individual sites. Sites located in floodplain contexts are found associated with faunal remains that are often poorly preserved or in a badly comminuted state, while 'channel' sites are almost completely devoid of any associated fauna. Table 2 shows quite distinct differences in the stone artefact assemblage composition at sites in these different sedimentary contexts.

Isaac has suggested that in some instances the nature, composition and spatial configuration of an archaeological occurrence with preserved bone specimens found in association with stone artefacts, could be interpreted as a 'home base' (Isaac 1967). In the case of the Karari sites, a large proportion of the 'home base' localities are found in river floodplain contexts. The fauna recovered from the excavations represent:

(a) riverine habitat—hippopotamidae, crocodylidae and fish;
(b) woodland/savanna habitat—rhinocerotidae, proboscidae, bovidae, suidae, equidae.

The remains found in association at home base sites show that the choice of location was affected by the accessibility of a whole set of resources. In the case of the Karari sites, the scatter of fragmentary bone specimens at these sites represents hunting, scavenging and/or collecting behaviours in which both riverine and open woodland/savanna habitats were exploited. The exploitation of two diverse habitats may have led to a subdivision of the social unit into smaller groups for subsistence activities, i.e. division of labour. The implication is either that the social unit ('band') was exploiting two habitats on at least two different occasions, or that there was a division of responsibilities within the unit and sub-groups were exploiting different habitats at the same time or on different occasions. Whatever the social organisation was at this time, however, the products of their efforts were brought back to a central location. The concept of the home base locus also implies the distribution and the sharing of food.

By contrast, accumulations of stone artifacts found in or on the banks of stream channels more closely approximate the interpretation of a single activity site, i.e. factory site. However, a note of caution must be sounded in the interpretation of any activity 'facies' as one must take into consideration factors that may bias the assemblage composition such as the hydraulics of streams and the presence, absence or partial representation of associated remains.

We have shown that stone artifact assemblages and associated remains vary and differ in composition from one site to another. We have tentatively suggested that at some sites the particular composition may be accounted for as due to a specific activity. However, a loose network of band or tribal affiliations with the idiosyncratic differences of one craftsman to another together with the idea that many of the implements were multi-purpose tools, could easily account for a part of the variation.

Seasonal occupation of sites

Evidence from sites located on the alluvial plain suggest that they may have been occupied seasonally. The remains of riverine faunas, especially crocodile and fish, found at these sites suggest that hominid occupation took place during that part of the annual cycle when water was flowing in the adjacent stream channels. Ephemeral streams and rivers draining the volcanic uplands, flowing across the plains during periods of precipitation, would have enabled crocodiles and fish, usually found in the lake, to migrate to higher reaches of the rivers.

The idea of seasonal occupation is further borne out by what appears to be a short-term reoccupation of FxJj 20 East, located on floodplain silts. Here the vertical scatter of stone artifacts and comminuted bone fragments demonstrates two stratified archaeological horizons, separated by a very thin sterile horizon. Until we study the relationships between fauna found on the sites and the background densities in the beds away from the sites, these suggestions are tentative.

The distribution of sites indicates the exploitation of a wide range of microhabitats. Evidence of hominid activities at sites scattered over an ancient landscape may be indicative of the seasonal availability of food resources.

Relationships to hominids

There is now direct and indirect evidence for the association of hominid remains with archaeological sites of the Karari Industry. At FxJj 20 East, a hominid mandible, KNMER 3230, was recovered *in situ* by excavations in 1975 in tuffaceous floodplain silts of the Okote Tuff complex (Fig. 33:6). The mandible was partially crushed and broken at the symphysis, but the tooth row is complete and the tooth crowns well preserved. The specimen has been assigned to *Australopithecus* cf. *boisei* (Leakey & Harris 1975).

A preliminary analysis of the vertical distribution of stone artifacts and fossil bone specimens at the site shows that the hominid is stratigraphically located approximately 10–15 cms above the main artifact bearing horizon. However, there is a low density scatter of bone fragments and stone artefacts on the same horizon as the hominid. In light of the implications of the possible association of the hominid mandible with the stone artefacts, further excavations will be carried out in the future.

In area 130, a skull (cranium with associated mandible) was recovered *in situ* from channel deposits located stratigraphically below the Okote Tuff complex (Figs 33:2 and 33:8). The hominid, KNMER 1805, has been classified as belonging to the genus *Homo* sp. (Leakey, R. E. F. 1974). A large mandible, KNMER 1806, was found on the surface, 10 to 15 m distance from this specimen, and was identified as belonging to the genus *Australopithecus* cf. *boisei* (Leakey, R. E. F. 1974). Excavations were carried out at both of these localities and a well preserved fossil fauna and light scatter of stone artifacts was recovered, but no further hominid fragments were found. The site complex is one of the best examples at East Lake Turkana to demonstrate the sympatric coexistence of two hominid lineages during the Lower Pleistocene. Forty metres to the east of the hominid localities, on the same stratigraphic horizon, is an archaeological site FxJj 38 North West (Fig. 33:8). The character and composition of the stone assemblage from this site, which has a large proportion of choppers (70 per cent), would suggest affinities to the stone assemblages recovered from Olduvai Bed I (Table 3).

A wealth of hominid fossils, including specimens assigned to the genus

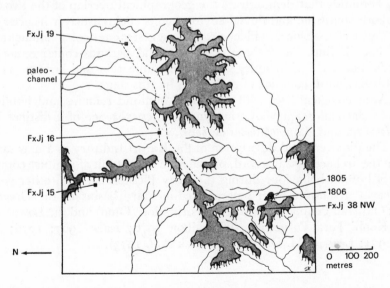

FIG. 33:8. Map of FxJj 15, 16, 19, 38 NW, Area 130, Karari Escarpment, in relation to present-day topography. ■ = archaeological site; ▲ = hominid site.

Australopithecus cf. *boisei* and the genus *Homo* have been recovered in beds of the Ileret Member (Leakey, R. E. F. 1971; 1972; 1973; 1974).

In summary, there are at least two hominid lineages present in the fossil record for this time at East Lake Turkana. The 'identity' of the toolmaker, however, must for the present time remain a matter of conjecture and speculation.

Conclusion

A preliminary description has been given of the sites and assemblages of the Karari Industry, located in a local sedimentary basin to the east of Lake Turkana.

1. An age of 1.2–1.6 million years for the sediments containing sites of the Karari Industry is based on isotopic dates, palaeomagnetic ages and palaeontological evidence.

2. Morphological characteristics of stone artefacts from 16 excavated localities show distinctive features which do not readily conform to those of other stone assemblages found in the Lower Pleistocene: Oldowan, Developed Oldowan and early Acheulian. A series of core scrapers form a very conspicuous set. However, the justification for creating a new taxonomic entity, the Karari Industry, stems from the distinctive character of the Upper Member material as a whole (Harris & Isaac 1976).

3. Detailed stratigraphic mapping and microstratigraphic studies undertaken in the East Lake Turkana area provide the basis for the reconstruction of features of an ancient landscape. Archaeological evidence of hominid activities is found in a mosaic of diverse microhabitats: the lake margin and inland at or adjacent to streams which flowed across a broad alluvial plain. These studies complement the research of Behrensmeyer on the palaeoecological relationships of early hominids that demonstrate the geographical overlap of the two hominid taxa in both lacustrine and fluvial environments (Behrensmeyer in press).

4. The archaeological evidence would suggest stone tool manufacture, meat eating, food sharing, the division of labour, and perhaps cohesive, cooperative band existence and rudimentary verbal communication had become part of the integral behavioural complex of early man by this stage.

5. Archaeological sites with associated hominid remains and hominid sites over the palaeolandscape show the sympatric coexistence of 2 distinct hominid taxa: *Homo* sp. and *Australopithecus* cf. *boisei*.

6. The discovery and description of the Karari Industry as a new taxonomic entity in the archaeological record, provides an opportunity for future comparative studies of human behaviour during the early Pleistocene, with other well-dated and excavated samples from separate sedimentary basins in the Gregory Rift Valley: Olduvai, Peninj, the Shungura Formation (Omo) and the Lower Member of the Koobi Fora Formation (Chavaillon 1976; Isaac 1967; 1976; Isaac & Curtis 1974; Leakey, M. D. 1971; Merrick *et al.* 1973).

Acknowledgements

This research was carried out as part of the archaeological investigations conducted under the direction of G. Isaac. His advice and guidance during the course of this study is gratefully acknowledged. The archaeological investigations are part of the interdisciplinary palaeoanthropological studies being carried out by the Koobi Fora Research Project.

We would like to thank Richard Leakey for his encouragement during this study and assistance provided by the National Museums of Kenya and the Government of Kenya.

R. Bainbridge, A. Behrensmeyer, B. Bowen, D. Burggrauf, I. Findlater, F. Fitch, H. Frank, C. Vondra, and H. White shared their views on the palaeogeography of the region and gave helpful comments on the microstratigraphy of the archaeological sites. J. M. Harris kindly identified the fauna recovered from the sites.

Field assistance has been given by J. Barthelme, D. Crader, D. Gifford, J. Gowlett, M. Mehlman, J. Onyango-Abuje, O. Odak & S. Wandibba. The excavations were carried out in a dedicated and skilled fashion by men of the WaKamba Tribe. John Kimengich, Kitibe Kimeu, Mukilya Magonka and Mutete Nume were site foremen.

Mary Leakey kindly made space available in her laboratory for the analysis of the Karari Industry. Her advice and helpful suggestions during the course of this study are gratefully acknowledged.

This study would not have been possible without the cooperation of all members of the Koobi Fora Research Project, especially Kimoya Kimeu, director of field operations, who organised logistical support.

R. Rodden read a draft of this paper and gave advice. Barbara Isaac has drawn Figs. 33: 1–5 and Judy Cohen compiled the tables and typed the final manuscript.

The archaeological investigations were carried out by grants from the National Science Foundation to G. Isaac as Principal Investigator. The project as a whole depends on financial assistance given to R. Leakey by the National Geographic Society, the L. S. B. Leakey Foundation and other donors.

References

BEHRENSMEYER, A. K. 1974. Late Cenozoic sedimentation in the Lake Rudolf Basin, Kenya. *Annals Geol. Soc. Egypt*, **IV**, 287–306.
——. In press. The habitat of Plio-Pleistocene hominids in East Africa; taphonomic and microstratigraphic evidence. In: Jolly, C. (ed.), *African Hominidae of the Plio-Pleistocene: Evidence, Problems and Strategies*. New York, Duckworth, Inc.
BOWEN, B. E. & VONDRA, C. F. 1973. Stratigraphical relationships of the Plio-Pleistocene deposits, East Rudolf, Kenya. *Nature, Lond.* **242**, 391–3.
BROCK, A. & ISAAC, G. LL. 1974. Palaeomagnetic stratigraphy and chronology of hominid-bearing sediments east of Lake Rudolf, Kenya. *Nature, Lond.* **247**, 344–8.
—— 1976. Reversal stratigraphy and its application at East Rudolf. In: Y. Coppens, F. C. Howell, G. Ll. Isaac and R. E. F. Leakey (eds). *Earliest Man and Environments in the Lake Rudolf Basin*. Chicago, University of Chicago Press, 148–62.

CERLING, T. E., BIGGS, D. L., VONDRA, C. F. & SVEC, H. J. 1975. Use of oxygen isotope ratios in correlation of tuffs, East Rudolf Basin, northern Kenya. *Earth and Planet Sci. Letters* **25**,2 91–6.

CHAVAILLON, J. 1976. Temoignage de l'Activité technique des hominidés du Pleistocene ancien Formation de Shungura, Basse Vallee de l'Omo. Ethiopie. In: Y. Coppens, F. C. Howell, G. Ll. Isaac and R. E. F. Leakey (eds), *Earliest Man and Environments in the Lake Rudolf Basin*. Chicago: University of Chicago Press, 565–73.

CLARK, J. D. 1970. *The Prehistory of Africa*. London: Thames and Hudson.

—— & HAYNES, C. V. 1969. An elephant butchery site at Mwanganda's village, Karonga, Malawi and its relevance for Palaeolithic archaeology. *World Archaeol.* **1**, 390–411.

CURTIS, G. H., DRAKE, R., CERLING, T. E. & HAMPEL, J. 1975. Age of the K.B.D. Tuff in the Koobi Fora Formation, East Rudolf, Kenya. *Nature, Lond.* **358**, 395–8.

FINDLATER, I. C. 1976. Tuffs and the recognition of isochronous mapping units in the Rudolf succession. In: Y. Coppens, F. C. Howell, G. Ll. Isaac, and R. E. F. Leakey (eds). *Earliest Man and Environments in the Lake Rudolf Basin*. Chicago: University of Chicago Press, 94–104.

FITCH, F. J., FINDLATER, I. C., WATKINS, R. T. & MILLER, J. A. 1974. Dating of the rock succession containing fossil hominids at East Rudolf, Kenya. *Nature, Lond.* **251**, 213–15.

—— & MILLER, J. A. 1970. Radioisotope age determination of Lake Rudolf artifact site. *Nature, Lond.* **226**, 226–8.

—— 1976. Conventional Potassium–Argon and Argon–40/Argon–39 dating of volcanic rocks from East Rudolf. In: Y. Coppens, F. C. Howell, G. Ll. Isaac and R. E. F. Leakey (eds). *Earliest Man and Environments in the Lake Rudolf Basin*. Chicago: University of Chicago Press, 123–47.

HARRIS, J. W. K. & ISAAC, G. LL. 1976. The Karari Industry: early Pleistocene archaeological materials from the terrain east of Lake Rudolf, Kenya. *Nature, Lond.* **262**, 102–7.

ISAAC, G. LL. 1967. The stratigraphy of the Peninj Group—early Middle Pleistocene formations west of Lake Natron, Tanzania. In: W. W. Bishop and J. D. Clark (eds). *Background to Evolution in Africa*. Chicago: University of Chicago Press, 229–58.

—— 1969. Studies of early culture in East Africa. *World Archaeol.* **1**, 1–28.

—— 1971. The diet of early man: aspects of archaeological evidence from Lower and Middle Pleistocene sites in Africa. *World Archaeol.* **2**, 278–99.

—— 1972a. Comparative studies of Pleistocene site locations in East Africa. In: P. J. Ucko, R. Tringham and G. W. Dimbleby (eds). *Man, Settlement and Urbanism*. London: Duckworth, Inc., 165–76.

—— 1972b. Chronology and tempo of cultural change during the Pleistocene. In: W. W. Bishop and J. A. Miller (eds). *Calibration of Hominoid Evolution*. Edinburgh: Scottish Academic Press, 381–430.

—— 1972c. Early phases of human behaviour: models in Lower Palaeolithic archaeology. In: D. L. Clarke (ed.). *Models in Archaeology*. London: Methuen & Co., 167–99.

—— 1975. Middle Pleistocene stratigraphy and cultural patterns in East Africa. In: K. W. Butzer and G. Ll. Isaac (eds). *After the Australopithecines*. The Hague: Mouton.

—— 1976. Plio-Pleistocene artifact assemblages from East Rudolf, Kenya. In: Y. Coppens, F. C. Howell, G. Ll. Isaac and R. E. F. Leakey (eds). *Earliest Man and Environments in the Lake Rudolf Basin*. Chicago: University of Chicago Press, 552–64.

——, LEAKEY, R. E. F. & BEHRENSMEYER, A. K. 1971. Archaeological traces of early hominid activities, east of Lake Rudolf, Kenya. *Science*. **173**, 1129–34.

—— & CURTIS, G. H. 1974. Age of early Acheulian industries from the Peninj group, Tanzania. *Nature, Lond.* **249**, 624–7.

——, HARRIS, J. W. K. & CRADER, D. 1976. Archaeological evidence from the Koobi Fora Formation. In: Y. Coppens, F. C. Howell, G. Ll. Isaac and R. E. F. Leakey (eds). *Earliest Man and Environments in the Lake Rudolf Basin*. Chicago: University of Chicago Press, 533–51.

KELLER, C. M. & ISAAC, G. LL. 1971. Reports of two short conferences of archaeologists working in Africa. In: J. D. Clark and G. Ll. Isaac (eds). *Fourth Bulletin of the Commission of Nomenclature of the Pan African Congress on Prehistory*. Berkeley: Mimeographed 1971.

LEAKEY, M. D. 1967. Preliminary survey of the cultural material from Beds I and II, Olduvai

Gorge, Tanzania. In: W. W. Bishop and J. D. Clark (eds). *Background to Evolution in Africa.* Chicago: University of Chicago Press, 417–46.

—— 1971. *Olduvai Gorge Volume 3: Excavations in Beds I and II, 1960–3.* Cambridge: Cambridge University Press.

LEAKEY, R. E. F. 1971. Further evidence of Lower Pleistocene hominids from East Rudolf, North Kenya. *Nature, Lond.* **231**, 241–5.

—— 1972. Further evidence of Lower Pleistocene hominids from East Rudolf, North Kenya, 1971. *Nature, Lond.* **237**, 264–9.

—— 1973. Further evidence of Lower Pleistocene hominids from East Rudolf, North Kenya, 1972. *Nature, Lond.* **242**, 170–3.

—— 1974. Further evidence of Lower Pleistocene hominids from East Rudolf, North Kenya, 1973. *Nature, Lond.* **248**, 653–6.

LEAKEY, R. E. F., BEHRENSMEYER, A. K., FITCH, F. J., MILLER, J. A. & LEAKEY, M. D. 1970. New hominid remains and early artefacts from northern Kenya. *Nature, Lond.* **226**, 223–30.

—— & HARRIS, J. W. K. 1975. Relationships of early Pleistocene hominids from Ileret and the Karari Escarpment to archaeological sites of the Karari Industry. Paper presented to the 44th Annual Meeting of the American Association of Physical Anthropologists, Denver, Colorado. Abstract, *Amer. J. Phys. Anthrop.* **42**, 313.

MAGLIO, V. J. 1972. Vertebrate faunas and chronology of hominid-bearing sediments east of Lake Rudolf, Kenya. *Nature, Lond.* **239**, 381–6.

MERRICK, H. V., DE HEINZELIN, J., HAESAERTS, P. & HOWELL, F. C. 1973. Archaeological occurrences of early Pleistocene age from the Shungura Formation, Lower Omo Valley, Ethiopia. *Nature, Lond.* **242**, 572–5.

VONDRA, C. F. & BOWEN, B. E. 1976. Plio-Pleistocene deposits and environments. In: Y. Coppens, F. C. Howell, G. Ll. Isaac and R. E. F. Leakey (eds). *Earliest Man and Environments in the Lake Rudolf Basin.* Chicago: University of Chicago Press, 79–93.

——, JOHNSON, G. D., BOWEN, B. E. & BEHRENSMEYER, A. K. 1971. Preliminary stratigraphical studies of the East Rudolf Basin, Kenya. *Nature, Lond.* **231**, 245–9.

34

DON C. JOHANSON, MAURICE TAIEB,
B. T. GRAY & YVES COPPENS

Geological framework of the Pliocene
Hadar Formation (Afar, Ethiopia)
with notes on paleontology
including hominids

Although detailed studies in the central Afar, in particular Hadar, have only recently begun, we feel that it is possible to present a provisional resumé of our research.

The Pliocene deposits of Hadar, about 160 m thick, represent a series of lacustrine and perilacustrine sediments with four major marker horizons. These deposits contain a rich and well preserved fossil fauna comprising over 70 species. The nature of the faunal assemblage especially the frequent remains of Nyanzachoerus pattersoni, Notochoerus euilus, Ceratotherium praecox, *early* Elephas recki, Hipparion *cf.* primigenium, *and the absence of* Equus *and tetraprotodont* Hippopotamus, *strongly suggests a biostratigraphic correlation with other sites in eastern Africa considered to be older than 3 million years. Fossil hominid remains, some uniquely complete, have been recovered from nine stratigraphic levels.*

In the north-east of Ethiopia, at about 9°N, the Ethiopian Rift, which represents the northern extension of the East African Rift System, is intersected by the Red Sea and Gulf of Aden Rifts, and widens out into the Afar Depression. All three systems have participated in the formation of the depression since the Miocene (Barberi *et al.* 1975; Black *et al.* 1975; Christiansen *et al.* 1975; Juch 1975; Zanettin & Justin-Visentin 1975; Mohr 1975).

Between 10° and 12°N, on the western margin of the Afar, a large sedimentary basin of Plio-Pleistocene age in the Hadar area (Fig. 34:1) has yielded a rich fauna of vertebrates including hominid remains (ca. 3 m.y.) From a structural viewpoint, this marginal basin is closely comparable with the Mio-Pliocene basin in the south-eastern Afar which is delimited by the occurrence of the Chorora Formation (Sickenberg & Schönfeld 1975). This formation belongs to the 'Lowermost Afar Series' of Christiansen *et al.* (1975) which is dated between 11 and 8 m.y. and outcrops along the foothills of the south-eastern escarpment. The continental sediments in these two basins are the oldest so far found in the Afar Depression. In contrast, sediments of Mio-Pliocene age appear to be completely lacking in the Ethiopian Rift, which was dominated by ignimbritic volcanism (Meyer *et al.* 1975; Morbidelli *et al.* 1975).

The Pliocene deposits of Hadar belong to the Afar Series. They overlie basalts in the Ethiopian escarpment region which are dated between 8 and 5 m.y. (Kunz *et al.* 1975). Basalts of this age are also exposed in the Millé area (11°30'N, 40°30'E), about 40–50 km east of the escarpment. On the western margin of the Afar the sediments contain only occasional intercalated basaltic flows dated around 3 m.y. whereas east of 40°30' stratoid basaltic activity has been more continuous between 3.6 and 2.2 m.y. (Varet 1973). Volcanism has been particularly intense along the present axial grabens of Issa and Tendaho, aligned respectively NE and NW, where the flows exceed several tens of metres in thickness. The marginal Plio-Pleistocene basin was tectonically controlled by the updoming and fissural volcanism occurring along the lines of these future grabens. The only sediments so far recognized from this stage in the evolution of the central Afar floor are dated between 3–4 and 1 m.y., and predate the inception of oceanic volcanism along

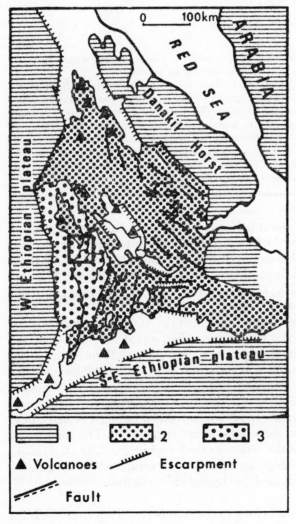

FIG. 34:1.

Location of the Plio/Pleistocene sedimentary basin in the Afar Depression:

1. Trap series and basement

2. Afar Series

3. Plio/Pleistocene lacustrine and fluvial deposits.

the axial ranges (Barberi *et al.* 1972). The base of these sediments has not been observed.

Travelling from west to east along 11°N from the Ethiopian escarpment towards the axial grabens, one observes the following succession over a distance of 80–100 km:

— a series of piedmont alluvial fan deposits

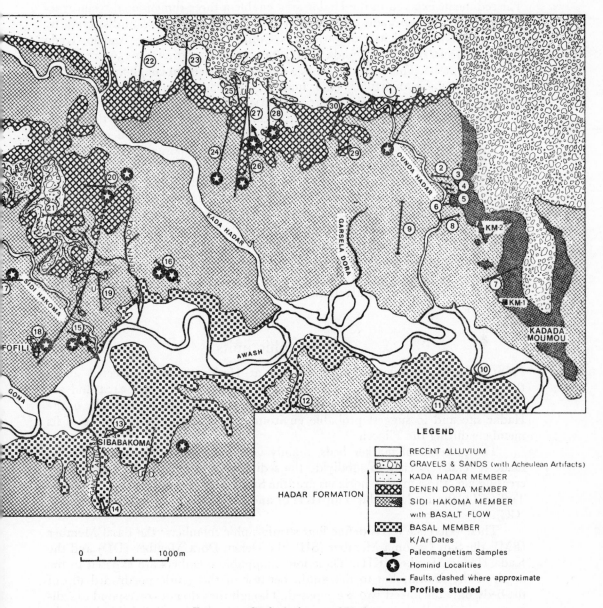

FIG. 34:2. Geological map of Hadar

— a series of delta, swamp and lake-margin sediments
— a series of typical lacustrine sediments
Only the second of these, represented at Hadar, has been studied in any detail
so far, and forms the subject of this article.

The Hadar area

The sediments exposed in the Hadar area enable a three-dimensional reconstruction to be made, especially around 11°10′N, 40°30′E where recent dissection by
the Awash River and its tributaries, the Gona, Gango Akidora, Andelo, Kada
Hadar, Garsela Dora and Ounda Hadar has cut down more than 150 m into the
Pliocene deposits. This incision has also affected the detrital conglomeratic sheets
containing Acheulean artifacts (Corvinus 1975) which channel and cap the whole
sequence. Along the Awash River a gravel and silt terrace at + 10 to 12 m represents
the most recent sedimentary unit in the area. The geomorphology of this region
of the Afar is thus relatively very recent. Furthermore, the earlier deposits show
very little tectonic disturbance in contrast to the contemporary basalt sequences
along the axial depressions of Tendaho & Issa (Taieb 1976). North of the Awash,
the sediments dip NNW, the angle of dip increasing eastwards from 2° to 4° near
Kadada Moumou. To the south of the Awash the dip is only about 2°. The sediments are also cut by a few short normal faults, aligned N to NE, with throws of
5 to 30 m which tend to diminish towards the south. For example the displacement
along the fault north of Kada Hadar, between profiles 24 and 27, which is 1.1 km
long, varies from 23 m to about 5 m N–S.

We can conclude that tectonic activity in this marginal zone of the Afar
Depression has been of relatively minor importance during the middle and late
Pleistocene.

Stratigraphic units

From some one hundred profiles studied at Hadar we have selected about
thirty of the most significant. They are shown in Figs 34:3 and 34:4 and are
located in Fig. 34:2. These sections enable us to establish the stratigraphy of the
Hadar site and to suggest probable environments of deposition for some of its
members during the Pliocene.

The presence of marker beds, mainly volcanic tuffs, throughout the area,
made it relatively easy to subdivide the sedimentary sequence. In order of decreasing age, these marker horizons are: the Sidi Hakoma Tuff (SHT), the Triple
Tuff (TT), the Kada Hadar Tuff (KHT) and a 1 m thick bed of green argillite
(CC).

These reference levels define four stratigraphic members: the Basal Member
(BM), the Sidi Hakoma Member (SH), the Denen Dora Member (DD) and the
Kada Hadar Member (KH). These four mappable members are exposed to the
north of the Awash, but to the south, because of the gentle northward dip of
the beds, only the lowest two are exposed. Though they do not correspond to sedimentary cycles, these members are in certain areas clearly subdivisible into sub-

members. Four submembers at most are to be found in the SH Member (SH–1, –2, –3, –4), three in the DD Member (DD–1, –2, –3) and four in the KH Member (KH–1, –2, –3, –4). The submembers distinguished in this way have been defined for the SH Member in the area between the Sidi Hakoma and the Denen Dora (profile 16); for the DD Member above the Denen Dora and for the KH Member in the northern part of the Kada Hadar (profile 22). Lateral variations in facies are often noted, as can be seen in Figs 34:3 and 34:4. Consequently, some of these submembers are not found throughout the area. Further correlations would therefore be premature at this stage and we prefer not to attempt detailed reconstructions of the paleogeography.

The Basal Member is incompletely exposed and only its upper strata are observed (about 15 m). It consists essentially of massive brown detrital clays and grey silts with small carbonate nodules, associated with thin sand horizons. To the south of the Awash the silts include two carbonate-cemented horizons overlying a well-developed sandstone. The exposed beds probably reflect sedimentation in a shallow basin, as indicated by the presence of detrital sands and carbonate horizons.

The SHT Volcanic Marker Bed outcrops close to the Awash River. It comprises at least two levels of white tuff, the lower being the thicker. Towards the Gango Akidora, to the south of the Awash, these volcanic ashes accumulated in a small basin, and attain a total thickness of about 10 m. On the left bank of the Awash (to the south of profile 16) the second tuff passes laterally into a channel fill where glass shards are coarser and exhibit very pronounced cross-bedding suggesting deposition in rapidly-flowing water. This channel is oriented 24°N and can be observed over a distance of about 125 m.

According to J. L. Aronson (Taieb *et al.* 1976) the K/Ar dates on the glass shards (3.1 to 5.5 m.y.) are inconsistent, due to an excess of radiogenic argon. The 4 m.y. average age of the SHT is therefore not meaningful.

The Sidi Hakoma Member in the west, spans about 40 to 50 m, reaching 90 m in thickness towards the Kadada Moumou in the east. Together with the overlying member it is the most widely exposed in the basin, and very numerous and complex lateral variations in facies are observed within it. From the base upwards, the most widespread units are as follows:

— the fossiliferous sandstone bed forming the base of the SH–2 submember. In the west, the sandstone caps numerous outliers (profiles 15 and 16) whereas to the east, they are surmounted by higher units (profiles 13, 14 and perhaps 19, 24).

— the gastropod-bearing level(s) in the SH–4 submember (profiles 7, 8, 9, 14, 18, 19, 24), constitutes shelly siltstone. It should be noted that it may form two distinct horizons, as can be seen towards the Sidi Hakoma (profiles 18 and 19).

The lower part of the SH Member (SH–1) consists of silty clays surmounted by two carbonate layers underlying the SH–2 sandstone. These silty clays are cut by lenses of fluvial deposits.

The intermediate part of the SH Member (SH–2 and SH–3) consists almost entirely of clayey silts with diffuse small carbonate concretions ('poupées')

The SH–4 submember is similar though carbonate may occur in more or less continuous beds of limited lateral extent (profiles 2, 3, 14 and 19). Furthermore, the silts are intersected by coarser detrital material (profiles 3, 14, 18 and 19) consisting of either conglomeratic channel fills or more commonly cross-bedded fluvial sands. These coarse detrital lenses have no constant stratigraphic position. They become less common towards the east, where the sediments are mainly silts and silty clays. Finally, the upper part of the SH–4 consists of compact, green-grey, finely laminated clays with extensive ferruginous mottling.

To the east, the middle part of the SH–4 submember includes a single basalt flow (2.5 km²) which forms the Kadada Moumou plateau. Its stratigraphic position has been clearly established by reference to a sand bed situated 10 m below the basalt, which can be followed along profiles 4 to 7 (Fig. 34:4). The basalt is therefore situated 21 m below the TT and approximately 60 m above the SHT. It is dated by two K/Ar determinations of 2.95 ± 0.2 m.y. and 3.0 ± 0.2 m.y. (Taieb *et al.* 1976). Six hominid levels have been found below it (three represented in profiles 15, 16, 24) and three above (profiles 20, 27, 28).

To summarize, the whole SH Member comprises facies of flood-plain, delta plain and delta margin type. Shallow lacustrine conditions are briefly manifested in SH–4 by the presence of the carbonate and gastropod-bearing horizons. These are underlain by dark reducing clays, rich in organic matter, which are of probable lagoonal origin. Finally, the very extensive mottled clays at the top of the SH Member most probably indicate a return to fluctuating shallow lacustrine conditions.

The Triple Tuff Marker Bed lies within the massive green-grey argillites with slickensides and small carbonate concretions. Immediately above the ferruginous horizon within the SH–4, three thin beds of whitish tuff, TT–1, TT–2 and TT–3, are to be found. Their thicknesses range from 1 to 10 cm, TT–2 being generally thicker and frequently reaching 5 to 6 cm. These ashes were deposited at the time of the maximum lacustrine extension seen at Hadar. Possible source areas for this explosive volcanism are rather distant such as the acidic volcanoes of Gurrale, Ida Ale and Gabillema, located 60 km N, 30 km NE and 80 km E of Hadar, respectively. These ashes were deposited quasi-uniformly over the whole Hadar region. An earlier ashfall is also evident 2 m below the ferruginous horizon. It is very thin (1–2 cm) and only found north of the Kada Hadar (between profiles 24 and 26). In that area it is therefore the TT–1, just above which determines the lower limit of the Denen Dora Member. Moreover, further to the west on the left bank of the Kada Hadar, the TT–3 occasionally reaches 20 to 30 cm in thickness, thinning towards the Kadada Moumou. The clays between the TT–1 and TT–2 levels include a 0.5–1 m thick layer rich in smooth-shelled ostracods. This has a constant stratigraphic position and towards the west includes fragments of gastropod shells. Such a high concentration of ostracods suggests a change in the lake-water chemistry. Future studies may enable us to determine whether the ostracod concentration was associated with the addition of chemical constituents of volcanic origin.

To sum up, the three volcanic horizons and the ostracod horizon represent an excellent marker level.

The Denen Dora Member exhibits similar facies over the whole Hadar site. Its thickness shows very slight variations, being about 30 m in the west and reaching 40 m in the east, near the Kadada Moumou. The lacustrine sedimentation observed in SH–4 continues in DD–1, although showing a trend towards fluvial conditions in the west. This is evident from the sand and sandstone beds in the Sidi Hakoma sector (profiles 17, 18, 19) and north of the Kada Hadar (profile 23). These fluvial incursions continued during the deposition of the middle part of the DD Member, whereas to the east, in the Kadada Moumou, essentially clayey facies continued to dominate (DD–1 and DD–2 submembers).

The lacustrine influence diminishes in the upper part of the intermediate submember where silts with carbonate concretions appear. Further evidence for this regressive phase is provided by the occurrence of fluvial deposits and abundant root casts. The latter are particularly common to the west and north-west (profiles 17, 20 and 26). The regression was probably characterized by swampy conditions with impeded drainage, which favoured the accumulation of carbonate.

At the top, the DD Member includes an ubiquitous and very fossiliferous sand-sandstone unit. These sands probably represent the deposits of an extensive distributary network which spread them over the whole Hadar site. Some of the channels carried coarser, pebble grade material.

To sum up, the DD Member is characterized by shallow-water lacustrine sedimentation passing up into swamp and even floodplain deposits. The hominid-bearing levels as well as the more or less complete faunal remains occur mainly within the intermediate and upper horizons.

The Kada Hadar Tuff and CC Marker Beds denote the boundary between the DD and KH Members. The KHT is a whitish-grey volcanic tuff consisting, like the SHT, of fine glass shards. Its thickness is about 20 to 40 cm but to the north of the Kada Hadar (profile 22) it passes laterally into a channel where the grain size is coarser. This channel, which is exposed over a distance of 325 m, trends N 77°E disappearing when it reaches the Kada Hadar.

The KHT is not a very good marker bed because it is only exposed in the NW and N of the Hadar site (profiles 20, 21, 22, 27, 28 and 30). South of the Sidi Hakoma, at Afofili, it passes into a beige-yellow tuffaceous clay. To the east, in the Kadada Moumou sector and beyond the Ounda Hadar it is no longer present. Its patchy distribution is probably related to the more localized sedimentation evident near the top of the DD Member and the base of the KH Member.

Nevertheless, about 10 m above the KHT, another marker bed, the 'Confetti Clay' (CC), with a thickness of 0.8 to 1 m, occurs throughout the whole Hadar site. It consists of finely laminated grey to green clays which in weathered exposures appear in the form of fine flakes similar to the ostracod-bearing horizon in the DD Member. This type of weathering results from the high percentage of swelling montmorillonitic clay and from the micro-layering. The CC Marker Bed proves that a second clearly lacustrine phase extended over the whole Hadar site. It was probably short and less important than the one observed at the base of the DD Member.

The lower limit of the KH Member therefore remains poorly defined in the areas where the KHT is missing. However, the occurrence of the CC enables us

Mammalia

    Artiodactyla

        Hippopotamidae
            Hippopotamus sp., *H.* cf. *imaguiculus*
            Hippopotamus sp.
        Suidae
            Sus sp.
            Notochoerus euilus
            Nyanzachoerus pattersoni
        Giraffidae
            Sivatherium maurusium
            Giraffa sp., *G.* cf. *jumae*
            Giraffa sp., *G.* cf. *gracilis*
            Giraffa sp., *G.* cf. *pygmaeus*
        Bovidae
            Ugandax sp.
            Tragelaphus sp., *T.* cf. *nakuae*
            Tragelaphini, gen. et sp. indet.
            Hippotragus sp.
            Kobus sp.
            Aepyceros sp.
            aff. *Parmularis* sp.
            Alcelaphini, gen. et sp. indet.
            Neotragini, gen. et sp. indet.
            Ovibovini, gen. et sp. indet.

    Perissodactyla

        Equidae
            Hipparion sp., *H.* cf. *primigenium*
            Hipparion sp.
         Rhinocerotidae
             Ceratotherium sp., *C.* cf. *praecox*
            Ceratotherium sp., *C.* cf. *simum*
            Diceros sp., *D.* cf. *bicornis*

    Proboscidea

        Elephantidae
            Elephas recki
            ?*Elephas* sp.
            Loxodonta sp., *L.* cf. *adaurora*
            Primelephas sp.

    Deinotheroidea

        Deinotheridae
            Deinotherium bozasi

    Rodentia

            Hystrix sp.
            Mastomys sp.
            Xenohystrix sp., *X.* cf. *crassidens*
            Xerus sp.
            Tachyoryctes sp.
            Pelomys sp.
            Oenomys sp.
            Tatera sp.

    Carnivora

        Canidae
            gen. et sp. indet.
        Hyaenidae
            Crocuta sp.
            Percrocuta sp.
            Euryboas sp.
         Mustelidae
            ?*Enhydriodon* sp.
            Poecilogale sp.
            Ictonyx sp.
         Felidae
            Megantereon sp.
            Homotherium sp.
            ?*Dinofelis* sp.
            Felis sp.
            Panthera sp.
         Viverridae
            Viverra sp.
            Herpestes sp.
            Galerella sp.

TABLE I:

Provisional faunal list from Hadar, Central Afar.

Primates

    Cercopithecidae
        Papio sp.
        Theropithecus sp.
        Colobus sp.

    Hominidae
        Australopithecus sp. *A.* aff. *africanus*
        Homo sp.

Chiroptera

    gen. et sp. indet.

Aves

    gen. et sp. indet.
    gen. et sp. indet.

Reptilia

    Chelonia
        Geochelone sp.
        Trionyx sp.
    Crocodilia
        Crocodilus sp.
    Ophidia
        gen. et sp. indet.
    Lacertilia
        gen. et sp. indet.

Osteichthyes

    Siluridae
        gen. et sp. indet.

Mollusca

    Gastropoda
        gen. et sp. indet.
    Bivalvia
        gen. et sp. indet.

Arthropoda

    Ostracoda
        gen. et sp. indet.

    Decapoda
        Potamidae
            gen. et sp. indet.

to locate it with considerable confidence. In addition, the silts and clayey-silts above and below the KHT are quite distinctive and in certain cases permit its stratigraphic position to be accurately determined.

The Kada Hadar Member is exposed only north of the Awash River, particularly on the left bank of the Kada Hadar, where it attains a thickness of 60–80 m. It has been studied only briefly thus far. The exposures upstream on the Kada Hadar, observed by J-J. Tiercelin & R. C. Walter and located in the NW of (the area represented in) Fig. 34:2, have not been reproduced in Fig. 34:4.

The KH Member consists primarily of slickensided brown and beige silty clays with carbonate nodules, cut by sands and conglomerates (profiles 21, 22, 25, 28). The correlation between the different sand lenses is still difficult to establish because most of them are not very extensive. Some are fossiliferous and fill channels which clearly truncate the underlying deposits (e.g., at the base of profile 27). The subdivisions of the KH Member are based on profile 22, and may have to be revised.

In the top of the upper KH which is partly represented in Fig. 33:3 two volcanic events are recorded, the former being characterized by two thin horizons of whitish ash and the latter by 0.6 m of beige tuffaceous sand. The upper KH Member includes a microconglomerate unconformably overlain by a cobble-gravel with Acheulian implements. These observations remain provisional and it would be premature to define more precisely the mode of sedimentation. However, a short lacustrine episode is evident from the CC argillite near the base. This lacustrine facies, as well as the swamp facies observed in the lower members, does not occur in the intermediate and upper parts of the KH Member where thin fining and coarsening-upwards sequences are seen. Brown silty clays without apparent stratification are also observed, which are very rarely fossiliferous.

We tentatively suggest that the deposits reflect rapid infilling of the Hadar basin by waters heavily loaded with fine suspended material, the area of deltaic sedimentation having probably moved to the east (profile 1 in Fig. 34:4) where a very fossiliferous sandstone bed similar to the sands in the DD–3 is observed.

Paleontology

Typical of the sites of eastern Africa, the mammalian fauna of the Hadar Formation is quite diverse (Table 1), although not as much as that of areas such as the lower Omo valley or the exposures east of Lake Turkana (Coppens *et al.* 1976). This is due to a longer period of field research and to the longer time spans represented by the Omo and East Turkana deposits. Many specimens are remarkably well preserved owing largely to what appears to be a rather unusual taphonomic setting (Taieb *et al.* 1976). The material is not uniformly distributed throughout the stratigraphic sequence, although the apparent absence of fauna above KH–2 is due primarily to a lack of intensive collection. The notably lower density in the SH–3 through the lower portion of the DD–1 submember is more real. Thus the collection may conveniently be broken down into three faunal units: lower (SH–1 and 2), middle (upper DD–1 and DD–2) and upper (DD–3 and KH–1, especially the basal sand horizon of the former). This subdivision will undoubtedly require future revision. While apparently largely homogenous, certain differences may be seen from the lower to the upper faunal units which may be attributed to evolutionary change and hence are of assistance in biostratigraphic correlation.

The artiodactyls are, as usual, the most common order represented in the collections. Among the suiformes are two common hippopotamids, both hexa-protodont. Suids are well represented by three genera. Notable is the fact that *Nyanzachoerus pattersoni* decreases markedly in its frequency above SH–2, correlating nicely with its apparent time of disappearance from the eastern African record and the K/Ar determinations for the basalt. Size increases may also be seen in the posterior dentition of *Notochoerus euilus* and *Sus* sp.

The bovids are at present the most diverse family, with *Tragelaphus* cf. *nakuae*, *Aepyceros* sp. and *Kobus* sp. dominating the assemblage. The former two are particularly abundant in the lower and upper units, while *Kobus* is the most common element known from the middle. Alcelaphines appear to increase in their frequency in the upper unit. Rather unusual but interesting forms include a neotragine, a more kudu-like tragelaphine and an ovibovine.

Perissodactyls are also common in the assemblage. The rhinocerotids are represented by both living genera, although *Diceros* by only two immature speci-mens. *Ceratotherium* is much more abundant, with the primitive *C. praecox* pre-dominating. Two equids are known, including *Hipparion* cf. *primigenium*, a very early form which is the largest known in eastern Africa. Some change can be seen in the equids, particularly in increased crown height and ectostylid development in the cheek teeth and reduction of I_3 through time.

The Proboscidea are remarkably frequent in the assemblage, due primarily to the fact that *Elephas recki* is the most common element in the lower and upper faunal units. It is also common in the middle unit, but this is overshadowed by the great abundance of *Kobus*. Some evolution may also be seen in the dentition of *E. recki*, commensurate with the approximate time range for the Hadar Forma-tion inferred from the Suidae. Perhaps distinctive is a separate form closely resembling *Elephas*, but which also shows some striking resemblances to *Mam-muthus*. A few unusual elephantid specimens have also been collected which suggest the existence of *Primelephas* in Hadar.

Other, less common forms are also represented by rodents, quite diverse in spite of the rather limited application of specialized techniques required for the recovery of abundant microfaunal material. Murids, hystricids and a primitive *Tachyoryctes* are present, among other forms. Notable carnivores include what are at present the earliest known *Crocuta* and one of the earliest canids. An unusual giant mustelid (*?Enhydriodon*), also known from the lower Omo valley, has been recorded. The non-mammalian fauna has not yet been studied in sufficient detail to allow additional remarks.

The presence of certain elements such as *Ceratotherium praecox*, *Hipparion primigenium*, *Primelephas*, early *Elephas* including *E. recki*, and *Nyanzachoerus*, as well as the notable absence of *Equus* and tetraprotodont hippos all suggest an early fauna, correlative with zone I in the Omo (Coppens & Howell 1976) and zone V of Coppens' (1972) African biozonation, datable to *ca.* 2.6–3.1 m.y. (Shuey, Brown & Croes 1974). In addition, the evolution observable in certain features of the equids, the suids and *Elephas*, as well as the strong decline in frequency of *Nyanzachoerus* above SH–2, are all consistent with the time range indicated by biostratigraphic comparisons as well as radiometric age data. Other striking frequency variations seen in various taxa through the sequence are likely due to ecological shifts which are broadly comparable to those recorded geologically.

TABLE 2: *Fossil Hominid Specimens from Hadar.*

Afar Locality (A.L.)	Date of Discovery	Description
	1973	
128–1	30 October	Left proximal femur fragment
129–1a	30 October	Right distal femur
129–1b	30 October	Right proximal tibia
129–1c	30 October	Right proximal femur fragment
166–9	11 December	Left temporal fragment
	1974	
188–1	16 October	Right mandibular corpus; M_2–M_3
198–1	18 October	Left mandibular corpus; C–P_4, dm, M_1–M_2
198–17a, b	5 December	Left I^1 and I^2
198–18	5 December	Left I_2
199–1	17 October	Right maxilla; C–M^3
200–1a, b	17 October	Complete maxilla; 16 teeth. Right M_1
211–1	20 October	Right proximal femur fragment
228–1	27 October	Diaphysis of right femur
241–14	22 December	Left lower molar
266–1	16 November	Mandibular corpus; left P_3–M_1, right P_3–M_3
277–1	19 November	Left mandibular corpus, C–M_2
288–1	24 November	Partial skeleton: occipital and parietal fragments; mandibular corpus with left P_3, M_3, right P_3–M_3, mandibular condyles; right scapula fragment; right humerus; proximal and distal left humerus; proximal and distal right and left ulnae; proximal and distal right radius; distal left radius; left capitate; 2 phalanges; 6 thoracic vertebrae and fragments; 1 lumbar vertebrae; sacrum; left innominate; left femur; proximal and distal right tibia; right talus; right distal fibula; numerous rib fragments.

The Hominidae

Intensive survey of a small portion of the Hadar Formation exposures has resulted in the recovery of an exceptional collection of hominid specimens. The current collection represents 12 individuals (Table 2) from nine stratigraphic horizons (Fig. 34:5). Characteristic of field research in eastern Africa, these discoveries were surface finds, but the lack of post-exposure alteration of the remarkably intact bone material supports the assumption that the specimens were recovered from situations closely associated with their *in situ* positions. In a number of instances it

was possible to match the matrix adhering to the specimen with a specific stratigraphic horizon located during detailed study of stratigraphic profiles.

Extensive and detailed anatomical, comparative, functional, statistical and
biomechanical investigations are currently in progress. A number of contributions
specifically concerning the hominid material are available (Johanson & Taieb
1976; Johanson & Coppens 1976; Johanson, Taieb & Coppens 1976; Johanson
et al. 1976) and our intention here is to simply provide the major aspects of our
preliminary assessment of the phyletic affinities and resemblances of the Hadar
hominid material with other Plio/Pleistocene hominid collections.

Recovery of fossil hominid specimens commenced in October, 1973, with
the fortunate discovery of four associated lower limb bone fragments (A.L. 128–1,
A.L. 129–1a, b, c). The proximal femoral fragments lack most of their necks and
heads but excellent surface anatomy provides details of muscle position and in-
sertion. The associated right distal femur and proximal tibia provide the only
knee joint (lacking a patella) of such antiquity. Anatomical aspects such as a high

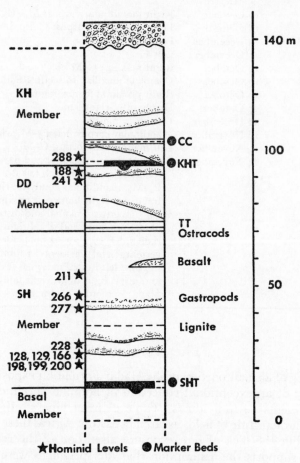

FIG. 34:5. Stratigraphic positions of hominid localities in the Hadar Formation.

preservation yielding such delicate items as crocodile and turtle eggs, rodent skulls and even crab claws.

Collection of a single hominid specimen (thus far the oldest most complete hominid ever recovered) composed of some 52 major osteological elements and numerous associated fragments is extraordinary. We are now in a position to more extensively comprehend the anatomy of an early hominid than was previously possible. Innumerable facets of articulations (joints), limb proportions, stature, biomechanical aspects, functional anatomy, etc. may be extensively pursued. In depth analysis will take considerable effort and time; however, it is appropriate to present some impressions and observations of 'Lucy's' skeleton.

Unfortunately most of the cranium is lost, but parietal and occipital fragments are clearly thin, with the sutures fused and lacking pronounced nuchal and temporal musculature.

The mandible is V-shaped, possesses a slight post-incisive planum with clear genioglossal fossae, a reduced incisor region and apparently relatively large canine (socket). The third premolars are small, have open anterior foveae, a well developed protoconid and virtual absence of a metaconid. Many of these features, especially when taken collectively are suggestive of a 'primitive' status for A.L. 288–1. A V-shaped mandible has been considered primitive (Leakey 1974) and is seen in KNM–ER 1482 (Leakey 1974) and Omo 18 (Arambourg & Coppens 1967).

The female status of A.L. 288–1 is suggested by a broad sciatic notch, an obtuse sub-pubic angle and a ventral arc on the pubis of the well preserved left innominate. Aspects of the sacrum confirm the female morphology of the innominate. In general, the innominate is similar to Sts 14 from Sterkfontein in the Transvaal; however, the ilium is higher with a straighter anterior border and the acetabulum is curiously shallow.

The left femur is complete, however the distal end requires some reconstruction due to crushing. Its length is about 280 mm and when compared to the humerus (235 mm long) a value of 83.9 is obtained for the humeral/femoral index.

For the moment the A.L. 288–1 partial skeleton is considered to have affinities to *A. africanus* (Sterkfontein *sensu stricto*) but because of certain distinctive aspects of the mandible, its dentition and of the pelvis and limb proportions it is not thought to be identical to *A. africanus*. We suggest that it is phylogenetically less advanced than the Sterkfontein material.

In review, then, the Hadar hominid collection may be composed of taxa which show strong resemblances to *Homo* on the one hand and with *A. africanus* on the other.

Appreciation of all Plio/Pleistocene hominid remains is constantly changing and we are cognizant of continued revision and re-evaluation concerning the earliest stages of human origins. We hope, however, that the hominids collected from the Hadar Formation will prove useful in unravelling the exciting puzzle of mankind's origins.

Acknowledgements

We very much appreciate the contributions of our colleagues and assistants of the International Afar Research Expedition, in particular: Ato Alemayehu Asfaw,

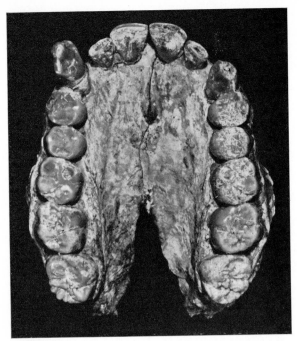

FIG. 34:6.

A.L. 200–1a, occlusal view

FIG. 34:7.

A.L. 266–1, occlusal view

FIG. 34:8.

A.L. 277–1, occlusal view

34:6

34:7

34:8

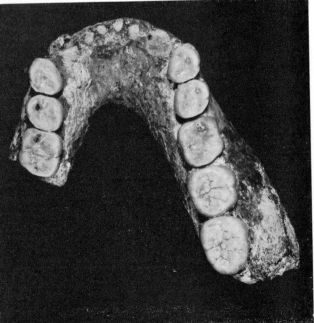

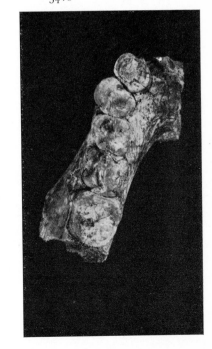

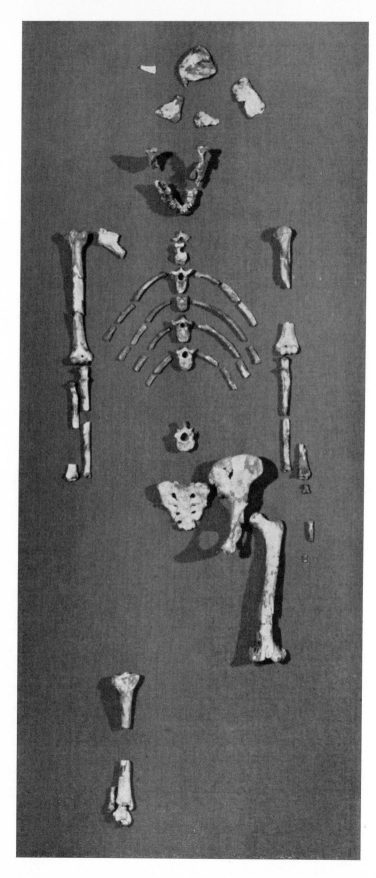

A.L. 288–1, anterior view

bicondylar angle, flattened and elongated (anteroposteriorly) lateral condyle
a raised lateral margin of the patellar groove are related to bipedal locor
activity. Recent biomechanical studies (Johanson *et al.* 1976) augment this
clusion by demonstrating external tibial rotation during extension and the cap
for hyperextension. Hence, it would appear that human bipedalism has a
siderable ancestry and by lowest Hadar Formation times the knee join
perfected for this unique locomotor mode.

Also from the lowest levels of the stratigraphic sequence, a complete
(A.L. 200–1a) (Fig. 34:6) and a demi-palate (A.L. 199–1) were collected
15 m of one another and may very well be derived from the same stratig
horizon.

The A.L. 200 specimen is interesting in that the incisors exhibit ribbe
wear. Also the canine tips show chippage and the buccal surfaces of the
premolars and the canines have suffered antemortem enamel loss. In contr
postcanine teeth do not have extreme occlusal wear.

Certain aspects of the A.L. 200 palate, such as the long sub-parallel ar
the well developed alveolar prognathism, are 'primitive'. However, the prese
large central incisors, large canines (separated from the lateral incisors
astemata) and large cheek teeth have parallels with more recent fossil m
such as KNM–ER 1590 (Leakey 1974) and *Pithecanthropus* IV from Java,
as to some older material from Laetolil (L.H.–3, –6) (Leakey *et al.* 1976
hominid material has been referred to the genus *Homo*.

The right demi-palate, relative to the A.L. 200 specimen, is smaller
but its similarities in nearly all details of the morphology with the larger
are provocative. It is our contention that these two specimens sample th
taxon and differences in size reflect normal variation. Similarities of the A
specimen are with KNM–ER 1813 (Leakey 1974) and with O.H. 13. Th
specimen has been assigned to *Homo habilis* (Leakey, Tobias & Napier 19
the former to *A. africanus* (Leakey 1974). It may not be unreasonable on t
of diagnostic features of the KNM–ER 1813 cranium (aside from its low
capacity) to consider its closest affinities to be with something like *Homo*

Mandibular material represented by a nearly complete specimer
266–1) (Fig. 34:7) and a fragmentary left lower jaw (A.L. 277–1) (Fig
exhibit clear U-shaped arches. Furthermore, aspects of occlusal morphol
tooth size of these specimens are reminiscent of material from Laetolil (I
Olduvai (O.H. 7) and Koobi Fora (KNM–ER 1802) which have been
to *Homo*.

A tantalizing fragmentary proximal femur (A.L. 211–1) is large an
affinities with O.H. 20, described by Day (1969) as *A. robustus*. Similaritie
size, lack of trochanteric flare, flattened femoral neck, weak intertrochant
and a hidden lesser trochanter when the specimen is viewed from the
aspect. For the moment, however, we prefer to reserve taxonomic identifi
A.L. 211–1 until more complete material is recovered.

Undoubtedly the most outstanding hominid thus far collected from
represented by a partial female skeleton (A.L. 288–1) (Fig. 34:9) k
'Lucy'. She was derived from a sandstone horizon characterized by sup

J. L. Aronson, M. Beden, R. Bonnefille, M. E. Bush, G. Corvinus, V. Eisenmann, C. Guerin, C. Guillemot, J.-J. Jaeger, J. C. Kolar, N. Page, G. Petter, P. Planques, T. J. Schmitt, B. A. Sigmon, H. Roche, J-J. Tiercelin, R. C. Walter & Ato Johannes Zeleke. We would like to thank (Miss) F. A. Street for her assistance and critical comments during the preparation and translation of the geological section.

Special gratitude is expressed to the Ethiopian Government, the Ministry of Culture, Youth and Sports Affairs and the Institute of Archaeology. This work has been supported financially by the Centre national de la recherche Scientifique, the National Science Foundation, the Harry Frank Guggenheim Foundation, the Cleveland Museum of Natural History, the National Geographic Society, the L. S. B. Leakey Foundation and the Singer-Polignac Foundation.

We express our gratitude to Dr Bishop for his invitation to contribute to a most exciting and useful conference.

References

ARAMBOURG, C. & COPPENS, Y. 1967. Sur la découverte dans le Pléistocène inférieur de la vallée de l'Omo (Éthiopie) d'une mandibule d'australopithecien. *C. R. Acad. Sciences*, Paris, **265**-D, 589–90.

BARBERI, F., BORSI, S., FERRARA, G., MARINELLI, G., SANTACROCE, R., TAZIEFF, F. & VARET, J. 1972. Evolution of the Danakil Depression (Afar, Ethiopia) in light of radiometric age determinations. *J. Geol.* **80**, 720–9.

——, FERRARA, G., SANTACROCE, R. & VARET, J. 1976. Structural evolution of the Afar triple junction. In: Pilger and Rösler (eds), Vol. **I**, pp. 38–54.

BLACK, R., MORTON, W. H. & REX, D. C. 1976. Block tilting and volcanism within the Afar in the light of recent K/Ar age data. In: Pilger and Rösler (eds), Vol. **I**, pp. 269–99.

CHRISTIANSEN, T. B., SCHAEFER, H.-U. & SCHÖNFELD, M. 1976. Geology of southern and central Afar, Ethiopia. In: Pilger and Rösler (eds). Vol. **I**, pp. 259–76.

COPPENS, Y. 1972. Tentative de zonation du Pliocène et du Pléistocène d'Afrique par les grands mammifères. *C. R. Acad. Sciences*, **274**-D, 181–4.

—— & HOWELL, F. C. 1976. Mammalian faunas of the Omo group: distributional and bio-stratigraphical aspects. In: Coppens, *et al.* (eds), pp. 177–92.

——, ——, ISAAC, G. LL. & LEAKEY, R. E. F. (eds) 1976. Earliest Man and Environments of the Lake Rudolf Basin. Univ. of Chicago Press, Chicago.

CORVINUS, G. 1975. Palaeolithic remains at the Hadar in the Afar region. *Nature, Lond.* **256**, 468–71.

DAY, M. H. 1969. A robust australopithecine fragment from Olduvai Gorge, Tanzania (Hominid 20). *Nature, Lond*, **221**, 230–3.

JOHANSON, D. C. & COPPENS, Y. 1976. A preliminary anatomical diagnosis of the first Plio-Pleistocene hominid discoveries in the central Afar, Ethiopia. *Am. J. Phys. Anthrop.* **45**, 217–34.

——, LOVEJOY, C. O., BURSTEIN, A. H. & HEIPLE, K. G. 1976. Functional implications of the Afar knee joint. *Am. J. Phys. Anthrop.* **44**, 188.

—— & TAIEB, M. 1976. Plio-Pleistocene hominid discoveries in Hadar, Ethiopia. *Nature, Lond.* **260**, 293–7.

——, —— & COPPENS, Y. 1976. Plio/Pleistocene hominid discoveries in Hadar, central Afar, Ethiopia. In: C. Jolly (ed.), *African Hominidae of the Plio-Pleistocene*, Duckworth, London. In press.

JUCH, D. 1976. Geology of the South–Eastern Escarpment of Ethiopia between 39° and 42° long. East. In: Pilger and Rösler (eds), Vol. **I**, pp. 310–15.

KUNZ, K., KREUZER, H. & MÜLLER, P. 1976. Potassium–Argon age determinations of the Trap basalt of the south-eastern part of the Afar Rift. In: Pilger and Rösler (eds), Vol. **I**, pp. 370–4.

LEAKEY, L. S. B., TOBIAS, P. V. & NAPIER, J. R. 1964. A new species of the genus *Homo* from Oldu-vai Gorge. *Nature, Lond.* **202**, 7–9.

LEAKEY, M. D., HAY, R. L., CURTIS, G. M., DRAKE, R. E., JACKES, M. K. & WHITE, T. D. 1976. Fossil hominids from the Laetolil Beds. *Nature, Lond.* **262**, 460–6.

LEAKEY, R. E. F. 1974. Further evidence of lower Pleistocene hominids from East Rudolf, North Kenya. *Nature, Lond.* 248, 653–6.

MEYER, W., PILGER, A., RÖSLER, A. & STETS, J. 1976. Tectonic evolution of the northern part of the Main Ethiopian Rift in southern Ethiopia. In: Pilger and Rösler (eds), Vol. **I**, pp. 352–61.

MOHR, P. A. 1976. Structural setting and evolution of Afar. In: Pilger and Rösler (eds), Vol. **I**, pp. 27–37.

MORBIDELLI, L., NICOLETTI, M., PETRUCCIANI, C. & PICCIRILLO, E. M. 1976. Ethiopian South–Eastern Plateau and related escarpment: K/Ar ages of the main volcanic events (Main Ethiopian Rift from 8°10′ to 9°00′ lat. North). In: Pilger and Rösler (eds), Vol. **I**, pp. 362–9.

PILGER, A. & RÖSLER, A. (eds) 1976. Afar Depression of Ethiopia. Inter-Union Commission on Geodynamics, Scientific Report No. **14**. 2 vols. Stuttgart: E. Schweizerbart'sche Verlags-buchhandlung.

SHUEY, R. T., BROWN, F. H. & CROES, M. K. 1974. Magnetostratigraphy of the Shungura Forma-tion, south-western Ethiopia: fine structure of the lower Matuyama polarity epoch. *Earth and Planetary Science Letters*, **23**, 249–60.

SICKENBERG, O. & SCHÖNFELD, M. 1976. The Chorora Formation—Lower Pliocene limnical sediments in the southern Afar (Ethiopia). In: Pilger and Rösler (eds), Vol. **I**, pp. 277–83.

TAÏEB, M. 1976. Evolution of Plio-Pleistocene Sedimentary basin of the Central Afar (Awash Valley, Ethiopia). In: Pilger and Rösler (eds), Vol. **II**, pp. 80–7.

——, JOHANSON, D. C., COPPENS, Y. & ARONSON, J. L. 1976. Geological and palaeontological background of Hadar hominid site, Afar, Ethiopia. *Nature, Lond.* **260**, 289–93.

VARET, J. 1973. Critères pétrologiques, géochemiques et structuraux de la genèse et de la dif-férenciation des magmas basaltiques: exemple de l'Afar. Thèse Doctorat d'Etat, Univ. Paris–Sud.

ZANETTIN, B. & JUSTIN-VISENTIN, E. 1976. Tectonical and volcanological evolution of the western Afar margin (Ethiopia). In: Pilger and Rösler (eds), **I**, 300–9.

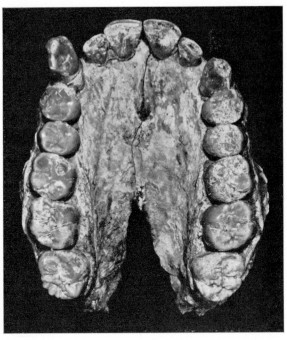

Fig. 34:6.

    A.L. 200–1a, occlusal view

Fig. 34:7.

    A.L. 266–1, occlusal view

Fig. 34:8.

    A.L. 277–1, occlusal view

34:6

34:7

34:8

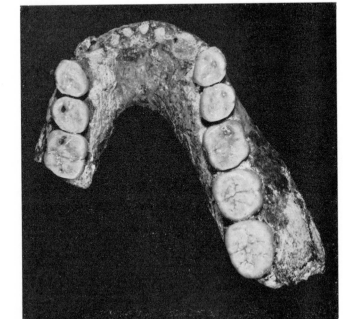

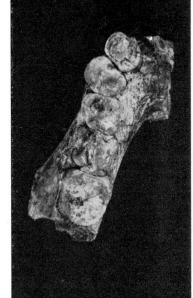

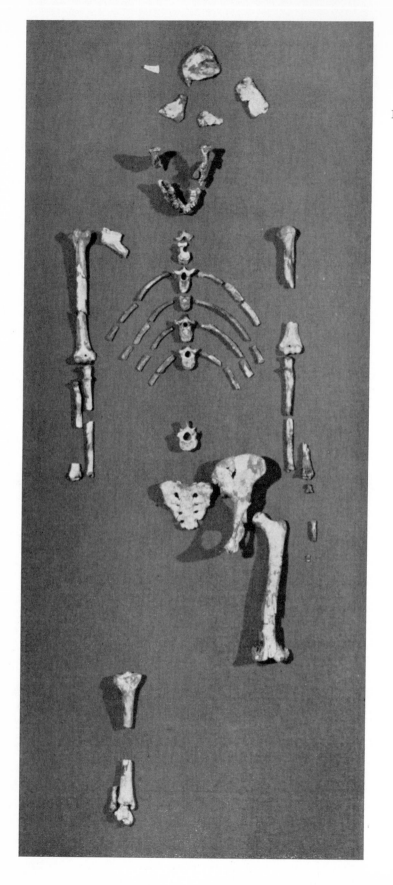

FIG. 34:9.

A.L. 288-1, anterior
view

J. L. Aronson, M. Beden, R. Bonnefille, M. E. Bush, G. Corvinus, V. Eisenmann, C. Guerin, C. Guillemot, J.-J. Jaeger, J. C. Kolar, N. Page, G. Petter, P. Planques, T. J. Schmitt, B. A. Sigmon, H. Roche, J-J. Tiercelin, R. C. Walter & Ato Johannes Zeleke. We would like to thank (Miss) F. A. Street for her assistance and critical comments during the preparation and translation of the geological section.

Special gratitude is expressed to the Ethiopian Government, the Ministry of Culture, Youth and Sports Affairs and the Institute of Archaeology. This work has been supported financially by the Centre national de la recherche Scientifique, the National Science Foundation, the Harry Frank Guggenheim Foundation, the Cleveland Museum of Natural History, the National Geographic Society, the L. S. B. Leakey Foundation and the Singer-Polignac Foundation.

We express our gratitude to Dr Bishop for his invitation to contribute to a most exciting and useful conference.

References

ARAMBOURG, C. & COPPENS, Y. 1967. Sur la découverte dans le Pléistocène inférieur de la vallée de l'Omo (Éthiopie) d'une mandibule d'australopithecien. *C. R. Acad. Sciences*, Paris, **265**–D, 589–90.

BARBERI, F., BORSI, S., FERRARA, G., MARINELLI, G., SANTACROCE, R., TAZIEFF, F. & VARET, J. 1972. Evolution of the Danakil Depression (Afar, Ethiopia) in light of radiometric age determinations. *J. Geol.* **80**, 720–9.

——, FERRARA, G., SANTACROCE, R. & VARET, J. 1976. Structural evolution of the Afar triple junction. In: Pilger and Rösler (eds), Vol. **I**, pp. 38–54.

BLACK, R., MORTON, W. H. & REX, D. C. 1976. Block tilting and volcanism within the Afar in the light of recent K/Ar age data. In: Pilger and Rösler (eds), Vol. **I**, pp. 269–99.

CHRISTIANSEN, T. B., SCHAEFER, H.-U. & SCHÖNFELD, M. 1976. Geology of southern and central Afar, Ethiopia. In: Pilger and Rösler (eds). Vol. **I**, pp. 259–76.

COPPENS, Y. 1972. Tentative de zonation du Pliocène et du Pléistocène d'Afrique par les grands mammifères. *C. R. Acad. Sciences*, **274**–D, 181–4.

—— & HOWELL, F. C. 1976. Mammalian faunas of the Omo group: distributional and bio-stratigraphical aspects. In: Coppens, *et al.* (eds), pp. 177–92.

——, ——, ISAAC, G. LL. & LEAKEY, R. E. F. (eds) 1976. Earliest Man and Environments of the Lake Rudolf Basin. Univ. of Chicago Press, Chicago.

CORVINUS, G. 1975. Palaeolithic remains at the Hadar in the Afar region. *Nature, Lond.* **256**, 468–71.

DAY, M. H. 1969. A robust australopithecine fragment from Olduvai Gorge, Tanzania (Hominid 20). *Nature, Lond*, **221**, 230–3.

JOHANSON, D. C. & COPPENS, Y. 1976. A preliminary anatomical diagnosis of the first Plio-Pleistocene hominid discoveries in the central Afar, Ethiopia. *Am. J. Phys. Anthrop.* **45**, 217–34.

——, LOVEJOY, C. O., BURSTEIN, A. H. & HEIPLE, K. G. 1976. Functional implications of the Afar knee joint. *Am. J. Phys. Anthrop.* **44**, 188.

—— & TAIEB, M. 1976. Plio-Pleistocene hominid discoveries in Hadar, Ethiopia. *Nature, Lond.* **260**, 293–7.

——, —— & COPPENS, Y. 1976. Plio/Pleistocene hominid discoveries in Hadar, central Afar, Ethiopia. In: C. Jolly (ed.), *African Hominidae of the Plio-Pleistocene*, Duckworth, London. In press.

JUCH, D. 1976. Geology of the South–Eastern Escarpment of Ethiopia between 39° and 42° long. East. In: Pilger and Rösler (eds), Vol. **I**, pp. 310–15.

KUNZ, K., KREUZER, H. & MÜLLER, P. 1976. Potassium–Argon age determinations of the Trap basalt of the south-eastern part of the Afar Rift. In: Pilger and Rösler (eds), Vol. **I**, pp. 370–4.

564   Don C. Johanson, Maurice Taieb, B. T. Gray & Yves Coppens

LEAKEY, L. S. B., TOBIAS, P. V. & NAPIER, J. R. 1964. A new species of the genus *Homo* from Oldu-
vai Gorge. *Nature, Lond.* **202**, 7–9.
LEAKEY, M. D., HAY, R. L., CURTIS, G. M., DRAKE, R. E., JACKES, M. K. & WHITE, T. D. 1976.
Fossil hominids from the Laetolil Beds. *Nature, Lond.* **262**, 460–6.
LEAKEY, R. E. F. 1974. Further evidence of lower Pleistocene hominids from East Rudolf, North
Kenya. *Nature, Lond.* 248, 653–6.
MEYER, W., PILGER, A., RÖSLER, A. & STETS, J. 1976. Tectonic evolution of the northern part of
the Main Ethiopian Rift in southern Ethiopia. In: Pilger and Rösler (eds), Vol. **I**, pp. 352–61.
MOHR, P. A. 1976. Structural setting and evolution of Afar. In: Pilger and Rösler (eds), Vol. **I**,
pp. 27–37.
MORBIDELLI, L., NICOLETTI, M., PETRUCCIANI, C. & PICCIRILLO, E. M. 1976. Ethiopian South-
Eastern Plateau and related escarpment: K/Ar ages of the main volcanic events (Main
Ethiopian Rift from 8°10′ to 9°00′ lat. North). In: Pilger and Rösler (eds), Vol. **I**, pp. 362–9.
PILGER, A. & RÖSLER, A. (eds) 1976. Afar Depression of Ethiopia. Inter-Union Commission on
Geodynamics, Scientific Report No. **14**. 2 vols. Stuttgart: E. Schweizerbart'sche Verlags-
buchhandlung.
SHUEY, R. T., BROWN, F. H. & CROES, M. K. 1974. Magnetostratigraphy of the Shungura Forma-
tion, south-western Ethiopia: fine structure of the lower Matuyama polarity epoch. *Earth
and Planetary Science Letters*, **23**, 249–60.
SICKENBERG, O. & SCHÖNFELD, M. 1976. The Chorora Formation—Lower Pliocene limnical
sediments in the southern Afar (Ethiopia). In: Pilger and Rösler (eds), Vol. **I**, pp. 277–83.
TAIEB, M. 1976. Evolution of Plio-Pleistocene Sedimentary basin of the Central Afar (Awash
Valley, Ethiopia). In: Pilger and Rösler (eds), Vol. **II**, pp. 80–7.
——, JOHANSON, D. C., COPPENS, Y. & ARONSON, J. L. 1976. Geological and palaeontological
background of Hadar hominid site, Afar, Ethiopia. *Nature, Lond.* **260**, 289–93.
VARET, J. 1973. Critères pétrologiques, géochemiques et structuraux de la genèse et de la dif-
férenciation des magmas basaltiques: exemple de l'Afar. Thèse Doctorat d'Etat, Univ. Paris-
Sud.
ZANETTIN, B. & JUSTIN-VISENTIN, E. 1976. Tectonical and volcanological evolution of the western
Afar margin (Ethiopia). In: Pilger and Rösler (eds), **I**, 300–9.

bicondylar angle, flattened and elongated (anteroposteriorly) lateral condyle and a raised lateral margin of the patellar groove are related to bipedal locomotor activity. Recent biomechanical studies (Johanson *et al.* 1976) augment this conclusion by demonstrating external tibial rotation during extension and the capacity for hyperextension. Hence, it would appear that human bipedalism has a considerable ancestry and by lowest Hadar Formation times the knee joint was perfected for this unique locomotor mode.

Also from the lowest levels of the stratigraphic sequence, a complete palate (A.L. 200–1a) (Fig. 34:6) and a demi-palate (A.L. 199–1) were collected within 15 m of one another and may very well be derived from the same stratigraphic horizon.

The A.L. 200 specimen is interesting in that the incisors exhibit ribbon-like wear. Also the canine tips show chippage and the buccal surfaces of the third premolars and the canines have suffered antemortem enamel loss. In contrast the postcanine teeth do not have extreme occlusal wear.

Certain aspects of the A.L. 200 palate, such as the long sub-parallel arch and the well developed alveolar prognathism, are 'primitive'. However, the presence of large central incisors, large canines (separated from the lateral incisors by diastemata) and large cheek teeth have parallels with more recent fossil material such as KNM–ER 1590 (Leakey 1974) and *Pithecanthropus* IV from Java, as well as to some older material from Laetolil (L.H.–3, –6) (Leakey *et al.* 1976); this hominid material has been referred to the genus *Homo*.

The right demi-palate, relative to the A.L. 200 specimen, is smaller in size but its similarities in nearly all details of the morphology with the larger palate are provocative. It is our contention that these two specimens sample the same taxon and differences in size reflect normal variation. Similarities of the A.L. 199 specimen are with KNM–ER 1813 (Leakey 1974) and with O.H. 13. The latter specimen has been assigned to *Homo habilis* (Leakey, Tobias & Napier 1964) and the former to *A. africanus* (Leakey 1974). It may not be unreasonable on the basis of diagnostic features of the KNM–ER 1813 cranium (aside from its low cranial capacity) to consider its closest affinities to be with something like *Homo habilis*.

Mandibular material represented by a nearly complete specimen (A.L. 266–1) (Fig. 34:7) and a fragmentary left lower jaw (A.L. 277–1) (Fig. 34:8) exhibit clear U-shaped arches. Furthermore, aspects of occlusal morphology and tooth size of these specimens are reminiscent of material from Laetolil (L.H.–4), Olduvai (O.H. 7) and Koobi Fora (KNM–ER 1802) which have been assigned to *Homo*.

A tantalizing fragmentary proximal femur (A.L. 211–1) is large and shows affinities with O.H. 20, described by Day (1969) as *A. robustus*. Similarities include size, lack of trochanteric flare, flattened femoral neck, weak intertrochanteric line and a hidden lesser trochanter when the specimen is viewed from the anterior aspect. For the moment, however, we prefer to reserve taxonomic identification of A.L. 211–1 until more complete material is recovered.

Undoubtedly the most outstanding hominid thus far collected from Hadar is represented by a partial female skeleton (A.L. 288–1) (Fig. 34:9) known as 'Lucy'. She was derived from a sandstone horizon characterized by superb fossil preservation yielding such delicate items as crocodile and turtle eggs, rodent skulls and even crab claws.

Collection of a single hominid specimen (thus far the oldest most complete hominid ever recovered) composed of some 52 major osteological elements and numerous associated fragments is extraordinary. We are now in a position to more extensively comprehend the anatomy of an early hominid than was previously possible. Innumerable facets of articulations (joints), limb proportions, stature, biomechanical aspects, functional anatomy, etc. may be extensively pursued. In depth analysis will take considerable effort and time; however, it is appropriate to present some impressions and observations of 'Lucy's' skeleton.

Unfortunately most of the cranium is lost, but parietal and occipital fragments are clearly thin, with the sutures fused and lacking pronounced nuchal and temporal musculature.

The mandible is V-shaped, possesses a slight post-incisive planum with clear genioglossal fossae, a reduced incisor region and apparently relatively large canine (socket). The third premolars are small, have open anterior foveae, a well developed protoconid and virtual absence of a metaconid. Many of these features, especially when taken collectively are suggestive of a 'primitive' status for A.L. 288–1. A V-shaped mandible has been considered primitive (Leakey 1974) and is seen in KNM–ER 1482 (Leakey 1974) and Omo 18 (Arambourg & Coppens 1967).

The female status of A.L. 288–1 is suggested by a broad sciatic notch, an obtuse sub-pubic angle and a ventral arc on the pubis of the well preserved left innominate. Aspects of the sacrum confirm the female morphology of the innominate. In general, the innominate is similar to Sts 14 from Sterkfontein in the Transvaal; however, the ilium is higher with a straighter anterior border and the acetabulum is curiously shallow.

The left femur is complete, however the distal end requires some reconstruction due to crushing. Its length is about 280 mm and when compared to the humerus (235 mm long) a value of 83.9 is obtained for the humeral/femoral index.

For the moment the A.L. 288–1 partial skeleton is considered to have affinities to *A. africanus* (Sterkfontein *sensu stricto*) but because of certain distinctive aspects of the mandible, its dentition and of the pelvis and limb proportions it is not thought to be identical to *A. africanus*. We suggest that it is phylogenetically less advanced than the Sterkfontein material.

In review, then, the Hadar hominid collection may be composed of taxa which show strong resemblances to *Homo* on the one hand and with *A. africanus* on the other.

Appreciation of all Plio/Pleistocene hominid remains is constantly changing and we are cognizant of continued revision and re-evaluation concerning the earliest stages of human origins. We hope, however, that the hominids collected from the Hadar Formation will prove useful in unravelling the exciting puzzle of mankind's origins.

Acknowledgements

We very much appreciate the contributions of our colleagues and assistants of the International Afar Research Expedition, in particular: Ato Alemayehu Asfaw,

NAMES AND ADDRESSES OF CONTRIBUTORS

ANNA K. BEHRENSMEYER
Department of Palaeontology
University of California
Berkeley, California 94720, U.S.A.

BRUCE E. BOWEN
Department of Earth Sciences
Iowa State University
Ames, Iowa 50010, U.S.A.

ANDREW BROCK
Department of Physics
University of Botswana
Lesotho & Swaziland
Roma, Lesotho, Africa

MAUREEN BROOK
Institute of Geological Sciences
Geochemical Division
64–78 Gray's Inn Road, London

FRANK H. BROWN
Department of Geology
University of Utah
Salt Lake City, Utah, U.S.A.

B. W. CERLING
Department of Geology & Geophysics
University of California
Berkeley, California 94720, U.S.A.

T. E. CERLING
Department of Geology & Geophysics
University of California
Berkeley, California 94720, U.S.A.

GREGORY R. CHAPMAN
Department of Geology
Bedford College
Regent's Park, London

YVES COPPENS
Musee de l'Homme
Palais de Chaillot 75116, Paris, France

SHIRLEY CAMERON CORYNDON
Department of Geology
University of Bristol
Bristol

GARNISS H. CURTIS
Department of Geology & Geophysics
University of California
Berkeley, California 94720, U.S.A.

PETER DAGLEY
Sub-Department of Geophysics
University of Liverpool
Liverpool

R. E. DRAKE
Department of Geology & Geophysics
University of California
Berkeley, California 94720, U.S.A.

G. G. ECK
University of Washington
Seattle, Washington, U.S.A.

IAN C. FINDLATER
Geology Department
Birbeck College
University of London
London

FRANK J. FITCH
Department of Geology
Birbeck College
University of London
London

ALAN W. GENTRY
British Museum
(Natural History)
Cromwell Road, London

JOHN A. J. GOWLETT
Department of Archaeology
University of Cambridge
Cambridge

B. T. GRAY
Department of Anthropology
Case Western Reserve University
Cleveland, Ohio 44106, U.S.A.

J. H. HAMPEL
Department of Geology & Geophysics
University of California
Berkeley, California 94720, U.S.A.

John W. K. Harris
Department of Anthropology
University of California
Berkeley, California 94720, U.S.A.

R. L. Hay
Department of Geology & Geophysics
University of California
Berkeley, California 94720, U.S.A.

Ingrid Herbich
Department of Anthropology
University of California
Berkeley, California 94720, U.S.A.

Andrew P. Hill
National Museums of Kenya
P.O. Box 40658, Nairobi, Kenya

Paul J. Hooker
Department of Geodesy & Geophysics
University of Cambridge
Cambridge

F. Clarke Howell
Department of Anthropology
University of California
Berkeley, California 94720, U.S.A.

Glynn Ll. Isaac
Department of Anthropology
University of California
Berkeley, California 94720, U.S.A.

M. K. Jackes
Department of Anthropology
University of Toronto
Toronto, Canada

D. Carl Johanson
Department of Anthropology
Case Western Reserve University
Cleveland, Ohio 44106, U.S.A.

Sir Peter Kent, f.r.s.
Chairman
Natural Environment Research Council
Alhambra House, 27–33 Charing Cross Road,
   London

M. Aftab Khan
Department of Geology
The University of Leicester
Leicester

Basil C. King
Department of Geology
Bedford College
Regent's Park, London

Mary D. Leakey
Centre for Prehistory & Palaeontology
P.O. Box 30239, Nairobi, Kenya

John A. Miller
Department of Geodesy & Geophysics
University of Cambridge
Cambridge

Alan E. Mussett
Oliver Lodge Laboratory
University of Liverpool
Liverpool

Ronald Oxburgh
Department of Geology & Mineralogy
University of Oxford
Parks Road, Oxford

H. C. Palmer
Department of Geophysics
University of Western Ontario
London, Ontario, Canada

Martin H. L. Pickford
Department of Geology
Queen Mary College
University of London
Mile End Road, London

Robert J. G. Savage
Department of Geology
University of Bristol
University Walk, Bristol

Robert M. Shackleton
School of Earth Sciences
The University
Leeds

Ralh T. Shuey
Department of Geology & Geophysics
717 Mineral Science Building
University of Utah
Salt Lake City, Utah 84112, U.S.A.

Christopher J. Swain
Department of Geology
The University
Leicester

Maurice Taieb
Centre National de la
Recherche Scientifique
Laboratoire de Bellevue
1 Place Aristide Briemd, 92190 Mendon,
France

Peter W. J. Tallon
5 Muswell Hill Road
London

T. D. White
Department of Anthropology
University of Michigan
Ann Arbor, Michigan 48104, U.S.A.

Laurence A. J. Williams
Department of Environmental Sciences
The University, Lancaster

Peter G. Williamson
Department of Geology
University of Bristol
University Walk, Bristol

Carl F. Vondra
Department of Earth Sciences
Iowa State University
Ames, Iowa 50010, U.S.A.

Bernard A. Wood
Department of Anatomy
Middlesex Hospital Medical School
Cleveland Street, London

Subject Index

Aberdare Mountains, 24, 37, 45, 50, 57
Acheulian artifacts, 141, 144, 145, 152, 154, 181, 194, 200, 202, 229, 232, 309, 316–18, 361, 371–3, 529, 544, 552, 557
  age of, 361, 373
Acheulian sites, 199, 200, 309, 322–4, 329–60
Aepyceros, 304, 305, 307, 431, 500, 552, 558
Afar Depression, 19, 20, 21, 23, 24, 25, 26
Afar Series, 549–64
Afar Triple Junction, 15, 71–3
Albertine Rift, *see* Western Rift
Alcelaphines, 294, 299, 300–7, 323, 431, 435–6, 500, 502, 503, 558
Algal deposits, 263, 264, 265, 276, 277, 312, 316, 361, 364, 365, 368, 369, 371, 385, 388, 401, 402, 403, 405, 406,
Allia Bay, 397, 399, 401, 403, 409, 411, 412, 419
Allometric growth, 125–38, 298
Allometry coefficient, 126, 131, 132
Alluvial deposits, 190, 192, 194, 197, 201, 313, 320, 364, 365, 375, 383, 387, 389, 390, 391, 397, 409, 422, 424, 539, 540, 542, 551
Ankaramitic basalt, 312, 313, 316
Antelopes, 201, 293, 297, 300, 301, 302, 303, 305, 307, 371, 500
Archaeological sites, 141–5
  Acheulian, 199, 200, 309, 322–4, 329–60
  Chesowanja, 141, 144, 319, 321, 323–5,
  East Lake Turkana, 141, 143, 144, 395, 406, 408, 415, 529
  Kapthurin, 324, 338, 350, 361–73
  Karari Industry, 141, 142, 529–47
  Kariandusi, 25, 127, 338, 350, 352, 253, 354
  KBS Industry, 529, 536, 541
  Kilombe, 232, 329–60
  Olduvai, 141–5, 151–5, 343, 352, 541, 543
  Olorgesailie, 174, 175, 180, 181, 199–203, 351, 353, 357
  Omo, 141–3, 504–5, 544
Archaeology, 139–47
Artifacts (*see also* Tools), 365, 395, 403–4, 406, 408, 463, 505
  Acheulian, *see* Acheulian artifacts
  dating, 446
  Developed Oldowan, 141–2, 154, 202, 309, 323, 324, 538, 544,

Karari, 533–8, 543–4
Levallois, 197
Oldowan, 143–4, 152–4, 309, 323, 324, 538, 544,
stone, 329–60, 533–42
Aterir Beds, 286, 287, 288, 289, 293, 304
Australopithecus, 309, 312, 322, 323, 504–5
  allometry, 132, 134
  bone collections, 93, 94
  teeth, 169
Australopithecus africanus, 504, 505, 561, 562
Australopithecus boisei, 153, 322, 529, 543, 544
Australopithecus robustus, 322, 505, 561
Awash River, 552, 553, 557

Baboon bones, 94, 126, 145, 201
Baringo Basin (*see also* Chesowanja, Kapthurin Formation, Kilombe, Lukeino Formation, Ngorora Formation), 23, 25, 207–373
  fossil bovids in, 293–308
  fossil hippopotamids in, 279–92
  geochronology, 207–23, 225–35, 443
  geological map, 208, 238, 251
  geology, 237–62, 361–73
  map, 226, 362
  palaeomagnetic stratigraphy, 225–35
  palaeontology, 263, 274–7
  sedimentation, 220–1, 280–2
  soil, 255
  stratigraphy, 207–23, 227, 230, 231, 234–5, 263–73, 277, 280–2
  structure, 210–11, 251, 252
  volcanism, 220–2
Baringo Lake, 25, 29, 36, 45, 47, 50, 59, 71, 77, 193, 251, 320, 361, 362, 368, 371
Baringo trachyte, 228, 364, 371, 372
Bartabwa Gap, 243, 251, 252, 253, 269, 270
Bartekero Valley, 361, 365, 366, 367, 371
BBS Tuff, *see* Okote Tuff
Bifaces, 202, 324, 337, 339, 340, 344, 345, 346–355, 357, 358, 529, 536, 538
Biostratigraphic correlation, 118–22, 195–6, 249–51, 279, 281–2, 421, 425–32, 434–7, 443, 473–95, 533, 549
Biostratigraphy, 103–18, 441–2, 445, 502–3, 517, 519–24,
Bird fossils, 164, 180, 255, 258, 276

Bison, 91, 92, 93
Bluck's conglomerate, 383, 386,
Bone tools, 89, 93, 94, 95, 96
Bones
  accumulations, 19, 87–101, 164, 174, 200,
    205, 259, 260, 534, 543
  burial in sediments, 97
  C-14 date, 152
  chemical alteration, 97, 98
  damage to, 91–6
  water-sorting, 96, 97
Boselaphines, 293, 294, 295–7, 299, 302, 304,
  307
Bovids, 91, 93, 151, 158, 165, 250, 256, 259,
  261, 293–308, 323, 325, 331, 338, 345, 371,
  427, 429, 430, 431, 434, 435, 436, 487–9,
  500, 502, 503, 556, 558
  correlation, 294, 430, 431, 434–5
Buluk, 377, 379, 388–9, 390–1, 392
Butchery, 88, 92, 93, 144, 200, 357, 534

Calderas, 25, 40, 55, 58, 59, 60, 61, 63, 65, 66,
  67, 71, 78, 79, 329
Caprinae, 293, 295, 297–8, 300, 307
Carnivora, 165, 239, 501, 502, 556
Cenozoic, 395–414, 473–93
  geochronology, 442–7, 473–98
  tectonics, 31
  volcanic rocks, 55–68, 207–23
Cephalines, 259, 261, 294, 296, 297
Cercopithecines, 164, 195, 263, 266, 501
Chalbi desert, 63, 68
Chari Tuff, 121–2, 399–403, 420, 425, 458,
  463, 478–80, 482, 483, 493, 494, 532, 533
Chemakilani Group, 217, 228, 361, 365
Chemeron Formation, 215–17, 228, 232, 264,
  268, 361, 362
  fossil bovids in, 293, 303–4
  fossil hippopotamids in, 279, 285–8, 290
Chemoigut Formation, 217, 230–1, 293, 305–7,
  309, 311, 312, 313, 316, 318, 320, 322, 323
  artefacts in, 312, 323, 324, 326,
  faunal list, 325
  fossils, 309, 311, 312, 316, 322, 323
Cheparchelon Plug, 265, 267, 277
Chesowanja, 230, 309–27
  archaeological site, 144, 319, 321, 323–5
  fossils in, 309, 311, 312, 316, 322, 323, 325
  geological maps, 310, 314, 315, 318
  geology, 309–23
  lithology, 313
  palaeomagnetic stratigraphy, 229–31, 322
  stratigraphic column, 311
  stratigraphy, 309, 312
  structure, 320–2

Chesowanja Formation, 229–30, 309–27
  age, 322, 324
  artefacts in, 141–4
  fossil hippopotamids in, 284, 285–7, 290
Choppers, 142, 153, 324, 343, 347, 348, 349,
  372, 529, 535, 538, 543
Chyulu Hills, 45, 64, 65, 66, 68
Cleavers, 144, 145, 174, 324, 329, 337, 338, 343,
  345, 347, 348, 372, 536, 538
Cobbles, 143, 153, 311, 325, 342, 343, 349,
  532, 534, 535, 536, 537, 538, 541
"Confetti Clay", 555, 557
Core tools, 141, 142, 143, 200, 324, 343, 344,
  347, 361, 371, 372, 529, 536, 537, 538, 544
Crater Highlands, 65, 68
Cretaceous peneplains, 20, 22, 23, 25, 26
Crocodile fossils, 257, 258, 325, 368, 380, 542
Crustacea, 257

Damota area, 155, 479
Death, 87, 90
Deinotherium, 165, 258, 323, 500, 502
Deltaic deposits, 143, 395, 397, 404, 405, 411,
  412, 415, 422, 424, 499, 532, 552, 554
Denen Dora Member, 552–5, 557
Developed Oldowan, 141, 142, 154, 202, 309,
  323, 324, 538, 544
Diatoms, 23, 183, 189, 191, 192, 194, 196, 197,
  203, 259, 263, 266, 311
Dida Galgalu lavas, 62, 63
Dinosaurs, 19, 378
Disarticulation, 91–3

East African Geological Research Unit, 30, 207,
  209, 279, 309
East African National Parks, 116–17, 120
East African Rift System (*see also* Gregory Rift
  Valley), 1–4, 7–18, 29–54, 375, 549
  geophysics, 7, 10–13, 15–17, 71–83
  lithosphere, 7, 9–10, 15–18
  map, 8
  palaeolakes, 507–27
  structural development, 19–28, 29–54
  tectonics, 7–17
  volcanism, 9, 20, 22, 24, 25, 26
East Lake Turkana (Rudolf) (*see also* Koobi
  Fora Formation, Omo Basin, Shungura
  Formation and Turkana Basin)
  artefacts, 141–4, 395, 406, 408, 529
  biostratigraphy, 473, 486, 489–95, 519–24
  correlation with other areas, 119, 121–2, 281,
    282, 290, 421–40, 458, 473, 476, 477, 488,
    489, 492–5
  fossil molluscs in, 399, 402, 403, 404, 405, 406,
    407, 408, 479, 519–26

fossils in, 97, 395, 399, 401, 402, 403–8, 463, 464
geological maps, 396, 415, 474
isochronous surfaces, 415–20
K-Ar ages, 419–20, 421, 463–9, 473, 483
lithofacies, 395–8
map, 423, 464
palaeoecology, 97, 431, 435–8, 489, 544
palaeoenvironments, 397, 404, 409–12, 489–492
palaeogeography, 409–12, 475–6
palaeomagnetic stratigraphy, 471, 473, 481, 482
sedimentation, 404–8, 411, 415–20
stratigraphy, 395–415, 446–7, 476–9
tectonics, 412
East Lake Turkana Tuffs (*see also* Chari Tuff, KBS Tuff, Koobi Fora Tuff and Okote Tuff), 415–19, 479–80
geochronology, 122, 420, 425, 434, 447, 455–458, 483
water-lain, 446–7
Eburu, 58
Edward Albert Rift, 508, 509, 523, 524, 526
El Moiti, 71, 78, 79
Elephant fossils, 3, 121, 165, 261, 331, 427, 429, 430, 464, 487, 489, 495, 500–3, 505, 533, 549, 558, 559
Elgeyo basalts, 21, 212, 234
Elgeyo escarpment, 21, 37, 78, 207, 210, 211, 212, 225, 234, 239, 240, 241, 244, 251, 252
Elgeyo fault, 34, 46, 50, 239, 243, 245, 246, 251
Elgon volcano, 22, 33, 39
Elmenteita Lake, 58, 59, 67, 193
Emuruangogolak, 61, 71, 78, 241
Equator monocline, 251
Equids, 151, 158, 164, 165, 249, 250, 258, 259, 260, 263, 270, 294, 430, 488, 499, 500, 502, 503, 549, 558
Erinei trachyte, 312, 313, 318, 320
Erosion surfaces, 31, 32, 33, 45, 63
Esayeti volcano, 24, 25
Etheria, 381, 385, 403, 407, 408, 503
Ethiopia (*see also* Afar, Hadar and Omo), 499–506, 549–64
Ethiopian dome, 8, 9, 15, 52, 63, 375, 391
Ethiopian Rift, 20, 21, 24, 25, 475
volcanism, 549–50
Evolution, 103–24, 132
mathematical model, 103–110
of hippopotamids, 283–5, 287, 288, 289–91
of hominids, 143, 145, 146
of mammals, 486–8, 499–506
of man, 87, 170
of Ngorara vertebrates, 249–51

Ewalel phonolites, 213, 216, 233, 237, 239, 240, 241, 243, 247, 248, 249, 250
Ewaso Ngiro, 47, 61, 63, 64
Explosion craters, *see* Maars

Fish fossils, 201, 239, 243, 257, 259, 264, 265, 276, 277, 278, 368, 380, 402, 405, 542
Fission-track dating, 444, 453
Flake tools, 141, 142, 145, 200, 324, 325–6, 335, 337, 338, 339, 343–51, 354, 371, 372, 534–8
Fluvial deposits, 200, 237, 241, 242, 244, 246, 264, 265, 266, 267, 276, 277, 278, 281, 329, 333, 362, 364, 395, 397, 404, 412, 415, 416, 422, 424, 436, 463, 476, 478, 493, 499, 510, 531, 532, 534, 541, 544, 550, 554, 555
Food debris, 88, 93–6
Fort Ternan, 249, 251, 279, 282, 294, 295, 297–300, 307, 392, 453
Fossil assemblages, 87–98
causes of death, 90, 93
evolutionary scaling, 103–24
Fossil bones, 88, 91–8, 152, 164, 174, 200, 201, 259, 534, 541, 542, 543,
Fossil man, 87–103
Fossil teeth, 331
bovid, 158, 295–307, 331
equid, 151, 158
hippopotamus, 279, 282–91
hominid, 132, 158, 165, 169, 542, 543, 561, 562
primate, 132–5, 169
Fossil vertebrate assemblages, 19, 87–101, 207, 215, 463, 473
Fossilization, 87–101
Fossils (*see also* individual entries)
loss of soft parts, 91
mammal, 152, 195–6, 242–3, 258, 276–8, 325, 331, 378, 379, 380, 381, 399, 464, 473, 484, 486–8, 499–506, 556, 558
plant, 242–3, 255–6, 259, 260, 264, 267, 274, 276, 287, 489–90, 501, 504
vertebrate, 19, 239, 249–51, 255, 257–61, 290, 294, 303, 309, 395, 405, 406, 407, 408, 463, 464, 473, 499–503, 549, 558, 559
wood, 61, 259, 260, 266, 274, 311, 378, 390, 501

Gadjingero valley, 158, 164
Galana Boi Beds, 396, 397, 403–4, 412, 478
Garusi Series, *see* Laetolil Beds
Garusi valley, 164, 169
Gazella, 301, 302, 303
Geochronology (*see also* Biostratigraphy, Palaeomagnetic stratigraphy and Radioisotopic

Geochronology—*cont.*
   dating), 207–23, 249–51, 281, 293, 311, 399, 400, 415–20, 421–22, 441–61, 473–98, 499
Geophysics, 3, 7–18, 30, 53, 71–83
Giraffes, 165, 259, 503
Gomphotheres, 249, 258, 388, 500
Gravity surveys, 3, 11, 12, 53, 71–83
Gregory Rift Valley (*see also* Baringo Basin, East African Rift System, East Lake Turkana, Kenya Rift, Kisumu Rift and Western Rift), 8, 9 21, 22, 23, 24, 25, 26
   biostratigraphy, 441–3, 445
   drainage, 32, 33, 44, 45, 46, 47, 55
   early exploration, 1–4
   faults, 41–5, 48–50
   geochronology, 207–23, 441–61
   geomorphology, 29, 31–3, 34, 41–6, 50, 52, 55
   geophysics, 3, 53, 71–83
   map, 31, 179, 180
   radioisotopic dating, 441–61
   section, 51
   sedimentatiom, 29, 33, 34, 38, 39, 40, 53, 55
   stone tools, 139, 141–5, 544
   structural history, 29–34, 41–5, 48–50, 52
   tectonics, 31, 32, 34, 41, 46, 55
   volcanism, 9, 20–6, 29, 33–41, 44, 52, 53, 55–69
Growth, 125–38
Guomde Formation, 397, 398, 403, 406, 412, 478, 479, 480, 495

Hadar
   geological map, 551
Hadar Formation, 549–64
   faunal list, 556
   fossil hominids in, 170, 559–62
   stratigraphic units, 552–7
   volcanism, 550
Handaxes (*see also* Acheulian artefacts), 144, 145, 152, 154, 174, 201, 324, 329, 337, 338, 343, 344, 347, 348, 349, 350, 538,
Hannington Lake area, 29, 41, 42, 45, 47, 50, 72, 193, 207, 218, 219, 237, 251, 256, 260, 312, 313, 371
Hasuma Tuff, 419
Hipparion, 151, 165, 249, 250, 259, 260, 294, 430, 500, 502, 549, 558
Hippopotamids, 92, 164, 200, 201, 259, 261, 263, 270, 274, 276, 277, 279–92, 368, 429, 436, 500, 502, 549, 558
   evolution, 283–91, 487
Hominid fossils, 153, 157–70, 251, 263, 276, 309, 312, 322, 323, 361–73, 395, 402, 463, 549, 554, 559–62

allometry, 130–5,
chronology, 103–24, 158, 361, 373, 445–6
mandibles, 133, 165–8, 361
palaeoenvironments, 499–506, 539–44
taphonomy, 87–101
taxonomic differences, 130, 131, 132
teeth, 132, 158, 165–9, 542, 543, 561
Hominid sites, 529–47
Hominids
   behaviour, 139–46, 200, 538–47
   evolution, 125, 143, 145, 146, 499–506, 561
   feeding, 92, 93, 143, 144, 200
Homo, 132, 165, 504–5, 529, 543, 544, 562
   teeth, 132, 169, 561, 562
Homo erectus, 153, 361, 372, 505
   teeth, 169
Homo habilis, 153, 505, 561
Homo sapiens, 151–2
Hope Fountain Industry, 203, 344
Horn cores, 91
Hyaenas, 93, 95, 239
Hyracoids, 239, 249, 258, 502

Ignimbrites, 23, **24**, 40, 444, 454, 549
Ileret, 397, 400, 401, 402, 403, 406, 409, 411, 412, 483, 484, 487, 530, 532, 533, 534, 535, 538, 544
Invertebrate fossils (*see also* Molluscs), 90, 113, 115–17, 195, 257, 259, 260, 479, 503, 507–527, 554, 555
Isochronous surfaces, 415–20
Isometric growth, 126–8

Kabarnet trachytes, 24, 40, 214–17, 220, 233, 254, 255, 264, 266, 269–71, 273, 274
   age, 233, 250
Kabarsero, 239, 243, 245, 247–9, 253, 255, 257, 258
Kada Hadar Tuff, 552–5, 557
Kagilip Fault, 251, 253
Kajong, 375, 377, 381, 383
Kajong Formation, 381–92
Kalimale, 253, 257
Kamasia Range (*see also* Tugen Hills), 24, 37, 38, 41, 43, 44, 46, 47, 48, 207
   K-Ar ages, 213–17, 220, 221
   stratigraphy, 210, 212, 220
   structure, 210–11
Kampi-ya-Samaki Beds, 365, 366, 367
Kanapoi, 24, 283, 288, 476
Kaparaina Basalt, 37, 215, 216, 220, 228, 232, 264, 268–70, 273, 277, 312, 362
Kaperyon Formation, 215–17
Kapitan Fault, 253, 254
Kapkiamu, 239, 245, 247, 248, 255, 257, 259

Kapikiamu Graben, 241, 243, 245, 254
Kapthurin Basin
  geological map, 363
  geology, 361–73
Kapthurin Formation, 217, 290
  age, 217, 228, 372
  artefacts and hominids in, 324, 338, 350, 361–373
  faunal list, 370
  palaeoemvironments, 367–71
  stratigraphic succession, 361–7
Kaption, 241, 243, 244, 245, 246, 248, 254
Karari escarpment, 400–7, 473, 480, 482, 530–533, 538, 543
Karari Industry, 141, 142, 529–47
Karari Tuff, 399–402, 458, 463, 483, 494, 531, 532, 544
Karau Formation, 229, 230, 232, 313, 316, 318, 320, 322, 324
Kariandusi Archaeological Site, 25, 127, 338, 350, 352–4
Karmosit Beds, 288, 293, 304–5, 307
Kavirondo Block, 22
Kavirondo Rift, *see* Kisumu Rift
Kavirondo Trough, 30, 48, 50, 73, 76, 81, 211, 282
KBS Industry, 529, 536, 541
KBS Tuff, 400, 401, 404, 411, 412, 419, 478, 480–2, 531, 532
  age, 420, 421, 424, 425, 432–3, 441, 455–8, 463–9, 471, 473, 483–5, 487, 494, 529, 532, 533
  artefacts in, 142–3, 400, 401, 529
  fauna, 121–2, 402, 406, 473, 532
Kedong Depression, 67, 179, 191, 192
Kedong Embayment, 58, 59
Kenya (*see also* Baringo, East Lake Turkana, Gregory Rift Valley, Lukeino Formation, Ngorora Formation, Olorgesailie and Turkana)
  geological map, 171
  gravity anomaly map, 71, 73, 75, 77, 81
  gravity survey, 71–83
Kenya Plateau Phonolites, 36–8, 210, 213, 239, 240, 249, 441, 453, 455
Kenya Rift (*see also* Baringo Basin), 57–65, 309–60
  archaeological sites, 329–60
  drainage, 64, 66, 68
  fossil bovids in, 293–308
  geomorphology, 57, 60, 63–7
  sedimentation, 59, 61, 62, 66–8
  structures, 57, 59, 60, 61, 65
  tectonics, 33, 55, 60, 61, 63, 65, 66
  volcanism, 55–68

Kenyan dome, 8, 9, 15, 26, 50, 52, 71, 74, 81, 375, 387, 388, 391, 392
Kerimasi, 65
Kerio River, 45, 46, 47, 71, 81, 207, 211, 217, 233, 239, 251
Kibingor, 232, 239, 247
Kibish Formation, 478, 499
Kilimanjaro, 33, 65, 66, 67, 68
Kilombe archaeological site, 329–60
  age, 329, 334, 337, 358
  excavation and artefacts, 357–60
  fossils, 331, 338, 345
  geology, 329–36
  map, 330, 339
  occupation surface, 337, 354–8
  palaeoenvironment, 335
  palaeomagnetism, 232
  stratigraphy, 331
Kilombe volcano, 58, 329, 330, 334
Kinangpop tuffs, 24–5
Kisingiri volcano, 20, 22, 39
Kisitei anticline, 270, 272, 273
Kisumu Rift, 20, 21, 22, 23, 30, 392
Kitale surface, 22, 23
Kito Pass Fault, 78, 211, 241–3, 245, 251, 253, 270, 272, 273
Kobuluk sill, 265, 267, 273, 277
Kobus, 323, 371, 429, 436, 502, 503, 558
Kokoi horst complex, 397, 400, 401, 403, 412, 530
Koobi Fora, 397, 401–12, 422
Koobi Fora Formation (*see also* KBS Tuff, Koobi Fora Tuff, Suregei Tuff and Tulu Bor Tuff), 396, 398–403, 406, 532
  age, 424, 425, 433, 463–9, 473
  archaeology, 529, 530
  correlation with other areas, 121–2, 424–38, 473–9, 484, 487, 490, 521, 523, 544
  faunal zones, 426–8
  magnetostratigraphy, 424, 433–4, 480–3
Koobi Fora Tuff, 121–2, 323, 398, 400, 401, 419–20, 533
Koora Graben, 179, 181, 183, 189–92, 194, 197, 199
Kubi Algi Formation, 396, 397–9, 403, 406, 409, 478, 484, 488, 494, 519
Kwaibus basalts, 228–9
Kyoga surface, 22

Lacustrine deposits, 23, 24, 25, 40, 60, 61, 64, 115, 193, 194, 197, 237, 242, 243, 244, 245, 246, 259, 263, 264, 267, 276, 278, 281, 316, 318, 362–5, 367, 368, 390, 395, 397, 399, 404, 411, 412, 415, 417, 419, 422, 424, 436, 463, 475–9, 481, 482, 485, 488, 493,

Lacustrine deposits—*cont.*
   494, 499, 500, 510, 544, 549, 550, 552, 554, 555, 557
Lacustrine molluscs, 405, 476–8, 507–27, 554
Laetolil, 160
Laetolil Beds
   age, 157–63
   fossils, 157–70, 294, 300, 301, 561
   stratigraphy, 158–62
Laga Bura Hasuma, 397, 399, 400, 402, 403, 412
Lahars, 241, 244, 248
Laikipia area, 21, 23, 24, 37, 43, 45
Laikipia escarpment, 207, 211, 217–20, 241, 244, 274
   faults, 43, 45, 46, 50, 211, 239, 240, 251, 252
Lake Rudolf, *see* East Lake Turkana
Lakes (*see also* Palaeolakes and individual lakes), 26, 193
   recent molluscs in, 513
Langebaanweg, 294, 296, 297, 299, 301
Lapurr Range, 377, 378, 390
Lava artefacts, 142, 143, 343, 344, 350, 353, 446, 535, 537
Legemunge Beds, *see* Olorgesailie Formation
Limuru trachyte, 24, 25
Listriodon, 250, 251, 259, 261, 381
Loiengalani Formation, 381, 385, 387, 388
Lokichar River, 71, 81
Lokitaung, 19, 377–8
Lomi, 60, 71, 78, 79
Londiani, 58, 71, 77, 338
Loperot, 20, 377, 378–80, 388, 390, 392
Loriu plateau, 81
Losokweta River, 312, 313, 316, 320
Lothagam, 24, 270, 276, 283–8
Lothidok, 79, 377, 379, 380, 390, 392
Loxodonta, 165, 331, 500, 502, 503
Loxodonta africana zone, 121, 427, 429, 487, 489, 495, 533
Lubur Series, 377, 378, 380
"Lucy", 561–2
Lukeino Block, 268, 270, 272–4, 277
Lukeino Formation, 215, 217, 263–78
   age, 264–5, 269–70, 293
   correlation with other areas, 264–5, 270
   faunal lists, 275
   fossil bovids in, 256, 293, 302–3, 307
   fossil hippopotamids in, 283–8
   fossil plants in, 256, 264, 267, 274, 277
   fossils in, 263, 264, 267, 270, 274, 275–7
   geological map, 262
   palaeoenvironments, 263, 264, 274, 277
   palaeogeomorphology, 263, 273, 277
   palaeomagnetism, 232–3

sedimentation, 263–8, 273, 277
stratigraphy, 265–9
structure, 263, 266, 268, 270–3, 277
taphonomy, 276–7
volcanic history, 265, 266, 268, 277

Maars, 55, 61, 63, 65, 68
Magadi area, 25, 171, 173, 174, 176, 189, 192, 193, 194, 197
Magnetostratigraphy, *see* Palaeomagnetostratigraphy
Makapansgat Cave, 93, 94, 96
Mammals (*see also* Fossil mammals), 490–2
Mandibles, 133, 134, 135, 165–9, 361, 372, 542, 543, 561
Mantle plume tectonics, 7, 12–13, 15, 17
Manuports, 142, 337, 345, 356, 357, 358, 534
Marabou stork, 255, 258
Maralal, 36, 45,
Marigat trachyte, 228, 229 371
Marsabit, 36, 45, 61–3, 68, 71, 73, 74
Masek Beds, 153, 154
Masek Lake, 158, 159
Mau, 57, 58, 59, 66
Meganthropus africanus, 157
Melanoides, 368, 385, 403, 405, 503, 508, 512, 522, 523, 524
Melka Kunture, 142, 144
Mellivora, 249, 255, 261
Membrane tectonics, 7, 12, 14–17
Menelikia, 429, 436, 503
Menengai, 26, 58, 59, 71
Meru, 65, 66, 68
Mesochoerus, 436, 437, 500, 503
Mesochoerus limnetes, 427, 429, 430, 500, 505
Mesochoerus limnetes zone, 115, 121, 427, 487–9, 495, 500
Mesozoic tectonics, 19–20, 26, 32, 375
Metridochoerus, 322, 323, 500, 503
Metridochoerus andrewsi zone, 121, 427, 487–9, 495, 500
Microfossils, 113–15, 127
Microvertebrates, 502
Miocene
   deposits, 44–5, 263–78, 309, 311, 375–94, 454
   fossils, 19–20, 98, 263–5, 270, 274–9, 293–308
   tectonics, 15, 21–4, 26, 29, 32, 33
Mollusc fossils, 180, 195, 196, 257, 259, 260, 265, 325, 368, 370, 380, 385, 399, 401, 402, 403, 404, 405, 406, 407, 408, 479, 503, 508, 512, 514–26, 553, 554, 555,
   lacustrine, 405, 476, 477, 478, 507–27
Molluscs, recent, 514
Moruarot, 373, 379, 380, 390, 391

evolutionary scale, 107–15
magnetostratigraphy, 424, 434, 477, 479, 480–4,
tools in, 141, 142, 544
Sidekh, 234, 240, 241, 244–7, 253, 254, 274
Sidekh phonolites, 212–14
Sidi Hakoma Tuff, 552–5, 558
Silali, 59, 60, 71, 78, 79, 207, 241
Simopithecus, 145, 195, 201
Siwalik Hills, 237, 249, 270, 293, 294, 297, 302, 307
Skeletons, disarticulation of, 91, 92, 193
Skulls, 126, 132
Statistical techniques, 103–10
Stephanie Arch, 395, 412
Stephanie Lake, 145, 479
Stephanie Rift, 21, 475, 492
Steppe Limestone, 182
Stone Age burial, 151–2
Stone blocks, 337, 345, 351, 357
Stone industries, 142, 143, 145, 152, 153, 154, 309, 323–5, 337–60
Stone tools (*see also* Artefacts), 139–47, 534–538
Stratigraphic palaeontology, *see* Biostratigraphy
Suguta, 36, 40, 44–6, 50, 60, 61, 71, 78, 81, 207, 277
Suids, 115, 121, 164, 165, 201, 239, 250, 251, 259, 263, 270, 291, 312, 323, 371, 427, 429, 434, 436, 437, 486–9, 492, 495, 500, 502, 503, 505, 549, 558, 559
Suregei Cuesta, 397, 400, 401, 409, 411, 531
Suregei Tuff, 398–400, 406, 409, 420, 458, 478, 479, 488, 493–5, 519
Suswa, 58, 59, 67, 71, 191

Tanganyika Rift, 33, 53
Tasokwan trachytes, 220–1
Tanzania (*see also* Laetolil, Olduvai Gorge and Serengeti), 47, 55, 65–7, 73, 152, 157
Taphonomy, 87–101, 259–60, 276–7, 437, 557
Tectonics, 12–17, 19–28, 391, 392, 412
mid-plate, 7–17
Teeth, *see* Fossil teeth
Tertiary environments, 97
Tertiary tectonics, 20–1, 26, 32
Theropithecus, 164, 165, 195, 196, 429, 505
Tiati Grit Series, 380
Tiati volcano, 39, 217, 239, 240, 241, 273
Tiim phonolites, 211–13, 233, 237, 239–41, 247–51, 253
Tinderet volcano, 21, 22, 24
Tools (*see also* Artefacts, Bifaces, Choppers, Cleavers, Core tools, Flake tools and Handaxes)

bone, 89, 93, 94, 95, 96
Palaeolithic, 95
stones used for, 142, 143, 323, 325, 335, 343, 344, 350, 353, 371, 372, 446, 535, 537
Trachyte artefacts, 343, 371, 372
Tragelaphines, 164, 165, 259, 261, 274, 294, 296–7, 299, 301–7, 429, 430, 431, 500, 558
Trilobites, 107, 109, 113–15, 117
Triple Tuff, 552, 554
Tuffs (*see also* Chari Tuff, Karari Tuff, KBS Tuff, Koobi Fora Tuff, Okote Tuff, Omo Tuff, Sidi Hakoma Tuff, Suregei Tuff, Triple Tuff and Tulu Bor Tuff), 446–7, 478–88
Tugen Hills (*see also* Kamasia Range), 238–47, 251, 252, 261, 263–8, 272, 273, 277, 293, 361, 371
palaeomagnetic survey, 225–35
Tulu Bor Tuff, 398, 400–2, 404, 406, 409, 411, 478, 479, 488, 493, 494
age, 400, 420, 433, 455, 463, 483, 485
Turkana Basin (*see also* East Lake Turkana, Omo and Shungura Formation), 19, 21, 23, 38, 40, 46, 47, 193
erosion surfaces, 391, 392
fossil mammals in, 378, 379, 392
fossils in, 91, 375, 377, 378–81, 385, 388–90, 392
geochemistry, 479–80
gravity anomalies, 71, 72, 74, 78–92
hominids in, 473–98
map, 376, 423, 474
modern animals, 92–3
palaeogeography, 475–6
sedimentary history, 375–94
stratigraphy, 476–9
tectonics, 375, 391, 392
Turkana sandstone, 19, 20, 33, 79, 377, 378, 380

Uasin Gishu, 22, 37, 39, 45, 210, 212, 219, 234
Usno Formation, 118–19, 476–8, 483, 499, 502, 503

Vogel River Series, *see* Laetolil Beds
Volcanic rocks, 207–23, 444, 446–7, 454–9
Volcanism, 9, 22, 24, 25, 26, 33–41
Quaternary, 55–69
Tertiary, 55, 56, 59, 60, 61, 65
Volcanoes, 20–6, 33, 35, 39, 40, 55, 58–61, 63, 65–8, 71, 78, 79, 161, 239, 240, 241, 244, 273, 321, 334, 446–7

Waril, 245, 248, 257, 259

Water–lain volcanic tuffs, 446–7
Western Rift, Valley, 2, 3, 21, 29, 30, 31, 33, 53, 523, 524, 526

Yatya anticline, 269, 270, 272, 273

Yatta plateau, 23, 27, 39

Zinjanthropus, 501
Ziphid whale, 379, 380
Zoogeography, 115–18

Author Index

Adam, W., 508, 527
Aguirre, E., 239, 261, 283, 291
Allen, J. R. L., 405, 413
Anderson, O. L., 15, 17
Andrews, F., 175
Ansell, W. F. H., 294, 308
Arambourg, C., 19, 26, 294, 297, 308, 372, 373, 377, 378, 380, 392, 393, 562, 563
Armstrong, R., 10, 17
Aronson, J. L., 553, 564
Ashton, E. H., 138
Atwater, T., 13, 18
Azzaroli, A., 19, 26

Backhouse, R. W., 82
Bagdasaryan, G. P., 444, 459
Bailey, O. K., 11, 12, 13, 17
Baker, B. H., 8, 9, 10, 12, 17, 19, 20, 21, 22, 23, 24, 25, 26, 27, 31, 33, 36, 52, 53, 54, 55, 63, 64, 69, 171, 174, 176, 177, 178, 191, 192, 197, 204, 206, 212, 213, 215, 219, 222, 311, 326, 375, 376, 381, 387, 391, 392, 393, 444, 459
Baker, S. W., 1, 3
Bambach, R. K., 492, 498
Banks, R. J., 73, 82
Barberi, F., 20, 24, 27, 549, 551, 563
Bard, Y., 105, 110, 123
Beadle, L. C., 527
Beaux, O. de., 490, 495
Beden, M., 121, 123, 165, 430, 438, 486, 495
Beerbower, J. R., 99
Behrensmeyer, A. K. 95, 96, 97, 98, 99, 122, 278, 400, 413, 414, 421–40, 463, 469, 475, 476, 482, 487, 489, 495, 530, 544, 545, 546, 547
Beinhart, G., 100
Bell, P., 152
Beloussov, V. V., 54, 222
Bequaert, J., 385, 394, 514
Berggren, W. A., 249, 261
Bibus, E., 132, 136
Bigalke, R. C., 489, 490, 495
Biggs, D. L., 413, 546
Binford, L. R., 202, 204
Binge, F., 24, 27
Bishop, W. W., 21, 22, 23, 24, 27, 31, 37, 40, 54,

69, 97, 98, 99, 123, 142, 144, 146, 154, 195, 203, 204, 205, 206, 212, 213, 215, 217, 222, 223, 225, 228, 229, 230, 233, 234, 239, 249, 261, 262, 263, 270, 278, 279, 282, 288, 292, 293, 294, 297, 305, 308, 309–27, 329–36, 338, 344, 345, 354, 356, 359, 393, 439, 444, 453, 454, 460, 461, 469, 495, 496, 527, 546
Black, C. A., 413
Black, R., 21, 27, 549, 563
Bluck, B. J., 383, 386, 393
Blundell, D. J., 82
Boaz, N., 97, 99
Bonatti, E., 27
Bonjer, K. -P., 83
Bonnadonna, F. P., 196, 204
Bonnefille, R., 436, 438, 490, 504
Bordes, F., 142, 146
Borsi, S., 563
Boswell, P. G. H., 152, 154
Bourguignat, 508
Bourlière, F., 204
Bowden, P., 69
Bowen, B. E., 395–414, 415, 418, 419, 420, 422, 424, 438, 469, 476, 482, 483, 485, 490, 492, 495, 498, 530, 531, 532, 545, 547
Boyd, F. R., 10, 17
Brace, C. L., 125, 136
Brain, C. K., 93, 94, 95, 96, 99
Breuil, H., 94, 99
Briggs, I. C., 74, 82
Brock, A., 122, 123, 232, 235, 322, 425, 433, 438, 471, 480, 482, 483, 495, 529, 530, 533, 545
Brook, M., 207–23, 228, 233, 250, 269
Brothwell, D. R., 95, 99, 101
Brown, D. S., 508, 527
Brown, F. H., 103–24, 424, 425, 433, 434, 439, 444, 457, 458, 460, 469, 473–98, 558, 564
Brown, L., 490, 498
Buckland, W., 93, 99
Bull, W. B., 387, 393
Bullard, E. C., 3, 72, 82
Burke, K., 15, 17
Burstein, A. H., 563
Butzer, K. W., 97, 99, 145, 146, 154, 193, 195, 204, 205, 206, 294, 308, 360, 424, 436, 439, 476, 478, 492, 496, 546

Carmichael, C. M., 450, 461
Carmichael, I. S. E., 469, 475, 496
Carney, J. N., 43, 170, 222, 229, 232, 235, 309, 312, 313, 322, 327, 444, 460
Carr, C. J., 490, 496
Cerling, B. W., 463
Cerling, T., 123, 400, 401, 413, 420, 460, 463–469, 483, 485, 496, 523, 532, 546
Champion, A. M., 378, 393
Champy, C., 126, 136
Chapman, G. R., 36, 37, 38, 40, 48, 54, 207–223, 225, 226, 228, 233, 234, 235, 239, 249, 250, 261, 263, 269, 270, 278, 282, 292, 308, 311, 326, 327, 460
Chavaillon, J., 141, 142, 144, 146, 505, 544, 546
Cheetham, A. H., 105, 123
Chessex, R., 21, 27
Chmielewski, W., 357, 359
Choubert, G., 308
Christiansen, T. B., 23, 27
Churcher, C. S., 95, 100, 380, 393
Clark, G., 146
Clark, J., 95, 99
Clark, J. D., 69, 92, 96, 99, 145, 146, 154, 174, 202, 204, 205, 206, 223, 327, 344, 354, 355, 357, 359, 372, 373, 393, 461, 527, 534, 546, 547
Clark, W. E. Le Gros, 137, 378, 380, 393
Clarke, D. L., 204, 205, 359, 546
Clarke, G. C., 69
Clarke, R. J., 170
Clifford, T. N., 8, 17
Cloud, P. E., 369, 373
Cole, G. H., 205, 359
Cole, S., 142, 146, 175, 204
Cook, H. E., 117, 123
Cook, S. F., 357, 359
Cooke, H. B. S., 115, 121, 122, 123, 175, 195, 196, 204, 263, 270, 278, 288, 292, 322, 380, 427, 430, 437, 439, 464, 469, 486, 488, 496, 497
Coppens, Y., 98, 107, 118, 123, 124, 137, 146, 196, 203, 204, 285, 289, 292, 294, 308, 323, 327, 395, 413, 414, 422, 438, 439, 460, 469, 471, 496, 497, 498, 499–506, 545, 546, 547, 549–64
Corruccini, R. S., 130, 131, 136, 137
Corvinus, G., 552, 563
Coryndon (Savage), S. C., 206, 263, 270, 279–292, 394, 429, 439, 487, 488, 496
Cox, A., 159, 426, 433, 439, 471, 481, 495, 496
Crader, D., 546
Craig, G. Y., 90, 99, 100
Croes, M. K., 115, 120, 124, 424, 434, 439, 480, 482, 498, 558, 564
Crossley, R., 48

Crusafont-Pairo, M., 239, 261
Cunnington, 508
Curtis, G. H., 122, 123, 141, 144, 147, 153, 154, 157–70, 194, 195, 205, 218, 337, 359, 419, 420, 444, 452, 453, 457, 458, 460, 461, 463–9, 483, 484, 487, 488, 496, 533, 544, 546, 564
Dagley, P., 215, 222, 225–35, 250, 269, 334, 336, 358, 359, 372
Dalrymple, G. B., 12, 17, 464, 465, 469
Damon, P. E., 450, 451, 461, 469
Dandelot, P., 117, 123
D'Arcy Thompson, W., 126, 129, 137
Darracott, B. W., 10, 11, 17, 73, 82, 83
Dart, R. A., 93, 94, 96, 99
Davidson, A., 20, 27
Davies, J. C., 27
Davis, D. H., 489, 496
Dawson, J. B., 65, 69, 461
Day, M. H., 153, 154, 170, 563
De Heinzelin, J., 107, 124, 147, 422, 424, 439, 476, 477, 494, 497, 505, 547
Delaloye, M., 27
Della Giustina, G., 175
De Lumley, H., 346, 360
Deraniyagala, P. E. P., 380, 393
De Vore, I., 100, 205
Dewey, J. F., 17
Dickson, G. O., 235
Dietrich, W. O., 151, 154, 157, 170, 300, 308
Dimbleby, G. W., 99, 205, 359, 546
Dixey, F., 20, 22, 23, 27, 30, 32, 54, 378, 393
Dodson, P., 91, 92, 96, 99
Dodson, R. G., 60, 69, 79, 83, 191, 206, 375, 377, 378, 380, 387, 393, 394, 412, 413, 414
Donaldson, A. C., 405
Dorst, J., 117, 123
Dove, G., 158
Downie, C., 66, 69
Drake, R., 123, 157–70, 420, 460, 463–9, 496, 546, 564
Draper, N. R., 118, 123
Eck, G. G., 103–24, 473–98, 501
Edwards, A. B., 114, 117, 123
Efremov, I. A., 88, 99
Eisenmann, V., 121, 123, 488, 497
Erk, F. C., 495
Eugster, H. P., 193, 204
Evans, A. L., 444, 460
Evernden, J. F., 153, 154, 194, 205, 218, 444, 452, 453, 460, 461
Ewer, R. F., 263, 270, 278
Fairhead, J. D., 10, 17, 18, 59, 69, 72, 82, 444, 460

Faure-Muret, A., 308
Ferrara, G., 27, 563
Findlater, I. C., 124, 415–20, 424, 460, 469, 478, 497, 521, 522, 523, 530, 531, 546
Fischer, G., 1
Fitch, F. J., 27, 54, 69, 122, 124, 141, 212, 222, 235, 261, 294, 308, 326, 389, 393, 400, 401, 413, 419, 420, 424, 425, 432, 433, 434, 439, 441–61, 463, 464, 467, 469, 471, 475, 483, 484, 485, 488, 489, 492, 494, 497, 529, 530, 533, 535, 546, 547
Fleischer, R. L., 453, 461
Flinn, R. M., 138
Flint, R. F., 175, 191, 197, 203, 205
Fois, V., 19, 26
Folk, R. L., 362, 373
Forster, S. C., 460
Frazer, L. N., 10, 18,
Freedman, L., 132, 136
Frick, H., 132, 136
Fuchs, K., 83
Fuchs, Sir V. E., 3, 79, 82, 152, 377, 378, 393, 508, 527

Gale, W. A., 197, 205
Galileo, 125
Gass, I. G., 8, 12, 15, 17, 18
Gautier, A., 91, 99, 381, 393, 477, 479, 497, 498, 503, 508, 523, 524, 527
Gentner, W., 461
Gentry, A. W., 196, 249, 256, 261, 293–308, 323, 429, 430, 431, 439, 487 488, 497
Gerasimovskiy, V. I., 459
Gibson, A. B., 378, 394
Giles, E., 131, 136
Ginsburg, R. N., 373, 394, 413
Girdler, R. W., 10, 11, 17, 18, 20, 27, 30, 54, 72, 73, 82
Giustina, G. della, 175
Glass, B., 495
Gleadow, A. J. W., 459, 461
Glob, P. V., 91, 99
Golden, M., 36, 37
Goldstein, M. S., 95, 99
Goodall, J. van Lawick, 140, 146
Gould, S. J., 125, 126, 127, 128, 129, 130, 132, 136, 137, 138
Gowlett, J. A. J., 323, 324, 325, 329, 331, 333, 336, 337–60
Grant, F. S., 74, 82
Grasty, R. L., 394, 461
Gray, B. T., 549–64
Gregnanin, B., 28
Gregory, J. W., 2, 3, 29, 54, 173, 174, 177, 205, 210, 222

Griffiths, D. H., 72, 78, 82
Griffiths, P. S., 222, 312, 313, 327, 361
Grimwood, I. R., 490, 497
Gumper, F., 10, 18, 72, 82
Gunther, 508
Guyasyan, R. K., 459

Haesaerts, P., 107, 124, 147, 439, 476, 477, 494, 497, 547
Hall, S. A., 17
Hamblin, W. K., 393
Hampel, J., 123, 420, 460, 463–9, 496, 546
Harris, J., 165
Harris, J. M., 121, 124, 196, 322, 327, 388, 393, 427, 430, 431, 439, 488, 497
Harris, J. W. K., 141, 322, 323, 327, 529–47
Harris, P. G., 12, 18
Harrison, C. G. A., 25, 27
Hay, R. L., 65, 69, 121, 124, 142, 152, 153, 154, 157–70, 175, 189, 193, 197, 203, 205, 433, 444, 452, 453, 460, 471, 483, 496, 564
Hay, W. W., 124
Hayatsu, A., 450, 461
Haynes, C. V., 359, 534, 546
Hazel, J. E., 105, 123
Heinzelin, J. de., 107, 124, 147, 422, 424, 439, 476, 477, 494, 497, 505, 547
Hein, S. M., 10, 17
Heiple, K. G., 563
Heirtzler, J. R., 233, 235
Heizer, R. F., 359, 361
Hemmer, H., 132, 136
Hendey, Q. B., 294, 308
Herbich, I., 141, 529–47
Herron, E. M., 235
Hersh, A. H., 131, 132, 136
Higgs, E., 99, 100
Hill, A., 87–101, 146, 170, 222, 235, 261, 278, 292, 308, 309–27, 460
Hillhouse, J., 433, 471, 495
Hinde, R. A., 146
Hobley, C. W., 2
Hodgson, W. D., 413
Höhnel, L. von., 1, 3, 381, 393
Holmes, A., 33, 54
Hooijer, D. A., 107, 239, 249, 259, 261, 263, 270, 278, 378, 380, 393, 488, 497
Hooker, P. J., 441–61
Hopwood, A. T., 152, 154, 170
Houston, R. S., 98, 100
Houtz, R. E., 123
Howell, F. C., 98, 103–24, 137, 142, 146, 147, 174, 201, 204, 205, 308, 323, 327, 359, 395, 413, 422, 439, 469, 471, 473–98, 505, 545, 546, 547, 558, 563

Hoyt, J. H., 413
Hughes, A. R., 93, 100
Hughes, R. Murray., 377, 378, 394
Hur.ord, A. J., 459, 461, 469
Hussain, S. T., 294, 308
Huxley, J. S., 126, 127, 128, 131, 132, 136, 137

Iles, W., 235
Irving, E., 71, 82
Isaac, B., 171, 537
Isaac, G. L., 95, 98, 100, 117, 122, 123, 124,
    137, 139–47, 154, 171, 173–206, 308, 327,
    337, 346, 347, 348, 351, 352, 353, 357, 359,
    360, 395, 413, 414, 419, 422, 425, 433, 438,
    440, 469, 471, 480, 482, 483, 495, 496, 497,
    498, 505, 529, 530, 531, 533, 534, 538, 540,
    544, 545, 546, 547, 563

Jack, R. N., 469
Jackes, M. K., 157–70, 564
Jackson, E. D., 17
Jaeger, J. J., 164, 259, 297, 436, 439, 502
Jennings, D. J., 24, 27, 64, 69
Johanson, D. C., 170, 549–64
Johnson, G. D., 405, 414, 424, 439, 469, 498,
    547
Johnson, R. W., 67, 69, 191, 206
Jolicoeur, P., 127, 128, 130, 136, 137
Jolly, C., 99, 545, 563
Jones, N,. 344, 360
Jones, W. B., 329, 331, 334, 335, 336, 338
Jordan, T. H., 10, 18
Joubert, P., 20, 27, 375, 378, 380, 393
Juch, D., 549, 563
Justin-Visentin, E., 20, 21, 28, 549, 564

Kamau, C. Washbourn-, 496
Kandindi, E., 166
Kavanagh, A. J., 127, 129, 137
Kay, R. F., 132, 137
Keast, A., 495
Keay, R. W. J., 489, 497
Keller, C. M., 538, 546
Kelley, J. C., 100
Kendall, C. G. St. C., 405, 413
Kennett, J. P., 123
Kent, Sir P., 1–4, 152, 158, 159, 170, 294, 308,
    460
Kenyon, W. A., 95, 100
Kew, H. W., 509, 527
Khan, M. A., 71–83
Kietzke, K., 99
King, B. C., 12, 18, 29–54, 207, 209, 222, 261,
    309, 311, 327
King, E. W., 100

King, R. F., 82
Kingdon, J., 491, 497
Kinsman, D. J. J., 9, 18
Kinzey, W. G., 132, 137
Klatt, B., 126, 137
Kleindienst, M. R., 201, 202, 205, 206, 347,
    354, 359, 360
Koenigswald, G. H. R. von., 452, 461
Kohl-Larsen, L., 157, 158, 170
Kolata, G. B., 104, 106, 124
Kreuzer, H., 563
Kronberg, P., 50, 54
Kruskal, J. B., 106, 124
Kruuk, H., 90, 100
Kuester, J. L., 105, 124
Kunz, K., 550, 563
Kurten, B., 104, 106, 124, 132, 137
Kuss, S. E., 95, 100

Lajoie, K., 424, 439, 444, 460, 469, 483, 485,
    496
Langdale Brown, I., 489, 498
Lanphere, M. A., 464, 465, 469
Larsen, L. Kohl, 157, 158, 170
Laughton, A. S., 15, 18
Lavocat, R., 294, 308
Leakey, L. S. B., 3, 151, 152, 153, 154, 155,
    157, 158, 170, 171, 173, 174, 175, 183, 195,
    196, 202, 206, 292, 294, 308, 378, 380, 393,
    452, 455, 461, 501, 561, 564
Leakey, M. D., 94, 100, 141, 142, 147, 151–5,
    157–70, 174, 195, 200, 201, 202, 203, 205,
    206, 294, 308, 323, 327, 329, 337, 343, 346,
    347, 348, 349, 360, 361, 413, 461, 469, 501,
    533, 534, 537, 541, 544, 546, 561, 564
Leakey, M. G., 121, 124, 164, 196, 206, 217,
    222, 324, 327, 338, 360, 361, 371, 372, 373,
    429, 439
Leakey, R. E. F., 98, 121, 123, 124, 137, 146,
    170, 195, 196, 204, 206, 308, 361, 373, 381,
    395, 397, 400, 413, 414, 422, 467, 469, 471,
    488, 495, 496, 497, 498, 505, 529, 530, 543,
    544, 545, 546, 547, 561, 562, 563, 564

Le Bas, M. J., 54
Le Blanc, R. J., 413
Lee, R. B., 100, 205
Le Gros Clark, W. E., 137
Lehrman, D. S., 146
Leopold, L. B., 413
Le Pichon, X., 235
Lerner, I. M., 127, 136
Lind, E. M., 489, 498
Lippard, S. J., 36, 38, 39, 54, 213, 222, 239,
    241, 251, 261

Lippolt, H. J., 461
Livingstone, D. A., 196, 206
Logan, B. W., 368, 373, 385, 394, 405, 413
Logatchev, N. A., 33, 38, 54, 222
Long, R. E., 72, 82
Lovejoy, C. O., 563
Lowe, C. van Riet, 372, 373
Lumer, H., 132, 137
Lumley, H. de., 346, 360

Maasha, N., 10, 18
McArthur, R. H., 512, 527
McCall, G. J. H., 23, 26, 27, 73, 79, 83, 206, 215, 217, 218, 222, 312, 327
McClenaghan, M. P., 38, 213, 223, 239, 251, 262, 263, 264, 270, 278
McConnell, R. B., 8. 18
McCown, E., 440
McElhinny, M. W., 15, 18
McHenry, H. M., 130, 137
MacInnes, D., 152, 195, 196, 394
Macintyre, R. M., 65, 69, 444, 461
McKenzie, D. P., 18
Madden, C. T., 380, 381, 394
Maglio, V. J., 115, 121, 122, 123, 124, 157, 170, 239, 249, 258, 262, 263, 278, 294, 308, 378, 380, 394, 395, 414, 426, 427, 428, 430, 439, 464, 469, 484, 486, 487, 488, 496, 497, 498, 533, 547
Maguire, P. K. H., 72, 82
Makris, J., 72, 73, 83
Mansfield, J., 73, 78, 82
Marinelli, G., 563
Martyn, J. E., 37, 38, 43, 48, 54, 215, 223, 225, 226, 235, 239, 248, 251, 255, 262, 263, 264, 270, 274, 278, 361, 365, 373
Mason, P., 378, 394
Matheson, F. J., 176, 206
Matsuda, R., 127, 137
Mead, J. G., 378, 380, 394
Medawar, P. B., 129, 137
Meester, J., 308
Mégnin, P., 91, 100
Mellor, D. W., 450, 461
Menzel, H., 83
Merrick, H. V., 141, 142, 147, 505, 544, 547
Meyer, W., 24, 25, 27, 549, 564
Mezacasa, G., 28
Milanovsky, E. E., 54, 222
Miller, J. A., 27, 54, 69, 122, 123, 124, 141, 170, 194, 195, 204, 205, 206, 212, 214, 216, 218, 222, 235, 250, 261, 262, 269, 278, 292, 294, 297, 308, 322, 326, 327, 359, 372, 389, 393, 400, 401, 413, 419, 420, 424, 425, 432, 433, 434, 439, 441-61, 463, 464, 467, 469,

471, 483, 484, 485, 494, 496, 497, 529, 530, 532, 533, 546, 547
Miller, T. G., 461
Mise, J. H., 105, 124
Mitchell, J. G., 69, 176, 178, 191, 192, 197, 204, 460, 461
Mohr, P. A., 17, 20, 25, 26, 27, 54, 393, 549, 564
Molnar, P., 13, 18
Moore, J. E. S., 508, 527
Moore, J. M., 27
Morbidelli, L., 20, 27, 549, 564
Moreau, R. E., 489, 498
Morgan, W. J., 7, 12, 17, 18
Morgan, W. T. W., 206
Morris, S. F., 239, 262
Morrison, M. E. S., 489, 498
Mortelmans, G., 359, 360
Morton, W. H., 27, 563
Mosimann, J. E., 128, 130, 137
Müller, A. H., 92, 100
Müller, P., 563
Muller, J., 27
Muluila, M., 166
Mungai, J. M., 170, 469
Muoka, M., 166
Muret, A. Faure-., 308
Murray-Hughes, R., 377, 378, 394
Mussett, A. E., 214, 216, 222, 225-35, 250, 269, 336, 359, 394, 450, 461

Naeser, C. W., 459, 461
Napier, J. R., 561, 564
Nash, W., 424, 425, 439, 479, 480, 483, 484, 496
Ndombi, J., 433, 471, 495
Needham, J., 127, 136
Nenquin, J., 359, 360
Neumann, O., 490, 498
Nicoletti, M., 27, 564
Nielsen, K. Perch-, 114, 117, 123

Oakley, K. P., 142, 147
Oertel, G., 90, 99
Olson, E. C., 90, 100
Osborn, F., 131
Osmaston, H., 255, 259
Oswald, F., 2, 4,
Ottey, P., 73, 82
Oxburgh, E. R., 7-18
Oxnard, C. E., 128, 137, 138

Pairo, M. Crusafont-, 239, 261
Palmer, A. R., 113, 114, 124

Palmer, H. C., 222, 225–35, 250, 269, 336, 359,
Parmentier, E. M., 12, 13, 18
Patterson, B., 270, 278, 378, 395, 414
Payne, J. A., 91, 100
Payne, S., 97, 100
Pelseneer, 508,
Penrose, L. S., 130, 137
Perch-Nielsen, K., 114, 117, 123
Perkins, P. C., 15, 17,
Petrucciani, C., 27, 564
Pezard, A., 126, 137
Piccirillo, E. M., 27, 28, 564
Pickering, R., 159 170
Pickford, M. H. L., 23, 27, 146, 213, 215, 217,
    222, 223, 228, 232, 233, 235, 237–62, 263–78,
    282, 292, 293, 294, 297, 307, 308, 309–27
Pilbeam, D., 125, 132, 137
Pilger, A., 27, 28, 563, 564
Pilgrim, G. E., 298, 300, 308
Pilsbry, H. A., 385, 394, 514
Pitman, W. C. III., 235
Polyakov, A. I., 459
Pomeroy, P. W., 10, 18, 72, 82
Posnansky, M., 175, 183, 202, 206, 344, 360
Powell, D. G., 65, 69
Price, P. B., 461
Protsch, R., 152, 154
Pulfrey, W., 22, 23, 27, 33, 54

Raja, P. K. S., 394, 461
Randel, R. P., 61, 69
Rattray, J. M., 489, 498
Reck, H., 151, 152, 154
Reeve, E. C. R., 126, 137
Reeves, C. C., 511, 527
Reilly, T. A., 381, 394, 444, 461
Rex, D. C., 20, 27, 563
Reyment, R. A., 128, 137
Reynolds, R. G. H., 424, 439
Rezak, R., 373, 394, 413
Rhemtulla, S., 81, 83
Rhoads, S. N., 490, 498
Richards, O. W., 127, 129, 137
Richardson, A. E., 196, 206
Richardson, J. L., 189, 196, 206, 496
Richter, R., 88, 100
Rigby, J. K., 393
Rix, P., 64, 69
Robbins, L. H., 395, 414
Robinson, J. T., 170
Roe, D. A., 347, 360
Rösler, A., 27, 28, 563, 564
Rohlf, F. J., 127, 137
Rosiwal, A., 393
Rutherford, E., 441

Saggerson, E. P., 19, 22, 25, 27, 35, 54, 63, 64,
    69
Sandison, A. T., 95, 99, 101
Sansom, H. W., 192, 206
Santacroce, R., 27, 563
Savage, R. J. G., 206, 292, 308, 375–94, 423
Savage, (Coryndon), S. C., 107, 121, 124
Sceal, J. S. C., 223
Schaefer, H.-U., 27, 563
Schäfer, W., 91, 92, 100
Schaller, G. B., 90, 100
Schönfeld, M., 23, 27, 549, 563, 564
Schopf, T. J. M., 104, 124
Schultz, A. H., 132, 137
Schumann, H., 91, 99
Sclater, J. G., 18
Searle, R. C., 11, 18, 53, 54, 73, 83
Searle, R. E., 73, 83
Setzer, H. W., 308
Shackleton, R. M., 19–28, 50, 54, 64, 69, 171,
    173–206, 219, 223
Shafiqullah, M., 450, 451, 461, 469
Shaw, A. B., 106, 111, 113, 114, 124
Shaw, E., 146
Shuey, R. T., 103–124, 424, 429, 434, 439, 477,
    480, 482, 488, 489, 496, 498, 558, 564
Sickenberg, O., 23, 549, 564
Sieveking, G. de G., 146
Sikes, H. L., 192, 206
Sill, W. D., 278, 414
Silver, E. A., 17
Simons, J. W., 93, 94, 100
Simpson, G. G., 104, 106, 109, 124, 294, 308,
    428, 429, 440
Skinner, N. J., 228, 235
Skipwith, P. A., 405, 413
Smith, E. A., 508, 527
Smith, H., 118, 123
Smith, N. D., 414
Sneath, P. A., 105, 124, 129, 130, 137
Snell, O., 126, 137
Snelling, N. J., 250, 269
Sørensen, H., 17
Sokal, R. H., 105, 124
Southam, J. R., 103, 124
Spence, T. F., 138
Stets, J., 27, 564
Stewart, D. R. M., 490, 498
Stewart, J., 490, 498
Stieldjes, L., 27
Styles, P., 20, 27
Suess, E., 393
Sunderalingham, K., 82
Sutcliffe, A. J., 93, 95, 100
Sutherland, D. S., 35, 54

Svec, H. J., 4, 13, 546
Swain, C. J., 71–83
Szalay, F. S., 136, 322, 327

Taieb, M., 170, 549–64
Tallon, D., 329
Tallon, P., 217, 223, 228, 329, 338, 361–73
Tarling, D. H., 71, 82
Taylor, M. E., 117, 123
Tazieff, R., 563
Teissier, G., 126, 127, 136, 137
Thomas, O., 490, 498
Thomas, P., 305
Thompson, A. O., 191, 206
Thompson, B. W., 192, 206
Thompson, W. D'Arcy, 126, 129 137
Thomson, J., 1, 4
Thurber, D. L., 294, 308, 478, 496
Tiercelin, J. -J., 557
Tobias, P. V., 153, 155, 170, 372, 373, 561, 564
Toots, H., 92, 95, 97, 100
Torrance, K. E., 18
Toula, F., 393
Trapnell, C. G., 489, 498
Trendall, A. F., 22, 23, 24, 27, 31, 54
Tringham, R., 99, 205, 359, 546
Truckle, P., 40, 79
Turcotte, D. L., 7, 9, 10, 14, 15, 18
Turekian, K. K., 235

Ucko, P. J., 99, 205, 359, 546
Urban, E. K., 490, 498

Valentine, J. W., 512, 527
Van Couvering, J. A., 239, 249, 261, 262, 294, 297, 308, 428, 429, 440, 444, 461
Van Couvering, Judy, 428, 429, 440
Van Damne, D., 381, 393, 477, 479, 498
Van Gerven, D. P., 130, 137
Van Lawick, J. Goodall, 140, 146
Van Riet Lowe, C., 372, 373
Varet, J., 27, 550, 563, 564
Varne, R., 22, 28
Verdcourt, B., 196, 381, 394
Visentin, E. Justin-, 20, 21, 28, 549, 564
Vondra, C. F., 395–414, 415, 418, 419, 420, 422, 424, 438, 469, 475, 476, 478, 483, 485, 490, 492, 495, 497, 498, 529, 530, 531, 532, 545, 546, 547
Von Höhnel, L., 1, 3, 381, 393
Von Koenigswald, G. H. R., 452, 461
Voorhies, M. R., 95, 96, 97, 100

Walker, A., 95, 97, 100, 170, 222, 235, 322, 327

Walker, A. C., 460, 469
Walker, K. R., 492, 498
Walker, R. M., 461
Walsh, J., 21, 23, 27, 79, 83, 206, 212, 214, 217, 222, 223, 239, 262, 264, 278, 377, 378, 380, 394, 412, 414, 461
Walter, R. C., 557
Washbourn-Kamau, C., 496
Washburn, S., 140
Waterman, T. H., 137
Watkins, R. T., 124, 388, 389, 393, 420, 458, 460, 469, 497, 546
Wayland, E. J., 2, 152
Weaver, S. D., 40, 54, 211, 223, 316
Webb, P. K., 38, 40, 54, 211, 223, 241, 262
Weidmann, M., 27
Weigelt, J., 88, 91, 92, 101
Weinert, H., 157, 170
Wells, C., 95, 101
Wendorf, F., 359
Wesselman, H., 436, 439, 502
Wheat, J. B., 93, 101
White, J. F., 128, 138
White, S., 152
White, T. D., 157–70, 323, 327, 564
Whiteman, A. J., 17
Whitworth, T., 196, 249, 262, 380, 394
Wiessner, P., 358, 360
Wilkinson, A., 259, 262
Wilkinson, A. F., 380, 394
Wilkinson, P., 66, 69
Williams, L. A. J., 17, 26, 27, 35, 36, 54, 55–69, 176, 206, 212, 222, 223, 326, 393, 459, 460
Williamson, K., 73, 83
Williamson, P. G., 375–94, 507–27
Willis, B., 3, 4
Wilson, E. O., 512, 527
Wilson, J. T., 15, 17
Wohlenberg, J., 19, 22, 27, 83
Wolff, R. G., 19, 26, 377, 378, 393
Wolman, M. W., 413
Wolpoff, M., 125, 137
Wood, B. A., 125–38, 170
Worsley, T. R., 124
Wright, R., 175, 197
Wright, R. V. S., 140, 147

Yonge, C. M., 508, 527
York, D., 449, 451, 461

Zanettin, B., 20, 21, 28, 549, 564
Zimmermann, J., 83
Zuckerman, S., 126, 130, 131, 138